T0408310

CROP SUSTAINABILITY AND INTELLECTUAL PROPERTY RIGHTS

CROP SUSTAINABILITY AND INTELLECTUAL PROPERTY RIGHTS

Edited by
Soumya Mukherjee
Piyali Mukherjee
Tariq Aftab

First edition published 2024

Apple Academic Press Inc.
1265 Goldenrod Circle, NE,
Palm Bay, FL 32905 USA

760 Laurentian Drive, Unit 19,
Burlington, ON L7N 0A4, CANADA

CRC Press
6000 Broken Sound Parkway NW,
Suite 300, Boca Raton, FL 33487-2742 USA

4 Park Square, Milton Park,
Abingdon, Oxon, OX14 4RN UK

© 2024 by Apple Academic Press, Inc.

Apple Academic Press exclusively co-publishes with CRC Press, an imprint of Taylor & Francis Group, LLC

Library and Archives Canada Cataloguing in Publication

Title: Crop sustainability and intellectual property rights / edited by Soumya Mukherjee, Piyali Mukherjee, Tariq Aftab.
Names: Mukherjee, Soumya, editor. | Mukherjee, Piyali, editor. | Aftab, Tariq, editor.
Description: First edition. | Includes bibliographical references and index.
Identifiers: Canadiana (print) 20220479976 | Canadiana (ebook) 20220480028 | ISBN 9781774913192 (hardcover) | ISBN 9781774913208 (softcover) | ISBN 9781003383024 (ebook)
Subjects: LCSH: Intellectual property (International law) | LCSH: Plant varieties—Patents. | LCSH: Plant varieties—Protection. | LCSH: Crops—Effect of stress on. | LCSH: Sustainable agriculture. | LCSH: Agricultural biotechnology. | LCSH: Farm produce. | LCSH: Traditional ecological knowledge.
Classification: LCC K3876 .C76 2023 | DDC 346.04/8—dc23

Library of Congress Cataloging-in-Publication Data

Names: Mukherjee, Soumya, editor. | Mukherjee, Piyali, editor. | Aftab, Tariq, editor.

Title: Crop sustainability and intellectual property rights / edited by Soumya Mukherjee, Piyali Mukherjee, Tariq Aftab.

Description: First edition. | Palm Bay, FL, USA : Apple Academic Press, 2023. | Includes bibliographical references and index. | Summary: "This book, Crop Sustainability and Intellectual Property Rights, merges the concepts of traditional agriculture, crop sustainability, and intellectual property rights associated with plant protection and agricultural products. This volume provides a comprehensive account of the perspectives of traditional agriculture, crop sustainability, and various strategies associated with crop tolerance to adverse environmental conditions. It also highlights the role of agricultural intellectual property rights, along with the implications for plant patents, protection of farmers' rights, and geographical indication in plant products to provide a broader outlook toward strategies for sustainable agriculture and global food security associated with IPR. The chapters provide an overview of sustainable crop cultivation in a variety of ways, with both traditional agriculture as well as with new biotechnological approaches. The volume explores several stress resilience strategies and issues for crops, considering how to mitigate the effect of increased carbon dioxide concentration, heavy metal pollution, over-salinized soils, and cold spells. It also discusses how to make desert farming more efficient; how to increase abiotic stress tolerance of crops with grafting, seed soaking/priming, soil amendment, and foliar applications; and more"-- Provided by publisher.

Identifiers: LCCN 2022055239 (print) | LCCN 2022055240 (ebook) | ISBN 9781774913192 (hardback) | ISBN 9781774913208 (paperback) | ISBN 9781003383024 (ebook)

Subjects: LCSH: Intellectual property (International law) | Plant varieties--Patents. | Plants varieties--Protection. | Sustainable agriculture. | Crops--Effect of stress on. | Agricultural biotechnology.

Classification: LCC K3876 .C765 2023 (print) | LCC K3876 (ebook) | DDC 346.04/86--dc23/eng/20230111

LC record available at https://lccn.loc.gov/2022055239
LC ebook record available at https://lccn.loc.gov/2022055240

ISBN: 978-1-77491-319-2 (hbk)
ISBN: 978-1-77491-320-8 (pbk)
ISBN: 978-1-00338-302-4 (ebk)

About the Editors

Soumya Mukherjee, PhD, is currently affiliated with the faculty in the Department of Botany, Jangipur College, University of Kalyani, West Bengal, India. He was formerly with Ramjas College, University of Delhi, India. He has also worked as a CSIR research fellow in abiotic stress physiology of plants. He has published both research and review articles in various peer-reviewed international journals (SCOPUS and SCI indexed). He has contributed various chapters (plant physiology, abiotic stress signaling) and popular articles to various books published by national and international publishers. He has been publishing edited volumes on plant signaling, communication (Springer), and root biology (Springer). He has authored e-learning modules in plant physiology and biochemistry published as an effort of the National Mission on Education through ICT (an MHRD Project undertaken by the University of Delhi). He has acted as a subject reviewer of several peer-reviewed international journals on plant science. He has presented brief research works at international and state-level conferences in India and abroad. Currently he is focused on undergraduate teaching to botany (Hons.) and life science students in various fundamental aspects areas of plant physiology. He continues to work with research interests in salt-stress physiology associated with the aspects of hormone signaling and biomolecular crosstalk in plants. Dr. Mukherjee earned his PhD at the Department of Botany, University of Delhi, India.

Piyali Mukherjee is an advocate affiliated with the West Bengal Bar Council, Kolkata, West Bengal, India, since 2017. She has earned a postgraduate diploma in forensic sciences and cyber forensics from the International Forensic Science Institute, Pune, India. She has gained her specialization in intellectual property rights and patent cooperation treaty from the World Intellectual Property Organization (WIPO), Geneva, Switzerland. She has gathered experience in the field of fingerprinting, document examination, and cyber investigations from the Sherlock Institute of Forensic Sciences, New Delhi, India. To her credit she has developed various course content and published book chapters related to law, humanities, management, and legal perspectives associated with commercially important plants.

Tariq Aftab, PhD, is currently Assistant Professor in the Department of Botany at Aligarh Muslim University, India, from which he earned his PhD. He formerly worked as Visiting Scientist at Leibniz Institute of Plant Genetics and Crop Plant Research (IPK), Gatersleben, Germany, and in the Department of Plant Biology, Michigan State University, USA. He is a member of various scientific associations in India and abroad. He has edited seven books with international publishers, including Elsevier, Springer Nature, and CRC Press (Taylor & Francis Group); co-authored several book chapters; and published over 60 research papers in peer-reviewed international journals. His research interests include physiological, proteomic, and molecular studies on medicinal and aromatic plants. He is the recipient of a prestigious Leibniz-DAAD fellowship from Germany, Raman Fellowship from the Government of India, and Young Scientist Awards from the State Government of Uttar Pradesh (India) and Government of India. After completing his doctorate, he has worked as a Research Fellow at the National Bureau of Plant Genetic Resources, New Delhi, and as Post-doctorate Fellow at Jamia Hamdard, New Delhi, India.

Contents

Contributors

Aditya Banerjee
Department of Biotechnology, St. Xavier's College (Autonomous), Kolkata, West Bengal, India

Sunanda Bharti
University of Delhi, Delhi, India

Atanu Bhattacharjee
Department of Pharmaceutical Sciences, Royal School of Pharmacy, Assam Royal Global University, Guwahati, Assam, India

Saransh Chaturvedi
Associate, Khurana & Khurana, Advocates and IP Attorneys

Chumki Chowdhury
Department of Botany, Jangipur College, University of Kalyani, West Bengal, India

Rashmita Dasgupta
Calcutta High Court, Kiran Sankar Roy Road, Kolkata, West Bengal, India

Subhashis Debnath
Department of Pharmaceutical Sciences, Royal School of Pharmacy, Assam Royal Global University, Guwahati, Assam, India

Shagun Danda
Department of Botany, Daulat Ram College, University of Delhi, Delhi, India

Hany G. Abd El-Gawad
Department of Horticulture, Faculty of Agriculture, Ain Shams University, Cairo, Egypt

Mohamed F. M. Ibrahim
Agricultural Botany Department, Ain-Shams University, Cairo, Egypt

Ranjan Dutta Kalita
Department of Biotechnology, Royal School of Biosciences, Assam Royal Global University, Guwahati, Assam, India

Sandeep Kaushik
Department of Environmental Science, Indira Gandhi National Tribal University, Amarkantak, Madhya Pradesh, India

Hala B. Khalil
Department of Genetics, Faculty of Agriculture, Ain Shams University, Cairo, Egypt

Anil Kurmi
Plant protection Division, Krishi Vigyan Kendra, Indira Gandhi National Tribal University, Amarkantak, Madhya Pradesh, India

Pooja Jha Maity
Department of Botany, Ramjas College, University of Delhi, Delhi, India

Mohamed M. El-Mogy
Vegetable Crops Department, Faculty of Agriculture, Cairo University, Giza, Egypt

Piyali Mukherjee
Department of Law, Brainware University, Kolkata, West Bengal, India

Yutika Nath
Department of Serology, Directorate of Forensic Sciences, State Forensic Science Laboratory, Guwahati, Assam, India

Soma Halder Paul
Department of Botany, Asutosh College, Kolkata, West Bengal, India

Manoj K. Rai
Department of Environmental Science, Indira Gandhi National Tribal University, Madhya Pradesh, India

Rishav Ray
School of Law, Christ (Deemed to be University) Bangalore, Karnataka, India

Aryadeep Roychoudhury
Department of Biotechnology, St. Xavier's College (Autonomous), Kolkata, West Bengal, India

Reetu Sharma
Department of Botany, Ramjas College, University of Delhi, Delhi, India

S. A. Shehata
Agricultural Botany Department, Ain-Shams University, Cairo, Egypt

S. Shweta
Department of Botany, Guru Ghasidas Vishwavidyalaya, Koni, Bilaspur, Chattisgarh, India

Ankur Singh
Department of Biotechnology, St. Xavier's College (Autonomous), Kolkata, West Bengal, India

B. Subramaniam
Retired Chief Scientist and Coordinator, Traditional Knowledge Digital Library Project, CSIR-National Institute of Science Communication and Information Resources, New Delhi, India

Bibin George Varughese
T. A. Pai Management Institute, Manipal, Karnataka, India

Muhammad A. Zayed
Botany and Microbiology Department, Menoufia University, Shebin El-Kom, Egypt

Abbreviations

Aβ	amyloid beta
ABA	abscisic acid
ABPs	actin-binding proteins
ABRE	ABA-responsive elements
ABS	access and benefit-sharing
AD	Alzheimer's disease
ADF	actin-depolymerizing factor
AF	actin filaments
AMF	arbuscular mycorrhizal fungi
AP	ascorbate peroxidase
APP	amyloid precursor protein
APX	ascorbate peroxidase
AREBs	ABA-responsive element-binding protein
AsA	ascorbic acid
BA	6-benzyaldenine
BACE	The beta site of the APP cleavage enzyme
BAP	benzylaminopurine
BC	biochar
BL	brassinolide
BRs	brassinosteroids
CaM	calmodulin-dependent protein kinases
CAT	catalase
CBL	calcineurin B-like proteins
CDF	cation diffusion facilitators
CK	cytokinins
CL	compulsory license
CML	calmodulin-like proteins
CNGCs	cyclic nucleotide-gated channels
CO_2	carbon dioxide
CPK	calcium-dependent protein kinases
CSIR	Council for Scientific and Industrial Research
CRT	calreticulin
CTR/COPT	copper transporter
Ca^{2+}	calcium

C99	99 amino acid fragment
DACCs	depolarization-activated calcium channels
DAG	diacylglycerol
DGK	diacylglycerol kinase
DNA	deoxyribonucleic acid
DUS	distinct, uniform, and stable
EC	peroxidases
EDV	essentially-derived variety
EPO	European Patent Office
ER	estrogen receptor
FA	fulvic acid
FAO	Food and Agriculture Organization
FYM	farm yard manure
GB	glycine betaine
GHG	greenhouse gas
GI	geographical indications
GIAHS	globally important agricultural heritage systems
GLRs	glutamate-receptor-like channels
GMOs	genetically modified organism
GPOD	guaiacol peroxidase
GPoX	guaiacol peroxidase
GPX	glutathione peroxidase
GR	glutathione reductase
GSH	glutathione
HA	humic acids
HACCs	hyperpolarization-activated calcium channels
HLH	helix-loop-helix
HM	heavy metals
HMA	heavy metal ATPase
HMT-PGP	heavy metal tolerant and plant growth promoting
H_2O_2	hydrogen peroxide
IAA	indole-3-acetic acid
IBA	Indole-3-butyric acid
IPO	Indian Patent Office
IP	intellectual property
IPRs	intellectual property rights
IRRI	International Rice Research Institute
IRT	iron-regulated transporter
ISO	isoorientin

JA	jasmonic acid
Kin	kinetin
LEAs	late embryogenesis abundant
MAPK	mitogen-activated protein kinases
MAP2Ks	MAPK kinases
MAP3Ks	MAPK kinase kinases
MAPs	microtubule-associated proteins
MCAs	mechanosensitive channels
MDA	malondialdehyde
Mn-SOD	manganese SOD
MT	microtubules
MT	metallothioneins
NAA	naphthaleneacetic acid
NO	nitric oxide
NPOP	National Program for Organic Production
NRAMP	natural resistance-associated macrophage protein
OPT	oligopeptide transporters
O_3	ozone
PA	phosphatidic acid
PAs	polyamines
PAL	phenylalanine ammonia lyase
PCs	phytochelatins
PCD	programmed cell death
PDR	pleiotropic drug resistance
PGPB	plant growth promoting bacteria
PGPR	plant growth promoting rhizospheric
PLD	phospholipase D
PBRO	Plant Breeder's Right Office
PBRs	Plant Breeder's Rights
PH	potassium humate
PPo	polyphenol oxidase
PPV and FR	Protection of Plant Varieties and Farmers' Rights
Pro	proline
PS	potassium silicate
PVPA	Plant Variety Protection Act
PVPAT	Plant Variety Protection Appellate Tribunal
RuBisCO	ribulose-1,5-bisphosphate carboxylase/oxygenase
ROS	reactive oxygen species
SA	salicylic acid

sAPP'A	soluble APP'A
SD	superoxide dismutase
SDGs	Sustainable Development Goals
SL	strigolactone
SOD	superoxide dismutase
SWE	seaweed extract
TA	titratable acidity
TBY-2	Tobacco Bright Yellow-2
TCE	traditional cultural expressions
TDZ	Thidiazuron
TFs	transcription factors
TK	traditional knowledge
TPCs	two-pore channels
TRIPS	The Agreement on Trade-Related Aspect of Intellectual Property Rights
TSS	total soluble solids
UPOV	International Union for the Protection of New Varieties of Plants
UPR	unfolded protein response
UPS	ubiquitin proteasome system
USPTO	United States Patent and Trademark Office
VAM	vesicular arbuscular mycorrhiza
V-ATPase	vacuolar proton-ATPase
V-Ppase	vacuolar proton pyrophosphatase
WIPO	World Intellectual Property Organization
WTO	World Trade Organization
WUE	water use efficiency
X	antioxidants
YSL	yellow stripe like
ZIP	ZTR/IRT-related protein
$^1O_2$	singlet oxygen

Preface

The last few decades have shown radical advances in biotechnology, which has revolutionized agriculture and trading practices related to plant-based products. Climate change, extensive anthropological activity, and environmental pollution are the major menaces that affect agricultural productivity worldwide. Land fragmentation and changes in land pattern usage often limit the area of crop cultivation. Biodiversity losses are responsible for a reduction in the number of ecosystem services, which affects crop productivity, thus resulting in food shortages. Despite the advancements brought about by Green Revolution, only 12% of the world's land surface is used for crop production. Therefore, the per capita production needs to be enhanced in the next few decades. Changes in the lifecycle pattern and diversity of pollinators adversely affect seed seta in major crop plants. More than 75% of the world's population relies on 12–15 plant species as sources of their major staple food, among which rice, wheat, and maize serve for 60% of the food value. Investigations have surged in plant sciences to produce resilient crops toward various biotic and abiotic stress factors like diseases, salinity, drought, and heavy metal contamination. With the advent of biotechnology and decision from the US Supreme Court, private sector companies have invested in agricultural research programs in order to patent genetically modified organisms. Nonetheless, conventional breeding practices are still required to be improved with exploitation of wild genes.

Simultaneous with agricultural development, there have been remarkable changes in the implication of intellectual property rights (IPR) in plant sciences. The emergence of greater globalization of trade has led to a more extensive use of IP on agricultural inventions. Thus, there are various concerns associated with the use of technology in developing countries and the legal rights to use improved germplasms and breeding programs. A higher percentage of people in the developing world focus on agriculture as their major livelihood. In this present situation, there is a need to enhance IP knowledge and set up more active programs in order to protect proprietary rights associated with plant varieties, agricultural biotechnology, and other plant-derived products, especially in the rural areas of developing countries.

The present volume collates various aspects of crop sustainability associated with traditional agriculture, physiological stress tolerance of crops, and exploitation of plant varieties. Furthermore, this book highlights the potential role of IP in commercialization of traditional knowledge, and its implication in plant patents, protection of farmer's rights, and geographical indication in plant products. We hope that the book will be beneficial to plant biotechnologists, breeders, corporate officials, and policymakers working in various aspects of crop management and IP in agriculture.

Best wishes
Editors

Introduction

"The average person is still under the aberrant delusion that food should be somebody else's responsibility until I'm ready to eat it."

– Joel Salatin

Human lifestyle and agriculture have coevolved on Earth across millions of years. The transition of humans from their nomadic to social lifestyle has been accompanied by domestication and diversification of various crop plants. Various edible and nonedible crops provide important ecosystem services to mankind and other organisms in diverse geographical areas of the world. Apart from the benefit of crops in attaining livelihood and food security, sustainable agriculture is imperative toward maintaining ecological balance, harboring pollinators, and developing environment-friendly strategies for cultivation. Plant tolerance to abiotic and biotic stress involves a plethora of physiological and molecular changes that involve rapid and slow signaling events. Various biotechnological strategies have been developed to multiply elite variety of crops and ornamental plants associated with disease tolerance, high yield, and reproductive success.

Improvement of plant varieties is important for the development of high-yielding cultivars with better quality and resistance to biotic and abiotic stress. Conventional plant-breeding methods are carried out for successive generations and desirable characters with superior traits are selected for release of high vigor cultivars. Presently, with the advent of molecular biology and chromosome mapping, marker-assisted breeding has appeared fruitful toward better screening of genetic background of crops. Conservation of germplasm for various indigenous (medicinal and commercially important) plants in various diverse geographical areas can be regulated by rights of protection and benefits shared by local people. In early 1990s, the term "biopiracy" was coined by governmental and nongovernmental organizations that indicate illegal and unauthorized exploitation of biological resources associated with traditional knowledge. Furthermore, biopiracy also involves patenting of knowledge and/or resources without providing any compensation. The extension of intellectual property rights in various communities or global area can facilitate in protection of farmer's rights and help in the socioeconomic evolution of various rural communities by establishing their role in national and international dealings. However, there are often conflicts between the intellectual property rights (IPR) establishers and opponents where community-based rights are replaced by individual-based property right systems. The extension of IPR in agriculture has made substantial progress in the last two decades. In the earlier period of 1900s, plant-based natural resources were considered as common resources without the ownership of any community or local people. However, in the latter half of the 19th century (1930s), patents on plants were granted by the United States followed by Europe. Followed by the establishment of International Union for the Protection of New Varieties of Plants (UPOV) in 1961, patents and breeder's rights were the two main forms of IPR in agriculture, which provided rights for the protection of seeds, crops, and other plants in a community. However, later on, various NGOs and farmers organized movements for promoting farmer's rights as an alternative to breeder's rights. The idea of implementation of farmer's rights was to enable farmers to save and exchange seeds and provide them benefits of ownership for development and conservation of plant varieties. The Convention on Biological Diversity (CBD) was signed at the Earth Summit held in Rio de Janeiro in 1992, which brought about a sovereign control over the natural resources. In the year 2000, the World Intellectual Property Organization (WIPO), has implemented the Intergovernmental Committee on Intellectual Property and Genetic Resources, Traditional Knowledge, and Folklore, which aims to interpret farmer's and community rights, including prospecting biodiversity, assigning patent, and ensuring benefit/ownership rights to farmers.

The present volume aims to provide a comprehensive account of the perspectives of traditional agriculture, crop sustainability, and various strategies associated with crop tolerance to adverse environmental conditions. Furthermore, the book highlights the role of intellectual property rights associated with plant protection, rights associated with agricultural products among indigenous communities and global food security. This book shall enable the readers to obtain a broader outlook toward strategies for sustainable agriculture and global food security associated with IPR.

Editors
S. Mukherjee, P. Mukherjee, and T. Aftab

PART 1
SUSTAINABLE CROP CULTIVATION

CHAPTER 1

Sustainable Crop Cultivation: A Comprehensive Update

CHUMKI CHOWDHURY

Department of Botany, Jangipur College, University of Kalyani, West Bengal, India

ABSTRACT

Nowadays, sustainable crop production is an important aspect of agriculture to attain and maintain the rising demand for food security throughout the world. Sustainable farming is one of the best ways to overcome the climate change-related shortcomings of crop production. Sustainable agricultural production may lead to the way to ensure food safety and food quality without harming the environment. With the increasing population of developing countries, the food demand is projecting to double over the next two to three decades. For the last few decades, permaculture, biodynamic farming, agroforestry, polyculture, crop rotation, etc., were the popular and helpful practices for sustainable crop production to mitigate the increasing demand for food. But, recently, integrated farming in saline soil or mangrove-covered areas in different parts of Asia has gained importance in sustainable crop production and food security. In West Asia, the integrated farming of livestock and crop cultivation has proven fruitful to crop production by using saline water. Indian farmers also have successfully implemented integrated farming with multicrop farming, livestock-crop farming, and aquaculture, along with crop cultivation in the saline soil of the Indian Sunderbans. Thus,

Crop Sustainability and Intellectual Property Rights. Soumya Mukherjee, Piyali Mukherjee & Tariq Aftab (Eds)
© 2023 Apple Academic Press, Inc. Co-published with CRC Press (Taylor & Francis)

they are now able to overcome the harsh effect of repeated storm surge on crop cultivation in coastal West Bengal.

1.1 INTRODUCTION

1.1.1   *FOOD SECURITY AND CLIMATE CHANGE*

The different steps of present-day food system like production, transport, processing, packaging, and storage are serving the livelihood of 1 billion people every day. According to the 2018 report of the UN Food and Agriculture Organization (FAO), 50% increase in food production will be needed by 2050 in order to feed the increasing world population (FAO, 2018; Searchinger et al., 2019). Although, scientists and activists had the assumption of the increasing demand for food security long back when the industrial revolution had just started. To solve this problem, the *green revolution* came into act in the 1960s. From 1961, agricultural sectors started to use nitrogen fertilizers and water resources for irrigation in a greater scale (increase of about 800% in case of fertilizers and 100% for water use) and achieved an increased GDP of more than 30% per capita (Anibaldi, 2021). But till now, after such a drastic increase, approximately 821 million people are presently undernourished, 151 million children under five are stunted, 613 million women and girls have iron deficiency, and approximately 2 billion adults have unhealthy weight gain (https://www.unicef.org, September, 2018). These figures easily indicate toward the limited success of the green revolution. Total greenhouse gas emissions from agriculture is likely to be increased further by about 30–40% by 2050, due to continually rising demand of the on-growing population (Searchinger et al., 2019), income growth, and dietary change (IPCC, 2019). So it is obvious that the conventional methods of farming and food production in a large scale are leading to severe climate change-related issues (Pannell et al., 2006). Availability, access, utilization, and stability of food security and their interactions are affected by climate change as well (FAO, 2018).

In the last few decades, investigations are going on in terms of maximization of profit from agricultural land without harming the environment. To reach this goal, a balance between ecological and economical demands on agricultural lands (Pannell et al., 2006) is the main discussable aspect. Sustainable agriculture can be defined in many ways. But, the main concept of these definitions is that sustainable farming comprises ranching and cultivation practices in specific land types designed to overcome the continuous need for food, fiber, energy, and ecosystem services including soil conservation, clean water, biodiversity, and many more. Sustainable farming encourages production of food and other organic consumables that are profitable and environment friendly and will serve the better livelihood for both farmers and the public (Narayan, 2012). Before introducing sustainable farming, more and more scientific studies are needed from which we can estimate and compare the cost of sustainable and conventional farming and also the advantages over disadvantages of sustainable farming so that it can be chosen as an alternative way of agriculture (Anibaldi, 2021). Technology may have a prominent role in the sustainable farming as well. To establish the sustainable crop production system, few initiation steps are needed among which the most important step is to reduce the negative effects of current system of agriculture while maintaining the high yield and sustainability.

Awareness about the agroecosystems is a very important step. Ecological units like soil, water, air, wildlife, insects, pathogens, plants, humans, and interactions among them are the different aspects of agroecosystem (Borremans et al., 2018). So the decision about the agricultural management have to be taken in such a way that the harmony among crops, livestock, beneficial organisms, pests, and the physical environment will be maintained properly. Apart from the environmental concern, the policies should consider the socioeconomic impacts as well. For sustainable farming, artificial agents used for high yield like artificial fertilizers and pesticides may harm the environment. The people related to farming and production must have the knowledge of ecological processes such as nutrient cycling, crop–weed competition, host–parasite and predator–prey relationships for estimating crop yields, and system sustainability.

The goal of sustainable farming can be a way to maximize crop yields along with the supply of clean water and air, the conservation of wildlife and other organisms valued by society, recreation, etc. Thus, to achieve sustainability, all participants including farmers, laborers, policy makers, retailers, consumers, and researchers should come across to work together.

Keeping the aforementioned points in mind, the United Nations (UN) Sustainable Development Goals (SDGs) for 2030 aimed toward:

- Large-scale production and consumption of food,
- The protection of land and water resources and change in land use,
- Proper framing for demand and supply for achieving food security by 2050, through agricultural development along with reducing

greenhouse gas (GHG) emissions as well as land and water degradation from farming (Anibaldi, 2021).

According to the IPCC's special report in 2019, potential harmful effect of climate change and its mitigation pathways may lead to harmful or beneficial alteration on agricultural sectors.

There are several ways of incorporating sustainable crop production system. Apart from minimizing the time gap between the yields, adoption of sustainable land management programmes along with the harmony of ecological functions such as conservation of agriculture and community adaptations with proper support and market access are two important footsteps which are to be taken to materialize the idea of sustainable food and crop production (Mbow et al., 2014a). Introduction to agroecology and agroforestry may be beneficial for sustainable crop production.

1.1.2 SUSTAINABLE CROP PRODUCTION

The transition to sustainable agriculture normally requires a series of small, realistic steps, like organic farming, which includes use of organic manures, vermicomposting, selection of crop for crop rotation, intercropping and companion cropping, integrated pest management, exploitation of organic nitrogen fixation, etc. (Anibaldi et al., 2021). Socioeconomic status and personal goals influence the transition from conventional farming to organic and sustainable farming. The decisions made in the basic levels control the whole system and the advancement of that sustainability of system depends on those decisions. To attain the success in sustainable agriculture, active participation of all participants in the system, including farmers, laborers, policymakers, researchers, retailers, and consumers is needed. The work plans contain three separate, but inter-related, areas of concern like farming and natural resources; practices of plant and animal production (Narayan, 2012); and the economic, social, and political issues (Pannell et al., 2006).

1.1.3 EFFICACY OF SUSTAINABLE FARMING FOR "FEEDING THE WORLD"

Sustainable crop production includes ecological and biological principles and has significantly lower costs and less environmental damage. Organic farming is one of the best alternatives that has been adopted throughout the world. Biotech industries of the world promotes agriculture with artificial inputs while organic agriculture is being promoted by the United Nations Food & Agriculture Organization (FAO) and described as a holistic production system which promotes and enhances agro-ecology, along with biodiversity and natural processes. In fact, according to a number of studies, organic farming methods can produce higher yields than conventional methods if proper implementation is there. Moreover, a worldwide conversion to organic farming may lead to increased food production levels and can reduce the degradation of agricultural soils by increasing soil fertility and soil health.

1.2 ECO-FRIENDLY METHODS IN SUSTAINABLE CULTURE

Considering both nutritional security and environmental sustainability, it is inevitable to adopt organic or eco-farming system, whereas integrated farming served the purpose of covering many more other aspects of livelihood in the remote places. Eco-farming implies a farming system that primarily aims at cultivating land and raising crops under ecologically favorable condition with a target of wild and native varieties of crops. In spite of using chemical fertilizers or pesticides, this method relies more on an integrated approach of crop management practices by making use of cultural, biological, and natural inputs. Use of organic manures such as farm yard manure (FYM) (Dey et al., 2018), recycling of organic wastes via composting, green manures, and biological inputs like vermicomposts and biofertilizers serve as the major plant nutrient input in eco-farming (Narayana, 2012). Bio-agents such as natural predators of pests are used in pest management which causes no harm to the environment. Agronomic practices such as crop rotation with proper selection of crops, intercropping and companion cropping, stubble mulching, and use of resistant varieties (Iqbal et al., 2019) are among the important factors of organic farming and sustainable farming (Kassam, 2020). Some of the eco-friendly methods are listed as follows:

1.2.1 USE OF EARTHWORM FOR PRODUCTION OF VERMICOMPOST

Decomposable organic wastes get composted naturally by earthworms present in the soil. This process is called vermicomposting. The vermicompost is a mixture of worm casts, macro- and micronutrients (N, P, K, Mn., Fe, Mo, B, Cu, and Zn), and some plant growth hormones like gibberellins and auxins. This mixture also contains some bacterial community like Azospirillum, Actinomyces, Phosphobacillus, etc. Vermicompost contains 1-1.5%N, 0.6-0.8% P, and 1.2–1.5% K, and is nutrient-rich because earthworms convert organic matter into available forms of nutrients through their physiological processes (Narayan, 2012). Vermicomposting helps to improve the soil fertility by better aeration and water-holding capacity as a result of the churning and turning of soil by earthworms. The excreta of earthworms are rich source of plant nutrients in the soil. A study says that soils with casts contain 5 times more nitrogen, 7 times more phosphorus, 11 times more potash, 2 times more magnesium, and 7–8 times more Actinomyces population in soils than the soil without earthworm casts. Using earthworms to enhance soil fertility is a completely natural means of soil management, which perfectly goes with integrated plant nutrient management strategy for sustainable agriculture.

1.2.2 BIOLOGICAL NITROGEN FIXATION—WAY OF SUSTAINABLE AGRICULTURE

Seventy-eight percent nitrogen is available in atmosphere but cannot be up taken by living beings and need to be fixed biologically. Biologically fixed atmospheric nitrogen in the soil can act as a potential nitrogen source for plants. This is an economic and eco-friendly source of nutrient supply to the soil (Nawaz and Farooq, 2021). Biologically fixed nitrogen retains longer in the soil than fertilizer nitrogen and may act as future source of readily mineralizable organic nitrogen in soil. Biological fixation of nitrogen is mediated by archaebacteria, eubacteria, cynobacteria, actinomyces, and other organisms categorized as diazotrophs. Diazotroph can convert atmospheric elemental nitrogen ($N2$) to ammonia ($NH3$) in the presence of nitrogenase enzyme system. Rhizobium, Azotobacter, and some cynobacteria are well known for high nitrogen fixation in fields. By reducing the use of artificial fertilizer, biologically fixed nitrogen restores the microflora of soil and makes the way for sustainable farming.

1.2.3 CHOICE OF CROP FOR CROP ROTATION, INTERCROPPING, AND ASSOCIATE CROPPING

To implement sustainable farming, rotational crop cultivation is an important aspect (Nawaz and Farooq, 2021). Proper utilization of plant nutrient from different parts of soil largely depends on the choice of crop. Intercropping helps to resist crop failure during the period of uncertain rainfall. It also helps to minimize the infestation of pest and diseases. For example, intercropping of sorghum and red gram at 2:1 row ratio gave 70% higher yield over the individual cropping of both the crops (Narayan, 2012). Intercropping reduces the infection of wilt diseases of gram. Red gram fixes atmospheric nitrogen, which is up taken by the sorghum crop. Associate cropping of three rows of gram followed by one row of linseed was proved fruitful in case of pod borer in gram (Narayan, 2012). It was helpful for reduction of fly invasion in linseed plants. Production of both the crops was increased by about 14% after the introduction of intercropping over single-crop cultivation (Narayan, 2012).

1.2.4 INTEGRATED PEST MANAGEMENT

Prolonged use of chemical pesticides in the field results in the deterioration of soil quality and also is a potential source of environmental pollution. So, nowadays, integrated pest management has gained a lot of importance to make the agricultural system eco-friendly. The combination of physical, chemical, biological, and cultural methods is very much useful for pest control in agricultural field with least ecological damage. To introduce natural pest management, modifications of controlling environment of crop cultivation to restrict pest infection is necessary. Cultivation of resistant transgenic plant varieties is the preferable way toward integrated pest management. The methods of biocontrol, like allelopathy, (Cheema et al., 2013) and use of natural predators suppress the parasites and microbial agents which are potential threat for crops. Conservation of natural enemies can replace the need of other control automatically.

1.2.5 DIFFERENT WAYS OF SUSTAINABLE CULTURE

1.2.5.1 MULTICROPPING

Multicropping system is a very useful and most common way toward sustainable agriculture compared to monoculture system (Francis and Porter, 2011). There are different types of multicropping practices. In Table 1.1, different types of multicropping system have been listed.

TABLE 1.1   Types of Multicropping System.

Name of the system	Definition	Advantage	Disadvantage
1. Monocropping	Continuous culture of one species in a field over a sequence of growing season	Conventional type of culture	Repeated cycle of pest invasion, disease, and growth of weed
2. Sequential cropping	Crops cultivated one after another but no overlapping phase	Cultivation of rain-fed cereals or paddy along with oilseed or legumes in same field but in a different period of time	Growing period is lengthy Chances of pest infection and disease spread are high Second crop may face drought as the amount of residual moisture is low
3. Multicropping	Two or more crops cultivated in a same piece of land with overlapping lifecycle	Large number of crops can be grown in same field	Crops sharing the same field can come to the exposure of each other in one of their growth phases, which may lead to spreading of diseases or may promote negative interactions between the crops
4. Intercropping	Multicrop system involve two or more annual crops growing in a same field at a same time	Maximum utilization of land in a single time	Any of the simultaneously growing species may retain in the field
5. Mixed intercrops	Multicrop system where different plant species are randomly arranged	Better utilization of land Different crops can be grown without affecting the yield of main crop	Soil quality may deteriorate
6. Row intercrop	Multicropping in arranged rows with different species	Different crops can be sown and harvested separately	Demand of soil nutrient is high
7. Agroforestry	Along with tree or shrubs, annual or perennial crops are cultivated	Multiproduct harvesting at a time Better utilization of resources	Resources can become limited in a short period of time Proper guarding system is needed to protect the crop

1.2.5.2   *INTEGRATED FARMING OF LIVESTOCK AND CROP*

In developing countries, the revenue generated from livestock farming is a strong base for agricultural economy (Singh et al., 2011). The integrated farming practice of livestock and crop is important both ecologically and economically as it is resource saving and sustainable way of agricultural production. In integrated livestock and crop cultivation, both the system shares the same field and resources (Seufert et al., 2019). Moreover, the waste from one system can be supplied as manure for the other system. In the African countries, cattle grazing in the fields after harvesting of crops serves as 40% of forage energy source (Stotz, 1983). This system is especially useful for smallholders in remote places for their day-to-day income and household supply of food. Integrated and comprehensive method of aquaculture, agriculture, and livestock rearing in a same time and space is another important aspect of sustainable agriculture. It is a very ancient practice in some of the countries of Asia. These integrated farms are mostly community controlled rather than being any farmer's personal property. These farms are mostly found in the agricultural lowlands with water resources taken from rivers, pools, etc. (Yang et al., 2001). The farms grow mostly Gramineae species (mostly paddy). Naturally grown aquatic plants are used to feed the fishes and other livestocks.

1.3   SCENARIO OF SUSTAINABLE AGRICULTURE IN ASIA

Organic agriculture is a rapidly growing sector, and 141 countries of the world have already taken the initiatives to incorporate organic farming. The survey data from certified organic farming throughout the world showed that 32.2 million hectares of agricultural lands are managed organically by more than 1.2 million producers, including smallholders.

Approximately 2.9 million hectares of agricultural land in Asia are under organic farming. This constitutes 9% of the world's organic agricultural land. About 230,000 farmers are now involved in this practice. The leading countries in organic farming are China (1.6 million hectares) and India (1 million hectares). The total shares of organic fields of all agricultural land are 7% in India and China (Narayan, 2012). Shrimp and fish farming in different parts of China, Vietnam, Malaysia, and other developing countries have become an important part of sustainable culture.

Saline soil is the most challenging area for agricultural production. The total mangrove coverage in the world is near about 15 million hectares where the farmers face a great challenge as tides make the soil saline and nonfertile in a regular basis. To overcome this problem, the farmers have initiated integrated farming where the crop production and livestock culture are practiced side-by-side. The aquaculture farms in Bang La, Vietnam, are the classic example of integrated farming along with agriculture, mangrove plantation, and mangrove conservation. Here, the main household income comes from aquaculture (40.3%) along with agriculture (24.4%) and due to sustainable farming and mangrove plantation, people managed to survive the devastation of the storm hit in 2005 in the Northern coast of Vietnam (Dat & Yoshino, 2013).

1.4   SUSTAINABLE AGRICULTURAL DEVELOPMENT AND ORGANIC FARMING IN DEVELOPING COUNTRIES

1.4.1   SUSTAINABLE AGRICULTURE

China and India share a large portion of World's population. To establish food security, both the countries are continually working to improve their agricultural sector and succeeded in that venture. In this section, we will discuss how China, India, and other Asian countries have adopted sustainable agriculture. China is one of the leading nations to introduce sustainable agriculture in their agricultural sector. From 2005 to 2018, five-fold increase in organically cultivated land was reported from China. In the export of organically cultivated crops, China is in fourth position in the world. The use of compost, cover crop and intercropping, and avoiding the use of chemical pesticides are the major steps taken by the Chinese toward sustainable agriculture. Almost US$65 billion of agrifood is exported from China and China's most important markets are Japan, Europe, and United States (https://theprint.in. April, 2020). According to the experts, the Chinese Government, along with the farmers and common people, must ensure food safety along with good food quality. Good-quality organic food production will also secure better domestic and export markets. Large-scale use of pesticides caused the rejection of exported agricultural products from China. So it was advisable that the Chinese government should enforce conversion of use of chemical pesticide toward organic ones. In the present days, many of the laws related to pesticide and other agrochemical manufacture, certification, and use have been implemented by the Chinese government (https://www.wilsoncenter.org, 2003).

Apart from China, India has started sustainable agriculture on a large-scale basis. In the context of sustainable agriculture, another term "alternative agriculture" is being used recently. In most of the cases, alternative agricultural practices are effective ways of agricultural sustainability. Any food or fiber production that has a more thorough incorporation of natural processes, reduced use of off-farm inputs with less harm to environment, and consumers are categorized as alternative agriculture. The farmers and policy makers should keep in mind that the degree of sustainability in agriculture varies with the types of crops, environment, and socioeconomic status. A sustainable system will be more resource conservative and needs less external inputs. In India, a new approach in farming system has been incorporated in the areas where the environment is challenging like estuarine islands, coastal areas, hill areas, etc.

To overcome the food demand of a very large population, the goal of Indian agriculture was always to attain the high production of crops without the concern of environment. From 1970 to 1980, India underwent significant rise of production, while in 1990, it started declining. Agricultural sector faced negative growth rate from 2000 to 2003 (Cook et al., 2016). In the time of high production, introduction of chemical fertilizers and pesticides boosted crop production but after long use of those harmful chemicals, the fertility and the quality of soil deteriorated drastically (Narayan, 2012). So, it was evident that, to achieve long-term productivity of the soil, the country must go for eco-friendly ways. By adopting organic farming method, India can not only achieve food security but also can draw attention in international market. To reach this goal, the Indian Government started the National Program for Organic Production (NPOP). Starting from providing the means to evaluate certification programmes for organic agriculture and products, deploying different policies for the development, and endorsement of organic products, formulating the National Accreditation Policy and Programme are among the main agendas of this program to encourage the development of organic farming and organic processing (Narayan, 2012; Vyas, 2003). In October 2000, the Ministry of Agriculture developed an organic agriculture task force that was separate from the NSCOP. The objectives of this separate task force include—advising the Ministry of Agriculture on all aspects of organic agriculture; promoting general awareness of organic agriculture; and developing a map for organic agriculture. State governments have initiated a number of projects throughout the 26 states of India. The setting of organic farm models, guidance about certification, knowledge about composting, and other practices relevant to arranging conferences, providing subsidies, and providing training are some of the steps taken by governments (Vyas, 2003). Organic farmers are getting financial support from government during the period of transformation (3 years) and after that. The farmers are rewarded US$228 (Rs.10,000) per hectare approximately for adopting organic agriculture. But, this programme is very limited in scale and can be pursued under particular schemes only. The total support cost to India is now well over US$ 80,000 (Rs. 3.5 million) per year (until 2006–2007; Narayana, 2102). Financial support is targeted to compensate for potential losses, to promote organic agriculture, to support and develop infrastructure, to conduct feasibility studies, and to prepare guidelines for "package of practices." Other specific incentives exist for organic farmers too.

Another Asian country, Bangladesh, also has agri-based economy. Most of the land in the country are low lying and salt water intrusion, flood, soil erosion drought, and other natural calamities often take a disastrous form. The agricultural sector of Bangladesh suffers climate-driven difficulties due to unplanned and unorganized agricultural practices. To solve this problem, a special wing of World Bank has started sustainable agricultural practices in Bangladesh. They have started promoting stress-tolerant high-yielding seeds and are educating farmers to manage climate friendly agricultural methods. Women farmers have been included and educated to take part in integrated farming. Promotion of organic products and new market set up were initiated for better distribution of products and revenue generation. Bangladesh Government has also taken many fruitful steps to promote sustainability in the agricultural sector (https://www.ifc.org, 2014).

Both in Malaysia and Vietnam, sustainable agriculture has occupied a large share. Malaysia has adopted sustainable agriculture in the form of precision agriculture. Wide use of advanced technologies, like remote sensing, different navigation and positioning system, drones, and robots are used for agricultural purpose in the areas where human-mediated agricultural activity is difficult. According to a survey, the on-growing sectors that showed excellence under precision agriculture are rubber production, oil palm production and forestry, and timber production (Khorramnia et al., 2014).

Vietnam is one among the world's leading countries of exporting agricultural products. The immediate goal is to reach the export ratio of 2019. In 2018, the country achieved excellence in export of agriculture and forestry products with a record turnover of over 40 billion USD (Doung et al., 2020). The country reached a goal of earning revenue of 48.6 billion USD by exporting agricultural and agro-forestry products by the end of 2021 which is nearly 15% increase compared to 2020.

1.4.2 POSSIBLE ACTIONS

There is a strong need of some initiatives that should be taken by the government and common people to improve the status of sustainable agriculture in India. Not only production-related issues but market access and transport of agricultural product are equally important. Approval of loans in a large scale and other financial support should be given to the needy people even from remote places. Propaganda of detail methods and awareness for sustainable farming through different media will be needed to make the system popular to common mass. Government of some states in India have started these awareness programmes for the sustainable culture.

1.4.3 INTEGRATED FARMING IN MANGROVES—SPECIAL REFERENCE TO MANGROVE FORESTS

High-yielding rice cultivation started in Indian Sundarbans villages in the early eighties. As a result of use of urea, soil quality had fallen drastically. Cultivating high-yielding varieties was practiced still the end of 1994–1995. During that time, due to the high cost of fertilizers, some of the farmers started showing interest in organic farming. From 1996 to 1997 onward, organic farming started in the Sunderbans. After the devastation caused by cyclone Aila, farmlands of Sundarbans suffered a lot of damage and the farmers realized the importance of integrated farming. Still, going organic was a time-taking process. Some NGOs are still trying to build different models where food security and market returns are integrated through organic farming, and can set a good example of adaptation to climate-friendly agriculture.

1.4.3.1 CLIMATE-ADAPTIVE FARMING

For the promotion of climate-friendly farming in the Sundarbans, the NGOs have set up a training module and a market-connected strategy with the hope that it will help the farmers to switchover to organic farming of rice, turmeric, pulses, and other items. They used online marketing to sell the products.

A survey in the year 2017 found that 70% of farmers in the area practised agriculture which was mainly dependant on chemicals, while only 10% used organic inputs exclusively. It takes a lot of effort of the NGOs to convince people for organic farming. According to the local people, the overuse of fertilizers spoils the quality of soil. Large amount of unused fertilizers, such as urea by crops, releases unused ammonia, which pollutes the environment. Only 40–50% of the applied urea can be utilized by a plant. Nitrate toxicity is another outcome of application of urea. Phosphate fertilizers cause arsenic pollution, which is the cause of chronic kidney disease in human beings.

1.4.3.2   ECOSYSTEM MANAGEMENT

In financial year 2017–2018, 100 farmers were trained to use organic farming for over 80 acres. Next year, the target was to train 300 farmers for using organic farming in 165 acres. Introduction of sustainable agriculture in the Indian Sundarbans included growing vegetables in the adjacent spaces of their houses during the extra hot summers for their own food demand, and honeybee culture in the fields was encouraged to facilitate pollination of the crop plants and production of honey for exportation and domestic consumption.

1.4.3.3   LOCAL RICE VARIETIES

The use of salt-tolerant crop varieties in different countries in Asia has brought a great success in the crop cultivation in saline soil (Iqbal et al., 2019). Farmers of Sunderban have the proof that local varieties of rice are hardier and more adapted to survive through the climatic ups and downs. About 140 local varieties have been identified in the farming area. Adoption of better farming method helped to reduce the farming costs. Today, the expense of organic rice variety in one *bigha* (0.33 acres) of land is INR 3700, while the equivalent for high-yielding rice is INR 5500; profit for organic rice is about INR 3300, while for the high-yielding varieties, it is about INR 2500 (indiaclimatedialogue.net, 2018). Indian Sunderbans undergo frequent storm surge and increasing salinity of soil and water due to sea level rise and declining river flow as the rapid climate change occurs. As a consequence, these calamities affect approximately the lives of 4 million people living here, like in other Asian deltas. To ensure the food security and livelihood of the inhabitant of the Sunderban area, cultivation of salt-tolerant crops mainly rice is the most useful way. The scientists of Central Soil Salinity Research Institute (CSSRI) have taken an initiative to help the farmers to implement new farming techniques. Use of improved salt-tolerant rice varieties along with proper water harvesting proved to be successful to overcome the climate-driven difficulties in farming in different areas of Sunderbans (https://cssri.res.in, 2018). It allowed the farmers to grow a second crop in the same field and earn profit in lakhs. Super cyclones like Aila and Amphan in Indian Sunderban area helped the local farmers realize the benefit of cultivation of salt-tolerant rice varieties that were native to that soil. Increase of one unit of salinity after a certain limit causes the drop of 12% in the crop yield for rice. Increase in sea level and repeated flood led to excess salt covering in agricultural fields. To cope up with the saline soil, scientists try to develop advanced variety salt-tolerant rice species. International Rice Research Institute (IRRI) had put efforts to get suitable tolerant varieties, which ran a project for Stress Tolerant Rice for Africa and South Asia from 2007 to 2019 and distributed 1 million tons of stress-tolerant rice seeds to 30 million farmers in Africa and South Asia (https://cssri.res.in, 2018). India's CSSRI has developed 120 stress-tolerant rice varieties till date, 40 of which are highly salinity resistant. According to the head of CSSRI regional field station at Canning Town in the Sundarbans of West Bengal, the available rice varieties are suitable for cultivating in monsoon and dry season. CSSRI also produces salt-tolerant seeds in accordance with different types of land and rice consumption preferences. Seeds tolerant to multiple stressors like salinity as well as drought are also necessary for saline soil rice cultivation. Scientists have developed different rice varieties, like Canning 7, which can tolerate higher salinity levels, while Amal Mana and Swarna Sub 1 can tolerate many days of flooding (https://cssri.res.in, 2018).

Dams like Farakka constructed in the Bhagirati river have drastically reduced the freshwater flows. Matla, one of the main rivers of Sunderban, is now largely fed by the monsoon rain water and tides, thus lacking fresh water in the dry seasons. Reduced fresh water flow fuels the salinity increase in the associated water and causes imbalance in the ecosystem. Population of many mangrove species along with Sundari is decreasing day-by-day as a result of drastic change in the salinity regime. After the devastation by Aila in 2009, the farmers were forced to leave their fields abandoned for salt encrustation. Some of the farmers in Gosaba block of Sunderban said in 2019 that the salts remaining in the agricultural fields made them unable to cultivate profitable vegetables like chilli even after 10 years of Aila. Another study in Gosaba block by Debnath (2013) said that the production of paddy, wheat, sugarcane, chilli, and pulses was heavily destroyed after Aila in 2009. The average production of Boro Paddy reduced to 20,833 kg/hectare from 34,671 kg/ hectare, and Aman paddy production reduced to 14,525 kg/hectare from 28,004kg/ hectare after Aila. Boro paddy production suffered maximum loss in Rangabelia mouza and Hentalbari mouza underwent maximum loss of Aman paddy production. According to an expert's opinion, use of organic fertilizer, proper soil management, and proper irrigation system are needed to overcome this problem and to establish a sustainable system in the Sundarbans (Debnath, 2013). Then, there were Amphan in 2020 and Yaas in 2021. Both of these super-cyclones made a remarkable devastation in Sunderbans. Though after Aila, the local farmers in Sagar Island revived heirloom rice varieties that lost importance after the Green Revolution. After Aila, these rice varieties flourished in newly saline soil where seeds with high yields failed. CSSRI also collected 63 traditional varieties and created new varieties. Traditional varieties are hardy but low in yield. Moreover, the grain shape and texture is often not lucrative. So, it is still a challenge for scientists to create seeds that satisfy diverse requirements. In Gosaba, farmers need was tall, long-duration rice for the monsoon. Yield and texture were not a matter of concern as monsoon rice is mainly cultivated for home consumption. In the dry season, farmers prefer to sow high-yielding salt-tolerant, long fine grained breeds called Laliminikit for the demand and high market value of the variety. Gosaba farmers are able to sow a second rice crop due to availability of enough fresh water but neighboring island of Bali does not (www.ipsnews.net/2015). Most important fact is these rice varieties dosen't require too much fertilizer or pesticides, which is beneficial for the farmers who are below poverty level. To fulfil these needs, CSSRI and NGOs like the Tagore Society for Rural Development are promoting water harvesting, drip irrigation, soil management, and crop diversity, alongside better seeds. In Rangbelia, some farmers have started growing grass pea, or *lathyrus* sp, along with rice in the dry season. They have used small ponds to store water for irrigation and cultivate fresh-water fish that brought some extra money. Dry straw mulching is used in some fields to protect soil moisture. CSSRI is promoting a new technique, which is new to the farmers of this area. In this technique, the land with ridges and furrows structure is used (https://cssri.res.in, 2018). The furrows solve the purpose of drainage of extra water and the less-saline ridges are used to grow vegetables. Thus, sustainable agriculture shows a ray of hope for economic growth and the battle against poverty in Sunderban. According to the report from five villages of Sagar block by Roy (2016), existing agricultural practices in Sunderban need to be modified and advanced method of sustainable rice cultivation should be introduced.

1.4.3.4 INTEGRATED FARMING IN MANGROVE

Gradual sea level rise, frequent storm surge, and regular flood with saline water cause lack of fresh water and make the situation more difficult for farmers to gain success in agricultural production in the Sunderban. So, farmers of Sunderban have chosen integrated farming method to survive the climate-related difficulties in this delta region. The farmers use their agricultural fields to cultivate different species throughout the year. The number of species grown in same field sometimes crosses more than 25 in a year. The lowlands adjacent to the paddy field, which have sufficient water, are transferred to aquaculture ponds. The integrated farming method in the Sunderban region can enhance the production amount to upto 200% higher than conventional farming. Success in the integrated farming can be achieved if the government implements policies to help the small-scale and landless farmers. It is also suggested that if the farmers work together to introduce integrated farming in large fields, the production will be more (www.ipsnews.net/2015). A report of 46 villages in North and South 24 Parganas says that after the Aila devastation in 2009, Development Research Communication and Service Center started working and introduced integrated farming in the said villages. Apart from paddy, they have started growing many vegetables in the fields throughout the year, Moreover, they have fish ponds to look after and other livestock like chicken, etc., along with a biogas plant, which run with the excreta from their livestock rearing. The group of farmers working there also managed to build a grain bank with the support of Area Resource Training Center of Development Research Communication and Service Center. In spite of making personal approach, the villagers are working here as a group or community and this venture got immense success in last few years (www.indiawaterprtal,org. 2015). Another group of farmers are working in integrated farming method in villages Patharpratima block of Sunderbans with the help of Development Research Communication and Service Center and have shown success in that approach.

1.5   FUTURE PERSPECTIVE

Implementation of sustainable agriculture is largely dependent on combined and integrated action of the government, public sectors, and the common people of the country. The size of agricultural land may be another important aspect for introducing sustainable methods in farming. According to the global survey data, many countries have adopted sustainable farming and play an increasingly important role. Asian countries, like China, have implemented sustainable agriculture successfully with the help of technologies and institutional innovations. But agriculture–environment agreement on a more sustainable track is facing great challenges today to ensure food security. The future of sustainability in the agricultural sector lies on some key points, which may be pointed out as follows:

1.   Improvement in agricultural technology
2.   Implementation of performance-based environmental policies
3.   Maintaining food quality
4.   Minimizing food waste

KEYWORDS

- **sustainable agriculture**
- **crop rotation**
- **food security**
- **climate change**
- **organic farming**

REFERENCES

After cyclone Aila, Farming Nurtures Food, Faith in Sunderbans, Oct, 2015. https://www.indiawaterportal.org

Borremans, L.; Marchand, F.; Visser, M.; Wauters, E. Nurturing Agroforestry Systems in Flanders: Analysis from an Agricultural Innovation Systems Perspective. *Agric. Syst.* **2018,** *162,* 205–219.

Cheema, Z. A.; Farooq, M.; Wahid, A. *Allelopathy: Current Trends and Future Applications*; Springer-Verlag: Heidelberg, 2013.

Cook, S.; Henderson, C.; Kharel, M.; Begum, A.; Rob, A.; Piya, A. Collaborative Actions on Soil Fertility in South Asia: Experiences from Bangladesh and Nepal, 2016.

Dat, P. T.; Yoshino, K. Comparing Mangrove Forest Management in Hai Phong City, Vietnam Towards Sustainable Aquaculture. *Procedia Environ. Sci.* **2013,** *17,* 109–118.

Debnath, A. Condition of Agricultural Productivity of Gosaba C.D. Block, South24 Parganas, West Bengal, India after Severe Cyclone Aila. *IJSRP,* **2013,** *3* (7), 1–4.

Dey, P.; Srivastava, S.; Lenka, N. K.; Shinogi, K. C.; Vishwakarma, A. K.; Patra, A. K, *Integrated Nutrient Management for Improving Soil Health and Crop Productivity, A SAARC Training Manual*; ICAR-Indian Institute of Soil Science: Bhopal, India, 2018; p 124.

Duong, D. T. Sustainable Development for Vietnam Agriculture, E3S Web of Conferences 175, 01015 (2020), *INTERAGROMASH 2020.* https://doi.org/10.1051/e3sconf/202017501015.

Exploring Sustainable Agriculture in China. https://www.wilsoncenter.org (accessed Mar 12, 2003).

FAO. *The Future of Food and Agriculture: Alternative Pathways to 2050*; Food and Agriculture Organization of the United Nations: Rome, Italy, 2018; p 228.

Fracis, C.; Porter, P. Ecology in Sustainable Agriculture Practices and Systems. *Cirt. Rev. Plant Sci.* **2011,** *30* (1–2), 64–73. DOI: 10.1080/07352689.2011.554353.

Global Hunger Continues to Rise. Press Release, Rome, Italy, September, 2018. https://www.unicef.org

Integrated Farming: The Only Way to Survive a Rising Sea. Inter Press Service, Jan 2015. https://www.ipsnews.net/2015.

Intergovernmental Panel on Climate Change (IPCC). Summary for Policymakers. Climate Change and Land: An IPCC Special Report on Climate Change, Desertification, Land Degradation, Sustainable Land Management, Food Security, and Greenhouse Gas Fluxes in Terrestrial Ecosystems; Intergovernmental Panel on Climate Change (IPCC): Geneva, Switzerland, 2019.

Iqbal, S.; Basra, S. M.; Afzal, I.; Wahid, A.; Saddiq, M. S.; Hafeez, M. B.; Jacobsen, S. E. Yield Potential and Salt Tolerance of Quinoa on Salt-Degraded Soils of Pakistan. *J. Agron. Crop Sci.* **2019,** *205,* 13–21.

Kassam, A. Advances in Conservation Agriculture Volume 2: Practice and Benefits; Burleigh Dodds Science Publishing: Sawston, 2020; p 498.

Khorramnia, K.; Shariff, A. R. M.; Abdul Rahim A.; Mansor, S. Toward Malaysian Sustainable Agriculture in 21st Century, IOP Conference Series: Earth and Environmental Science, Volume 18, 8th International Symposium of the Digital Earth (ISDE8), 26–29 August **2013,** Kuching, Sarawak, Malaysia.

Making the Agricultural Sector More Sustainable and Resilient in Bangladesh, Oct 2014. https://www.ifc.org

Mbow, C.; Smith, P.; Skole, D.; Duguma, L.; Bustamante, M. Achieving mitigation and adaptation to climate change through sustainable agroforestry practices in Africa. *Curr. Opin. Env. Sust.* **2014,** *6,* 8–14.

Organic Farming Gathers Steam in Indiaan Sunderban, May 2018. https://indiaclimatedialogue.net.

Roy, C. The Sustainability of Rice Cultivation: Evidence from Sunderbans in India. *Int. J. Humanit. Soc. Sci.* **2016,** *4* (5), 160–164.

Seufert, V.; Ramankutty, N. Many Shades of Gray—The Context-Dependent Performance of Organic Agriculture. *Sci. Adv.* **2017,** *3* (3), e1602638. DOI: 10.1126/sciadv.1602638.

Sharma, P. C.; Singh, A.; Choudhary, M. ICAR-Central Soil Salinity Research Institute. Annual Report, Karnal, Haryana, India, 2018–19.

Sharma, P. C.; Singh, A, ICAR-Central Soil Salinity Research Institute, Annual Report, Karnal, Haryana, India, 2016–17.

Singh, Y. P.; Naresh, R. K.; Singh, P. K.; Singh, H. L.; Rathi, R. C.; Singh, S. V. Crop-Livestock Interaction to Food Security and Sustainable Development of Agriculture for Small and Marginal Farmers in North West of India. *J. Rural Agric. Res.* **2011,** *11* (2), 6–13.

Stotz, D. *Production Techniques and Economics of Smallholder Livestock Production Systems in Kenya. Farm Management Handbook of Kenya,* 4th ed.; Ministry of Agriculture Republic of Kenya: Nairobi, 1983.

Veerapa, N. K.; Chien, T.M. Sustainable agricultural practices in Vietnam: A Supply Chain Perspective of Organic Vegetable Production in Hanoi Agricultural Practises in Vietnam: A Supply Chain Perspective of Organic Vegetable Production. *Acta Hortic.* **2012, 2958,** 35–42. DOI:10.17660/ActaHortic.2012.958.2

Vyas, V. S. *India's Agrarian Structure, Economic Policies and Sustainable Development;* Academic Foundation Publishers: New Delhi, 2003.

Why China Is Emerging as a Leader in Sustainable and Organic Production? 13th April, 2020. https://theprint.in

Yang, J.; Zhang, J.; Wang, Z.; Zhu, Q.; Wang, W. Remobilization of Carbon Reserves in Response to Water Deficit During Grain Filling of Rice. *Field Crop Res.* **2001,** *71* (1), 47–55.

CHAPTER 2

Traditional Agriculture: A Sustainable Approach Toward Attaining Food Security

S. SHWETA[1], B. SUBRAMANIAM[2], MANOJ K. RAI[3], SHAGUN DANDA[4], ANIL KURMI[5], and SANDEEP KAUSHIK[3]

[1]*Department of Botany, Guru Ghasidas Vishwavidyalaya, Bilaspur, Chhattisgarh, India*

[2]*Retired Chief Scientist and Coordinator, Traditional Knowledge Digital Library Project, CSIR-National Institute of Science Communication and Information Resources, New Delhi, India*

[3]*Department of Environmental Science, Indira Gandhi National Tribal University, Amarkantak, Madhya Pradesh, India*

[4]*Department of Botany, Daulat Ram College, University of Delhi, Delhi, India*

[5]*Plant protection Division, Krishi Vigyan Kendra, Indira Gandhi National Tribal University, Amarkantak, Madhya Pradesh, India*

ABSTRACT

The challenge confronting human beings in the near future is to enhance crop productivity and simultaneously mitigating the adverse environmental impacts arising out of such an agricultural practice. The present-day

Crop Sustainability and Intellectual Property Rights. Soumya Mukherjee, Piyali Mukherjee & Tariq Aftab (Eds)

© 2023 Apple Academic Press, Inc. Co-published with CRC Press (Taylor & Francis)

agricultural practices meet the ever-increasing food requirements, which is causing depletion of the genetic resource base and relying chiefly on a small number of varieties meant for intensive cultivation. This has also led to a severe loss of agricultural biodiversity along with a near to or total disappearance of several traditional varieties leading to a reduction in the diversity of plant species under cultivation regime. High input or industrialized agriculture has also been a cause of concern for the changes it brings out in the soil profile, soil microbiota, loss of pollinators, and insect predators besides causing severe damage to the soil, water bodies, and air. Interestingly, traditional agricultural practices such as agroforestry, intercropping, crop rotation, ley farming in arid zones, use of organic manures, mulching practices, use of trash lines, combined raised beds and canals, use of retention ditches, composting and mulching, use of plant-based insecticides are eco-friendly approaches are likely to address some of the above-mentioned issues such as loss of agrobiodiversity, nutrient loss, water deficit, etc. Besides traditional crops, the use of diverse underexploited crop plants and land races existing with the tribal people that lack formal crop improvement will provide a possible pathway toward attaining the food security in a more sustainable way. The agricultural heritage systems that are eco-friendly sustainable systems, preserve the agrobiodiversity, and meet the food and nutritional requirements of the local people needs to be preserved in their entirety along with their life support systems. The current chapter focuses on the traditional practices which can be employed to combat the problems posed by modern agricultural practices and paving way toward fulfilling the food preferences and dietary needs of the people, a food-secure nation not only inculcating physical, social, and economic access to sufficient, safe, and nutritious food but also conserving the genetic resources, agrobiodiversity, and an overall sustainable development.

2.1 INTRODUCTION

Agriculture started about 10,000 years ago that enabled the settlement of nomadic people (Smith, 1997; Diamond and Bellwood, 2003; Balter, 2007). Coupled with the beginning of agriculture, domestication of crop species began. The developments in agriculture were primarily related to enhance productivity and improve nutritive value of the crop produce. Crop species that were native to the region were identified and selected for cultivation by the local people. Crop produce that were acceptable by the people were preferred for cultivation and this paved their ways to the market. With advent of time and ever-increasing population, the food security became a major concern. Crop improvement programmes were subsequently taken up by the agricultural scientists on a few selected desirable crop species. This involved genetic improvements through conventional breeding for enhancing productivity and nutritive value, for resistance to pests and diseases, tolerance to various edaphic factors, varieties with adaptability to different agro-climatic zones; these efforts resulted in release of genetically elite varieties. The use of agrochemicals such as commercial chemical fertilizers and pesticides was liberal for enhancing agricultural output.

High input agriculture or industrialized agriculture has been necessary for meeting the food requirements of the burgeoning population. However, high input agriculture has made an adverse impact on the environment. It has led to climate change, food unsafety, biodiversity loss, and environmental pollution (Zhang et al., 2017). The rampant use of chemical fertilizers and commercial insecticides/pesticides has led to accumulation of the same in soil, air, and the water bodies, loss of soil microbial diversity, decline in the population of natural pollinators and predators, rapid evolution of insect pests, reduction of the carbon reserves of the soil, secondary salinity, decreased soil fertility, loss of top layer of soil due to erosion, increased cost of production, eutrophication of water bodies (Ram, 2003; Sharma, 2005), and has also led to an increase in use of fossil fuels by the agro-industries emitting greenhouse gases.

India has a wealth of genetic resources that is used in traditional agriculture chiefly by the tribal people. Land races of the cultivated crops form a valuable resource since these are cultivated without inputs of agrochemicals, and also have better nutrient profile and yield features compared to the most of the varieties under cultivation at present. Several traditional agricultural practices exist in India, and since these are sustainable, this knowledge could be made use of in present-day agricultural practice.

Reliance on few crop species for meeting most of the energy requirements of human beings in present-day agricultural scenario has reduced the diversity of crop species under cultivation. Most of the arable lands of the world (~80%) grow only a few crops such as wheat, rice, corn, sugarcane, soybean, etc. (Heinemann et al., 2013; Altieri et al., 2015). Intensive monoculture of high-yielding varieties adapted to local conditions has also led to loss of genetic variability within the crop species (Evenson and Gollen, 2003; Sardaro et al., 2016). High-yielding varieties that are genetically identical have replaced the local varieties and this has led to a loss of about 75% of world's crop diversity (FAO, 2009a, 2009b, 2009c; Gonzalez, 2011).

Traditional farming systems, however, have played a key role in preserving and cultivating diverse crop species and land races on subsistence farming basis for meeting the local food requirements. In fact, traditional farmers are the custodians of traditional crop varieties that can tolerate environmental stresses (Chhatre and Agrawal, 2008; Gonzalez, 2011). They are also using the age-old traditional agricultural practices generations after generations for raising the crops, which is dynamically evolving based on the local conditions and requirements. Traditional agroecosystems that have been practiced for thousands of years (Pulido and Bocco, 2003) are time tested and have continued as sustainable agriculture systems (Ellis and Wang, 1997). Their crops have not undergone any formal crop improvements, instead the farmers retain their own seed material as planting material from previous crops. They also do not use expensive commercial agrochemicals as it is used by the modern day farmers. The diverse crop species used by the traditional farmers have remained largely neglected and still occupy a very small area under cultivation. These underexploited traditional "poor man's crop species" and land races are being viewed as the crops for the future, since they are able to sustain harsh environmental conditions and are climate resilient crops, produce reasonable yield even under adversities of soil and weather conditions, are nutritionally superior to most of the present day cultivars, have fair amount of resistance to pests and diseases, do not require the use commercial fertilizers since the age-old agricultural practices followed by the traditional farmers maintains required soil fertility and minimize the loss of soil due to erosion. Hence, the underexploited crops have an ability to withstand challenging environmental conditions and can be grown on marginal lands and have a potential to augment total agricultural produce.

Integration of traditional agricultural practices is likely to play a vital role in alleviating the environmental crisis, which has arisen due to modern-day agricultural practice. Though high input agriculture is the common practice in many regions of the world but traditional systems of agriculture are also meeting the food requirements of the people and perhaps could play a major role in food security in a sustainable way in times to come. This chapter examines different traditional agricultural practices, agricultural heritage systems that preserve the agrobiodiversity and traditional eco-friendly agricultural practices, underexploited crops that have been cultivated for thousands of years but now are cultivated in few pockets, and have potential to provide food and nutrition security in a sustainable way.

2.1.1 PRACTICES ADOPTED IN TRADITIONAL SYSTEMS OF AGRICULTURE

There is a need to adopt climate-smart approaches where the food production can be enhanced sustainably, coupled with climate adaptation and resilience besides reducing the emission of greenhouse gases (FAO, 2010). Traditional agricultural systems are considered to be climate-smart approaches for sustainable food production (Singh and Singh, 2017). Several organic farming practices have been followed traditionally that are sustainable, for example, intercropping, mixed cropping, crop rotation, ley cropping, cover cropping, organic composting, crop-livestock farming, use of trash lines, retention ditches, use of ridges, combined raised beds and canals, mulching practices, agroforestry, and use of botanicals in insect pest controls are likely to increase the crop productivity, at the same time mitigating climate change (Sharma, 2005; Singh and Singh, 2017). Some of the traditional systems of agricultural patterns and practices are discussed below.

2.2 TRADITIONAL AGRICULTURAL CROPPING PATTERNS

2.2.1 INTERCROPPING AND MIXED CROPPING

Intercropping refers to a cropping system where, in the same piece of land two or more crops are simultaneously cultivated in specific row pattern with the objective of enhancing total crop productivity and enhancing agrobiodiversity (Shava et al., 2009; Hu et al., 2017). Intercropping is done to optimally utilize the space in the agricultural land by cultivating another crop in the space left between the rows of the main crop. Intercrops provide stability to the ecosystem and have an ability to optimally utilize the resources such as water, nutrients, light, land, etc. to give more yield (Mushagalusa et al., 2008; Ning et al., 2017). In India, intercropping is an age-old traditional practice and several intercrop patterns are followed in different agroecosystems. For example, cereal pulse intercrops are common such as sorghum and pigeon pea that are grown as intercrops (Wang et al., 2010).

Mixed cropping on the other hand is a cropping system where two or more crops are cultivated in the same piece of land simultaneously to reduce the risk of crop failure due to adverse climatic conditions (Kaushal et al., 2016). Mixed cropping is done to obtain at least one crop and it does not follow any specific planting pattern. Mixed cropping includes diversification of food besides pest and disease suppression (Sauerborn et al., 2000). In India, millets are often cultivated in mixed cropping systems as they are climate resilient crops and drought tolerant (Kaushal et al., 2016), and assure reasonable crop productivity even in case of adverse environmental conditions.

Tropical soils are often poor in available nitrogen, hence the productivity is lower since nitrogen is a limiting factor in crop productivity. Intercropping or mixed cropping with legume crops, integrating legume trees in agroforestry system or crop rotations with legume crops, which could be grain legumes or forage crops as practiced in traditional agriculture would ideally overcome the soil limitations with respect to availability of available nitrogen.

2.2.2   CROP ROTATION

An agricultural practice where different crops are cultivated in sequence on the same piece of land and it is a practice commonly followed by traditional farmers (Hobbs et al., 2008). Crop rotation is a sustainable system where there is increase in agricultural productivity and reduces soil erosion (Huang et al., 2003). It is a practice adopted to mitigate agroecological problems such deterioration of soil quality and climate change resulting from monocropping (Liu et al., 2016). It is also considered to be effective in carbon sequestration (Triberti et al., 2016). The use of leguminous crops enhance the soil nitrogen status reducing use of commercial fertilizers, which in turn are synthesized using fossil fuels, hence there is reduction in carbon dioxide emissions (Wang et al., 2010). Cowpea is one-grain legume used in crop rotation programmes, which is able to grow in nutrient-poor soils and is an efficient nitrogen fixer (Sanginga et al., 2000; Singh et al., 2003).

2.2.3   TRADITIONAL LEY FARMING IN DRYLANDS

Traditional ley farming is practiced in the drylands of India. Ley farming is growing of grass or pasture legumes in rotation with grain or tilled crops on the marginal lands. It not only conserves the soil but also provides fodder and edible grains. Ley farming has been one of the best practices for restoration of fertility in the dry regions of India (Muthana et al., 1985). The structure of the marginal soils as well as the microbiotic activities in the soil improves enabling enhanced yields. Ley farming in arid zones improves retention of moisture and rates of infiltration. Forage production in ley farming also enhances viable VAM spores, organic matter content, nitrifying bacteria, and total nitrogen besides improvement in various enzymatic activities in the soil. There is also an increase in yields of pearl millet grain and straw (Rao et al., 1997).

2.2.4   COVER CROPPING

Cover cropping is a sustainable approach practiced by traditional farmers. Cover crops are plants that are cultivated to reduce the soil erosion and nutrient loss (Dabney et al., 2001), improves soil health by enhancing the soil fertility, increase the soil carbon reserves once ploughed into the field, increase the soil microbes, increase carbon sequestration, weed, and pest control, help in nutrient recycling, and also help in conserving soil moisture (Schipanski et al., 2014; Frasier et al., 2016; Pinto et al., 2017). Use of leguminous crops adds to available soil nitrogen status thereby reducing the greenhouse gas emission in case the same quantity of nitrogen is commercially synthesized using fossil fuels. Some plants that are used as cover crops are pea, clover, beans (legumes), sorghum, rye, brassicas, etc. (Singh and Singh, 2017).

2.3   TRADITIONAL AGRICULTURAL PRACTICES

2.3.1   TRADITIONAL ORGANIC COMPOSTING

Traditional agricultural practices depend on the organic manures inputs that are made locally. The process of composting is the activity of microorganisms, which are able to decompose organic matter and form organic manure with the release of bound nutrients for their use by the plants. Organic composts have been used for several centuries (Oudart et al., 2015) and has been a sustainable and eco-friendly approach. It is a climate-smart approach to enhance soil fertility (Singh and Singh, 2017). The use of organic composts has several advantages such as it improves the soil organic matter content thereby builds up carbon reserves in the soil, provides ideal conditions for enhancement of soil microbiota and invertebrate communities, which help in nutrient recycling, facilitates absorption and retention of soil moisture, reduces soil erosion, improves cation exchange capacity, reduces risk of attack by pests and diseases (Peacock et al., 2001; Magdoff and Weil, 2004; Sun et al., 2004; Kremen and Miles, 2012; Ge et al., 2008; Zhang et al., 2012; Liu et al., 2013). It also reduces the greenhouse gas emissions indirectly since in case the commercial fertilizers are used it would lead to greenhouse gas emission since fossil fuels are required for making fertilizers (Wang et al., 2017). The use of commercial fertilizers would also lead to decline in the soil microbial diversity and its activities (Ding et al., 2017) particularly the nutrient recycling. In India, the use of composted farm yard manure is a common practice adopted in the traditional agriculture (Gopinath et al., 2008). Farm yard manure in the form of compressed cake is prepared by the farmers in Jharkhand using plant material and flowers. Some of the commonly used plant materials such as *Azadirachta indica, Madhuca longifolia* var. *latifolia* (syn. *Madhuca indica),* and *Derris indica* are also used to make the farm yard manure (Dey and Sarkar, 2011).

2.3.2   CROP-LIVESTOCK FARMING SYSTEMS

Traditional farming systems commonly integrates crop-livestock combinations, which form an interdependent system and provides agroproducts as well as livestock-based foods such as milk, meat, etc. Livestock feed is made available from the leaves of cultivated trees commonly used in agroforestry or grasses, whereas the wastes of animals form the organic manures for enhancing the fertility of the soil. Thus, there is enhancement of agrobiodiversity, diversification of the food obtained and at the same time maintaining the fertility of soil, organically improving the soil carbon content. Integration of crop and livestock is an approach that reduces the dependence of traditional farmers on external inputs such as fossil fuels, fertlizers, and pesticides (Naylor et al., 2005; Anex et al., 2007). It is one of the ideal method of land resource management and resilience of agroecosystems to changing climate (Singh and Singh, 2017). In Kerala, a mixture of cow's urine and dung is frequently applied to enhance the soil fertility.

2.3.3   TRASH LINES, RETENTION DITCHES

Trash lines are often used in traditional agriculture. Crop residues of *Sorghum*, millets, and clovers are placed in lines across fields in rows. They prevent the flow of water permitting infiltration of water besides preventing the soil erosion. Crop residues are those that takes a longer time for decomposition, however when decomposed release organic matter and nutrients to the soil. Crop residues also help in adsorption of moisture. It is also effective in enhancing the soil carbon reserves. Generally the crop residues used are less palatable to cattle (Eyong, 2007).

Retention ditches are made on steep slopes in mountainous areas. The purpose of making retention ditches is to prevent the runoff and assist in infiltration of water so that the crop roots get sufficient water to absorb. Besides retention ditches help in preventing soil erosion. Terraces are also built using stones and soil with similar purpose (Eyong, 2007)

2.3.4   USE OF RIDGES

Ridges are often made in traditional agricultural systems for cultivating certain crops such as groundnuts, potatoes, maize, radish, sweet potatoes, legumes, etc. Ridge top may be made flat or rounded (Eyong, 2007). Making ridges enhance the surface area of soil that helps in retaining moisture for longer period of time and better exchange of gases. It also helps in preventing soil erosion.

2.3.5   COMBINED RAISED BEDS AND CANALS

In traditional agriculture, raised beds and canals have temperature moderation effects for the growing season of the crop and leads to higher productivity (Altieri, 1999a, 1999b; Altieri and Koohafkan, 2008).

2.3.6   ORGANIC MULCHING PRACTICES

Organic mulching is commonly practiced in the traditional agricultural systems. In organic mulching, dried leaves of the plants are spread in the field. The advantages of mulching are: (i) conserves the soil moisture by preventing direct evaporation of moisture from the soil (Naskar, 2009), (ii) prevents the direct impact of hail and rain drops on the soil thereby helping in preventing soil erosion (Altieri and Koohafkan, 2008), (iii) reduces runoff losses (Telkar et al., 2017) and suppresses weed growth, (iv) biodegradation of the leaves contributes to the enrichment of the soil with the organic matter and nutrients (Eyong, 2007), (v) very effective in enhancing the soil carbon reserves, and (vi) regulates soil temperature (Altieri and Koohafkan, 2008; Naskar, 2009).

2.3.7   AGROFORESTRY PRACTICES IN TRADITIONAL FARMING SYSTEMS

Agroforestry practice of inclusion of tree crops with other agricultural crops in multistorey approach, ensures higher productivity per unit area and provides multiple agricultural products. Multipurpose tree species particularly legume trees would be desirable since these will not only provide diverse agroproducts but will also enrich the soil with available nitrogen. Trees serve as a wind break and shelter belts (Lasco et al., 2014). The trees also reduce evaporation besides reducing the impact of raindrop and hail. Hence, agroforestry is a strategy to mitigate microclimate variability by traditional farmers (Altieri and Koohafkan, 2008). The farmers create agroforestry designs and grow trees to provide shade cover to prevent the crop plants from extreme heat (Altieri and Koohafkan, 2008). Deep and well-spread roots of trees can absorb water and nutrients from large area of soil, thus enabling them to adapt to drought conditions (Verchot et al., 2007). Agroforestry optimizes space usage, reduce infestation by pests and diseases, prevents soil erosion, sustains soil fertility (Doddabasawa and Chittapur, 2020), provides food security (Nair et al., 2009; Coulibaly et al., 2017) and also conserves agrobiodiversity. Agroforestry is a climate-smart practice and has been recognized as a mitigation option for greenhouse gas emissions (Smith et al., 2007; FAO, 2009a, 2009b, 2009c; Singh and Singh, 2017) since they act as carbon sinks due to their high ability to sequester carbon thereby reducing atmospheric loads thus buffering against climate variability and alleviates climate change (Nowak and Crane, 2002; Montagnini and Nair, 2004; Altieri and Koohafkan, 2008; Nair et al., 2009; Obsertein et al., 2010). It also enhances the carbon density besides enriching the carbon reserves in the soil and improve water retention in soil (Obsertein et al., 2010; Zomer et al., 2016; Paul et al., 2017; Doddabasawa and Chittapur, 2020). The diversity that gets enhanced by agroforestry has the ecological role of replenishing the litter layer to protect soil and cycling of nutrient (Elevitch et al., 2014). It has been observed that in agroforestry systems, farmers suffer less due to weather-associated crop failures (Altieri and Koohafkan, 2008). Livestocks managed along with farming systems not only provide milk and meat but also provide manure and enhance carbon sequestration (Altieri, 1999a, 1999b).

In Kerala, traditional farmers who cultivate perennial cash crops carry out intercropping to enhance the productivity per unit area and also to prevent from devastation by pest or disease. In Arecanut plantations, bananas are commonly intercropped. Arecanuts are also intercropped with turmeric, cardamom, ginger, and chillies as cover crops. Black pepper climbers are trained on the arecanut plants. Sorghums, yams, pineapples are also inter-cropped with turmeric. Cowpeas are also grown that contribute to the soil nitrogen through biological nitrogen fixation.

There are different categories of agroforestry practices adopted in hill farming system in Sikkim that are briefly discussed below.

2.3.7.1   FARM-BASED AGROFORESTRY

Farm-based agroforestry is an agrosilvicultural practice that uses multipur-pose trees on terrace edges in and around the cultivable lands. Wastelands are often converted into the agroforestry systems that also reclaims the waste-lands. It is an important land use system on the slopes that maintains soil fertility by recycling of organic matter. Multipurpose trees provide timber, fuel and fodder, and other products besides preventing soil erosion. Legume trees are preferred since they also contribute to soil enrichment because of their ability to fix atmospheric nitrogen into usable form of nitrogen. Fruit tree crops act as wind breaks alongwith terrace paddy fields and grasses such as *Agrostis, Thysanolaena* cultivated for use as fodder for the livestock that are essential part of the agroecosystems. Terraces also include a number of lesser known crops that meets the protein and other nutrient requirements. Livestock is an integral component and management of fodder species is critical (Sharma and Dhakal, 2011; Sharma et al., 2009).

2.3.7.2   FARM FOREST-BASED AGROFORESTRY

The farm forest-based agroforestry practices include a number of multipur-pose species that meets their requirements of fodder for livestock, fuelwood, timber, etc. Bamboo grooves are also common in these agroforestry systems. Rice crop is also cultivated in the valleys. Organic matter, nutrients, and litter from this agroforestry system are used in the farms and thus supports the cultivated systems. Terraced slopes that are cultivated are protected by agro-forests on all the sides. Several nontimber products are also collected for use by the locals such as medicinal plants, fodder, tubers, wild vegetables, wild edible fruits, mushrooms, etc. Remarkably, the local people have stabilized the barren slopes by vegetation using the multipurpose species and also the fodder grasses for the livestock (Sharma and Dhakal, 2011; FAO SHA).

2.3.7.3 HIGH-VALUE CROP-BASED TRADITIONAL AGROFORESTRY

This traditional agroforestry having cardamom is known as high-value crop-based agroforestry practice. The large cardamom (*Amomum subulatum*) and Mandarin oranges (*Citrus reticulata*) grown in this agroforestry system are high-value cash crops. In this agororestry system, a variety of multipurpose keystone species that conserve the biodiversity grow such as *Artocarpus lacucha, Ficus, Erythrina variegata (*syn. *Eleusine indica)*, *Dendrocalamus, Thysanolema latifolia (*syn. *T. maxima)*, etc. However, alders (*Alnus nepalensis*) are most commonly grown. Alders associate with the actinomycetes *Frankia* and fix atmospheric nitrogen and are able to enhance the soil fertility. Their nutrient-rich litter also contributes to soil nitrogen and organic matter. *Alnus* agroforestry system has greater carbon stocks in comparison to other agroforestry systems (Rai and Sharma, 2004). Other multipurpose trees that are grown in agroforestry practice on terraced croplands are *Albizia* species, which is also a nitrogen fixing species. *Alnus* and *Albizia* have an ability to occupy degraded ecosystems and reclaim the ecosystems due to their nitrogen fixing abilities and speed up nutrient recycling in agroecosystems. Mandarin oranges-based agroforestry systems are also common. Besides Mandarin oranges and large cardamom, ginger is also a cash crop. Along with the fruit trees such as Mandarin oranges (*C. reticulata*), guava (*Psidium guajava*), *Morus alba, Artocarpus lakoocha* are grown in multistory, the farmers carry out understory mixed intercroppings of cereals such as maize, millets such as finger millet, pseudocereals such as buckwheat, grain legumes, vegetables such as taro, yam, spices such as ginger and oil seeds. The agroforestry system is ecofriendly, self sustainable, and highly productive (Sharma and Dhakal, 2011; Sharma et al., 2009).

2.3.8 USE OF BOTANICALS IN INSECT PEST CONTROL IN TRADITIONAL AGRICULTURE

Indiscriminate use of commercial chemical insecticides to control the insect attack on crop plants has resulted in rapid evolution of insect resistance in pests, increase in residual toxicity besides causing environmental pollution (Raghavendra et al., 2016).

Botanical insecticides are toxic to insects or repellent or antifeedant against insects damaging crop species. These natural, cost effective, easily available, biodegradable, specific to target insect pests are safe, environmental friendly and sustainable without causing pollution, and are less persistent in the environment (Devlin and Zettel, 1999; Narayanasamy, 2002; Pavela, 2007; Dar et al., 2014; Geda, 2015; Raghavendra et al., 2016). Most of the traditional botanical insecticides have been used for control of insect pests before the use of synthetic insecticides (Saxena, 1987; Dar et al., 2014).

In India, neem tree (*A. indica*) has been reported in archaic Sanskrit manuscript, the Veda dated at least 4000 years ago (Philogene et al., 2005). Neem has been used as traditional insecticide for many years both for crops in field and also for stored grains. Extract of neem leaf is used as against sucking pests and defoliators. Neem leaf, seed, and oil contain the active principles azadirachtin, nimidin, salanin, meliantrol, and terpenoids, which deters locusts from feeding and interferes with life cycle of insects (Saxena, 1987; Debashri and Tamal, 2012; Beaulieu, 2013; Dar et al., 2014; Raghavendra et al., 2016).

Some of the other commonly used traditional botanical insecticides used are leaf extract of *Justicia adhatoda* (syn. *Adhatoda zeylanica*) against defoliators and sucking pests; extract of garlic against fruit borer, leaf eating caterpillar; plant extract of *Datura* against thrips, jassids, tea mosquito bug, aphids; extract of garlic-chilli is used against fruit borer, red-headed hairy caterpillar, leaf eating caterpillar, brinjal fruit, and shoot borer; leaf powder of *Lantana* against aphids; and leaf extract of *Lantana* against defoliators, leaf miners, beetles (Raghavendra et al., 2016). All the botanicals for example neem, *Datura*, *Lantana*, garlic, *Adhatoda* are easily available and therefore, the use of these plants for insect control is less expensive.

The insecticidal properties of the botanicals are due to the presence of the active principles. These active principles also known as allelochemicals are secondary metabolites and act as self-defense molecules for the plants synthesizing them (Harborne, 1989; Regnault-Roger and Philogene, 2008). Some of the examples of the active principles synthesized by the insecticidal plants used in traditional insect control along with their plant parts are anonine present in leaf, semi-ripe fruits of custard apple (*Annona squamosa*) is an insecticide; caffeic acid, alkaloids, and quercetin present in seed, bark, leaf, pod of wild sirissa (*Albizia lebbeck*) are insecticidal; phenolic compounds present in shell oil of cashew nut (*Anacardium occidentale*) are insecticidal; saponins present in seed oil of *M. longifolia* var. *latifolia* are repellent and insecticidal; meliacin present in fruit and seed of common bead (*Melia azedarach*) are antifeedant and insecticidal; and karanjin present in seed and seed oil of India beech (*Pongama pinnata*) are repellent and insecticidal

(Saxena, 1987; Dar et al., 2014). Different traditional agricultural practices and patterns adopted are summarized in Figure 2.1.

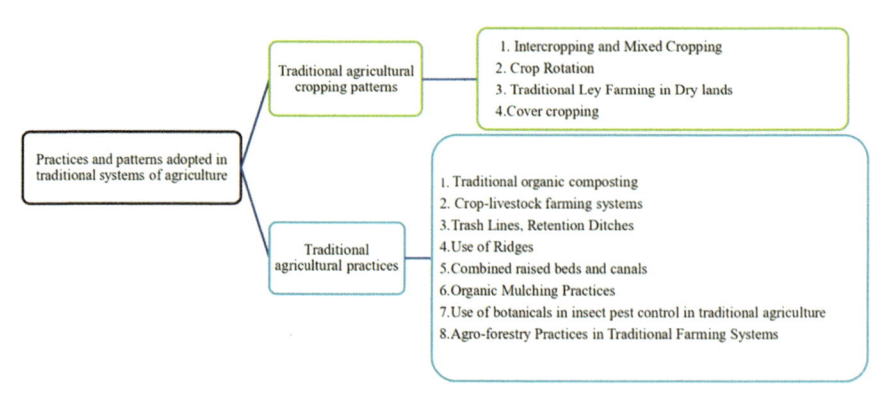

FIGURE 2.1　Summary of practices and patterns adopted in traditional systems of agriculture.

2.4　GLOBALLY IMPORTANT AGRICULTURAL HERITAGE SYSTEMS (GIAHS)

GIAHS was initiated by Food and Agricultural Organization of the United Nations in 2002 with the objectives of recognizing the traditional agriculture practiced for centuries around the world, which are sustainable and provide food security, conserve and manage the local agrobiodiversity, preserve the traditional agricultural practice and knowledge associated with it, and provide support so that the traditional agricultural practice along with the associated genetic resource and traditional knowledge are sustained. GIAHS plays a key role in safeguarding world agricultural heritage sites. Focus of GIAHS is also to preserve the resilient ecosystems and valuable cultural heritage (FAO, 2018). GIAHS recognize particular sites around the world, 62 GIAHS-designated sites of traditional agricultural practices have been recognized so far in 22 countries (FAO, 2021). Subsistence farming is carried out in these regions and provide multiple agricultural products and ecosystem services, and at the same time meeting the livelihood security for several farmers (FAO, 2018; Sood, 2012). Different aspects and services of agricultural heritage systems are depicted in Figure 2.2. Three agricultural heritage systems which are saffron heritage system from Pampore region of Kashmir, traditional agriculture system in Koraput region of Odisha and sub-sea level farming system in Kuttanad region of Kerala from India have been recognized as GIAHS in the years 2011, 2012 and 2013, respectively. Besides the above, ten other agricultural systems viz. traditional Ladakh agricultural system, Ladakh; Raika pastoralists of the Thar desert, Rajasthan; Korangadu silvo-pastoral system, Tamil Nadu; Soppina Bettas systems, Western Ghats; Grand Anicut (Kallanai) farming system, Tamil Nadu; tribal agricultural systems in Sethamphat and Srikakulam, Andhra Pradesh; Apatani rice fish culture system, Arunachal Pradesh; Sikkim Himalaya agriculture, Sikkim; Catamaran fishing system, Tamil Nadu; and Darjeeling system in the Himalayas are also high priority agricultural systems of India. The agricultural heritage sites are unique, have individuality and originality, and follow traditional eco-friendly agricultural practice, preserve and cultivate agrobiodiverse crops and land races, provide nutrition and food security and grow climate resilient crops. These agricultural heritage systems need to be preserved in their entirety along with their life support systems.

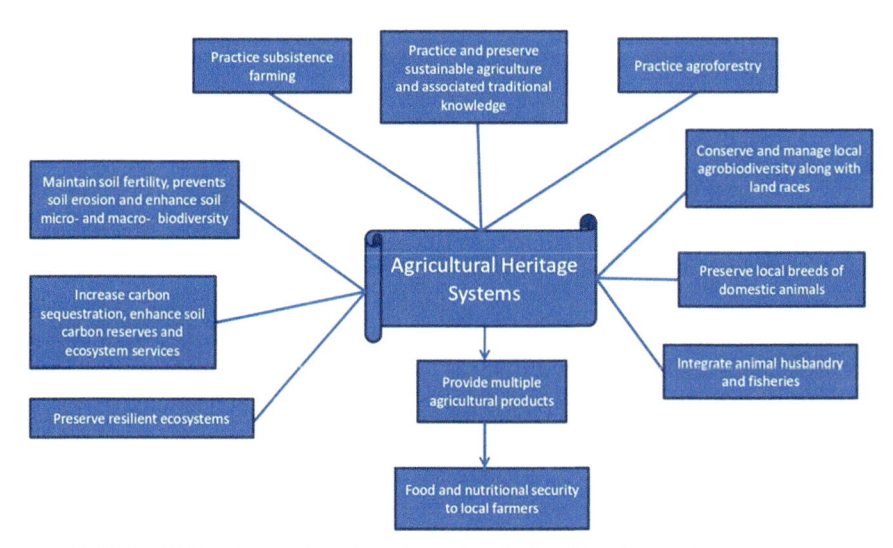

FIGURE 2.2   Different aspects and services of agricultural heritage systems.

Some of these systems are discussed below.

2.4.1 SUBSEA LEVEL FARMING IN KUTTANAD

One of the sustainable system of traditional agriculture practiced in Kuttanad, Kerala is the subsea level farming, which is practiced for the last one and half centuries (Kerala tourism, 2013; Swaminathan, 2013). The farming practice plays an important role in conserving the ecosystem and maintaining biodiversity (Kerala tourism, 2013). (Swaminathan, 2013; Kerala tourism, 2013). Paddy cultivation is carried out 2–3 m below the sea level. This traditional paddy cultivation is carried out in the Ramsar site in Kuttanad delta, which is fed by six rivers of the western Ghats (Swaminathan, 2013; Sreeja et al., 2015). Kuttanad is also known as the rice bowl of Kerala since it is the major crop of Kuttanad and chief contributor of rice produced in Kerala. Bio-bunds are created using coconut stems (Swaminathan, 2013; Joseph, 2002). Such subsea level farming system would be an example to follow in case of there is raise in water level as a result of the greenhouse effect (Swaminathan, 2013). This traditional farming practice not only contributes to the production of rice but plays an important ecological role such as mitigation of the impact of the climate change, prevents soil erosion, recharging of the ground aquifer systems, conservation of biodiversity (Zedler and Kercher, 2005), aquaculture (Joseph, 2016) and suppresses growth of weeds. Another additional advantage is that diazotrophic Cyanobacteria can grow for a longer period under such conditions. Not only the diazotrophic Cyanobacteria will contribute to additional availability of nitrogen but will also contribute to the carbon reserves in such ecosystems due to their ability to carry out photosynthesis.

2.4.2   TRADITIONAL AGRICULTURAL SYSTEM OF KORAPUT IN ODISHA

Local tribal communities practice sustainable traditional Agriculture integrated to its environment (FAO-GIAHS 2) in the Koraput region. They also have their distinct cultural traditions and have contributed to invaluable ecological services (FAO-GIAHS). Their traditional agriculture knowledge includes seed viability assessment prior to sowing, sustain the soil fertility besides preserving the valuable landraces of rice and other crops (FAO-GIAHS 2). The tribal communities have conserved in situ about 340 land races of rice (FAO-GIAHS 2) many of which have distinct aroma, taste, resistance to stress both biotic and abiotic (FAO-GIAHS 3) and are desirable resource for genetic upgrading of the present day cultivated rice. The tribal communities chiefly depend on rice and also cultivate and preserve minor millets, pulses, oilseeds, fibrous plants and vegetables (FAO-GIAHS 2, Sood, 2012) besides maize that ensures food security to the tribal communities. They also maintain sacred grooves, which are biological heritages and hence play a significant role in the conservation of forests (Sood, 2012) and their associated biodiversity. The communities play an important role in sustainably managing the forests and conserving about 2500 species of flowering plants, gymnosperms, and ferns, which includes the 79 endemic flowering plants and a gymnosperm species (FAO-GIAHS 2; Sood, 2012; FAO, 2008), and about 1200 species are also used in traditional systems of medicine meeting their health requirements (Anon 2; Sood, 2012; FAO, 2008). Many endemic species have been preserved by the tribal communities. The traditional agricultural practices, land races need to be preserved under changing environmental conditions and anthropogenic influences (FAO, 2008).

2.4.3   TRADITIONAL LADAKH AGRICULTURAL SYSTEM

Ladakh has traditional agricultural practice that uses the environmentally compatible organic farming, which is its strength. About 70% of the population is directly or indirectly dependent on agriculture (LAHDC, 2019). Mixed farming is practiced since the traditional farming systems includes livestocks and crop plants where the crops get the manure from the livestock (Dolker, 2018; LAHDC, 2019). The main crops cultivated are wheat, barley, and peas. Alfalfa is cultivated as a fodder crop (Dolker, 2018). Vegetables such as potatoes, onion, cabbage, peas, kidney beans, cauliflower, carrot, radish, beans, and cucurbits are also cultivated (LAHDC, 2019). With respect to fruits, apple, and apricots are the major crops.

Seabuckthorn [*Elaeagnus rhamnoides* (L.) A. Nelson, syn. *Hippophae rhamnoides L.*], a member of the family *Elaeagnaceae* grows naturally in Ladakh (LAHDC, 2019) and is a multipurpose species (Li and Schroeder, 1996). There is a great demand for seabuckthorn fruits since these are rich in minerals and vitamins particularly vitamin C. Hence, fruits contain nutrients constitute a good source of nutraceuticals as well as cosmoceutical (Bal et al., 2011). The fruits are used in making jams, jellies, juices, sauces, etc. and are used in traditional systems of medicine. The wood finds use as a fuelwood. The berries have anti-proliferation property besides boosting immune system (Olas et al., 2018). Seabuckthorn oil contains omega-fatty acids (Marsinach and Cuenca, 2019).

Sea buckthorn plants form root nodules by associating with the actinomycetes, *Frankia*, and play a key role in fixing atmospheric nitrogen and enhancing the available soil nitrogen. The plants prevent soil erosion by binding to the soil. The plant can be used for the restoration of degraded ecosystems since the plant is tolerant to drought and salinity. Since there is a great demand for the fruits in traditional systems of medicine, there is a high potential for cultivating seabuckthorn in Ladakh as the plants do not require any agrochemicals, since it grows naturally in Ladakh and can be grown with least amount of human care (LAHDC, 2019).

The high-altitude cold desert region of Ladakh poses a great challenge to the farmers. However, the farmers have been able to evolve methods by which they are able to cultivate crop plants even under these harsh conditions (FAO, 2008). It is unique since the agriculture is practiced in one of the tough environments due to low temperature, undulating landscapes, and dry soil conditions. The farmers have built stone terraces and utilized the glacial-fed rivers in order to carry out agriculture. Trees are planted to prevent soil erosion and soil sedimentation (FAO, 2008). They practice resource efficient organic composting that is their traditional method for producing organic manure (LAHDC, 2019). They have also played an important role in conserving the land races of crops such as the land races of alfalfa (FAO, 2008).

2.4.4 RAIKA PASTORALISTS OF THE THAR DESERT, RAJASTHAN

By tradition, Raikas maintain camel and also cultivate crops during summer depending on the rains as a source of water for rearing of the crops. They are representatives of dromedary pastoralism (Kohler-Rollefson, 1992). Besides camel, Raikas also keep goats, sheep, and water buffaloes. These animals represent the sustenance of the rural people (Kohler-Rollefson, 1992). The Raikas are agro-pastoralists and cultivate crops during the rainy season. Subsequently during autumn, winter, and spring season, which are dry months they become pastoralists. One of the remarkable contribution of the Raikas is that have outstanding variety of indigenous breeds including 11 Bikaneri sheep breeds. Raikas, therefore, are custodians of unique agro-biodiversity which they have protected for centuries inspite of the harsh challenges such as Drought conditions, climate change, fodder shortages, and fluctuating harvests of the crops (FAO, 2008).

2.4.5 KORANGADU SILVOPASTORAL SYSTEM, TAMIL NADU

Korangadu is a semiarid region that is rich in grazing land and includes the districts Karur, Dindigul, Erode, and Coimbatore in west Tamil Nadu

(FAO, 2008; Singh and Rana, 2019). It is a time-tested model maintained by the local farmers (Akila and Bharathy, 2016) and is one of the agricultural heritage sites with rich biodiversity. For centuries, the grazing land has been conserved and managed in a sustainable way since it provides fodder for the cattle, which is a means of sustenance for the local communities (FAO, 2008). They do not use any agrochemicals and plants are manured using cattle dung. The available nitrogen in the soil is enhanced by the hardy legumes, *Vigna trilobata* and *Vigna unguiculata* (syn. *Dolichos biflorus*) (Udhaya and Kumar, 2019) which also provides protein-rich feed for cattle. The cultivation of shrubs, trees, and fodder plants in multistory fashion is a typical agroforestry approach. The lower tier consists of fodder plants that includes grasses such as African foxtail grass (*Cenchrus ciliaris*) and legumes, upper tree species include White Bark Acacia (*Acacia leucophloea*) (Udhaya and Kumar, 2019). They also cultivate the thorny shrub or a small tree that is Indian bdellium-tree (*Commiphora berryi*) for biofencing which forms the middle tier, a plant adapted to the semi-arid conditions of the tropics (Udhaya and Kumar, 2019). In addition to it, the plant is a multipurpose species and finds use in traditional Indian systems of medicine and also yields a gum.

The local communities are also known for breeding cattle native to the region known as "Kangeyam" cattle, which includes local breeds of bullocks, buffaloes, sheeps, and goats (FAO, 2008; Singh and Rana, 2019). Interestingly, this region possseses calcareous soil is a rich source of calcium for these sturdy cattles (Udhaya and Kumar, 2019). However, the maintenance of the traditional grazing land and the livestock has become difficult in recent times due to several factors (FAO, 2008) such as erratic rainfall (Udhaya and Kumar, 2019), introduction of Jersey cows that are gradually bred with the native Kangeyam cows to enhance milk production, loss of species and ecosystem, etc. (Singh and Rana, 2019).

2.4.6   SOPPINA BETTAS SYSTEMS, WESTERN GHATS

Soppina Bettas system is in Karnataka, where 16 local varieties of paddy are cultivated along with other plants (FAO, 2008). It is one of the agricultural heritage sites which is sustainable where local farmers have evolved a mechanism for producing composted organic manure using leaves and litter, which is used for the paddy crop (FAO, 2008). For their foliage requirements, they chiefly use the trees *Memecylon umbellatum, Aporosa cardiosperma (*syn. *Aporosa lindleyana), Syzygium cumini*, and *Syzygium* *caryophyllatum* (Nayak et al., 2000). They do not use chemical pesticides but rely on botanical pesticides for controlling insect pests. They use plant resources of the region for medicine and timber, etc. for their daily needs (FAO, 2008) such as using firewood from the forest to burn areca nuts before drying (Nayak et al., 2000). They practice agroforestry having areca nut intercropped with citrus fruits, banana, coffee, pepper, vanilla, etc. However, The Soppina Bettas system is in constant threat due to land conversion, overuse, etc. (FAO, 2008).

2.4.7   GRAND ANICUT (KALLANAI) FARMING SYSTEM

Grand Anicut farming system lies in the eastern part of Tamil Nadu. Widespread network of irrigation of six lakh hectares of chiefly rice crop has been developed due to engineering talents of the local communities and past rulers in the delta with the support of Grand Anicut dam. The livelihood of 60% population is agriculture (FAO, 2017). However, besides agriculture, the local communities maintain livestocks and carry out fishing in a sustainable manner. The local communities have played a great role in conserving genetic diversity of crop plants particularly rice and minor millets (FAO, 2017). Interestingly, three crops of rice are raised enabled by water supply through Cauvery river. Black or green gram seeds are pushed into the soil by feet when the previous crop gets harvested. Irrigation is not needed since the soil has residual moisture (FAO, 2017). Other crops that are cultivated include sesame, maize, groundnut, banana, sugarcane, etc. However, the threats to the Kallanai farming system enabled-biodiversity are due to expanding agriculture, decrease in rainfall, etc. (FAO, 2017).

2.4.8   TRIBAL AGRICULTURAL SYSTEMS (SETHAMPHAT AND SRIKAKULAM, ANDHRA PRADESH)

Based on the type of use of the land, slope, elevation, and ecology, the region has been divided into three-tier agricultural system. This comprises forests in upland, slash-and-burn agriculture practiced in the mid-elevation lands, and crops cultivated by tribal farmers in the plains (FAO, 2009a, 2009b, 2009c; Kaushal et al., 2016). Lentils, millets, and oilseeds are cultivated in the mid-elevation lands along with cashew and rice is cultivated in the plains chiefly comprising the wetlands (Kaushal et al., 2016). The traditional approach of mixed cropping pattern and cultivation of millets ensures food security since millets are able to withstand dry conditions and are suitable for dry land agriculture, can grow on marginal soils and are not attacked by insect pests (Kaushal et al., 2016). The local tribals of Sethamphat have grain reserves of maize, lentils, millets, sorghum, etc. for meeting their food requirements for about a year. This ensures food security in times of adverse climatic conditions. Interestingly the grains stored by the local communities are resistant to postharvest pest attacks. The tribals also collect edible parts of plants from the forests (Kaushal et al., 2016), which supplements their food requirements.

Traditional manuring is done using cow dung and its urine, fodder, compost, etc. The tribal farmers follow an integrated farming approach wherein they maintain goat, sheep, poultry, dairy, fishery, sericulture, duckery, apiary, etc. also besides the crops they cultivate depending on the agro-climatic conditions (Rao et al., 2019).

The sustainable traditional system of agriculture is threatened by a number of factors such as monocropping, loss of land races particularly of millets and pulses, loss of traditional agricultural practices, water contamination, etc. (Kaushal et al., 2016).

2.4.9 APATANI RICE FISH CULTURE SYSTEM, ARUNACHAL PRADESH

The Apatani Plateau is located in a hilly plateau in lower Subansiri district of Arunachal Pradesh (Saikia and Das, 2008) and the Apatani tribe inhabits the Ziro valley (Ministry of Culture-GOI, 2021). It is one of the world heritage sites tentatively added by UNESCO due to its high productivity and its sustainability. Apatani is a rice-based fish farming system (Giri, 2009). Rice cultivation is done at wet valley bottoms at the base of the hill slopes that forms a saucer-shaped structure (FAO, Agroecology Knowledge Hub). Rice-based fish farming is a low-cost and highly productive sustainable farming system and meets the livelihood of marginal farmers (Saikia and Das, 2008). The subsistence farmers have combined the rice cultivation and fish culture in a creative sustainable system. Apatani is a self-sustaining agroecosystem where rice and fish have a symbiotic relationship (Giri, 2009). The valley receives good rainfall and the water flows in from the surrounding mountains providing ideal conditions for the cultivation of the rice crop along with fish farming. Annually, two rice crops are grown in the valley. In betel nut groves, a novel approach of bamboo pipes is used for watering (Agarwal and Narain, 1997) and the supply of water is carried out from the water stored during the monsoon season (Giri, 2009). Chemical fertilizers and pesticides are not used. Millets are also cultivated in mixed cropping system meets the varied needs of people (Rai, 2005). Animal wastes and crop residues provide the required nutrients to the rice crop (Tangjang and Nair, 2015), which is a remarkable system of effective cycling of nutrients. Bunds are made in the rice fields that prevents the flow of water out of the field and it also prevents the escape of fishes. Bamboos base is made for creating the bunds. Millets such as *Eleusine coracana* and vegetables such as pumpkin, cucumber, beans, cauliflower, tomatoes, brinjal, radish, cabbage, chillies, etc. are cultivated on the bunds (Vikaspedia; Ministry of Culture-GOI, 2021; Kumar and Ramakrishnan, 1990). The Apatani rice-fish culture system can be a good model, which can be adopted in wetlands since the net productivity in rice-fish farming system is higher than cultivation of rice alone. The unique Apatani agroecosystem is a good example of an eco-friendly agricultural system in the north-eastern hill region of India (Kumar and Ramakrishnan, 1990).

2.4.10   SIKKIM HIMALAYA AGRICULTURE

Sikkim Himalaya agriculture is rich in agro-biodiversity and is managed using the traditional ecological knowledge by ethnic communities. The managed agro-biodiversity can broadly be divided into three different multifunctional vertically segregated agro-ecological zones that are (i) alpine plateaus with agropastoral ecosystems that forms the grazing ground for yak, sheeps, goats, etc., and has been the life support system for centuries providing wool, milk, cheese, meat, fat and other products (FAO, 2014); (ii) traditional agrofor-estry, such as the alder-cardamon and farm-based systems in temperate zone where subsistence farming is carried out which includes crops such as wheat, potato, barley, maize, cabbage, apple, peas, beans, etc. and practice organic agriculture; and (iii) rice system grown in the terraces or valleys in the lower subtropical zones (FAO, 2014) where a large number of land races of rice having characteristic aroma, fine quality of grains and having medicinal properties are cultivated (Sharma and Liang, 2006) besides managing farm based agroforestry (Sharma and Dhakal, 2011). Some other plants cultivated in the crop production systems includes legumes, pseudo-cereals such as buckwheat, fruits and nuts, vegetables, etc. These are traditional sources of food and contributes to the food security of the ethnic communities (FAO, 2014). Spices, medicinal, or aromatic plants constitutes cash crops in the agroforestry system. Mandarins are commonly cultivated in the farm-based agroforestry systems. Legumes such as peas and beans cultivated along with rice enhance soil fertility due to their ability to fix atmospheric nitrogen. Rice cultivation along with agroforestry along the slopes is an unique traditional ecological management system in the mountains of Sikkim (FAO, 2014).

The ethnic community preserves the germplasm by cultivating the traditional varieties and are owners of the genetic diversity of crop plants (FAO, 2014). They have rich genetic resource of land races, medicinal plants, underutilized crop plants, wild relatives of crop species, and crops unique to the region (Sharma and Dhakal, 2011). For example, the wild allies of cardamom, citrus, and yams exist in the wild. The communities have preserved the diverse crop species in seed banks (FAO, 2014). Land races of finger millets, colocasia, yams, oil seeds, leafy vegetables are cultivated and preserved as seed banks. Buckwheat, beans, pulses, and wild relatives of citrus are also conserved (FAO, 2014).

Agroforestry in Sikkim Himalaya agriculture has been practiced for centuries where multipurpose tree species are grown along with fruit trees and other crops in multistory method (Sharma and Dhakal, 2011). The fruit trees cultivated include *P. guajava, M. alba, A. lakoocha,* besides *Juglans regia,* a dry fruit and *Tamarindus indica* yielding edible fruits that are edible (FAO, 2014) and used as a cuisine. The agroforestry system acts as carbon sinks and plays significant role in reducing greenhouse gases (Sharma et al., 2000).

Multipurpose species are important components of the agroforestry system. Nepalese Alder, *A. nepalensis,* a valued keystone species forms root nodules by associating with the actinomycetes, *Frankia* and enriches soil by contributing to the available soil nitrogen up to 125 kg/ha/year (Ramakrishnan, 1992). The agro-ecosystem has remarkable potential for ecological resilience with respect to climate change and contributes to the environmental services. The communities play a key role in protecting the biodiversity of the adjacent forests, which provide them with the several plant products such as fruits, medicines, fibers, tubers, mushrooms, other edible plant parts, etc. (Sharma and Dhakal, 2011), which are used by the communities or sold in the market. Sea buckthorn (*Hippophae tibetana*) is a multipurpose species found in the forests of Sikkim. Besides providing nutritive fruits, the wood finds use as firewood, the plant has medicinal properties in Indian systems of medicine, root nodules are formed by associating with *Frankia* and contributes to available soil nitrogen. A medicinal fungi belonging to Ascomycetes, *Cordyceps sinensis* is also collected that is used to boost immunity, and it shows activity against lung and skin cancer and shrinks tumor size. The forests play an important role in providing valuable ecological services and hence forests have been preserved by the ethnic communities. Sikkim Himalayas are a part of biodiversity hotspots and is a cradle of evolution of novel gene pools (Takahashi, 2009) with several endemic species that have also been conserved by the local communities.

Traditional practices of inclusion of multipurpose nitrogen fixing trees such as *Erythrina, Albizia, Hippophae*, and *Alnus* enhance soil fertility and reduces soil erosion (Sharma and Dhakal, 2011). The ethnic communities have played an important role in conservation and maintenance of agrobiodiversity in Sikkim Himalaya. Wild edible plants and honey obtained from wild provide food security. The tribals are able to generate cash from ginger, oranges, herbs of medicinal value, orchids, and other ornamental plants (Takahashi, 2009). However, the Sikkim Himalaya agriculture has several issues with respect to the management of the traditional sustainable agricultural system such as market access, poverty, ecological sensitivities (FAO, 2014), loss of biodiversity and medicinal plants, climate change, problems due to invasive exotic species in farmlands, agroforests, wetlands, forests, etc. (Sharma and Dhakal, 2011, FAO, 2014).

2.5   UNDEREXPLOITED TRADITIONAL CROPS FOR PROVIDING FOOD SECURITY

The cultivation of underexploited traditional crops that have an adaptability to grow on marginal lands having high nutrient value can be expanded since these will provide both the food and nutritional security. This also enhances the diversity of crop species under cultivation. These traditional crops could be used as supplementary diet locally. Since these crops do not require much attention of farmers in terms of farm inputs, the crop produce would be more affordable for a larger sections of the society. Customer acceptable commercial food products from such crops having higher nutritional value needs to be popularised as supplementary foods particularly for infants, invalids, and pregnant women for meeting nutrient deficiencies and thus would also alleviate from the hidden hunger of the people besides providing food security. Also enhancing the shelf life of the acceptable products without loss of nutrient value would be beneficial for storing the food products for a longer period. Strategies for wider marketing and distribution network of such agroproducts would also enhance the reach of such nutrient-rich traditional food products. Choice of traditional crops that are early maturing, hardy species, climate resilient would be desirable. Needless to mention that cultivating such crops would be sustainable agriculture and would provide food security particularly in places which face periodic food shortage chiefly due to inclement weather conditions. Potentials of underexploited crops to provide food and nutritional security is illustrated in Figure 2.3.

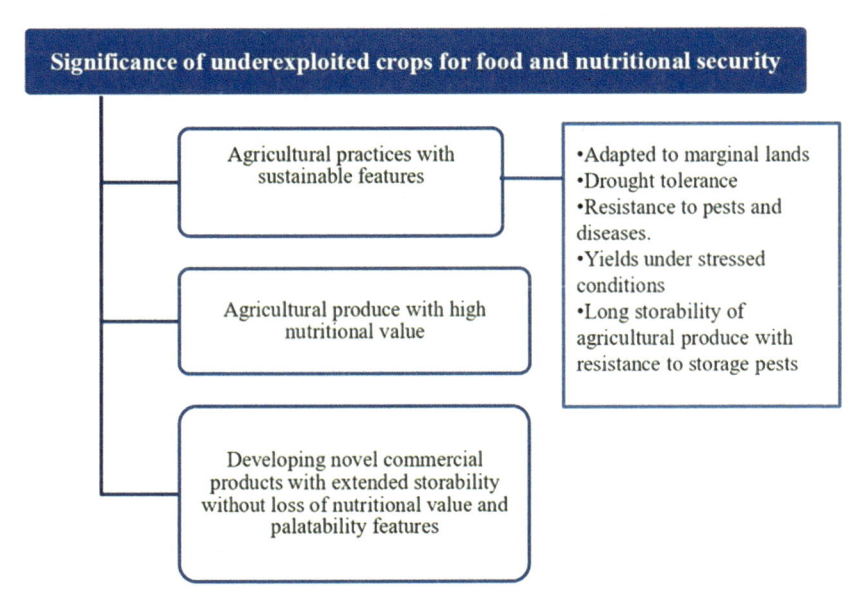

FIGURE 2.3 Potential of underexploited crops to provide food and nutritional security.

Some of the underexploited traditional crops cultivated on a limited areas which have remarkable potential to alleviate hunger and nutritional deficiencies are given below.

2.5.1 *GRAIN AMARANTHS* (*Amaranthus hypochondriacus* L., *A. cruentus* L., *A. caudatus* L., Family: Amaranthaceae)

Grain amaranths are ancient crop native to the Andean region of South America including Argentina, Bolivia and Peru (FAO, 2021), cultivated for thousands of years in Latin America (Guzmán-Maldonado and Paredes-Lopez, 1998) and domesticated as early as 4000 BC (Rosentrater and Evers, 2018). It perhaps met 80% of the caloific needs of the Aztecs (Rosentrater and Evers, 2018). However, they are mostly getting lost to present-day cultivation (Das, 2016). The seeds do not contain gluten and are known as pseudograin since the flavor and cooking characteristics resemble grains (FAO, 2021). Three species *Amaranthus caudatus, A. hypochondriacus*, and *A. cruentus* constitute the grain amaranths since they are cultivated for their grains. The two-grain amaranths, *A. hypochondriacus* and *A. cruentus,* are native to Mexico and Guatemala whereas *A. caudatus* was domesticated in the Andes of Peru (Stetter et al., 2016; Bhatia, 2005; Rosentrater and Evers, 2018). *A. edulis*, a race of *A. caudatus* is native to Peru and Andean region. *A. caudatus*, constituted the staple crop of the ancient civilizations Aztec, Mayan, and Incas. The grain amaranths are cultivated on limited scale chiefly in the hill farming systems in India and grown in Himalayas and South Indian hills. However, it is also becoming popular crop in Gujarat (Bhatia, 2005). Grain amaranths grow vigourously and are C_4 plants (Sage, 2004; Rosentrater and Evers, 2018). They are short-duration crops maturing in about 60–70 days (King and Ravi, 2012).

Grain amaranths are highly tolerant to heat and drought (Wu and Blair, 2017) as well as salinity and resists pests (Bhatia, 2005). The plant can be cultivated on marginal lands and adapted to poor soil fertility (Saucedo et al., 2017) with little care of the human beings. Amaranth grains contain 12.5–17.6% proteins (Teutonico and Knorr, 1985; Caselato-Sousa and Amaya-Farfan, 2012) having good balance of amino acids with high amounts of essential amino acid lysine 0.73–0.84% of the total protein (Becker et al., 1981; Bressani, 1994; Rastogi and Shukla, 2013; De Ron et al., 2017) and are also having high amounts of methionine and cysteine (Caselato-Sousa and Amaya-Farfán, 2012; Bressani et al., 1987). Amaranths also contain good quantities iron, β carotenes, and folic acid (Chauhan and Singh, 2013).

Amaranth grains are small approximately 1 mm in diameter and are yellow to red. Seed flour is used to make bread, noodles, cookies, pancakes, etc. Seed flour mixed with wheat or rice flour is used to make bread and other preparations (Bhatia, 2005). The seeds are popped and made into pancakes or sweetmeat balls and also can be cooked and eaten. Seeds are also used for making porridge. Enhancing the acreage of grain amaranths that can be eaten as supplementary or alternate food will not only ensure food security but will also alleviate the suffering of people from nutrient deficiencies. Red pigment obtained from the plant is used in food industry (Teng et al., 2015; Bhatia, 2005).

2.5.2   *QUINOA* (*Chenopodium quinoa* Willd., Family: Chenopodiaceae)

Quinoa, a pseudocereal, is also known as "the golden grain of the Andes." It originated in Andes and cultivated in the Andean region for thousands of years (Bhargava et al., 2006). Using their traditional knowledge, the Andean people have protected and preserved this valuable food crop (FAO, 2021).

Quinoa has adaptability to wide range of altitudinal from sea level to 3800 m. It has deep root system and is able to draw minerals and water from deeper layers of soil. It has an ability to tolerate extremes of ecological conditions such as saline and drought conditions and can withstand frost (Bhargava et al., 2006) and has been considered as a wonder grain.

Quinoa would have met most of the food requirements of Andean people. Several nutritive preparations can be made from grains, they can be cooked and eaten like a cereal or made into a pasta, roasted seed flour can be made into bread, biscuits (Bhargava et al., 2006; Rane et al., 2019), and other diverse edible products such as sponge cakes, noodles, quinoa flakes, and puffed quinoa (Angeli et al., 2020). Quinoa can be fermented to produce alcoholic drinks and can be used to obtain starch, protein, and saponin by the industries (Jacobsen, 2003). Quinoa starch can be put to diverse industrial uses since it has properties such as good water-holding capacity, gelation, emulsifying, and foaming. Seed oil contains good proportion of omega-6 and vitamin-E (Sharma et al., 2015). Quinoa is a health food free from gluten, grains are rich in proteins (11.1–18.1%), carbohydrates (48.6–68.1%) and fats (4.0–7.9%) (Angeli et al., 2020). Grains also contain good quantities of the essential amino acids lysine (2.8–5.4 g/100 g crude protein), leucine (3.7–7.2 g/100 g crude protein), and methionine (1.1–2.2 g/100 g crude protein) (Angeli et al., 2020). Seeds are also rich in vitamins and minerals and are considered as an important nutraceutical.

The year 2013 was declared as the "International Year of Quinoa" by the United Nations General Assembly. FAO expects that sustainable production of traditional crops such as quinoa can contribute to alleviation of the food security issues of the present day world (FAO, 2021).

Since quinoa has an adaptability to different agro-environmental conditions, it is an appropriate choice for cultivation in marginal soils and can be an ideal crop for meeting the food and nutritive requirements for regions suffering from food insecurity (FAO, 2021). It is considered that quinoa has a potential to become a significant industrial and food crop in the present century (Chavan et al., 2019). Cultivation can be diversified in Himalayas, North Indian plains, semiarid regions of Telangana (Bhargava et al., 2006; Rathore et al., 2019; Ramesh et al., 2019), Maharashtra (Rane et al., 2019), Rajasthan, and Andhra Pradesh.

2.5.3 BUCKWHEAT [*Fagopyrum esculentum* Moench, *F. tataricum* (L.) Gaertn., Family: Polygonaceae]

Buckwheat is native to Western China and Tibet and Eastern India (FAO, 2021) and was domesticated possibly in 6000 BC (Ahmad et al., 2018). It is an annual having fast growth rate. The plant has short growing season (2–4 months) (FAO, 2021) and hence can be grown in high-altitude temperate zones (Shah, 2013) where available growing period is short. It acts as a life support for people living in cold deserts of Himalayas since it is the only crop that can be cultivated at an altitude of 4500 m (Joshi, 1999; Tundup et al., 2016). Two species of buckwheat namely, *Fagopyrum esculentum* (common buckwheat) and *F. tataricum* (Tartary buckwheat) are cultivated in the Himalayas (Tundup et al., 2016).

Buckwheat has been cultivated on marginal and fairly unproductive land where it forms a subsistence crop. Tartary buckwheat has better tolerance to frost hence is generally cultivated at the higher altitudes compared to common buckwheat that is grown at the lower altitudes (Campbell, 1997).

All parts of plant are useful. Leaves and tender shoots are used as vegetable, grains as pseudocereal (FAO, 2021) is ground into flour, mixed with wheat flour is used to make biscuits, bread, pancakes, noodles, porridge, and soup. It is a staple food in many hilly areas where bread is made out of seed flour. Buckwheat is also used to make alcoholic drink. Buckwheat has also been used to make green manure crop for lands with low productivity. Interestingly, due to its ability to smother weeds, it has been used for control of broad-leaved weeds (Ahmad et al., 2018). The plant acts as a cover crop, binds soil, and checks soil erosion (FAO, 2021).

The grains of common buckwheat contains carbohydrates varying from 59% to 70% of dry matter. Most commonly grown cultivars yield seeds with 11–15% protein on whole seed basis. Seeds of common buckwheat contain 1.5–3.7% total lipids. The composition of some of the essential amino acids are lysine 5.9; methionine 3.7, leucine 5.8, and tryptophan 1.4% of total proteins (Campbell, 1997; Ahmad et al., 2018).

2.5.4 DRUMSTICK (*Moringa oleifera* Lam., Family: Moringaceae)

Drumstick tree is indigenous to North-west India and grows wild in sub-Himalayan tracts (Ramachandran et al., 1980). The tree is widely cultivated and used in India.

Drumstick is fast growing and is drought tolerant (FAO, 2021) since the tree is adapted to dry and hot climates (Pandey et al., 2011). The trees are used as windbreaks and reduces soil erosion (FAO).

The tree is a multipurpose species. All parts of plant namely bark, tender fruits, leaves, seeds, roots, and flowers are edible (FAO). Young leaves are cooked and eaten, and also made into soups and salads. The plant leaves are a rich source of vitamins, copper, calcium, potassium, magnesium, iron, manganese, and zinc (Chorage et al., 2020).

It has importance in medicine having several health benefits, nutritional, and industrial applications (Pandey et al., 2011). Seed oil is sweet, nonsticking, nondrying, resists rancidity and seed cake used to purify potable water.

Leaves are rich in protein, fatty acids, vitamins, and minerals that form part of its quality as superfood. Leaves form excellent source of vitamin A, B1, B2, B3, C, nicotinic acid, pyridoxine, ascorbic acid and folic acid and minerals such as iron, calcium, magnesium, and phosphorus. A substantial amount of essential amino acids are found in the pods. Leaf powder is used to treat malnourishment in children. Leaves constitue excellent nutrient food especially for pregnant and nursing mothers as well as young children (Pandey et al., 2019)

2.5.5 TEFF [*Eragrostis* tef (Zucc.) Trotter, Family: Poaceae]

The origin and diversification of teff have taken in Ethiopia where it is endemic (Ketema, 1997). In Ethiopia, it is cultivated as a staple crop. Teff can be cultivated under varied cropping systems such as intercropping (FAO, 2021) and can form supplementary feed and can overcome food crisis providing food security and a safeguard against famine (Ruskin, 1996).

The plant can grow under various stress conditions (Ketema, 1997). It is cold tolerant and is drought resistant and can grow in difficult conditions that is not ideal for most of the other cereals (Ruskin, 1996). The plant can regenerate rapidly after a moderate water stress (Miller, 2009). In Ethiopia, teff is grown even when other crops such as wheat, maize etc. give good yield. In case, the key crops such as wheat, maize, or sorghum fail, the early maturing teff is cultivated as a backup crop for sustenance.

Teff, also called "lazy man's crop" since the seeds are scattered in moist soil and the farmers return on around 45th day to collect the grains (Ruskin, 1996). Teff does not require much attention, its rapid growth smothers weeds and can grow on marginal soils. Early maturing varieties take 45–60 days. It is a C_4 plant (Miller, 2009). Grains can be stored in the traditional store-houses for many years since they are not attacked by storage insect pests (FAO, 2021, Ruskin, 1996).

Teff is a promising crop under dry farming systems in India and inter-cropped with other crop plants. This will enable providing food security in case of failure of the main crop plant. It has been found that plants with red and brown seeds are hardier, mature fast, and grown easily. The usage of agropesticides is not required since the plant is resistant to diseases and pests. It can also be grown in heavy soils (Ruskin, 1996).

Teff grains are white, red, or brown. Rich flavor is found in the darker seeds (Ruskin, 1996; FAO, 2021). In Ethiopia, a flat, spongy, and slightly sour bread called injera is made from teff. Spongy texture is retained for 3 days (Ruskin, 1996; Ketema, 1997; FAO, 2021). The seeds are low in glycemic index and are gluten-free and can also be used for making gruel, soups, stews, gravies, muffins, biscuits, cookies, puddings, porridge or pancakes, baked into cakes, and made into a sweet bread or beverages, and also can be used for making alcoholic drinks.

Teff grains are rich in proteins. Seeds contain proteins, 9–11%; carbohydrates, 73 g; and fats 2.0 g. It is rich in vitamins and minerals. It has 88 mg of vitamin C, niacin 2.5 mg, iron, 5.8 mg, magnesium 180 mg, calcium 159 mg, phosphorus 378 mg, and potassium 401 mg (Ruskin, 1996). Seeds are also rich in lysine. Essential amino acid composition in seeds include lysine 3.68, leucine 8.53, phenylalanine 5.69, and valine 5.46 g/16 gN (Jansen et al., 1962; Ketema, 1997).

It is a useful fodder crop. In South Africa, teff is also used as a nurse crop since it covers the soil rapidly and nurtures the establishment of other perennial grasses. It prevents soil erosion (Ruskin, 1996).

2.5.6   *YAM BEANS OR POTATO BEAN* [*Pachyrhizus erosus* (L.) Urb., Family: Leguminosae, Sub-family: Fabaceae]

The plant is a climber and is native to Mexico and Central America (FAO, 2021; Naskar, 2009). In India, it is cultivated in West Bengal, North Bihar,

Uttar Pradesh, Jharkhand, Chhattisgarh, Orissa and Assam (Naskar, 2009). Yam beans adapt to hot humid climates or subtropical or warm temperate zones. The plant matures in 110–140 days (Naskar, 2009). The plant is high yielding even in arid climates. The plant produces root nodules and is able to fix atmospheric nitrogen and contribute to soil fertility. Yam is intercropped with maize and beans (FAO, 2021). Seeds or sprouted roots are used for propagation. It is also used in crop rotation programmes and enhances the soil fertility.

An underutilized tuber crop, the large tuberous roots which are 0.5–2.5 kg of Yam beans are edible (FAO, 2021; Naskar 2009) and are rich in ascorbic acid. In India, crisp tender tubers are eaten raw. It contains proteins 1.5%, starch 10% and sugar 5–6%. The mature tubers yield starch. Tender pods are edible, whereas the mature pods are toxic.

2.5.7 BREADFRUIT [*Artocarpus altilis* (Parkinson ex Zorn) Fosberg, Family: Moraceae]

Breadfruit is believed to have originated in Oceania (FAO, 2021). Breadfruit is similar to freshly baked bread hence the common name. It is staple crop cultivated in Oceania, Central Africa, and Central America (Huang et al., 2020). Breadfruit has been providing food security and agricultural sustainability for centuries in Hawaiin islands (Elevitch et al., 2014). It is ideal crop for hot, humid, and tropical lowlands areas (Sikarwar et al., 2014). Breadfruit is cultivated in the hotter parts of India, however, commonly grown in Kerala and Karnataka.

A tree crop highly valued for its fruits and is known to be a prolific yielder. The plant is adapted to grow on sandy, sandy loam, loam, coastal sand, or saline soils (Mohanty and Pradhan, 2015). The plant requires very little inputs, it is easy to grow and produces fruits during its lifespan of 50 years or more (Roberts-Nkrumah, 2015). Hence, breadfruit has potential to provide food security in the developing world (Huang et al., 2020). It is one of the sought after plant in agroforestry and can be intercropped with other crops such as coconut, kava, taro, yam, banana, cacao, coffee, Citrus, pineapple, papaya, tomato, eggplant, etc. (Elevitch et al., 2014; FAO, 2021). It has an ability to sequester atmospheric carbon dioxide, hence has an extended environmental benefit (Elevitch et al., 2014). Breadfruit is a multipurpose species. A single tree produces about 100–200 fruits per year (Elevitch et al., 2014). The soft and sweet fruits are generally round, oval, and oblong in shape measuring 9–20 cm, and weigh between 0.25 and 6 kg (Sikarwar et al., 2014).

Breadfruit flour is a gluten-free flour and is rich in vitamin A, C, and potassium (Turi et al., 2015). Edible portion of the breadfruit contains: Protein 4 g, carbohydrate 31.9 g, fat 0.2 g, calcium 16.8 mg, potassium 376.7 mg, magnesium 34.3, phosphorus 43.1 mg/100 g of edible portion (Elevitch et al., 2014). Proteins present in breadfruit is easier to digest compared to wheat protein. Higher growth rate and body weight in mice fed the breadfruit diet has been reported when compared with standard diet-fed mice (Liu et al., 2020). Breadfruit flour is a complete protein option for present-day food which is gluten-free, low in glycemic index and nutrient dense (Liu et al., 2020). Breadfruit contains good quantities of essential amino acids and are especially rich in phenylalanine, leucine, isoleucine, and valine (Liu et al., 2015).

The flour of breadfruit is a healthy alternative to other starches for modern foods (Liu et al., 2020). Breadfruit flour can be a component in a number of food products such as pasta, instant baby food, bread, cake, cookies, snacks, pastries, chips, biscuits, crackers, and meat (Esparagoza and Tangonan, 1993; Akubor and Badifu, 2014; Malomo et al., 2011; Ajani et al., 2012; Sikarwar et al., 2014; Elevitch et al., 2014; Nochera and Ragone, 2019; Huang and Bohrer, 2020). The fruits are low in fat.

2.5.8   FINGER MILLET [*Eleusine coracana* (L.)Gaertn., Family: Poaceae]

Finger millet is an annual cereal that was domesticated 5000 years ago in Ethiopia to Uganda (FAO, 2021) in African continent (Gupta et al., 2017). The crop was introduced into India about 3000 years ago (FAO, 2021). Finger millet is an important staple crop in several semiarid and tropical regions of the world (Gupta et al., 2017). Finger millet is cultivated in tropical and subtropical regions of India in nutrient-poor soils (Maitra et al., 2020) chiefly in Karnataka, Tamil Nadu, Andhra Pradesh, Orissa, Jharkhand, Uttaranchal, Maharashtra, and Gujarat (Maitra et al., 2020).

Since India represents secondary center of gene diversity (FAO, 2021), finger millet shows numerous variations such as early and vigorous growth, large panicle size, increased finger number, and branching as well as high-density grains. Varieties that are water-efficient and high carbon dioxide fixation are also known (Gupta et al., 2017).

Finger millet is a climate resilient crops. Finger millet ensures food security even during harsh environment (Gupta et al., 2017). It can tolerate various abiotic stresses (Gupta et al., 2017) and grow under high temperature, low moisture, and poor or marginal soils (Gull et al., 2014; Gupta et al., 2017). The crop is hardy and strongly resist the conditions of drought (Chandrashekar, 2010; Lal et al., 2017), hence it is preferred in dry farming system. Small farmers cultivate it in subsistence farming (Maitra et al., 2020). Finger millet can be intercropped with maize, sorghum, or legumes (FAO, 2021). The crop is free from major pests and diseases.

Gluten-free seeds have excellent making qualities hence can be easily processed in the food industry (Chandrashekar, 2010). Seeds are white, light brown, or dark brown and can be popped or made into bread or biscuits made from milled flour. Grains after malting are used for the making fermented beverages (Chandrashekar, 2010). Malted grains used as weaning food have high energy levels (Chandrashekar, 2010). Porridge and alcoholic drinks are also made from seeds. Finger millets are good food for diabetics since they have low glycaemic index. The grains have new applications such as decortication, puffing, extrusion, and expansion (Chandrashekar, 2010). Development of novel food products and value-added food products based on finger millet, have the potential to achieve nutritional food security.

Seeds have excellent nutraceutical properties (Gupta et al., 2017). Finger millet is an important diet for pregnant and lactating mothers, and children (Gupta et al., 2017). Finger millet or Ragi (*Eleusine coracana* L.) is a rich source of protein and nutrients (Poornima et al., 2021). Seeds contain crude protein 7.30%, carbohydrates 68.10% (Chandra et al., 2019). Finger millet contains high levels of methionine and lysine (Mbithi-Mwikya et al., 2000). Finger millets are also rich in manganese, phosphorus, zinc, and iron as and other minerals and vitamins (Vadivoo et al., 1998; Tripathi and Platel, 2010; Shashi et al., 2007; Poornima et al., 2021). The grains are good source of calcium and contain 450 mg/100 g of calcium (Gupta et al., 2011; Kumar et al., 2014).

The plant has inherent ability to ward off pests and pathogens (Chandrashekar, 2010). Finger millet is known as "famine crop" since the seeds are resistant to storage insect pests for about 10 years, and can meet the food requirements even in case of crop failures (Mgonja et al., 2007). Resistance of seeds of finger millet to storage insect pests is due to hardness of the grains (Audilakshmi et al., 1999), presence of seed phenols (Kavitha and Chandrashekar, 1992; Seetharam and Ravikumar 1993; Chethan and Malleshi, 2007), formation of phytoalexins (Snyder et al., 1991) and also the formation of prolamins (Gupta et al., 2011).

2.5.9   *INDIAN BARNYARD MILLET* (*Echinochloa frumentacea* Link, Family: Poaceae)

Barnyard millet is an underutilized crop and has a potential crop in contributing to food and nutritional food security (Renganathan et al., 2020). It is an annual robust crop (Denton, 1987). It is also called Indian barnyard millet, Sawa millet, or billion dollar grass and has its origin probably in Central Asia wherefrom it has spread to Europe and America (Prabu et al., 2020). Parallel line of evolution has also been suggested for Barnyard millet both in India and Africa (Sood et al., 2015).

Barnyard millet is an important minor millet crop of warm and temperate regions of the world and cultivated in India, China, Japan, Korea, Nepal, and Pakistan (Renganathan et al., 2020). The crop is chiefly cultivated in the Himalayan region of Uttarakhand and Deccan plateau region of Tamil Nadu.

The crop has extensive adaptability and is grown in marginal rainfed areas as it is fast growing and a short duration crop (Sood et al., 2015). It has an ability to grow harsh environmental conditions such as drought or flooded situation and is an option during famine years (Sood et al., 2015; Renganathan et al., 2020). They are less susceptible to biotic and abiotic stresses (Singh et al., 2010; Renganathan et al., 2020). It does not require agro-inputs. A crop having C_4 photosynthetic pathway has been identified for climate-resilient agriculture (Bouhache and Bayer, 1993; Sood et al., 2015). It is an ideal crop for subsistence farming and can meet the food requirements in case of failure of monsoons (Gupta et al., 2009). Inspite of all good qualities, the crop is considered as poor man's food (Sood et al., 2015).

The millet is cultivated for human consumption and livestock feed (Renganathan et al., 2020). The grains are staple diet of some communities and are eaten after cooking in water or can be taken after boiling with milk and sugar and made into porridge (Sood et al., 2015). The flour is used for making baby foods, snacks, and dietary foods (Anju and Sarita, 2010; Surekha et al., 2013). The millet flour can be combined with other flours for making value-added products (Veena et al., 2004; Surekha et al., 2013). Barnyard millet is used in traditional snacks. It has also been used to make noodles, rusk, biscuits, sweets, ready mix, popped products, etc. (Veena et al., 2004; Sood et al., 2015). Beer can also be made after fermentation. The crop produces huge quantities of fodder due to its rapid growth (Sood et al., 2015).

Barnyard millet is a functional food crop due to its nutritional qualities coupled with its antioxidant properties (Sood et al., 2015; Prabu et al., 2020). The millet contains good quantities of protein (5–8.5%) and carbohydrate

(57–66%) and fat (3.5–4.6%) (Monteiro et al., 1987; Renganathan et al., 2020), and also micronutrients such as iron, calcium, magnesium, phosphorus, and zinc (Singh et al., 2010; Saleh et al., 2013; Chandel et al., 2014; Chandra et al., 2019; Renganathan et al., 2020). The iron content in barnyard millet grain is about 15.6–18.6 mg/100 g (Saleh et al., 2013; Renganathan et al., 2020).

2.5.10   LITTLE MILLET (PANICUM SUMATRENSE ROTH EX ROEM. AND SCHULTZ, FAMILY: POACEAE)

Little millet is considered to have originated in India (Maitra and Shankar, 2019) and is one of the oldest crops domesticated in India (Kamatar et al., 2013). It is an underutilized crop (Vetriventhan and Upadhyay, 2016). It is chiefly cultivated in Andhra Pradesh, Odisha, Tamil Nadu, Karnataka, Madhya Pradesh, Gujarat, Chhattisgarh, and Maharashtra (Maitra and Shankar, 2019). It is cultivated by the resource-poor tribal farmers in South India but there is a gradual decline in the cultivation of this millet (Arunachalam et al., 2005). Little millet enhance the agro-biodiversity (Maitra and Shankar, 2019). It is a short-duration crop and can be integrated into different farming systems due to its short life cycle (Maitra and Shankar, 2019). It can grow in different soil and environmental conditions (Vetriventhan and Upadhyay, 2016). It can grow in marginal lands with restricted quantities of water and constitutes an important minor millet grown in dry lands (Kamatar et al., 2013). It produces high biomass and good yield of grains. Not only it yields food but also fodder for the livestock (Vetriventhan and Upadhyay, 2016).

Little Millet constitutes a staple food for a large number of tribal people in several parts of the world (Salini et al., 2014). Grains are nutritious food (Maitra and Shankar, 2019). Little millet is a nutri-cereals and contains carbohydrates 65.5 g, proteins 10.1 g, fat 3.89 g, calcium 16.1 mg, phosphorus 130 mg, magnesium 91 mg, zinc 1.8 mg, iron 1.2 mg, thiamin 0.26 mg, riboflavin 0.05 mg, niacin 1.3 mg and folic acid 362 µg in 100 g of the grains (Bhat et al., 2018). Little millet provides nutritional and food security to marginal farmers in drylands (Maitra and Shankar, 2019). Processing of little millets increases the total phenolic, flavonoid, and tannin contents (Pradeep and Guha, 2011). Straw is used as fodder. They are useful to prevent soil erosion and can sequester carbon (Maitra and Shankar, 2019).

2.5.11   *ITALIAN OR FOXTAIL MILLET* [*Setaria italica* (L.)P.Beauv., Family: Poaceae]

Foxtail or Italian millet, an annual tufted grass is also known as German millet. It is grown in China since 5000 BC (Brink, 2006). Perhaps after its first domestication in central China, it got spread to India and Europe (Brink, 2006). It is at present cultivated almost throughout the world (Brink, 2006). In India, foxtail millet is cultivated in Andhra Pradesh, Tamil Nadu, Gujarat, Maharashtra, and Karnataka (Khare, 2007).

Foxtail millet is tolerant to herbicide (Zhu et al., 2006) and also to drought and salinity (Jayaraman et al., 2008; Krishnamurthy et al., 2014; Sudhakar et al., 2015). Foxtail millet (*Setaria italica)* is a C_4 plant and has improved water and carbon use efficiencies (de Oliveira et al., 2016). Foxtail millet produces 1 g of dry mass at the cost of 257 g of water, which is considerably low compared to maize and wheat (Li and Brutnell, 2011). Foxtail millet is also an N- and P-efficient crop (Nadeem et al., 2020).

In India and China, it is an important crop. Husked grains are used as food after cooking. Grain flour is used for making bread either alone or by mixing with wheat flour. It can also be made into noodles after mixing with wheat flour. The flour can also be used for making other products such as cakes, porridges, puddings, mini crisp chips, millet crisp rolls, and flour for baby foods. It can also be cooked after mixing with pulses. It is a nutritious food and can be given to elderly and for pregnant women. Beer and alcohol can also be made from foxtail millet (Brink, 2006).

The foxtail millet grain contains: protein 11.2 g, fat 4.0 g, carbohydrate 63.2 g, Ca 31 mg, Fe 2.8 mg, thiamin 0.6 mg, riboflavin 0.1 mg, and niacin 3.2 mg/100 g edible portion (FAO, 1995). The essential amino-acid constituents are threonine 328 mg, valine 728 mg, leucine 1764 mg, tryptophan 103 mg, lysine 233 mg, methionine 296 mg, phenylalanine 708 mg, and isoleucine 803 mg/100 g grains (FAO, 1970).

2.5.12   *BROWNTOP MILLET* [*Brachiaria ramosa* (L.) Stapf, syn. *Panicum ramosum* L., *Urochloa ramosa* (L.) Nguyen, Family: Poaceae]

The browntop millet is native to India (Ashoka and Sunitha, 2020). It is an ancient crop cultivated in India as a subsistence crop in Neolithic India (Boivin et al., 2008). The millet is still cultivated in India in dry regions of

Karnataka and Andhra Pradesh for its grains as well as fodder (Madella et al., 2016; Ashoka and Sunitha, 2020).

Browntop millet is a heavy yielder and grows in warm season. The millet has adaptability to grow in diverse soils and climates. The grains are edible and have a variety of uses such as fodder, for controlling soil erosion, production of hay, etc. Brown top millet, known as miracle crop or positive crop due to its ability to grow in dry and rainfed situations. The millet contains all essential nutrients. It produces quick forage. Along with plantation crops, the browntop millet is grown as a cover crop and controls soil erosion. The plant has biological control property in suppressing root-knot nematode. This millet can be used for providing nutritional security (Ashoka and Sunitha, 2020).

The plant is heat tolerant and can grow both in drought as well as flooded conditions. It can be cultivated as mixed crop along with redgram. It is a quick maturing crop, maturing in 90 days (Ashoka and Sunitha, 2020). The crop finds use as a catch crop, cover crop, or nurse crop. The use of millet as a cover crop under agroforestry system controls soil erosion (Sheahan, 2014). As a nurse crop, it is used for the establishment of perennial grass cover (Sheahan, 2014)

Grain is used as a boiled whole grain, porridge, gruel, or for making bread (Nesbitt, 2005). The millet can also be cultivated as a catch crop between commodity crops (Sheahan, 2014) or nurse crop for the establishment of perennial grass cover (Ashoka and Sunitha, 2020).

Millet is small greenish grain and is gluten free. It has high protein content 11.5 g/100 g of millet (Ashoka and Sunitha, 2020). It is also rich in iron, calcium, potassium, magnesium, zinc, phosphorus, protein, and vitamin B complex.

Browntop millet is used for biological control by suppressing root-knot nematode populations in tomato and pepper crops (McSorley et al., 1999). The browntop millet is a bioaccumulator and used for soil remediation since it accumulates lead and zinc in shoot and root (Lakshmi et al., 2013).

2.5.13 *PROSO MILLET/BROOMCORN MILLET/COMMON MILLET*
(*Panicum miliaceum* L., Family: Poaceae)

Proso millet is one of the oldest grain crops. Proso millet (*Panicum miliaceum* L.) was domesticated in the semiarid regions of China approximately 10,000 years ago (Santra et al., 2009, Baltensperger, 1996) from where it spread to other parts of the world (Prabhakar et al., 2017). Proso millet is commonly cultivated in tropics and subtropics of India, Sri Lanka, Japan, Pakistan, Nepal, Myanmar, China, Russia, Egypt, Arabia, and Western Europe (Baltensperger, 1996; Prabhakar et al., 2017;). It is chiefly cultivated in Madhya Pradesh, Uttar Pradesh, Bihar, Tamil Nadu, Maharashtra, Andhra Pradesh, Karnataka, and Uttarakhand (Prabhakar et al., 2017).

Proso millet is a short-duration crop maturing in 60–100 days (Baltensperger, 2002) and thus is able to evade drought (Prabhakar et al., 2017). It is tolerant to drought and heat and ideal crop for dryland farming (Habiyaremye et al., 2017). It grows in mixed cropping systems with cowpea or soybeans (Schonbeck and Morse, 2006). It could be useful in low-input sustainable systems (Sheahan, 2014).

The millet grows up to an altitude of 3500 m in India (Baltensperger, 2002). Proso millet is a C_4 plant with a low transpiration ratio; its high water use efficiency allows it to grow in water-limited environments (Baltensperger, 1996; Lyon et al., 2008). It is able to carry out photosynthesis even in drought, high temperatures, and limited nitrogen and CO_2 (Habiyaremye et al., 2017). The plant can be cultivated on a variety of soil types such as slightly acidic, saline, sandy loam, or low-fertility soils (Riley et al., 1989; Changmei and Dorothy, 2014) and can give reasonable yield in degraded soils and harsh climatic conditions (Prabhakar et al., 2017). Proso millet cultivation post wheat crop controls weed growth, conserves stored moisture (Krishna, 2013), prevents the loss of organic matter, and reduces disease pressure (Santra et al., 2009).

Millet has high nutritional content and is a major source of energy and protein for African and Asian countries (Amadou et al., 2013). Grains of Proso millet are gluten free with a low glycemic index and contain carbohydrates and variety of fatty acid. It also contains minerals such as manganese, phosphorus, magnesium, and amino acids (Prabhakar et al., 2017). The millet is used for producing ethanol.

The millet can be grown as a catch crop in sorghum or corn stubble fields.

2.5.14 *KODO MILLET* (*Paspalum scrobiculatum* L., Family: Poaceae)

The millet is found widely distributed in the tropics of the old world. In India and West Africa, the grains are harvested from wild. It is an ancient millet and was domesticated about 3000 years ago in India. The millet is cultivated in India, Pakistan, Vietnam, Thailand, Philippines, Indonesia, and

West Africa. In India, it is chiefly cultivated in Tamil Nadu, Madhya Pradesh, Gujarat, Karnataka, Maharashtra, Odisha, Kerala, Rajasthan, Chhattisgarh, Uttar Pradesh, West Bengal, Andhra Pradesh, and Himalayas (de wet et al., 1983; Deshpande et al., 2015).

The seeds though small are rich in nutrients. The plant can be cultivated in marginal soils has low input requirements and grows in a range of climatic conditions (Biodiversity International, 2021).

Growth period ranges from 120 to 180 days. The crop can be cultivated in semiarid regions since it is a drought-tolerant crop (Deshpande et al., 2015). The millet can also be grown as intercrops with pigeon pea, green gram, and soybean (Prabhakar et al., 2017) due to its shallow root system. It provides food security in case of water stress and assist as "low risk" crop reducing threats of poor yield due to biotic or abiotic stress. The millet has the highest productivity amongst millets and has potential yield of 2000 kg/ha. It also provides food security since the grains can be stored for a long period of time and meeting the food requirements during food shortages (Biodiversity International, 2021).

The millet is a nutraceutical and functional food containing fat 1.1–3.4 g and carbohydrates 66.6 g/100 g of grain. Grains contain protein 8% and minerals 2.6–4.9% (Kalpana, 2015; Deshpande et al., 2015). Kodo millet is a good source of phosphorus and iron. It is rich in proteins, sulfur-containing amino acids, methionine and cysteine, vitamins such as niacin, pyridoxine and folic acids (Biodiversity International, 2021). Grains are light brown to a dark grey (Deshpande et al., 2015). In rural India, the grains are eaten traditionally as health and vitality foods (Hegde and Chandra, 2005). Kodo millet is free from gluten and can be cooked and eaten. Seeds can be popped or puffed to make snack items. Kodo flour can be made into bread or pudding, soup, pasta, etc. (Biodiversity International, 2021; Deshpande et al., 2015). It can be included to make different Indian dishes, biscuits, and cutlet that are acceptable. They have the potential to produce novel traditional foods (Kalpana, 2015). It is cultivated on hillsides for providing famine food and to prevent soil erosion (Deshpande et al., 2015).

2.6 SUMMARY

Traditional agricultural practices are low-input agriculture that are eco-friendly and preserve the agrobiodiversity. The time-tested farming methods meet the food and nutrient demands in subsistence farming. The agricultural lands are maintained fertile through the traditional agropratices. Agricultural heritage systems includes systems which practice traditional agriculture and have individuality and originality, preserve and cultivate agrobiodiverse climate resilient crops and land races, provide nutrition and food security to the local people. These unique agroecosystems needs to be preserved in their entirety along with their life support systems. Traditional farming systems have also played a key role in preserving the underexploited crop species such as the pseudocereals, millets, drumsticks, teff, breadfruit, yam beans, etc., which have potential to augment the food resources particularly in areas that face food shortages. Popularization of the underexploited crops is required that will give boost to undertake cultivation of these crops on marginal lands since these are easy to cultivate, do not need the use of expensive agrochemicals, fairly resistant to pests and diseases, and give reasonable yield even under stressed environmental conditions. The popularization of these crops can be done by developing new products, which are acceptable by people, have a longer shelf life, have wider and efficient distribution via marketing networks of these products so that the nutrient-rich food can easily reach pregnant women, invalids, and nutrient deficient young and adults. These underexploited crops have great potential to provide food and nutrient security in a sustainable way.

KEYWORDS

- **traditional agriculture**
- **subsistence farming**
- **globally important agricultural heritage systems (GIAHS)**
- **underexpolited crops**
- **climate resilient crops**
- **agrobiodiversity**

REFERENCES

Agarwal, A.; Narain, S. Dying Wisdom: Rise, Fall and Potential of India's Traditional Water Harvesting Systems. *State India Environ.* **1997,** *4,* 404.

Ahmad, M.; Ahmad, F.; Dar, E. A.; Bhat, R. A.; Mushtaq, T.; Shah, F. Buck Wheat (*Fagopyrum esculentum*)—A Neglected Crop of High Altitude Cold Arid Regions of Ladakh: Biology and Nutritive Value. *Int. J. Pure App. Biosci.* **2018,** *6* (1), 395–406.

Ajani, A. O.; Oshundahunsi, O. F.; Akinoso, R.; Arowora, K. A.; Abiodun, A. A.; Pessu, P. O. Proximate Composition and Sensory Qualities of Snacks Produced from Breadfruit Flour. *Glob. J. Sci. Front. Res. Biol. Sci.* **2012**, *12*, 1–8.

Akila, N.; Bharathy, N. SWOC Analysis of Korangadu Pasture Land for Sheep in Tamil Nadu, a Time Tested Pasture Land Model. *North East Vet.* **2016**, *16* (3), 14–16.

Akubor, P. I.; Badifu, G. I. O. Chemical Composition, Functional Properties and Baking Potential of African Breadfruit Kernel and Wheat Flour Blends. *Int. J. Food Sci. Tech.* **2014**, *39*, 223–229.

Altieri, M. A. Applying Agroecology to Enhance the Productivity of Peasant Farming Systems in Latin America. *Environ. Dev. Sustain.* **1999a**, *1* (3), 197–217.

Altieri, M. A. The Ecological Role of Biodiversity in Agroecosystems. *Agric. Ecosyst. Environ.* **1999b**, *74* (1), 19–31.

Altieri, M. A.; Koohafkan, P. *Enduring Farms: Climate Change, Smallholders and Traditional Farming Communities*; Vol. 6; Third World Network: Penang, 2008; p 37.

Altieri, M. A.; Nicholls, C. I.; Henao, A.; Lana, M. A. Agroecology and the Design of Climate Change-Resilient Farming Systems. *Agron. Sustain. Dev.* **2015**, *35* (3), 869–890.

Anex, R. P.; Lynd, L. R.; Laser, M. S.; Heggenstaller, A. H.; Liebman, M. Potential for Enhanced Nutrient Cycling Through Coupling of Agricultural and Bioenergy Systems. *Crop. Sci.* **2007**, *47*, 1327–1335.

Angeli, V.; Silva, P. M.; Massuela, D. C.; Khan, M. W.; Hamar, A.; Khajehei, F.; Grae-Hönninger, S.; Piatti, C. Quinoa (*Chenopodium quinoa* Willd.): An Overview of the Potentials of the "Golden Grain" and Socio-Economic and Environmental Aspects of Its Cultivation and Marketization. *Foods.* **2020**, *9*, 216.

Anju, T.; Sarita, S. Suitability of Foxtail Millet (*Setaria italica*) and Barnyard Millet (*Echinochloa frumentacea*) for Development of Low Glycemic Index Biscuits. *Malays. J. Nutr.* **2010**, *16*, 361–368.

Aquaculture in Apatani Plateau in Arunachal Pradesh. In *Vikaspedia*. https://vikaspedia.in/ agriculture/fisheries/fish-production/culture-fisheries/types of aquaculture/aquaculture in apatani plateau in arunachal pradesh (retrieved May 16, 2021).

Arunachalam, V.; Rengalakshmi, R.; Raj, M. S. K. Ecological Stability of Genetic Diversity Among Landraces of Little Millet (*Panicum sumatrense*) in South India. *Genet. Res. Crop. Evol.* **2005**, *52*, 15–19.

Ashoka, P.; Sunitha, N. H. Review on Browntop Millet—A Forgotten Crop. *J. Exp. Agric. Int.* **2020**, *42* (7), 54–60.

Audilakshmi, S.; Stenhouse, J. W.; Reddy, T. P.; Prasad, M. V. R. Grain Mold Resistance and Associated Characters of Sorghum Genotypes. *Euphytica.* **1999**, *107*, 91–103.

Bal, L. M.; Meda, V.; Naik, S. N.; Satya, S. Sea Buckthorn Berries: A Potential Source of Valuable Nutrients for Nutraceuticals and Cosmoceuticals. *Food Res. Int.* **2011**, *44* (7), 1718–1727. DOI: 10.1016/j.foodres.2011.03.002.

Balak Ram B. Impact of Human Activities on Land Use Changes in Arid Rajasthan: Retrospect and Prospects. In *Human Impact on Desert Environment*; Narain, P., Kathju, S., Kar, A., Singh, M. P., Kumar P., Eds.; Scientific Publishers: Jodhpur, 2003; pp 44–59.

Balter, M. Seeking Agriculture's Ancient Roots. *Science* **2007**, *316*, 1830–1835.

Beaulieu, D. Neem Oil as Organic Insecticide. *Landsc. Newslett.* **2013**, 2.

Becker, R.; Wheeler, E.; Lorenz, K., Staffrod, A. E., Grosjean, O. K., Betschart, A. A.,; Saunder, R. M. A Compositional Study of Amaranth Grain. *J. Food Sci.* **1981**, *46*, 1175.

Bhargava, A.; Shukla, S.; Ohri, D. Chenopodium Quinoa—An Indian Perspective. *Industr. Crop. Prod.* **2006,** *23* (1), 73–87.

Bhat, B. V.; Rao, B. D.; Tonapi, V. A. *The Story of Millets* (Ed). Karnataka State Department of Agriculture, Bengaluru and ICAR-Indian Institute of Millets Research: Hyderabad, India, 2018; p 110.

Bhatia A. L. Growing Colourful and Nutritious Amaranths. *Nat. Prod. Radiance.* **2005,** *4* (1), 40–43.

Biodiversity International. NUS-Neglected and Underutilized Species Community: Kodo Millet. http://www.nuscommunity.org/nus/neglected-underutilized-species/kodo-millet/ (acessed April 19, 2021).

Boivin N.; Fuller D. Q.; Korisettar R.; Petraglia, M. First Farmers in South India: The Role of Internal Processes and External Influences in the Emergence and Transformation of South India's Earliest Settled Societies. *Pragdhara* **2008,** *18,* 179–200

Bouhache, M.; Bayer, D. E. Photosynthetic Response of Flooded Rice (*Oryza sativa*) and Three Echinochloa Species to Changes in Environmental Factors. *Weed Sci.* **1993,** *41,* 611–614

Bressani, R. *Composition and Nutritional Properties of Amaranth.* In *Amaranth: Biology, Chemistry and Technology*; Ed.; Paredes-López O.; CRC Press, 1994; pp 185–205.

Bressani, R.; Gonzalez, J. M.; Zuniga, J.; Breuner, M.; Elias, L. G. Yield, Selected Chemical Composition and Nutritive Value of 14 Selections of Amaranth Grain Representing Four Species. *J. Sci. Food Agric.* **1987,** *38,* 347–356.

Brink M. *Setaria italica* (L.) P. Beauv. Record from Protabase. In *PROTA (Plant Resources of Tropical Africa/Ressourcesvégétales de l'Afrique tropicale)*, Wageningen, Netherlands; Brink, M., Belay, G., Eds.; 2006.

Campbell, C. G. *Buckwheat: Fagopyrum Esculentum Moench. Promoting the Conservation and Use of Underutilized and Neglected Crops*, Vol. 19; Institute of Plant Genetics and Crop Plant Research, Gatersleben, International Plant Genetic Resources Institute: Rome, Italy, 1997; 95p.

Caselato-Sousa, V. M.; Amaya-Farfan, J State of Knowledge on Amaranth Grain: A Comprehensive Review. *J. Food Sci.* **2012,** *77,* 93–104.

Chandel, G.; Meena, R.; Dubey, M.; Kumar, M. Nutritional Properties of Minor Millets: Neglected Cereals with Potentials to Combat Malnutrition. *Curr. Sci.* **2014,** *107,* 1109–1111.

Chandra, S.; Saklani, S.; Semwal, R. B.; Semwal, D. K. Estimation of Nutritional and Mineral Contents of Eleusine Coracana and Echinochloa Frumentacea—Two Edible Wild Crops of India. *Curr. Nutr. Food Sci.* **2019,** *15* (4).

Chandrashekar, A. Finger Millet: *Eleusine coracana. Adv. Food Nutr. Res.* **2010,** *59,* 215–262.

Chauhan, A.; Singh, S. Influence of Germination on Physico—Chemical Properties of Amaranth (Amaranthus Spp.) Flour. *Int. J. Agric. Food Sci. Technol.* **2013,** *4* (3), 215–220.

Chavan, S. M.; Jain, N. K.; Kumar, V.; Jain, S. K.; Agarwal, C.; Wadhawan, N.; Kumar, A.; Mehta, A. K. Reviews on Agronomical and Functional Aspects of Quinoa. *Int. J. Chem. Stud.* **2019,** *7* (1), 2173–2178.

Chethan, S.; Malleshi, N. G. Finger Millet Polyphenols: Optimization of Extraction and the Effect of pH on Their Stability. *Food Chem.* **2007,** *105,* 862–870.

Chhatre, A.; Agrawal, A. Forest Commons and Local Enforcement. *Proc. Natl. Acad. Sci. U.S.A.* **2008,** *105,* 13286–13291.

Chorage C. A.; Solanke G. M.; Kalel, R. Processing of Moringa Oleifera Leaves to Develop Powder for Consumption: A Review. *Int. J. Sci. Res.* **2020,** *9* (5), 617–619.

Coulibaly, J. Y.; Chiputwa, B.; Nakelse, T.; Kundhlande, G. Adoption of Agroforestry and the Impact on Household Food Security Among Farmers in Malawi. *Agric. Syst.* **2017,** *155,* 52–69.

Dabney; S. M.; Delgado; J. A.; Reeves; D. W. Using Winter Cover Crops to Improve Soil and Water Quality. *Commun. Soil Sci. Plant Anal.* **2001,** *32,* 1221–1250.

Dar; S. A.; Dar; N. A.; Bhat; M. A.; Bhat; M. H. Prospects; Utilization and Challenges of Botanical Pesticides in Sustainable Agriculture. *Int. J. Mol. Biol. Biochem.* **2014,** *2* (1), 1–14.

Das, S.; Ed. Distribution and Maintenance of Amaranth Germplasm Worldwide. In *Amaranthus: A Promising Crop of Future*; Springer: Berlin, 2016; pp 99–106.

de Oliveira, D. M. C. G.; Orellana, C.; Gebbie, L.; Steen, J.; Hodson, M. P.; Chrysanthopoulos, P.; Plan, M. R.; McQualter, R.; Palfreyman, R. W.; Nielsen, L. K. Metabolic Reconstruction of *Setaria italica*: A Systems Biology Approach for Integrating Tissue-Specific Omics and Pathway Analysis of Bioenergy Grasses. *Front. Plant Sci.* **2016,** *7,* 1138.

De Ron, A. M.; Sparvoli, F.; Pueyo, J. J.; Bazile, D. Protein Crops: Food and Feed for the Future. *Front. Plant Sci.* **2017,** *8,* 105.

Debashri, M.; Tamal, M. A Review on Efficacy of *Azadirachtaindica* A. Juss. Based Biopesticides: An Indian Perspective. *Res. J. Recent Sci.* **2012,** *1* (3), 94–99.

Denton, D. C. Food Crops for Waterfowl. In *Fireside Waterfowler: Fundamentals of Duck and Goose Ecology*; Wesley, D. E., Leitch, W. G., Eds.; Stackpole Books: Mechanicsburg, PA, 1987; pp 1–352.

Deshpande, S. S.; Mohapatra, D.; Tripathi, M. K.; Sadvatha, R. H. Kodo Millet-Nutritional Value and Utilization in Indian Foods. *J. Grain Process. Storage* **2015,** *2* (2), 16–23.

Devlin, F.; Zettel, T. *Ecoagriculture: Initiatives in Eastern and Southern Africa*; Weaver Press: Harare, Zimbabwe, 1999.

Dey, P.; Sarkar, A. K. Revisiting Indigenous Farming Knowledge of Jharkhand (India) for Conservation of Natural Resources and Combating Climate Change. *Indian J. Tradit. Knowl.* **2011,** *10* (1), 71–79.

Diamond, J.; Bellwood, P. Farmers and Their Languages: The First Expansions. *Science* **2003,** *300,* 597–603.

Ding, J.; Jiang, X.; Guan, D.; Zhao, B.; Ma, M.; Zhou, B.; Li, J. Influence of Inorganic Fertilizer and Organic Manure Application on Fungal Communities in a Long-Term Field Experiment of Chinese Mollisols. *Appl. Soil Ecol.* **2017,** *111,* 114–122.

Doddabasawa; Chittapur, B. M. Agroforestry: An Ecosystem Approach in Farming for Sustainability. *J. Farm Sci.* **2020,** *33* (1), 1–7.

Dolker, P. An Overview of Transition in Traditional Agriculture of Ladakh. *J. Himal. Ecol. Sustain. Dev.* **2018,** *13,* 26–48.

Elevitch, C.; Ragone, D.; Cole, I. *Breadfruit Production Guide: Recommended Practices for Growing; Harvesting; and Handling*; 2nd ed.; Breadfruit Institute of the National Tropical Botanical Garden; Kalaheo; Hawaii and Hawaii Homegrown Food Network: Holualoa, Hawaii, 2014; pp 1–35.

Ellis, E.; Wang, S. M. Sustainable Traditional Agriculture in the Tai Lake Region of China. *Agric. Ecosyst. Environ.* **1997,** *61* (2–3), 177–193.

Esparagoza; R. S.; Tangonan; G. J. Instant Baby Food Using Banana and Breadfruit Flour as Food Base. *Univ. South Mindanao Coll. Agric. Res. J.* **1993,** *4,* 175–177.

Evenson, R. E.; Gollen; D. Assessing the Impact of the Green Revolution, 1960–2000. *Science* **2003,** *300,* 758–762.

Eyong; C. T. Indigenous Knowledge and Sustainable Development in Africa: Case Study on Central Africa. In *Indigenous Knowledge Systems and Sustainable Development: Relevance for Africa Emmanuel K. Boon; Luc Hens, Tribes and Tribals*; Special Vol. 1; 2007; pp 121–139.

FAO. *Amino-Acid Content of Foods and Biological Data on Proteins*; FAO Nutrition Studies No 24: Rome, Italy, 1970; pp 1–285.

FAO. *Sorghum and Millets in Human Nutrition*; FAO Food and Nutrition Series No 27; Food and Agriculture Organization: Rome, Italy, 1995; pp 1–184.

FAO. *Conservation and Adaptive Management of Globally Important Agricultural Heritage Systems (GIAHS). In PIMS 2050.Terminal Report.* Project Symbol: UNTS/GLO/002/GEF Project ID: 137561; Food and Agriculture Organization of the United Nations: Rome, Italy, Feb, 2008. http://www.fao.org/fileadmin/templates/giahs/PDF/GIAHS_B_terminalReport. pdf (acessed on May 19, 2021).

FAO. *Low Greenhouse Gas Agriculture: Mitigation and Adaptation Potential of Sustainable Farming Systems*; Food and Agriculture Organization of the United Nations: Rome, Italy, 2009a.

FAO. *International Treaty on Plant Genetic Resources for Food and Agriculture: First Fruits of Plant Gene Pact*; Food and Agriculture Organization of the United Nations: Rome, Italy, 2009b. http://www.fao.org/plant-treaty/news/news-detail/en/c/341439/ (acessed on April 23, 2021).

FAO. *Family Farming Knowledge Platform: Inventory and Documentation of Tribal GIAHS in India*; Food and Agriculture Organization of the United Nations (FAO): Rome, Italy, 2009c. http://www.fao.org/family-farming/detail/en/c/283072/ (accessed on April 05, 2021).

FAO. *Climate Smart Agriculture: Policies; Practices and Financing for Food Security. Adaptation and Mitigation*; Food and Agriculture Organization of the United Nations (FAO): Rome, Italy, 2010.

FAO. *Family Farming Knowledge Platform: Sikkim Himalaya-Agriculture: Improving and Scaling up of Traditionally Managed Agricultural Systems of Global Significance*, 2014. http://www.fao.org/family-farming/detail/en/c/283212/ (accessed on April 05, 2021).

FAO. *Template for GIAHS Proposal: Grand Anicut (Kallanai) and Associated Farming System in Cauvery Delta Zone of Tamil Nadu*; Food and Agricultural Organizations on the United Nations: Rome Italy, 2017; pp 52. http://www.fao.org/publications/card/en/c/ e8bebbc8-794f-4f48-bef2-ac1827d77c3c/ (acessed on April 23, 2021).

FAO. *Family Farming Knowledge Platform: Globally Important Agricultural Heritage Systems (GIAHS). Combining Agricultural Biodiversity; Resilient Ecosystems; Traditional Farming Practices and Cultural Identity*; Food and Agriculture Organization of the United Nations, 2018. http://www.fao.org/family-farming/detail/en/c/1147754/ (Accessed on April 23, 2021).

FAO. *GIAHS; Globally Important Agricultural Heritage Systems.* Food and Agriculture Organization of the United Nations, 2021a. http://www.fao.org/giahs/en/ (acessed on April 19, 2021).

FAO. *Traditional Crops*; Food and Agriculture Organization of the United Nations, 2021b. http://www.fao.org/traditional-crops/en/ (Accessed on April 23, 2021).

FAO. *Agroecology Knowledge Hub: Apatani Wet Rice Cultivation: An Example of a Highly Evolved Traditional Agroecosystem*; Food and Agriculture Organization of the United

Nations, 2021c. http://www.fao.org/agroecology/database/detail/en/c/442981/ (accessed on April 11, 2021).

FAO. *GIAHS: Asia and the Pacific: India-Koraput Traditional Agriculture, India*; Food and Agriculture Organization of the United Nations, 2021d. http://www.fao.org/giahs/giahsaroundtheworld/en/ (acessed on April 16, 2021).

FAO. *Family Farming Knowledge Platform: Kuttanad Below Sea Level Farming System*; Food and Agriculture Organization of the United Nations, 2021e. http://www.fao.org/family-farming/detail/en/c/283069/ (accessed April 16, 2021).

Frasier, I.; Quiroga, A.; Noellemeyer, E. Effect of Different Cover Crops on C and N Cycling in Sorghum NT Systems. *Sci. Total Environ.* **2016**, *562*, 628–639.

Ge, Y.; Zhang, J. B.; Zhang, L. M.; Yang, M.; He, J. Z. Long-Term Fertilization Regimes Affect Bacterial Community Structure and Diversity of an Agricultural Soil in Northern China. *J. Soils Sediment.* **2008**, *8*, 43–50.

Geda, M. K. Traditional Botanical Based Pesticides for Key Pest Animals Control in Dire Dawa Administration; Eastern Ethiopia. *Glob. J. Adv. Res.* **2015**, *2* (2), 388–399.

Giri, D. K. In India: Tribal Agricultural Heritage Systems. In *Proceedings of the Second International Forum on Globally Important Agricultural Heritage Systems*, Buenos Aires, Argentina, Oct 21–23, 2009; pp 67–68.

Gonzalez, C. G. Climate Change; Food Security; and Agrobiodiversity: Toward a Just; Resilient; and Sustainable Food System. *Fordham Environm. Law Rev.* **2011**, *22*, 493–522.

Gopinath, K. A.; Saha, S.; Mina, B. L.; Pande, H.; Kundu, S.; Gupta; H. S. Influence of Organic Amendments on Growth; Yield and Quality of Wheat and on Soil Properties During Transition to Organic Production. *Nutr. Cycl. Agroecosyst.* **2008**, *82* (1), 51–60.

Gull, A.; Jan, R.; Nayik, G. A.; Prasad, K.; Kumar; P. Significance of Finger Millet in Nutrition; Health and Value Added Products: A Review. *J. Environ. Sci. Comput. Sci. Eng. Technol.* **2014**, *3*, 1601–1608.

Gupta, A.; Mahajan, V.; Kumar, M.; Gupta, H. Biodiversity in the Barnyard Millet (*Echinochloa frumentacea* Link.Poaceae) germplasm in India. *Genet. Resour. Crop Evol.* **2009**, *56*, 883–889.

Gupta, N.; Gupta, A. K.; Singh, N. K.; Kumar; A. Differential Expression of *PBF Dof* Transcription Factor in Different Tissues of Three Finger Millet Genotypes Differing in Seed Protein Content and Color. *Plant Mol. Biol. Rep.* **2011**, *29*, 69–76.

Gupta, S. M.; Arora, S.; Mirza, N.; Pande, A.; Lata, C.; Puranik, S.; Kumar, J.; Kumar; A. Finger Millet: A "Certain" Crop for an "Uncertain" Future and a Solution to Food Insecurity and Hidden Hunger Under Stressful Environments. *Front. Plant Sci.* **2017**, *8*, 643.

Guzmán-Maldonado; S.; and Paredes-Lopez; O. Functional Products of Plants Indigenous to Latin America: Amaranth; Quinoa; Common Beans; and Botanicals. In *Functional Foods: Biochemical and Processing Aspects*; Mazza, G., Ed.; Technomic Publishing Co. Inc.: Lancaster, PA, 1998; pp 293–328.

Harborne, J. B. *Introduction to Ecological Biochemistry*; 3rd ed; Academic Press: London, 1989; pp 1–356.

Hegde, P. S.; Chandra, T. S. ESR Spectroscopic Study Reveals Higher Free Radical Quenching Potential in Kodo Millet (*Paspalum scrobiculatum*) Compared to Other Millets. *Food Chem.* **2005**, *92* (1), 177–182.

Heinemann, J. A.; Massaro, M.; Coray, D. S.; Agapito-Tenfen, S. Z.; Wen, J. D. Sustainability and Innovation in Staple Crop Production in the US Midwest. *Int. J. Agric. Sustain.* **2013**, *12* (4), 387–390.

Hillock, D.; Bolin, P. In Earth-Kind Gardening Series *Botanical Pest Controls*; Oklahoma Cooperative Extension Service. *HLA.* **2004,** *6433,* 1–4.

Hobbs; P. R.; Sayre; K.; Gupta; R. The Role of Conservation Agriculture in Sustainable Agriculture. *Philos. Trans. R. Soc. Lond. B. Biol. Sci.* **2008,** *363* (1491), 543–555.

Hu, F.; Feng, F.; Zhao, C.; Chai, Q.; Yu, A.; Yin, W.; Gan, Y. Integration of Wheat-Maize Intercropping with Conservation Practices Reduces $CO2$ Emissions and Enhances Water Use in Dry Areas. *Soil Till. Res.* **2017,** *169,* 44–53.

Huang, S.; Bohrer, B. M. The Effect of Tropical Flours (Breadfruit and Banana) on Structural and Technological Properties of Beef Emulsion Modeling Systems. *Meat Sci.* **2020,** *163,* 108082.

Huang, M.; Shao, M.; Zhang, L.; Li, Y. Water Use Efficiency and Sustainability of Different Long-Term Crop Rotation Systems in the Loess Plateau of China. *Soil Till. Res.* **2003,** *72* (1), 95–104.

Huang, S.; Roman, L.; Martinez, M. M.; Bohrer; B. M. Modification of Physicochemical Properties of Breadfruit Flour Using Different Twin-Screw Extrusion Conditions and Its Application in Soy Protein Gels. *Foods* **2020,** *9,* 1071.

Jacobsen, E. The Worldwide Potential for Quinoa (*Chenopodium quinoa* Willd.). *Food Rev. Int.* **2003,** *19* (1–2), 167–177.

Jansen, G. R.; DiMaio, L. R.; Hause; N. L. Amino Acid Composition and Lysine Supplementation of Teff. *J. Agric. Food Chem.* **1962,** *10,* 62–64.

Jayaraman, A.; Puranik, S.; Rai, N. K.; Vidapu, S.; Sahu, P. P.; Lata, C. cDNA-AFLP Analysis Reveals Differential Gene Expression in Response to Salt Stress in Foxtail Millet (*Setaria italica* L.). *Mol. Biotechnol.* **2008,** *40,* 241–251.

Joseph, E. A. Rice Cultivation in Major Wetlands of Kerala. *Int. J. Eng. Sci. Comput.* **2016,** *6* (8), 2136–2137.

Joshi, B. D. Status of Buckwheat in India. *Fagopyrum* **1999,** *16,* 7–11.

Kalpana, C. A. *Kodo Millet: A Boon to Food Industry and Health.* 7th Indo-Global Summit and Expo on Food and Beverages, New Delhi, India; Octr 08–10, 2015.

Kamatar, M. Y.; Hemalatha, S.; Meghana, D. R.; Talawar, S.; Naik; R. K. Evaluation of Little Millet (*Panicum sumatrense)* Land Races for Cooking and Nutritional Composition. *Curr. Res. Biol. Pharmaceut. Sci.* **2013,** *2* (1), 7–11.

Kaushal, R. K.; Das, G.; Suryakumari; Poornachander, E. *A Study of Indigenous Agricultural Practices Among the Tribals of Andhra Pradesh and Telangana—The Trajectory of Transition and Impacts on Livelihoods and Food Security.* Centre for People's Forestry: Secunderabad, 2016; pp 1–72.

Kavitha, R.; Chandrashekar; A. Content and Composition of Nonstarchy Polysaccharides in Endosperms of Sorghums Varying in Hardiness. *Cereal Chem.* **1992,** *69,* 440–443.

Kerala Tourism. FAO Heritage Status to Kuttanad. *Newsletter* **2013,** *241.*

Ketema, S. *Tef Eragrostis tef (Zucc.)Trotter.Promoting the Conservation and Use of Underutilized and Neglected Crops. 12.* International Plant Genetic Resources Institute: Rome, 1997; pp 1–50.

Khare, C. *Setaria italica* (Linn.) Beauv. In *Indian Medicinal Plants*; Khare, C., Eds.; Springer: New York, NY, 2007.

King; E. D. I. O.; Ravi; S. B. Documentation and Monitoring of Agrobiodiversity and Indigenous Knowledge on-Farm—Experiences from India. In *Proceedings of an International Conference on Farm Conservation of Neglected and Underutilized Species: Status, Trends and Novel Approaches to Cope with Climate Change*, Frankfurt, June 14–16,

2011; Padulosi, S., Bergamini, N., Lawrence, T., Eds.; Biodiversity International: Rome, 2012; pp 57–63.

Kohler-Rollefson, I. The Raika Dromedary Breeders of Rajasthan: A Pastoral System in Crisis. *Nomadic Peoples* **1992**, *30*, 74–83.

Kremen, C.; Miles, A. Ecosystem Services in Biologically Diversified Versus Conventional Farming Systems: Benefits; Externalities; and Trade-Offs. *Ecol. Soc.* **2012**, *17* (4), 40.

Krishnamurthy, L.; Upadhyaya, H. D.; Gowda, C. L. L.; Kashiwagi, J.; Purushothaman, R.; Singh, S. Large Variation for Salinity Tolerance in the Core Collection of Foxtail Millet (*Setaria italica* (L.) P. Beauv.) Germplasm. *Crop. Pasture Sci.* **2014**, *65*, 353–361.

Kumar, A.; Mirza, N.; Charan, T.; Sharma, N.; Gaur; V. S. Isolation; Characterization and Immunolocalization of a Seed Dominant CaM from Finger Millet *(Eleusine coracana* L. Gartn.) for Studying Its Functional Role in Differential Accumulation of Calcium in Developing Grains. *Appl. Biochem. Biotechnol.* **2014**, *172*, 2955–2973.

Kumar, A.; Ramakrishnan, P. S. Energy Flow Through an Apatani Village Ecosystem of Arunchal Pradesh in Northeast India. *Hum. Ecol.* **1990**, *18*, 315–333.

LAHDC. *Mission Organic Development Initiative of Ladakh; Policy; Strategy and Action Plan.* Ladakh Autonomous Hill Development Council: Leh, Ladakh, 2019; pp 1–18.

Lakshmi, P. M.; Jaison S.; Muthukumar T.; Muthukumar, M. *Assessment* of Metal Accumulation Capacity of *Brachiaria ramosa* Collected from Cement Waste Dumping Area for the Remediation of Metal Contaminated Soil. *Ecol. Eng.* **2013**, *60*, 96–98.

Lal, A.; Sircar, R.; Kumar; S. Super Food; Finger Millet (*Eleusine coracana* L.) a Source to Nutritional Security: A Review for Its Potential to Value Addition and Process Waste Utilization. *J. Nutr. Food Sci.* **2017**, *7* (5 Suppl).

Lasco, R. D.; Delfino, R. J. P.; Catacutan, D. C.; Simelton, E. S.; Wilson; D. M. Climate Risk Adaptation by Smallholder Farmers: The Roles of Trees and Agroforestry. *Curr. Opin. Environm. Sustain.* **2014**, *6*, 83–88

Li, P.; Brutnell; T. P. *Setaria viridis* and *Setaria italica*; Model Genetic Systems for the Panicoid Grasses. *J. Exp. Bot.* **2011**, *62*, 3031–3037.

Li; T. S.; Schroeder; W. R. Sea Buckthorn (Hippophae rhamnoides L.): A Multipurpose Plant. *Hort. Technol.* **1996**, *6* (4), 370–380. DOI: 10.21273/HORTTECH.6.4.370.

Liu, X.; Lehtonen, H.; Purola, T.; Pavlova, Y.; Rotter, R.; Palosuo; T. Dynamic Economic Modelling of Crop Rotations with Farm Management Practices Under Future Pest Pressure. *Agric. Syst.* **2016**, *144*, 65–76.

Liu, Y.; Brown; P. N., Ragone, D.; Gibson, D. L.; Murch; S. J. Breadfruit Flour is a Healthy Option for Modern Foods and Food Security. *PLoS One.* **2020**, *23*, 15 (7).

Liu, Y.; Duan, M.; Yu; Z. Agricultural Landscapes and Biodiversity in China. *Agric. Ecosyst. Environ.* **2013**, *166*, 46–54.

Liu, Y.; Ragone, D.; Murch; S. Breadfruit (*Artocarpus altilis*): A Source of High-Quality Protein for Food Security and Novel Food Products. *Amino Acids.* **2015**, *47* (4), 20–15.

Madella, M.; Lancelotti, C.; Garcia-Granero, J. J. Millet Microremains–An Alternative Approach to Understand Cultivation and Use of Critical Crops in Prehistory. *Archaeol. Anthropol. Sci.* **2016**, *8*, 17–28.

Magdoff, F.; Weil, R. Soil Organic Matter Management Strategies. In *Soil Organic Matter in Sustainable Agriculture*; Magdoff, F., Weil, R. Eds.; CRC Press, 2004; pp 269–283.

Maitra, S.; Reddy, M. D.; Nanda; S. P. Nutrient Management in Finger Millet (*Eleusine coracana* L. Gaertn) in India. *Int. J. Agric. Environm. Biotechnol.* **2020**, *13* (1), 03–21.

Maitra, S.; Shankar; T. Agronomic Management in Little Millet (*Panicum sumatrense* L.) for Enhancement of Productivity and Sustainability. *Int. J. Biores. Sci.* **2019**, *6* (2), 91–96.

Malomo, S. A.; Eleyinmi, A. F.; Fashakin, J. B. Chemical Composition; Rheological Properties and Bread Makingpotentials of Composite Flours from Breadfruit; Breadnut and Wheat. *Afr. J. Food Sci.* **2011**, *5*, 400–410.

Marsinach, M. S.; Cuenca, A. P. The Impact of Sea Buckthorn Oil Fatty Acids on Human Health. *Lipid. Health Dis.* **2019**, *18*, 145.

Mbithi-Mwikya, S.; Ooghe, W.; Van Camp, J.; Nagundi, D.; Huyghebaert, A. Amino Acid Profiles After Sprouting; Autoclaving; and Lactic Acid Fermentation of Finger Millet (*Eleusine coracana*) and Kidney Beans (*Phaseolus Vulgaris* L.). *J. Agric. Food Chem.* **2000**, *48*, 3081–3085.

McSorley, R.; Ozores-Hampton, M.; Stansly, P. A.; Conner; M. Nematode Management; Soil Fertility; and Yield in Organic Vegetable Production. *Nematropica* **1999**, *29*, 205–213.

Mgonja, M. A.; Lenne, J. M.; Manyasa, E.; Sreenivasaprasad, S. Finger Millet Blast Management in East Africa. Creating Opportunities for Improving Production and Utilization of Finger Millet; In *Proceedings of the First International Finger Millet Stakeholder Workshop, Projects R8030 & R8445UK Department for International Development—Crop Protection Programme* held 13–14 September 2005 at Nairobi Patancheru 502 324, Andhra Pradesh, India; International Crops Research Institute for the Semi-Arid Tropics, 2007; 196pp.

Miller, D. *Teff Grass: A New Alternative*; UC Davis: California, 2009.

Ministry of Culture, Government of India (GOI). *Parampara: Traditions & Practices-Apatani Paddy-Fish Cultivation*, 2021. https://paramparaproject.org/traditions_apatani-paddy-fish. html (accessed May 17, 2021).

Montagnini, F.; Nair, P. K. R. Carbon Sequestration: An Unexploited Environmental Benefit of Agroforest Systems. *Agrofor. Syst.* **2004**, *61*, 281–295.

Monteiro, P. V.; Sudharshana, L.; Ramachandra; G. Japanese Barnyard Millet (*Echinochloa frumentacea*): Protein Content; Quality and SDS-PAGE of Protein Fractions. *J. Sci. Food Agric.* **1987**, *43*, 17–25.

Mushagalusa, G. N.; Ledent, J. F.; Draye; X. Shoot and Root Competition in Potato/Maize Intercropping: Effects on Growth and Yield. *Environm. Exp. Bot.* **2008**, *64*, 180–188.

Muthana, K. D.; Sharma, S. K; Harsh; L. N. Study on Silvi-Pastoral System in Arid Zone. *My For.* **1985**, *21* (3), 233–238.

Nadeem, F.; Ahmad, Z.; Ul Hassan, M.; Wang, R.; Diao, X. and Li, X. Adaptation of Foxtail Millet (*Setaria italica* L.) to Abiotic Stresses: A Special Perspective of Responses to Nitrogen and Phosphate Limitations. *Front. Plant Sci.* **2020**, *11*, 187.

Nair, P. K. R.; Kumar, B. M.; Nair; V. Agroforestry as a Strategy for Carbon Sequestration. *J. Plant Nutr. Soil Sci.* **2009**, *172*, 10–23.

Nair, P. R.; Nair, V. D.; Kumar, B. M.; Haile; S. G. Soil Carbon Sequestration in Tropical Agroforestry Systems: A Feasibility Appraisal. *Environm. Sci. Policy* **2009**, *12* (8), 1099–1111.

Narayanasamy, P. Traditional Pest Control: A Retrospection. *Indian J. Trad. Knowl.* **2002**, *1* (1), 40–50.

Naskar, S. K. Progress and Status of Yam Bean Research in India. In *15th Triennial ISTRC Symposium, International Society for Tropical Root Crops*; Lima: Peru, 2009; pp 23–28.

Nayak, S. N. V.; Swamy, H. R.; Nagaraja, B.; Rao, U. Farmers' Attitude Towards Sustainable Management of Soppina Betta Forests in Sringeri Area of the Western Ghats, South India. *For. Ecol. Manag.* **2000,** *132* (2), 223–241.

Naylor, R.; Steinfeld, H.; Falcon, W.; Galloway, W.; Smil, V.; Bradford, E.; Alder, J.; Mooney; H. Losing the Links Between Livestock and Land. *Science* **2005,** *310,* 1621–1622.

Nesbitt, M. Grains. In *The Cultural History of Plants*; Prance, G., Nesbitt, M., Eds.; Routledge Press: New York, 2005; pp 45–60.

Ning, C.; Qu; J.; He, L.; Yang,R.; Chen, Q.; Luo, S.; Cai, K. Improvement of Yield; Pest Control and Si Nutrition of Rice by Rice-Water Spinach Intercropping. *Field Crop. Res.* **2017,** *208,* 34–43.

Nochera, C.; Ragone, D. Development of a Breadfruit Flour Pasta Product. *Foods* **2019,** *8,* 110.

Nowak, D. J.; Crane; D. E. Carbon Storage and Sequestration Byurban Trees in the USA. *Environ. Pollut.* **2002,** *116* (3), 381–389.

Obsertein, M.; Bottcher, H.; Yamagata; Y. Terrestrail Ecosystem Management for Climate Change Mitigation. *Curr. Opin. Environ. Sustain.* **2010,** *2,* 271–276.

Olas, B.; Skalski, B.; Ulanowska; K. The Anticancer Activity of Sea Buckthorn [*Elaeagnus rhamnoides* (L.) A. Nelson]. *Front. Pharmacol.* **2018.** DOI: https://doi.org/10.3389/fphar.2018.00232.

Oudart, D.; Robin, P.; Paillat, J. M., Paul; E. Modelling Nitrogen and Carbon Interactions in Composting of Animal Manure in Naturally Aerated Piles. *Waste Manag.* **2015,** *46,* 588–598.

Pandey, A.; Pradheep, K.; Gupta, R.; Nayar, E. R.; Bhandari; D. C. 'Drumstick tree' (*Moringa oleifera* Lam.): A Multipurpose Potential Species in India. *Genet. Resour. Crop. Evol.* **2011,** *58,* 453–460.

Pandey, V. N.; Chauhan, V.; Pandey, V. S.; Upadhyaya, P. P.; Kopp; O. R. *Moringa oleifera* Lam.: A Biofunctional Edible Plant from India; Phytochem. *Med. Propert. J. Plant Stud.* **2019,** *8* (1), 10–19.

Paul, C.; Weber, M.; Knoke, T. Agroforestry Versus Farm Mosaic Systems: Comparing Land-Use Efficiency; Economic Returns and Risks Under Climate Change Effects. *Sci. Total Environ.* **2017,** *587,* 22–35.

Pavela, R. Possibilities of Botanical Insecticide Exploitation in Plant Protection. *Pest Technol.* **2007,** *1,* 47–52.

Peacock, A. D.; Mullen, M. D.; Ringelberg, D. B.; Tyler, D. D.; Hedrick, D. B.; Gale, P. M.; White; D. C. Soil Microbial Community Responses to Dairy Manure or Ammonium Nitrate Applications. *Soil Biol. Biochem.* **2001,** *33,* 1011–1019.

Philogene, B. J. R.; Regnault-Roger, C.; Vincent, C. Botanicals: Yesteday's and Today's Promises. In *Biopesticides of Plant Origin*; Regnault-Roger, C., Philogene, B. J. R., Vincent, C., Eds.; Lavoisier: Paris, 2005; pp 1–15.

Pinto, P.; Long, M. E. F.; Pineiro; G. Including Cover Crops During Fallow Periods for Increasing Ecosystem Services: Is It Possible in Croplands of Southern South America? *Agric. Ecosyst. Environ.* **2017,** *248,* 48–57.

Poornima, R.; Prasad, K. N.; Jois; S. N. Growth; Yield and Nutritional Content of Finger Millet (*Eleusine coracana* L.) as Influenced by Pranic Energy Application. *J. Appl. Nat. Sci.* **2021,** *13* (1), 42–50.

Prabhakar; Ganiger, P. C.; Boraiah, B.; Bhat, S.; Nandini, C.; Kiran; Tippeswamy, V.; Manjunath, H. A. *Improved Production Technology for Kodo Millet.* Technical

Bulletin—2/2017-18.Project Coordinating Unit; ICAR-AICRP on Small Millets; GKVK: Bengaluru, 2017; pp 1–10.

Prabu, R.; Vanniarajan, C.; Vetrivanthan, M.; Gnanamalar, R. P.; Shanmughasundaram, R.; Ramalingam, J. Diversity and Stability Studies in Barnyard Millet [*Echinochloa frumentacea* (Roxb).Link.] Germplasm for Grain Yield and Its Contributing Traits. *Electr. J. Plant Breed.* **2020,** *11* (2), 528–537.

Pradeep, S. R.; Guha, M. Effect of Processing Methods on the Nutraceutical and Antioxidant Properties of Little Millet (*Panicum sumatrense*) Extracts. *Food Chem.* **2011,** *126* (4), 1643–1647.

Prakriti, M.; Sogani, R.; Gurung, N.; Rastogi, A.; Swiderska; K. *Smallholder Farming Systems in the Indian Himalayas; Key Trends and Innovations for Resilience.* Country Report. Natural Resource Group; International Institute for Environment and Development: London, 2018; p 68.

Pulido, J. S.; Bocco; G. The Traditional Farming System of a Mexican Indigenous Community: The Case of Nuevo San Juan Parangaricutiro; Michoacán; Mexico. *Geoderma* **2003,** *111,* 249–265.

Raghavendra, K. V.; Gowthami, R.; Lepakshi, N. M.; Dhananivetha, M.; Shashank; R. Use of Botanicals by Farmers for Integrated Pest Managementof Crops in Karnataka. *Asian Agric. History.* **2016,** *20* (3), 173–180.

Rai, S. C. Apatani Paddy-Cum-Fish Cultivation: An Indigenous Hill Farming System of North East India. *Indian J. Trad. Knowl.* **2005,** *4* (1), 65–71.

Rai, S. C.; Sharma, P. Carbon Flux and Land Use/Cover Change in a Himalayan Watershed. *Curr. Sci.* **2004,** *86* (12), 1594–1596.

Ramachandran, C.; Peter, K. V.; Gopalakrishnan, P. K. Drumstick (*Moringa oleifera*): A Multipurpose Indian Vegetable. *Econ. Bot.* **1980,** *34* (3), 276–283.

Ramakrishnan, P. S. *Shifting Agriculture and Sustainable Development: AnInterdisciplinary Study from North-Eastern India*; UNESCO-MAB Series, Paris; Parthenon Publ.: Carnforth,; Lancs. U. K. 1992; 424pp.

Ramesh, K.; Devi, K. B. S.; Gopinath, K. A.; Praveen, K. Geographical Adaptation of Quinoa in India and Agrotechniques for Higher Productivity of Quinoa. *J. Pharmacogn. Phytochem.* **2019,** *8* (3), 2930–2932.

Rane, J.; Pradhan, A.; Aher, L. K.; Singh, N. P. Quinoa: An Alternative Food Crop for Water Scarcity Zones of India. ICAR-NIASM Publications; ICAR-NIASM: Baramati, 2019; pp 1–12.

Rao, A. V.; Singh, K. C.; Gupta, J. P. Ley Farming—An Alternate Farming System for Sustainability in the Indian Arid Zone. *Arid Soil Res. Rehab.* **1997,** *11* (2), 201–210.

Rao, K. T.; Rao, M. M. V. S.; Nagarjuna, D. Tribal Farmer Success Story of Integrated Farming System in Andhra Pradesh. *Rashtriya Krishi* **2019,** *14* (1), 123–126.

Rastogi, A.; Shukla; S. Amaranth: A New Millennium Crop of Nutraceutical Values. *Crit. Rev. Food Sci. Nutr.* **2013,** *53,* 109–125.

Rathore, S.; Bala, M.; Gupta, M.; Kumar, R. Introduction of Multipurpose Agro-Industrial Crop Quinoa (*Chenopodium quinoa*) in Western Himalayas. *Indian J. Agron.* **2019,** *64* (2), 287–292.

Regnault-Roger, C.; Philogene, B. J. R. Past and Current Prospects for the Use of Botanicals and Plant Allelochemicals in Integrated Pest Management. *Pharm. Biol.* **2008,** *46* (1–2), 41–52.

Renganathan, V. G.; Vanniarajan, C.; Karthikeyan, A.; Ramalingam, J. Barnyard Millet for Food and Nutritional Security: Current Status and Future Research Direction. *Front. Genet.* **2020,** *11,* 500.

Roberts-Nkrumah, L. B. *Breadfruit and Breadnut Orchard Establishment and Management: A Manual for Commercial Production*; Food and Agriculture Organization of the United Nations: Rome, Italy, 2015.

Rosentrater, K. A.; Evers; A. D. Introduction to Cereals and Pseudocereals and Their Production. In *Kent's Technology of Cereals: An Introduction for Students of Food Science and Agriculture*; Elsevier, 2018; pp 1–76.

Ruskin, F. R. Lost Crops of Africa. In *Grains*; Vol. 1; The National Academic Press: Washington, D. C., 1996; pp 218–236.

Sage, R. F. The Evolution of $_{C4}$ Photosynthesis. *New Phytol.* **2004,** *161,* 341–370.

Saikia, S. K.; Das, D. N. Rice-Fish Culture and its Potential in Rural Development: A Lesson from Apatani Farmers; Arunachal Pradesh. *India J. Agric. Rural Dev.* **2008,** *6* (1), 125–131.

Saleh, A.; Zhang, Q.; Chen, J.; Shen, Q. Millet Grains: Nutritional Quality; Processing; and Potential Health Benefits. *Compr. Rev. Food Sci. Food Saf.* **2013,** *12,* 281–295.

Salini, K.; Kumari, A. N.; Senthil; N. Genetic Analysis of Yield and Its Components in Little Millet (*Panicum sumatrense* Roth ex Roem. and Schultz). In *2nd International Conference on Agricultural and Horticultural Sciences*, Hyderabad, India, Feb 03–05, 2014. *Agrotechnol 2*, 2–4. Omics Group Conferences.

Sanginga, N.; Lyasse, O.; Singh, B. B. Phosphorus Use Efficiency and Nitrogen Balance of Cowpea Breeding Lines in a Low P Soil of the Derived Savanna Zone in West Africa. *Plant Soil.* **2000,** *220* (1–2), 119.

Sardaro, R.; Girone, S.; Acciani, C.; Bozzo, F.; Petrontino, A.; Fucilli, V. Agro-Biodiversity of Mediterranean Crops: Farmers' Preferences in Support of a Conservation Programme for Olive Landraces. *Biol Cons.* **2016,** *201,* 210–219.

Saucedo, A. L.; Hernandez-Dominguez, E. E.; de Luna-Valdez, L. A.; Guevara-García, A. A.; Escobedo-Moratilla, A.; Bojorquez-Velazquez, E. Insights on Structure and Function of a Late Embryogenesis Abundant Protein from *Amaranthus cruentus*: An Intrinsically Disordered Protein Involved in Protection Against Desiccation; Oxidant Conditions; and Osmotic Stress. *Front. Plant Sci.* **2017,** *8,* 497.

Sauerborn, J.; Sprich, H.; Mercer-Quarshie, H. Crop Rotation to Improve Agricultural Production in Sub-Saharan Africa. *J. Agron. Crop. Sci.* **2000,** *184,* 67–72.

Saxena, K. Indigenous Crop Protection Practices in Sub-Saharan East Africa. In: *Database of Natural Crop Protectant Chemicals*, 1987.

Schipanski; M. E.; Barbercheck; M.; Douglas; M. R.; Finney; D. M.; Haider; K.; Kaye; J. P.; White; C. A Framework for Evaluating Ecosystem Services Provided by Cover Crops in Agroecosystems. *Agric. Syst.* **2014,** *125,* 12–22.

Sekhar, D.; Rao, K. T.; Rao, N. V. Studies on Integrated Farming Systems for Tribal Areas of Eastern Ghats in Andhra Pradesh. *Indian J. Appl. Res.* **2014,** *4* (10), 14–16.

Seetharam, A.; Ravikumar, R. L. Blast Resistance in Finger Millet-Its Inheritance and Biochemical Nature. In *Proceedings of the Second International Small Millets Workshop on Recent Advances in Small Millets; Bulawayo; Zimbabwe*; Riley, K. W., Gupta, S. C., Seetharam, A., Moshanga, J., Eds.; Oxford-IBH Publishing Company: New Delhi, 1993; pp 449–466.

Shah, R. A. First Report on Buckwheat (*Fagopyrum esculentum*) from HighAltitude Temperate Zone of North Western Himalayan Region. *Indian J. Hill Farming.* **2013,** *26* (1), 52–54.

Sharma; A. K. The Potential for Organic Farming in the Drylands of India. *Arid Lands Newslett.* **2005**, *58*.

Sharma, E.; Sharma, G.; Liang, L.; Subba, J. R.; Tanaka, K. Sikkim Himalayan-Agriculture: Improving and Scaling Up of the Traditionally Managed Agricultural Systems of Global Significance. *Resour. Sci.* **2000**, *31* (1), 21–30.

Sharma, G.; Dhakal, T. D. *Opportunities and Challenges of the Globally Important Traditional Agriculture Heritage of the Sikkim Himalaya*, 2011.

Sharma, G.; Liang, L. The Role of Traditional Ecological Knowledge Systems in Conservation of Agrobiodiversity: A Case Study in the Eastern Himalayas. In *Proceedings of International Policy Consultation for Learning from Grassroots Initiatives and Institutional Interventions*, May 27–29, 2006; Indian Institute of Management: Ahmedabad, India, 2006.

Sharma, V.; Chandra; S.; Dwivedi; P.; Parturkar; M. Quinoa (Chenopodium Quinoa Willd.): A Nutritional Healthy Grain. *Int. J. Adv. Res.* **2015**. https://www.journalijar.com/article/6138/quinoa-(chenopodium-quinoa-willd.):-a-nutritional-healthy-grain/.

Shashi, B. K.; Sunanda, S.; Shailaja, H.; Shankar, A. G.; Nagarathna; T. K. Micronutrient Composition; Antinutritional Factors and Bioaccessibility of Iron in Different Finger Millet (*Eleusine coracana*) genotypes. *Karnataka J. Agric. Sci.* **2007**, *20*, 583–585.

Shava, S.; O'Donoghue, R.; Krasny, M. E.; Zazu, C. Traditional Food Crops as a Source of Community Resilience in Zimbabwe. *Int. J. Afr. Renaiss. Stud.* **2009**, *4* (1), 31–48.

Sheahan; C. M. *Plant Guide for Browntop Millet (Urochloa ramosa)*. USDA-Natural Resources Conservation Service; Cape May Plant Materials Center: Cape May, NJ, 2014.

Sikarwar, M. S.; Hui, B. J.; Subramaniam, K.; Valeisamy, B. D.; Yean, L. K.; Balaji, K. A Review on *Artocarpus altilis* (Parkinson) Fosberg (Breadfruit). *J. Appl. Pharm. Sci.* **2014**, *4* (8), 91–97.

Singh, A. K.; Rana, R. S. Nationally Important Agricultural Heritage Systems in India: Need for Characterization and Scientific Validation. *Proc. Indian Natl. Sci. Acad.* **2019**, *85* (1), 229–246.

Singh, B. B.; Ajeigbe, H. A.; Tarawali, S. A.; Fernandez-Rivera, S.; Abubakar; M. Improving the Production and Utilization of Cowpea as Food and Fodder. *Field Crops Res.* **2003**, *84* (1), 169–177.

Singh, K. P.; Mishra, H. N.; Saha, S. Moisture-Dependent Properties of Barnyard Millet Grain and Kernel. *J. Food Eng.* **2010**, *96*, 598–606.

Singh, R; Singh, G. S. Traditional Agriculture: A Climate-Smart Approach for Sustainable Food Production. *Energ. Ecol. Environ.* **2017**, *2* (5), 296–316.

Smith, B. D. The Initial Domestication of *Cucurbita pepo* in the Americas 10,000 Years Ago. *Science* **1997**, *276*, 932–934.

Smith, P. D.; Martino, Z.; Cai, D.; Gwary, H.; Janzen, P.; Kumar, B.; McCarl, S.; Ogle, F.; O'Mara, C.; Rice, B.; Scholes, O.; Sirotenko. Agriculture. In *Climate Change (2007): Mitigation*. Contribution of Working Group III to the Fourth Assessment Report of the Intergovernmental Panel on Climate Change; Metz, B., Davidson, O. R., Bosch, P. R., Dave, R., Meyer, A., Eds.; Cambridge University Press: Cambridge; UK and New York; NY; USA.

Snyder, B. A.; Leite, B.; Hipskind, J.; Butler, L. G.; Nicholson, R. L. Accumulation of Sorghum Phytoalexins Induced by *Colletotrichum graminicola* at the Infection Site. *Physiol. Mol. Plant Pathol.* **1991**, *39*, 463–470.

Sood, J. UN Heritage Status for Odisha's Koraput Farming System. In *Down to Earth*, Jan 04, 2012.

Sood, S.; Khulbe, R. K.; Gupta, A. K.; Agrawal, P. K.; Upadhyaya, H. D.; Bhatt, J. C. Barnyard Millet—A Potential Food and Feed Crop of Future. *Plant Breed.* **2015,** *134* (2), 135–147.

Sreeja, K. G.; Madhusoodanan, C. G.; Eldho, T. I. In Climate and Landuse Change Impacts on Sub-Sea Level Rice Farming in a Tropical Deltaic Wetland. In *E-Proceedings of the 36th IAHR World Congress,* June 28–July 3, 2015; The Hague: The Netherlands, 2015.

Stetter, M. G.; Zeitler, L.; Steinhaus, A.; Kroener, K.; Biljecki, M.; Schmid, K. J. Crossing Methods and Cultivation Conditions for Rapid Production of Segregating Populations in Three Grain Amaranth Species. *Front. Plant Sci.* **2016,** *7,* 816.

Sun, H. Y.; Deng, S. P.; Raun, W. R. Bacterial Community Structure and Diversity in a Century-Old Manure-Treated Agroecosystem. *Appl. Environ. Microbiol.* **2004,** *70,* 5868–5874.

Surekha, N.; Ravikumar, S. N.; Mythri, S.; Rohini, D. Barnyard Millet (*Echinochloa frumentacea* Link) Cookies: Development; Value Addition; Consumer Acceptability; Nutritional and Shelf Life evaluation. *IOSR J. Environ. Sci. Toxicol. Food Technol.* **2013,** *7,* 01–10.

Takahashi; S. In Cherishing Our Agricultural Heritage Systems for Climate Change Adaptation and Mitigation. In *Proceedings of the Second International Forum on Globally Important Agricultural Heritage Systems Theme: Cherishing our Agricultural Heritage Systems for Climate Change Adaptation and Mitigation,* Buenos Aires; Argentina, Oct 21–23, 2009; pp 73–75.

Tangjang, S.; Nair, P. K. R. Rice + Fish Farming in Homesteads: Sustainable Natural-Resource Management for Subsistence in Arunachal Pradesh; India. *J. Environ. Sci. Eng.* A **2015,** *4,* 545–557.

Teng, X.-L.; Chen, N.; Xiao, X.-G. Identification of a Catalase-Phenol Oxidase in Betalain Biosynthesis in Red Amaranth (*Amaranthus cruentus*). *Front. Plant Sci.* **2015,** *6,* 1228.

Teutonico, R.; Knorr, D. Amaranth; Composition; Properties; and Applications of a Rediscovered Food Crop. *Food Technol.* **1985,** *39,* 49–60.

Telkar; S.; Singh; A. K.; Kant; K.; Pratap; S.; Solanki; S. P. S.; Kumar; D. Types of Mulching and Their Uses for Dryland Condition. *Biomol. Rep.* **2017,** *17* (6), 1–4.

Thomas, P. M. Problems and Prospects of Paddy Cultivation in Kuttanad Region: A Case Study of Ramankari Village in Kuttanad Taluk. In *Draft Report. A Project of Kerala Research Programme on Local Level Development,* Thiruvananthapuram, 2002.

Triberti, L.; Nastri, A.; Baldoni, G. Long-Term Effects of Crop Rotation; Manure and Mineral Fertilisation on Carbon Sequestration and Soil Fertility. *Eur. J. Agron.* **2016,** *74,* 47–55.

Tripathi, B.; Platel, K. Finger Millet (*Eleucine coracana*) Flour as a Vehicle for Fortification with Zinc. *J. Trace Elem. Med. Biol.* **2010,** *24,* 46–51.

Tundup, P.; Wani, M.; Hussain, S.; Dawa, S. Traditional Methods of Buckwheat (*Fagopyrum esculentum* Moench) Cultivation in High Altitudes Cold Desert Region of India. *Int. J. Agric. Sci. Res.* **2016,** *7* (1), 101–106.

Turi, C. E.; Liu; Y.; Ragone, D.; Murch, S. J. Breadfruit (*Artocarpus altilis* and Hybrids): A Traditional Crop with the Potential to Prevent Hunger and Mitigate Diabetes in Oceania. *Trends Food Sci. Technol.* **2015,** *45,* 264–272.

Udhaya, N. D.; Kumar, A. R. Korangadu—A Unique Silvi Pasture System. *Appro Poult. Dairy Vet. Sci.* **2019,** *6* (5).

Vadivoo, A. S.; Joseph, R.; Ganesan, N. M. Genetic Variability and Diversity for Protein and Calcium Contents in Finger Millet [*Eleusine coracana* (L.)Gaertn] in Relation to Grain Color. *Plant Foods Hum. Nutr.* **1998,** *52,* 353–364.

Veena, B.; Chimmad, B. V.; Naik, R. K.; Shantakumar, G. Development of Barnyard Millet Based Traditional Foods. *Karnataka J. Agric. Sci.* **2004,** *17,* 522–527.

Verchot, L. V.; Van Noordwijk, M.; Kandji, S.; Tomich, T.; Ong, C.; Albrecht, A.; Palm, C. Climate Change: Linking Adaptation and Mitigation Through Agroforestry. *Mitig. Adapt. Strat. Glob. Change* **2007,** *12* (5), 901–918.

Vetriventhan, M.; Upadhyaya, H. D. *Little Millet; Panicum sumatrense; An Under-Utilized Multipurpose Crop.* In *1st International Agrobiodiversity Congress,* New Delhi, India, Nov 06–09, 2016.

Wang, Q.; Li, Y.; Alva, A. Cropping Systems to Improve Carbon Sequestration for Mitigation of Climate Change. *J. Environ. Prot.* **2010,** *1* (03), 207.

Wang, Z. B.; Chen, J.; Mao, S. C.; Han, Y. C.; Chen, F.; Zhang, L. F.; Li, C. D. Comparison of Greenhouse Gas Emissions of Chemical Fertilizer Types in China's Crop Production. *J. Clean Prod.* **2017,** *141,* 1267–1274.

De Wet, J. M. J.; Prasada Rao, K. E.; Mengesha, M. H.; Brink, D. E. Diversity in Kodo Millet; *Paspalum scrobiculatum. Econ. Bot.* **1983,** *37* (2), 159–163.

Wu, X.; Blair, M. W. Diversity in Grain Amaranths and Relatives Distinguished by Genotyping by Sequencing (GBS). *Front. Plant Sci.* **2017.**

Yadav, P.; Mina, U. *Amaranthus*: Development Opportunity. *Indian Farm.* **2019,** *69* (4), 27–31.

Zedler, J. B.; Kercher, S. Wetland Resources: Status; Trends; Ecosystem Services and Restorability. *Annu. Rev. Environ. Res.* **2005,** *30,* 39–74.

Zhang, Q. C.; Shamsi, I. H.; Xu, D. T.; Wang, G. H.; Lin, X. Y.; Jilani, G.; Hussain, N.; Chaudhry, A. N. Chemical Fertilizer and Organic Manure Inputs in Soil Exhibit a Vice Versa Pattern of Microbial Community Structure. *Appl. Soil Ecol.* **2012,** *57,* 1–8.

Zhang, Y.; Min; Q.; Li; H.; He; L.; Zhang; C. and Yang; L. A Conservation Approach of Globally Important Agricultural Heritage Systems (GIAHS): Improving Traditional Agricultural Patterns and Promoting Scale-Production. *Sustainability* **2017,** *9* (2), 295.

Zhu, X. L.; Zhang, L.; Chen, Q.; Wan, J.; Yang, G. F. Interactions of Aryloxyphenoxypropionic Acids with Sensitive and Resistant Acetyl-Coenzyme a Carboxylase by Homology Modeling and Molecular Dynamic Simulations. *J. Chem. Inf. Model.* **2006,** *46,* 1819–1826.

Zomer, R. J.; Neufeldt, H.; Xu, J.; Ahrends, A.; Bossio, D.; Trabucco, A.; van Noordwijk, M.; Wang, M. Global Tree Cover and Biomass Carbon on Agricultural Land: The Contribution of Agroforestry to Global and National Carbon Budgets. *Sci. Rep.* **2016,** *6,* 29987.

CHAPTER 3

Plant Bioactive Compounds: Biotechnological Applications for Novel Molecules

ATANU BHATTACHARJEE[1,#], SUBHASHIS DEBNATH[1,#], YUTIKA NATH[2,#], and RANJAN DUTTA KALITA[3,#]

[1]*Department of Pharmaceutical Sciences, Royal School of Pharmacy, Assam Royal Global University, Guwahati, Assam, India*

[2]*Department of Serology, Directorate of Forensic Sciences, State Forensic Science Laboratory, Guwahati, Assam, India*

[3]*Department of Biotechnology, Royal School of Biosciences, Assam Royal Global University, Guwahati, Assam, India*

ABSTRACT

Medicinal plants based bioactive compounds have been utilized by the human civilization since ancient times. Bioactive compounds are secondary metabolites produced by plants for its self defense against various forms of stress and predators. These compounds are chemical complexes that are pharmacologically very active against a host of diseases inflicting human

[#]All authors contributed equally to this work.

Crop Sustainability and Intellectual Property Rights. Soumya Mukherjee, Piyali Mukherjee & Tariq Aftab (Eds)

beings. The declining status and over-exploitation of traditionally used medicinal plants have made the situation worse with the number of medicinal plants in the wild decreasing. Plant biotechnology-based techniques have opened up huge opportunities for rediscovering of the active ingredients that could be now manufactured in the laboratory. The present topic lays down the importance of medicinal plants, their uses in history, and how plant biotechnology led to a revolution in manufacturing of the bioactive components. The anticancer and anti-Alzheimer's plant molecules have also been elaborated to justify the necessity of plant-based bioactive compounds as an alternative form of treatment.

3.1   INTRODUCTION

Plants are nature's best gift to mankind. Of all the things nature provided, medicinal plants are the most important of all which are still in service of human civilizations since ages. For the past several centuries, medicinal plants have been used and well recorded in various written formats across the world. Written information on the application of plants in Mesopotamia dates back to 2600 BC. Such records state the use of more than 1000 plant-derived medicines (Atanas et al., 2016). The Ebers Papyrus that dates back to 1550 BC contains a record of 700 drugs of plant origin (Borchardt, 2002; Cragg and Newman, 2013). The Ayurveda of India dating back to the first millennium BC (Patwardhan, 2005) and the Chinese system of traditional medicine dating back to thousands of years (Unschuld, 1986) are two of the most extensive system of references in modern day containing the knowledge and applications of hundreds of plants. Use of plants in the West can be documented from the compendiums of Disocorides, the famous Greek physician of first century AD. The evidences left by the Romans such as Pliny the Elder and Galen in the first and second Century AD reveals a lot about the role of plants in the treatment of various symptoms in humans (Sneader, 2005). The Arabs kept documented evidences that are compilations of the Greco-Roman knowledge systems of the 5th–12th centuries and further complemented these with their own system of medicine including additions from the Indian and Chinese systems of traditional medicine (Cragg and Newman, 2013). Invention of the letterpress in the 15th century by Johannes Gutenberg led to the documentation and printing of several herbal medicinal books, which were widely distributed in Europe. Some of the notable books were *The Mainz Herbal; 1484, The German Herbal; 1485, Herbarium Vivae Eicones;1530, Kreütter Buch; 1546 and De Historia Stirpium; 1542*

(Sneader, 2005). The use of natural products has been reported since the day when history was recorded. The first uses of natural products can be found with the use of cedar oil, licorice oil, Cyprus oil, poppy juice, etc. Tapputi Beletekallim, from the Royal Palace of Mesopotamia, was one of the first chemists who documented all the raw materials used in the preparation of perfumes in the Royal Palace of Mesopotamia (Levey, 1956).

The various documented books of India reports hundreds of plants used in the treatment of different diseases. The Ayurveda lists about 1200–1800 plants while the Siddha system documents about 500–900 plants. Besides the *Astanga Hridaya, Sushruta Samhita*, and *Charak Samhita* reported approximately 901, 573, and 526 numbers of plants for treatment of symptoms related to a host of diseases (Sen and Chakraborty, 2017, 2015; Debnath et al., 2015; Shankar and Majumdar, 1997)..

Plant-based alternative medicines have always been popularized by the common populace and those people who live in places away from easy access to towns and cities and thus further away from modern health care. Besides the maladies of side effects of modern-day synthetic drugs too helped in the increase of the popularity of plant-based traditional medicines. Approximately 60% of the world's population uses alternative medicines as a primary source of health care (Ballabh and Chaurasia, 2007; Pandey et al. 2013). In India, about 8000 plants are believed to possess medicinal properties and 25,000 formulations derived from medicinal plants are used by the various ethnic and rural communities in India (Sen and Chakraborty, 2015).

Traditional folk medicines originating from plants are usually administered in the form of tinctures, teas, powders, poultices, or even other different formulations (Samuelsson, 2004b). Traditional medicines are popular because of the fact that the knowledge of such medicines have been passed down from generation to generation. Besides the safety issues, these plants find their place in the traditional folklore and have been used since ages and therefore the common people have their full faith and trust in using these plants (Sen and Chakraborty, 2017). The passage of information has also led to the conservation of these important plant species and with it the art of traditional or folk medicine (Jachak and Saklani, 2007).

Plant-based herbal medicines provide the basic support to modern medicines and new drug development (Sen et al., 2011). Natural product-based new molecule entities resourced from herbal sources can be used as a direct curative agent. These secondary metabolites play a promising role in the discovery of lead drug molecules targeting specific diseases. Thus renewed interest in research in medicinal plants could be seen in the last few decades after the realization of their health benefits. Also, safety and cost are the other two factors for increased dependency on medicinal plants (Ekor, 2014; Thomford et al., 2018; Anand et al., 2019).

India is a land with a rich biodiversity consisting of various endemic and rare species of plants, which are a true treasure for novel drug discovery. Of the 18 hotspot regions of the world, India is blessed in having 2 of them— The Eastern Himalayas including the Indo Burma region and the Western Ghats. India is also ranked 7th amongst 16 megadiverse countries of the world where 70% of all species of the world are found to occur (Jachak and Saklani, 2007). India consists of 8% of global biodiversity and has approximately 49,000 species of plants out of which approximately 5150 are endemic to the nation (Ramakrishnappa, 2002; Singh, 2007; National Biodiversity Authority, 2012). Approximately 65% population of India uses traditional medicine for daily requirements. Out of the approximately 49,000 plant species available in India, about 3520 species of plants have medicinal properties. Five hundred of these medicinally important plant species are used in the Ayurvedic industry (Ramakrishnappa, 2002; Singh, 2007).

Traditional medicine has found uses not only amongst the people in developing nations but also in various developed countries of the world. It has been reported that in Germany alone 40–50% people, 48% population of Australia, 49% French people, and 42% population of USA relies on traditional medicine. The herbal plant industry has been found to be a fast emerging industry gaining worldwide attention. The world trade in selected medicinal plants and secondary plant metabolites was approximately 60 billion in the year 2000 and expected to increase to USD 5 trillion by 2050 (Chandran et al., 2020).

A review of the pharmacopeias of the world revealed about 120 different molecules from plant sources, which are used as a component of various life-saving drugs. Extensive research contributed to a large number of natural product entities as drug molecules. In between 1981 and 2002, about 119 drugs were given approval, which was of natural product origin including plants (Gurib-Fakim, 2011).

The plant kingdom consists of a large number of species, which are responsible for producing a huge diversity of bioactive compounds having various chemical complexities. It has been reported that only 6% of existing plants have been investigated till now for their pharmacological properties and approximately 15% for their phytochemical components (Atanas et al., 2016). Recently available data indicated the study of nearly 100 natural product compounds in various clinical trials and about another 100 natural product molecules in the preclinical phase of drug discovery (Traditional Medicine Programme, 1998). Between 1981 and 2006, 24 natural products were the starting point for marketed drugs. Of these 24, 5 were of plant origin while the rest 19 were of microbial origin (Ganesan, 2008). The first plant-based compound to have found a place in modern medicine is morphine in the year 1827. It was first isolated by Friedrich Serturner from *Papaver somniferum*. Subsequently, various natural products were isolated from plants viz cocaine, quinine, nicotine, capsaicin, caffeine, codaine, etc. (Corson and Crews, 2007; Felter and Lloyd, 1898). Plant-derived anticancer molecules have been found to be of immense potential. Molecules like paclitaxel isolated from Taxus spp, vincristine and visblastine from *Catharanthus roseus* (Madagascar periwinkle) camptothecin from *Camptotheca acuminata* are some of the most notable drugs. Galanthamine isolated from *Galanthus nivalis* is another famous molecule used in the treatment of Alzheimer's disease. It acts by inhibition of the cholinesterase enzyme (Mashkovsky and Kruglikova-Lvova, 1951). A very important drug discovery is the antimalarial drug artemisinin isolated from the traditional Chinese medicinal plant *Artemisia annua* (Klayman et al., 1984). The first ever FDA-approved botanical drug isolated from *Camellia sinensis* Kuntze was Veregen, approved for treatment of genital and perianal warts (Ahn, 2017). The FDA approved another drug Fulyzaq, isolated from the blood red latex of *Croton lechlerii* Mull. Arg, a plant widely available in the South America. Fulyzaq was approved for the treatment of diarrhea in HIV patients (Ahn, 2017). Commercial production of plant bioactive compounds has been found to be less due to various regional and environmental problems (Yue et al., 2016). Besides it takes years for plants to develop the desired compounds and conventional methods are time-consuming process (Kolewe et al., 2008).

Plant tissue culture techniques provide an alternative and possible route for culture of plant cells in in vitro conditions and makes it easier for isolation of the bioactive compounds. Using such techniques, one can easily control the parameters for the production of specific compounds in a controlled condition in the laboratory throughout the year.

3.2   ROLE OF PLANT BIOTECHNOLOGY IN PRODUCTION OF PLANT METABOLITES

In the production of secondary metabolites, plant biotechnology has an immense role to play. Various parts of the plants, viz cells, organs, and tissues are used in the generation of bioactive metabolites. This exciting field also helps in the manipulation of the plant cells genetically to derive the targeted products. Plants are known to synthesize various natural products, which are basically secondary metabolites. These include flavonoids, alkaloids, phenolics, terpenoids, saponins, glycosides, steroids, tannins, and volatile compounds (Alamgir, 2018). These metabolites are heterogenous group of complex molecules that are highly diverse in structure (Rosenthal, 1991). These bioactive compounds, although not required by the plant for its normal growth and development, yet serve a major purpose in the defence of the plant against a host of insects and microbes (Ncube and Van Staden, 2015).

The secondary metabolites obtained from plants are divided into the following criteria-drugs, flavors, perfumes, pigments, and agrochemicals. It has been generally found to be difficult to obtain these metabolites in industrial setups as the structures are quite complex and these are synthesized using complicated biosynthetic pathways.

Thus plant biotechnology offers various efficient mechanisms to produce these compounds in the desired amounts. It uses a host of techniques that are in vitro plant cell culture processes. Basically, two major stages are present-Biomass aggregation and Synthesis of secondary metabolites. Four main approaches are utilized-callus cell culture, suspension cell culture, immobilized culture, and differentiated cell culture. The plant cells have a basic important criteria, namely, totipotency. It is because of this phenomenon that each cell in a culture are able to retain the complete genetic details and therefore produce a range of chemicals found in the parent plant (Rao and Ravishankar, 2002). The first stage is the callus culture process. Callus culture deserves particular importance because of the fact that they are useful in the production of secondary metabolites (Guerriero et al., 2018; Hussain et al., 2012). In this process, disorganized aggregate of cells is grown from a variety of plant explants in a nutritionally rich media consisting of various nutrients, vitamins, and hormones. The cells generated in the callus culture process are further required for the suspension cell culture process. In this step, the callus cells are added into a liquid medium consisting of similar nutrients, vitamins, and hormones. Suspension cell culture method is used widely in the culture and synthesis of various plant bioactive compounds. This process also enables to control the factors responsible for production of specific plant secondary metabolites, and is even faster than the original plant. The required compounds can be further produced all throughout the year by controlling the growth parameters and conditions (Alamgir, 2017). Callus culture has received wide impetus due to its reliability in the production of plant secondary metabolites and thus benefitting the process of conservation of rare and endemic medicinal plants (Efferth, 2019). Callus cultures also helps in the generation of micropropagated plants and in the development of single-cell suspension cultures. Micropropagation of plants helps in generation of homogenous plants in a very short time. Various hormones play an important role in the production of different metabolites during micropropagation. Cell suspension culture process is useful in the production of the preferred secondary compounds (Fischer et al., 1999; Xu et al., 2011). Manipulation of the biosynthetic machinery pathway of plants can also be carried out using callus and suspension culture. Apart from callus culture, other plant cell culture like differentiated organ culture of roots and shoots have also been found to be reputed for production of secondary metabolite production.

Apart from the common cell culture techniques, another process called the hairy root culture process can be used to produce a variety of plant metabolites if there is difficulty in production of the same by using the normal cell suspension culture method. The hairy root culture helps in the alternative production of compounds synthesized in the roots of the plants (Pence, 2011). Various tropane alkaloids, ginsenosides, artemisinin etc. have been found to be produced from plant hairy root culture. Hairy roots are preferred due to their various advantages such as high productivity, competency and constancy (Ha et al., 2016). This technique is also widely used in the biosynthesis of volatile compounds. A major advantage of the hairy root culture is the branched nature that grows faster than normal roots. They are able to synthesize compounds produced in the plant roots as well as in the acrial parts (Khan et al., 2021; Gounaris, 2010). A major positive point in hairy root culture is their stability and high productivity in hormone-free culture media (Alamgir, 2018).

Thus the plant biosynthetic machinery can be controlled by using various biotechnological techniques of using the cell, tissues, and organ culture. The molecules or chemical entities can be produced from a variety of explants such as leaves, stems, roots, etc. All these plant cells can be used for the production of the desired molecule having pharmacological functions and used for treatment of several important diseases (Karuppusamy, 2009).

The various cell and organ cultures have helped in the prevention of the threat of overharvesting and exploitation of plants and issues related to toxicity of plant constituents. Thus over the years, these techniques have paved the way for the production of different secondary metabolites utilized for the benefit of the humankind.

Table 3.1 highlights the various secondary bioactive compounds which have been produced using various plant biotechnology techniques.

Plant-based bioactive compounds have been extensively investigated for their anti-neural and anticancer properties. The mechanistic activities of various plant-based molecules have been deciphered against these diseases. Some of the plant-derived molecules as anti-Alzheimer's and anticancer agents are discussed in the following text.

3.2.1 PLANT DERIVED COMPOUNDS AGAINST NEURAL DISEASES

In neurodegenerative diseases like Alzheimer's disease (AD) loss of memory (dementia) along with psychological and behavioral changes has become a serious matter of concern. It is an irreversible and associated with memory and psychological abnormalities and causes changes in overall behavior (Amod et al., 2005). AD is associated with diminished level of Ach in brain (Colucci et al., 2012). As per WHO, dementia cases are increasing at rapid rate in developing nations and will soon become a global public health priority (Dwivedi and Singh, 1978; Zhang and Jiang, 2015; Anderson, 2002).

Formation of extracellular amyloid beta plaques with drastic neuronal and synaptic reduction in cholinergic system of brain is considered as major pathological hallmark of AD (Gupta and Bala, 2013; Azimi et al., 2017). According to "beta-amyloid cascade," deposition of amyloid beta peptide triggers neuro-inflammation resulted in neurodegeneration. Amyloid precursor protein (APP) upon degradation by beta and gamma-secretases yields Amyloid beta plaques. Initial cleavage by the membrane-immobilized aspartic protease beta secretase (BACE; the beta site of the APP cleavage enzyme) produces soluble N-terminal and membrane-bound C-terminal fragments. The C-terminal fragment undergoes proteolysis by gamma-secretase to produce amyloid beta peptide (Berrino, 2002; Skovronsky et al., 2006). BACE has emerged as a promising therapeutic target as it initiates the first steps in amyloid beta production (Brinton et al., 1998). In contrast, neuroinflammation in AD probably begins as a host defense response to the harmful effects of amyloid deposits in the brain. Therefore, anti-inflammatory drugs may be another potential therapeutic target to counteract the progression of AD (Barao et al., 2016). Medicinal plants are a large undeveloped reservoir of natural remedies and a potential source of anti-inflammatory agents, BACE, and cholinesterase inhibitors (Zhang and Jiang, 2015; Bachurin et al., 2017; Ben Halima et al., 2016). Therefore, it is worth investigating the

TABLE 3.1   Table Demonstrating the List of Compounds Produced Using Plant Biotechnology Techniques.

Compounds	Plant source	Plant hormone/growth regulator used	Plant Biotechnology technique	Reference
Pulegone and Menthofuran	*Mentha piperita*	4-indol-3-ylbutyric acid (IBA) and 6-benzylaminopurine (BAP).	Micropropagation	Santoro et al. (2013)
limonene, a-pinene, camphene, a-thujone, 1,8-cineole, borneol, camphor and nerol	*Ajuga bracteosa*	Thidiazuron (TDZ)	Micropropagation	Ali et al. (2018)
Monoterpenes and sesquiterpenes	*Lallemantia Iberica*	Thidiazuron (TDZ)	Micropropagation	Pourebad et al. (2014)
Gymnemic acid	*Gymnema sylvestre*	2,4-dichlorophenoxyacetic acid (2,4-D); kinetin (Kin); and 6-Benzylaminopurine (BAP)	Callus and cell cultures	Veerashree et al. (2012)
Diosgenin	*Helicteres isora L*	2,4-dichlorophenoxyacetic acid (2,4-D); kinetin (Kin); and 6-Benzylaminopurine (BAP)	Callus and cell cultures	Shaikh et al. (2020)
Thymol and p-cymene	*Carum copticum*	2,4-dichlorophenoxyacetic acid (2,4-D): benzyl amino purine (BAP)	Callus culture	Razavizadeh et al. (2020)
Rosmarinic acid	*Ocimum basilicum L. var. purpurascen*	Naphthaleneacetic acid (NAA)	Callus cultures	Nazir et al. (2020)
Quercetin	*Abutilon indicum L.*	2,4-dichloro phenoxy acetic acid (2,4-D) with indole-3-acetic acid (IAA)	Callus culture	Sajjalaguddam and Paladugu (2015)
Peonidin, Chicoric acid	*Ocimum basilicum*	Naphthaleneacetic acid (NAA)	Callus cultures	Nazir et al. (2020) and Jiao et al. (2018)

TABLE 3.1 *(Continued)*

Compounds	Plant source	Plant hormone/growth regulator used	Plant Biotechnology technique	Reference
Kaempferol	*Dysosma pleiantha*	Medium (B5) 2,4-dichlorophenoxyacetic acid (2,4-D); kinetin	Callus cultures	Karuppaiya and Tsay (2020)
Lignans	*Linum ussitatsimum L*	Naphthaleneacetic acid (NAA)	Cell suspension cultures	Nadeem et al. (2019)
Limonene	*Mentha pulegium*	2,4-D	Cell suspension cultures	Darvishi et al. (2016)
Isoorientin (ISO)	*Cecropia obtusifolia*	Naphthalene acetic acid (NAA); 2,4-dichlorophenoxyacetic acid (2,4-D); indole-3-butyric acid (IBA); indole-3-acetic acid (IAA); 6-benzylaminopurine (BAP)	Cell suspension cultures	Del Pilar Nicasio-Torres et al. (2012)
Gallic acid	*Barringtonia racemosa L.*	2,4-D and kinetin	Cell suspension cultures	Osman et al. (2018)
Scopoletin	*Spilanthes acmella Murr.*	6-benzyladenine; 2,4-dichlorophenoxyacetic acid	Cell Suspension cultures	Abyari et al. (2016)
Tyrosol	*Rhodiola crenulata*	6-benzyaldenine (BA); naphthalene acetic acid (NAA) and thidiazuron (TDZ)	Cell suspension cultures	Shi et al. (2013)
Wogonin	*Scutellaria lateriflora*	Phytohormone-free MS medium having sucrose and supplemented with antibiotic ampicillin and cefotaxim	Hairy root cultures	Wilczañska-Barska et al. (2012)
Saikosaponins	*Bupleurum falcatum L.*	Indole-3-butyric acid (IBA)	Root cultures	Kusakari et al. (2000)
Cynaroside, Rutin, Neochlorogenic acid	*Vitex agnus castus L.*	a-naphthaleneacetic acid (NAA); benzylaminopurine (BAP); gibberellic acid (GA3)	Agitated shoot cultures	Skrzypczak-Pietraszek et al. (2018)

usefulness of traditional medicines for the early prevention of AD, as traditional medicines are considered safe and economical by WHO. Therefore, systematic screening of such plants can provide valuable information when disclosing new drugs for the treatment of AD.

3.2.1.1   THE AMYLOID HYPOTHESIS

According to the amyloid hypothesis accumulation of insoluble forms of Aβ lead to oligomerization and formation of Aβ plaques and disrupt the neuronal network of brain (Dewachter and Van, 2002). These cause inflammatory mediator-induced hyperphosphorylation of tau protein and tangle formation in between neurons. This synchronization of neuronal network and diminish memory formation/retention in the AD patient (Choi et al., 2008a, 2008b; Jeon et al., 2005; Samuelsson, 2004a).

Amyloid beta (Aβ) is produced by (i) BACE-1 converts APP to soluble APPʹA (sAPPʹA) and a 99 amino acid fragment (C99), and (ii) C99 in the presence of γ-secretase produce toxic peptide AʹA42 (Bhattacharjee, 2018; Kinghorn, 2001; Balunas and Kinghorn, 2005; Gurib-Fakim, 2006; Huang and Mucke, 2012) (Fig. 3.1).

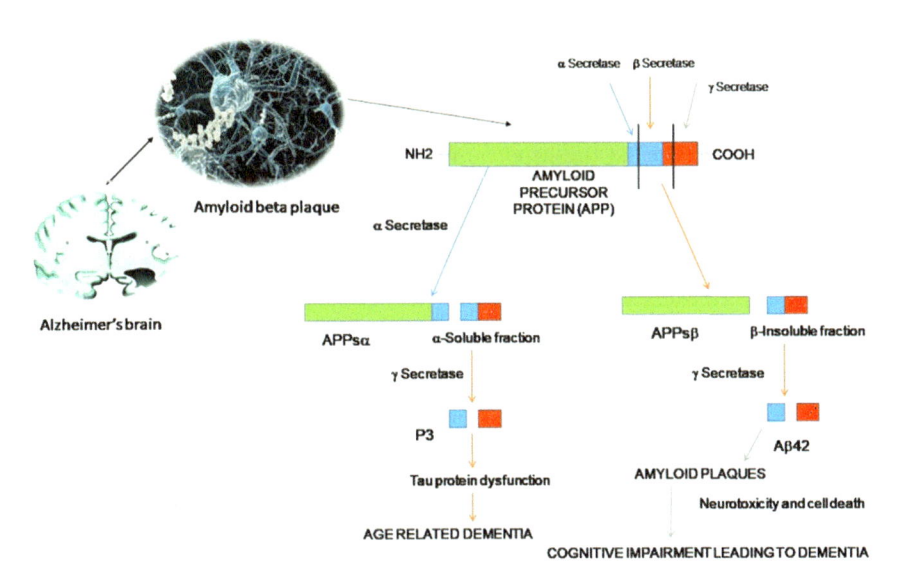

FIGURE 3.1   Mechanism of toxic amyloid B formation leading to cognitive impairment in AD.

Source: Adapted and modified from Bhattacharjee et al. (2020).

Recently, FDA approved a drug call Aducanumab (BIIB037) that possesses high-affinity toward human IgG1 monoclonal antibody against a conformational epitope found on Aβ. Aducanumab binds aggregated forms of Aβ and solubilize them (Crehan and Lemere, 2016). Despite the apparent success of drugs discovered from natural resources, the future is challenging and scientists are constantly working to identify new drugs to combat AD (Fenical and Jensen, 2006; Philomena, 2011). Natural resources are an integral part of the treatment of psychiatric disorders, with special consideration given to their ethno-pharmacological point of view (Perry et al., 1999; Vaidya, 1997). Phytoconstituents not only act synergistically but can also counteract the toxic effects of the compound from other plant species. A list of medicinal plants as BACE1 inhibitors is summarized below (Table 3.2).

3.2.1.2 PHYTOMOLECULES ACCESSED TO REVERSE AD PROGRESSION

3.2.1.3 PHENYL PROPANOIDS

Phenyl propanoids are widely distributed plant secondary metabolite formed from phenylalanine by the action of phenylalanine ammonia-lyase (PAL). It has been observed that they inhibit protein aggregation in nonstoichiometric fashion (Feng and Shoichet, 2006). Four bioactive compounds isolated from the fruit of *Cordia sebestena* (Hawaiian plant) ethanol extract showed BACE1 activity against the aspartic protease, which is a central blocker in the etiology of AD (Coan et al., 2009). New phenyl propanoids viz sebesteniods A-D (Fig. 3.2 (i)) had been isolated from bioactive fractions of the same plant and showed potent BACE1 inhibitory activity indose-dependent manner (Eglen, 2002). In-vitro studies revealed sebestenoids C and D were the most potent compounds with IC_{50} values of 20 and 22 μM, respectively. In addition, sebestenoids C and D were investigated using a surrogate system. These compounds were assayed against the serine protease chymotrypsin in a standard chemiluminescent assay with and without the addition of detergent. The IC_{50} values of sebestenoids C and D were strongly inhibit BACE1. These findings suggest that phenyl propanoid derivatives of sebestenoids A, B, C, and D are a potential agents for AD (Gunasekera et al, 2010; Youdim et al., 2004).

TABLE 3.2 Phytoextracts and Their Derivatives Accessed to Reverse AD Progression via BACE1 and AChE Inhibition.

Plant name	Part used	Extract	Lead compound(s)	Mechanism of action	Pharmacological study	References
Sophoraflavescens	Root	Chloroform extract	Lavandulyl flavanones (Saphoraflavones)	BACE1 inhibition in a dose-dependent manner	In-vitro study	Hwang et al. (2008)
Fructusgardeniae	Whole Plant	n-butanol extract	Geniposide and crocin	Geniposide exhibited AChE inhibitory activity on scopolamine-induced amnesic mice	In-vivo study	Nam and Lee (2013)
Paeonialactiflora—	Seeds	Ethanol extract	Resveratrol oligomer	BACE1 inhibition which was assessed by the Fluorescence Resonance Energy Transfer (FRET) assay using Rh-EVNLDAEFK as substrate.	In-vitro study	Choi et al. (2011)
Psoraleacorylifolia	Seeds	Methanol extract	Isoflavone, neocorylin	Baculovirus-expressed BACE1 inhibition	In-vitro study	Choi et al. (2008b)
Cordia sebestena	Fruit	Ethanol extract	Sebestenoids A-D	Demonstrated moderate inhibition of aspartic protease BACE1	In-vitro study	Dai et al. (2010)
Perillafrutescens	Leaves	Methanol extract	Luteolin and Rosmarinic acid	They inhibited BACE1 noncompetitively with a substrate in Dixon plots which suggest their binding with the regulatory site of BACE1.	In-vitro study	Choi et al. (2008a)

TABLE 3.2 *(Continued)*

Plant name	Part used	Extract	Lead compound(s)	Mechanism of action	Pharmacological study	References
Smilax china L.	Rhizomes	Methanol extract of	Resveratrol oxyresveratrol, Veraphenol, and cis-scirpusin A	BACE1 inhibition in a dose-dependent manner	In-vitro study	Smit (2004)
Sanguisorba officinalis	Roots	Methanol extract	1,2,3-trigalloyl-4,6-hexahydroxydiphenoyl-β-d-glucopyranoside and 1,2,3,4,6-pentagalloyl-β-d-glucopyranoside	BACE1 inhibition in a dose-dependent manner	In-vitro study	Lee et al. (2005)

Source: Adapted and modified from Bhattacharjee et al. (2020).

3.2.1.4 FLAVONOIDS

Flavonoids are one of the most widely distributed secondary metabolites found in a variety of fruits, vegetables, and flowers viz citrus, garlic, tea, etc. Flavanoids have the potential to cross blood-brain barrier and thereby promote neuro-protective effects particularly againstAβ plaques. They significantly reduced Aβ1-40 and Aβ1-42 formation by inhibiting BACE1 enzyme activity in-vitro and thereby decrease Aβ-induced neuronal cell death (Ross and Kasum, 2002; Commenges et al., 2000). The antioxidant activity of flavonoids inhibit Aβ oligomerization. Myricetin (Fig. 3.2 (ii)), an isolated flavanoid showed neuroprotective effect against Aβ-induced neuronal cell injury (Shimmyo et al., 2008). Methanol derivatives of *Moruslhou* stem bark extract exhibited significant BACE1 inhibitory activity with IC_{50} of 78.4 μg/mL (Ono et al., 2003). The bioactive fraction of ethanolic extract lead to the isolation of few compounds viz norartocarpetin, Kuwanon C, Morusin, Kuwanon A, Morusinol, Cyclomorusin, and Neocyclomorusin (Fig. 3.2 (iii–vii)). Each compound showed significant BACE1 reversible inhibition in a dose-dependent manner (Jeong et al., 2009; Kang et al., 2004).

3.2.1.5 NAPTHOQUINONEDERIVATIVES

Napthoquinone a water-soluble vitamin has been identified as a Hsp90 inhibitor. Hsp90 is aprotein that regulates homeostasis in response to stress and thus a potential target of newAD therapy (Muto ct al., 1987). The activity of BACE1 is sensitive to oxidative stress. There is convincing evidence that oxidative stress results in Aβ accumulation (Tauraite et al., 2009). Therefore, BACE1 inhibition by Napthoquinone could contribute to an improvement dissolution of Aβ (Montenegro et al., 2010). Naphthoquinone derivatives like Resveratrol (Fig. 3.2 (viii)) is a potential biomarker to encounter AD. Resveratrol oligomers isolated from the seed extract of *Paeoni alacti flora* showed potent BACE1 inhibitory activity in-vitro (Hadden et al., 2009).

Vitisin A, B (Fig. 3.2 (ix–x)) were found to be promising candidates for the design of new potential BACE1 inhibitors. They have been reported to decline amyloid-induced oxidative stress in PC12 cell lines. However, the mechanism has not yet been elucidated (Ko et al., 2005).

3.2.1.6 GLYCOSIDE DERIVATIVES

Glycosides isolated from different species of Aloes like Aloveroside A, Elgnica Dimer A, Elgnica dimer B, p-Coumaroylaloenin, and Aloenin (Fig. 3.2 (xi–xiv)) showed BACE1 inhibition in vitro. Assays were performed using different concentrations of isolated compounds. Amongst all the glycosides, coumaroyl aloenin demonstrated the most potent inhibitory activity with IC_{50} value of 68.3 µg/mL (Speranza et al., 1994).

In addtion, glycosides of *Nelumbo nucifera* stamens viz β-cyclogeraniol diglycoside, cycloartenol, ρ-hydroxy benzoic acid, vanilloloside, 5'-O-methyladenoside showed decline in Aβ25-35 fibril formation in-vitro. (Hyun et al., 1997; Solomon, 2019; Bhattacharjee et al., 2020).

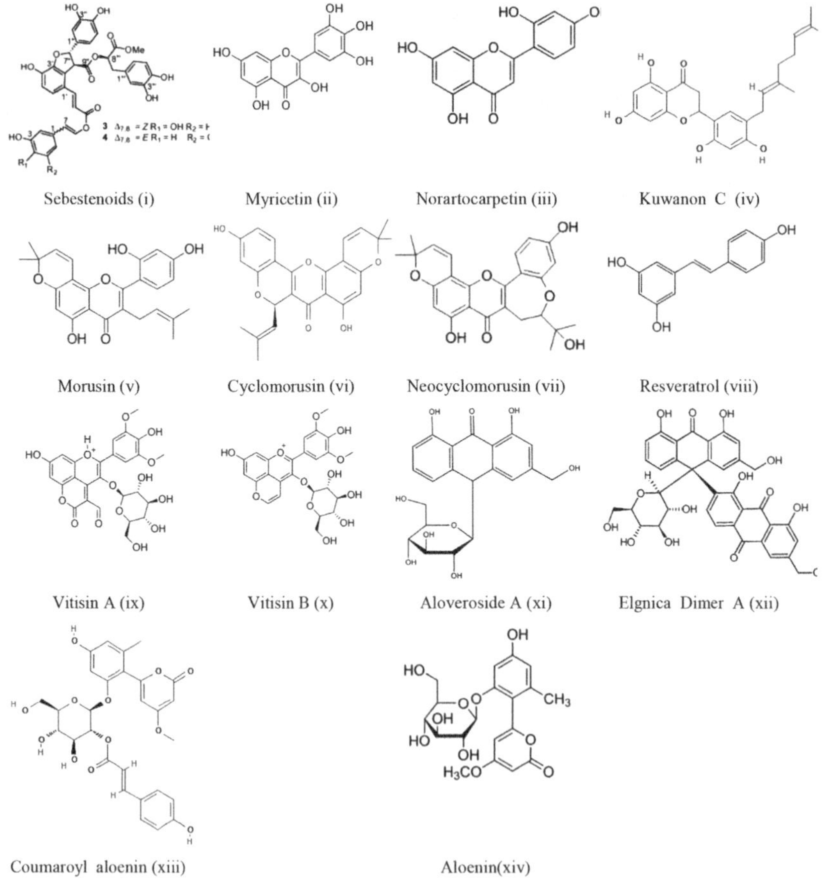

FIGURE 3.2   Structure of potential phytocompounds to encounter AD.

3.2.2   PLANT-DERIVED ANTI-CANCER MOLECULES

Cancer is a leading cause of death in many countries. It has challenged medical sciences for centuries. Amongst all noncommunicable diseases in India, cancer is responsible for 9% of the deaths (Prashant et al., 2020). In UK, alone 17 million cases were reported in 2018 with death resulting in 9.6 million people in 2018 (Ashraf et al., 2020). With the rising cause of cancer and the increasing use of synthetic drugs and chemotherapeutics, leading to further side effects, the role of plant-based novel molecules as anticancer agents have been investigated by various workers. Various plant-derived anticancer drugs are used to control the growth and spread of cancer (Debnath et al., 2010; Cragg and Newman, 2005) Some of them are explained below—

3.2.2.1   POLYPHENOLS

Curcumin, gallacatechins, flavonoids, and tanins are considered as poly-phenolic compounds and found to have anticancer activity. It has also been observed that polyphenols contain natural antioxidant and when included in the diet it can prevent cancer. Polyphenols are believed to have apoptosis-inducing properties thus demonstrating anticancer properties that can be used in controlling cancer. Polyphenols interferes the metabolism of copper ions that is bound to deoxyribonucleic acid (DNA) fragmentation and prevent growth of cancer. Resveratrol can cause DNA degradation in presence of Cu(II) (Tomeh et al., 2019). Polyphenols can also target and interact with the cancer cell proteins and prevents the growth of cancer cell. These agents may also alter polyphenol regulating acetylation and methylation by binding directly. It has also been observed that curcumin-treated cancer cells of different cells lines interact with various stimuli and suppresses the expression of Tumor Necrosis Factor (Nina et al., 2021).

3.2.2.2   FLAVONOIDS

Flavonoids are polyphenolic secondary metabolites of plants and consumed in diet. Chemically, it is having the general structure of a 15-carbon skeleton, and it contains 1 heterocyclic ring and 2 phenyl rings. Flavonoids are having nearly 10,000 known structures and they are physiologically active in plants and have various significant health benefits in human (Kopustinskiene et al., 2020).

Different plants have been studied to check their flavonoid content and their effect on cancer cell (Emanuele et al., 2017). Plants of various fern species and litchi leafs have been investigated for their chemical content and their effect on cancer. High contents of flavonoids like chalcones, flavones, and flavonols are found in the seeds. Researchers have studied anticancer activity of flavonoids from the fern species *Dryopteris erythrosora* and it was highly effective for lung cancer (Cao et al., 2013). Researchers have also studied cytotoxic effect of flavonoids on anticancer cell and they have observed high free radical scavenging activity. Several flavonoids that are purified are also used to prevent different types of cancer in human-like breast cancer, hepatoma, and cervical carcinoma. Cytotoxic effect of flavonoids present in the bark of *Erythrina suberosa* has also been evaluated on HL-60 cell and it was found to be very effective to prevent cancer (Ahmed et al., 2020). Alpinumisoflavone is also effective to treat cancer and it shows apoptosis activity through signaling pathway (both intrinsic and extrinsic pathway) (Fang et al., 2020). Reduction in mitochondrial membrane potential is observed because of the induction of apoptotic proteins. The cancer cell cannot survive after the damage of the mitochondria of the cells. Flavonoids extracted from the fern species showed anticancer activity even at a very low concentration.

3.2.2.3 BRASSINOSTEROIDS

Brassinosteroids are polyhydroxylated steroidal phytohormones found in plants. They have similar structures to animals' steroid hormones. It controls various physiological functions including development and growth of plants. It is a natural occurring substance that is having significant effect on control and cure against cancer (Malikova et al., 2008). 28-homocastasterone and 24-epibrassinolide are two natural occurring brassinosteroids found in plants having potential anticancer effect and effective at micromolar concentrations (Steigerova et al., 2010). It has been observed that cancer cells do not undergo natural apoptosis and they have characteristic cell multiplication. Cell growth responses were altered by brassinosteroids by alteration of the cell cycle that induces apoptosis. It is used to treat various cancer and effective against several cancer cell lines like cervical carcinoma HeLa and T-lymphoblastic leukemia CEM. Several receptor proteins like HER-2, *epidermal growth factor receptor* and Estrogen receptor (ER) are targeted to treat breast cancer as these proteins are present in breast cancer cell (Steigerova et al., 2010). Development of prostate cancer is highly effected by androgen receptor protein. Brassinosteroids having efficacy to bind with that receptors of these proteins and interfere with the growth of the cancer cell. It also blocks the cell cycle and prevent the growth of cancer. Researchers have also observed that breast cancer cell lines treated with 24-epiBL and 28-homoCS demonstrated reduction in the cyclin proteins involved in G_1 cell cycle phase. At this phase, cells in the cell cycle will either undergo repair or enter apoptosis. Treatment with brassinosteroids induces apoptosis at this stage, which cancer cells would not be able to do naturally without treatment. Brassinosteroids also shows various other activities in regular and tumor cells (Steigerova et al., 2010). During the treatment of cancer cell, it is very important to check that the active molecule do not affect the growth and functions of the normal cells. That means those agents should not be cytotoxic to normal cell and they should have specific action against cancer cell only. This is one of the unique property of the brassinosteroids and an effective way to treat cancer.

3.2.2.4 ANTICANCER PLANT-DERIVED DRUGS

Plant-derived anticancer drugs are natural and easily available. They can be consumed orally as part of the regular dietary intake of the patient suffering from cancer. Natural substances do not affect the normal cells and they are well tolerated by the body (Gurgul and Litynska, 2017). However, few plant derivatives like lectins, lignans, and taxanes can also effect the activity of the normal cells. Natural substances that are safe to the normal cells and able to kill cancer cells can play a lead role for the development of anticancer drug (Khazir et al., 2014; Subhashis et al., 2020). These plant-derived drugs can be classified under four categories:

a. Methytransferase inhibitors
b. Antioxidants or DNA damage preventive drugs
c. Histone deacetylases inhibitors, and
d. Mitotic disruptors.

Several plant-derived anticancer drugs with their origins and molecular mechanisms are explained in Table 3.3.

TABLE 3.3 Plant-Derived Drugs Having Anticancer Activity and Their Molecular Mechanism.

Anti-neoplastic agents	Derived from	Molecular mechanism	References
Curcumin	*Curcuma longa*	Reduces *matrix metalloproteinase-2 and interfere with metastasis and cell proliferation.*	Cragg and Newman (2005) and Tomeh et al. (2019)
Sulphoraphane	*Brassica*	Contains phase 2 detoxification enzymes and used to treat breast cancers.	Su et al. (2018)
Betulin, Betulinic acid	*Ziziphus nummularia*	It induces apoptosis through reactive oxygen species generation. Kills cancer cells more selectively.	Fulda (2008)
Epipodophyllotoxin	*Podophyllum peltatum* L	Pro-apoptotic effects	Ross et al. (1984)
Andrographolide	*Andrographis paniculata*	Helps to increases antioxidant enzymes Superoxide dismutases, Catalase, GST; decreases LDH and MDA	Rajagopal et al. (2003) and Khan et al. (2018)
Epigallacotechin-3-gallate	Green tea	Inhibit carcinogenesis induced chemically. Good antioxidant. Prevent DNA damage.	Kwak et al. (2015)
Roscovitine	*Raphanus sativus* L. (*Brassicaceae*)	It reduces cell cycle progression	Cicenas et al. (2015)
Asiaticoside	*Centella asiatica*	Inhibit synthesis of DNA and shows anticancer agent.	Yingchun (2019)
Vincristine	*Catharanthus roseus* G.	Have pro-apoptotic properties, act as anti miotic agent and microtubule inhibitor. Also induce cell cycle arrest.	Güthlein et al. (2002)
Vinblastine			Rtibi et al. (2017)
Vinflunine			Kruc zynski and Hill (2001) and Bachner and De (2008)

TABLE 3.3   *(Continued)*

Anti-neoplastic agents	Derived from	Molecular mechanism	References
Flavopiridol	*Dysoxylum binectariferum*Hook.f. (*Meliaceae*)	Anti-inflammatory; tyrosine kinase activyt; growth inhibitory effects, immunamodulatory activity;	Newcomb (2004) and Pinto et al. (2020)
Noscapine	Opium poppy (*Papaver somniferum*)	It is having antiproliferative properties and inhibit growth of cancer cell.	Mahmoudian and Rahimi (2009)
Triterpenic acids	*Boswellia serrata*	It inhibits DNA synthesis, apoptosis and topoisimerse I and II.	Petronelli et al. (2009)
Paclitaxel (Taxol)	*Taxus brevifolia* L	It blocks mitosis and disrupt formation of spindle.	Long (1994)
Pomiferin	*Maclura pomifera; Dereeis Malaccensis*	It is having antioxidant activity, prevent oxidative damage of DNA.	Yang et al. (2011) and Son et al. (2007)
Combretastatin A-4 phosphate	*Combretum caffrum*	It is having anti-angiogenic and tumor necrosis effect.	West and Price (2004)
Camptothecin	*Mappia foetida*	Effectively inhibits synthesis of nucleic acid and topoisomerase-1. Slowdown growth of human colon cancer cells.	Venditto and Simanek (2010)

Several plant-derived anticancer drugs like pomiferin, sulforaphane and isoflavones acts as HDAC inhibitors and controls cancer in human (Su et al., 2018; Yang et al., 2011; Son et al., 2007). These drug molecules are believed to alter the activity of carcinogenic proteins to control cancer. It has been observed that sulforaphane is very affective during breast cancer. A decreasein expression of estrogen receptor, human epidermal growth factor receptor 2, and *epidermal growth factor receptor* occurs due to HDAC inhibition caused by sulforaphane in the treatment of breast cancer cell lines (Su et al., 2018). In cancer cells, genes that are epigenetically-silenced and functional for chromatin acetylation are again reactivated by HDAC inhibitors and the cancer cells then enter into programmed cell death (apoptosis). Chemotherapeutic sensitivity in human cancers can be enhanced by inhibiting histone deacetylases by the plant-derived compounds.

Several derivatives of vinca alkaloids like vindesine, vincristine, vinblastine, and vinorelbine can inhibit the dynamics of the microtubules by binding to β-tubulin (Güthlein et al., 2002; Rtibi et al., 2017; Kruczynski and Hill, 2001; Bachner and De, 2008). These plant derived compounds causes apoptosis and cell cycle arrest and act against cancer. Paclitaxel stabilizes microtubules in the cells and decrease replication of cancer cells. One of the most important compounds derived from the plant is paclitaxel that have a great impact on the control of cancer (Long, 1994). On the other hand, vincristine and vinblastine were very important plant derivatives used in treatment of cancer. It has also observed that combination of drugs derived from various plant extract like camptotheca alkaloids, vinca alkaloids, and taxus diterpenes have improved anticancer activity. Extracts from various plants like *Urtica membranaceae* and *Origanum dayi Post* were tested for their anticancer activity in various organs and it has been observed that a combination of plant extract is more effective against cancer cell (Solowey et al., 2014; Stavri et al., 2005). Hence plant derivatives are most attractive anticancer agent then the synthetic drug that effects the normal cells and shows toxicity. Various plant derivatives' used in the treatment of cancer are listed below in the Figure 3.3.

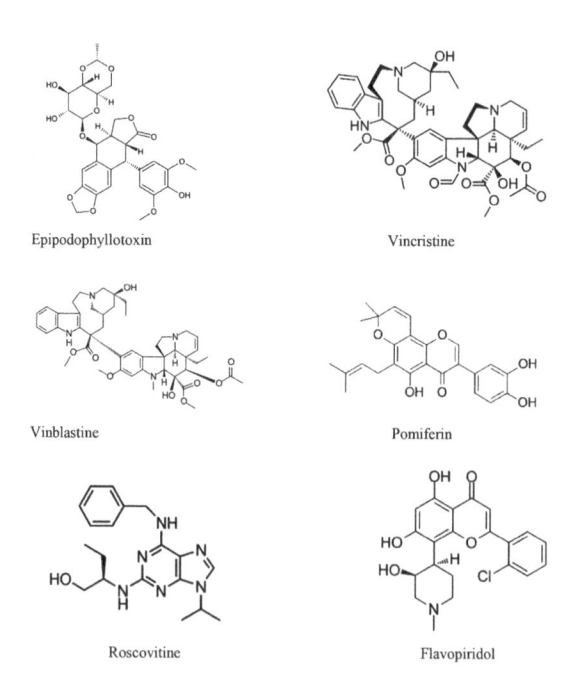

Epipodophyllotoxin

Vincristine

Vinblastine

Pomiferin

Roscovitine

Flavopiridol

FIGURE 3.3 Chemical structure of various plant-derived anticancer drugs.

CONFLICT OF INTEREST

The authors report no conflict of interest amongst themselves.

KEYWORDS

- **traditional medicine**
- **callus culture**
- **suspension culture**
- **secondary metabolites**
- **botanical drug**
- **medicinal plants**

REFERENCES

Abyari, M.; Nasr, N.; Soorni, J.; Sadhu, D. Enhanced Accumulation of Scopoletin in Cell Suspension Culture of Spilanthes acmella Murr. Using Precursor Feeding. *Braz. Arch. Biol. Technol.* **2016,** *59,* e16150533, 1–7.

Ahmed, Z.; Aziz, S.; Alauddin. S. In Vitro Cytotoxic and Antimicrobial Activities of Erythrina Suberosa (Roxb) Bark. *J. Pharm. Bioallied. Sci.* **2020,** *12* (2), 210–216.

Ahn, K. The Worldwide Trend of Using Botanical Drugs and Strategies for Developing Global Drugs. *BMB Rep.* **2017,** *50* (3), 111–116.

Alamgir, A. N. M. Therapeutic Use of Medicinal Plants and Their Extracts. Vol I, Pharmacognosy. Rainsford, K. D., Ed.; *Progress Drug Res.* **2017,** *73,* 403–426.

Alamgir, A. N. M. Biotechnology, In Vitro Production of Natural Bioactive Compounds, Herbal Preparation and Disease Management (Treatment and Prevention). Therapeutic Use of Medicinal Plants and their Extracts. *Progress Drug Res.* **2018,** *74* (2), 585–664. Springer International Publishing AG.

Ali, H.; Khan, M. A.; Kayani, W. K.; Khan, T.; Khan, R. S. Thidiazuron Regulated Growth, Secondary Metabolism and Essential Oil Profiles in Shoot Cultures of *Ajuga bracteosa*. *Ind. Crops Prod.* **2018,** *121,* 418–427.

Amod, P. K.; Laurie, A. K.; Girish, J. K. Herbal Complement Inhibitors in the Treatment of Neuroinflammation. *Ann. NY Acad. Sci.* **2005,** *1056,* 413–429.

Anand, U.; Jacobo-Herrera, N.; Altemimi, A.; Lakhssassi, N. A Comprehensive Review on Medicinal Plants as Antimicrobial Therapeutics: Potential Avenues of Biocompatible Drug Discovery. *Metabolites* **2019,** *9* (258), 1–13.

Anderson, R. N. Deaths: Leading Causes for 2000. *Natl. Vital Stat. Rep.* **2002,** *50* (16), 1–85.

Ashraf, U. Z.; Bhalerao, A.; Mikelis, C. M.; Cucullo, L.; German, N. A. Assessing the Current State of Lung Cancer Chemoprevention: A Comprehensive Overview. *Cancers* **2020,** *12* (5), 1265–1279.

Atanas, G.; Atanasov, A. G.; Waltenberger, B.; Wenzig, E. M. P.; Linder, T.;, Wawrosch, C.; Uhrin, P.; Temml, V.; Wang, L.; Schwaiger, S.; Heiss, E. H.; Rollinger, J. M.; Schuster, D.; Breuss, J. M.; Bochkov, V.; Mihovilovic, M. D.; Kopp, B.; Bauer, R.; Dirsch, V. M.; Stuppner, H. Discovery and Resupply of Pharmacologically Active Plant Derived Natural Products: A Review. *Biotechnol. Adv.* **2016,** *33* (8), 1582–1614.

Azimi, S.; Zonouzi, A.; Firuzi, O. Discovery of Imidazopyridines Containing Isoindoline-1,3-Dione Framework as a New Class of BACE1 Inhibitors: Design, Synthesis and SAR Analysis. *Eur. J. Med. Chem.* **2017,** *138*, 729–737.

Bachner, M.; De, S. M. Vinflunine in the Treatment of Bladder Cancer. *Ther. Clin. Risk Manag.* **2008,** *4* (6), 1243–1253.

Bachurin, S. O.; Bovina, E. V.; Ustyugov, A. A. Drugs in Clinical Trials for Alzheimer's Disease: The Major Trends. *Med. Res. Rev.* **2017,** *37*, 1186–1225.

Ballabh, B.; Chaurasia, O. P. Traditional Medicinal Plants of Cold Desert Ladakh-Used in Treatment of Cold, Cough and Fever. *J. Ethnopharmacol.* **2007,** *112* (2), 341–345.

Balunas, M. J.; Kinghorn, A. D. Drug Discovery from Medicinal Plants. *Life Sci.* **2005,** *78*, 431–441.

Barao, S., Moechars, D., Lichtenthaler, S. F. BACE1 Physiological Functions May Limit Its Use as Therapeutic Target for Alzheimer's Disease. *Trends Neurosci.* **2016,** *39* (3), 158–169.

Ben Halima, S.; Mishra, S.; Raja, K. M. P. Specific Inhibition of β-Secretase Processing of the Alzheimer Disease Amyloid Precursor Protein. *Cell. Rep.* **2016,** *14*, 2127–2141.

Berrino, F. Western Diet and Alzheimer's Disease. *Epidemiol. Prev.* **2002,** *26* (3), 107–115.

Bhattacharjee, A. Phytoextracts and Their Derivatives Affecting Neurotransmission Relevant to Alzheimer's Disease: An Ethno-Medicinal Perspective. In *Neurotransmitters in Plants: New Achievements and Perspectives*; Akula, R., Roshchina, V. V., Eds.; Taylor & Francis: USA, 2018; p 357

Bhattacharjee, A.; Saikat S.; Raja C. Plant-Based β-Secretase (BACE-1) Inhibitors: A Mechanistic Approach to Encounter Alzheimer's Disorder. In *Evidence Based Validation of Traditional Medicines*; Mandal, S. C., Chakraborty, R., Sen, S., Eds.; Springer Nature Singapore Pvt. Ltd.: USA, 2020; pp 257–266.

Borchardt, J. K. The Beginnings of Drug Therapy: Ancient Mesopotamian Medicine. *Drug News Perspect.* **2002,** *15*, 187–192.

Brinton, R. D.; Yamazaki, R. S. Advances and Challenges in the Prevention and Treatment of Alzeimer's Disease. *Pharm. Res.* **1998,** *15*, 386–398.

Cao, J.; Xia, X.; Chen, X.; Xiao, J.; Wang, Q. Characterization of Flavonoids from Dryopteris Erythrosora and Evaluation of Their Antioxidant, Anticancer and Acetylcholinesterase Inhibition Activities. *Food Chem. Toxicol.* **2013,** *51* (11), 242–250.

Chandran, H.; Meena, M.; Barupal, T.; Sharma, K. Plant Tissue Culture as a Perpetual Source for Production of Industrially Important Bioactive Compounds. *Biotechnol. Rep.* **2020,** *26*, e00450, 1–10.

Choi, S. H.; Hur, J. M.; Yang, E. J. Beta-Secretase (BACE 1) Inhibitors from *Perillafrutescens* var. acuta. *Arch. Pharmacal. Res.* **2008a,** *31*, 183–187.

Choi, Y. H.; Yon, G. H.; Hong, K. S.; Dae S. Y.; Chun, W. C.; Woo-Kyu, P; Jae, Y. K.; Young, S. K.; Shi, Y. R. In Vitro BACE1 Inhibitory Phenolic Components from the Seeds of *Psoraleacorylifolia*. *Planta Med.* **2008b,** *74*, 1405–1408.

Choi, C. W.; Choi, Y. H.; Cha, M. R., Young, S. K.; Gyu, H. Y.; Kyung, S. H.; Woo-Kyu, P.; Young, H. K.; Shi, Y. R. In Vitro BACE 1 Inhibitory Activity of Resveratrol Oligomers from the Seed Extract of *Paeonialactiflora*. *Planta Med.* **2011,** *77*, 374–376.

Cicenas, J.; Kalyan, K.; Sorokinas, A. Roscovitine in Cancer and Other Diseases. *Ann. Transl. Med.* **2015,** *3* (10), 135–148.

Coan, K. E.; Maltby, D. A.; Burlingame, A. L. Promiscuous Aggregate-Based Inhibitors Promote Enzyme Unfolding. *J. Med. Chem.* **2009,** *52,* 2067–2075.

Colucci, L.; Bosco, M.; Ziello, A. R. Effectiveness of Nootropic Drugs with Cholinergic Activity in Treatment of Cognitive Deficit: A Review. *J. Exp. Pharmacol.* **2012,** *4,* 163–172.

Commenges, D.; Scotet, V.; Renaud, S.; Jacqmin-Gadda, H.; Barberger-Gateau, P; Dartigues, J. F. Intake of Flavonoids and Risk of Dementia. *Eur. J. Epidemiol.* **2000,** *16,* 357–363.

Corson, T. W.; Crews, C. M. Molecular Understanding and Modern Application of Traditional Medicines: Triumphs and Trials. *Cell* **2007,** *130,* 769–774.

Cragg, G. M.; Newman, D. J. Natural Products: A Continuing Source of Novel Drug Leads. *Biochim. Biophys. Acta.* **2013,** *1830,* 3670–3695.

Cragg, G. M.; Newman, D. J. Plants as a Source of Anti-Cancer Agents. *J. Enthnopharmacol.* **2005,** *100* (2), 72–79.

Crehan, H.; Lemere, C. A. Anti-Amyloid-β Immunotherapy for Alzheimer's Disease. In *Developing Therapeutics for Alzheimer's Disease: Progress and Challenges*; Michael, S. W., Ed.; Academic Press, 2016.

Dai, J; Sorribas, A; Yoshida, W. Y.; Williams, P. G. Sebestenoids A. -D. BACE 1 Inhibitors from *Cordia sebestena. Phytochemistry* **2010,** *71,* 2168–2173.

Darvishi, E.; Kahrizi, D.; Bahraminejad, S.; Mansouri, M. In Vitro Induction of a-Pinene, Pulegone, Menthol, Menthone and Limonene in Cell Suspension Culture of Pennyroyal (Mentha pulegium). *Cell. Mol. Biol.* **2016,** *62,* 7–9.

Debnath, P. K.; Banerjee, S.; Debnath, P.; Mitra, A.; Mukherjee, P. K. Ayurveda—Opportunity for Developing Safe and Effective Treatment Choice for the Future. In *Evidence-based Validation of Herbal Medicine*; Mukherjee, P. K., Ed.; Elsevier Science Publishing Co Inc.: Amsterdam, 2015; pp 427–454.

Debnath, S.; Datta, D.; Babu, M. N.; Kumar, R. S.; Senthil, V. Studies on the Preparation and Evaluation of Chitosan Nanoparticles Containing Cytarabine. *Int. J. Pharm. Sci. Nanotechnol.* **2010,** *3* (3), 963–970.

Del Pilar Nicasio-Torres, M.; Meckes-Fischer, M.; Aguilar-Santamaría, L.; Garduño-Ramírez, M. L.; Chávez-Ávila, V. M.; Cruz-Sosa, F. Production of Chlorogenic Acid and Isoorientin Hypoglycemic Compounds in Cecropia Obtusifolia Calli and in Cell Suspension Cultures with Nitrate Deficiency. *Acta Physiol. Plant.* **2012,** *34,* 307–316.

Dewachter, I.; Van, L. F. Secretases as Targets for the Treatment of Alzheimer's Disease: The Prospects. *Lancet Neurol.* **2002,** *1,* 409–416.

Dwivedi, K. K.; Singh, R. H. A Clinical Study of Medhyarasayana Therapy in the Management of Convulsive Disorder. *J. Res. Ayurveda Siddha* **1978,** *13* (97), 105.

Efferth, T. Biotechnology Applications of Plant Callus Cultures. *Engineering* **2019,** *5* (1), 50–59.

Eglen, R. M. Enzyme Fragment Complementation: A Flexible High Throughput Screening Assay Technology. *Assay Drug Dev. Technol.* **2002,** *1,* 97–104.

Ekor, M. The Growing Use of Herbal Medicines: Issues Relating to Adverse Reactions and Challenges in Monitoring Safety. *Front. Pharmacol.* **2014,** *4* (177), 1–10.

Emanuele, S.; Lauricella, M.; Calvaruso, G.; DAnneo, A.; Giuliano, M. Litchi Chinensis as a Functional Food and a Source of Antitumor Compounds: An Overview and a Description of Biochemical Pathways. *Nutrients* **2017,** *9* (9), 992–1014.

Fang, M.; Liu, Y.; Liu, Q.; Qian, L. Alpinumisoflavone Inhibits Tumor Growth and Metastasis in Papillary Thyroid Cancer via Upregulating miR-141-3p. *Anat. Rec. (Hoboken)* **2020**, *303* (7), 1842–1850.

Felter, H. W.; Lloyd, J. U. *King's American Dispensatory*; Ohio Valley Co.: Cincinnatti, 1898. http://www.henriettes-herbcom/eclectic/kings/index/html.

Feng, B. Y.; Shoichet, B. K. A Detergent-Based Assay for the Detection of Promiscuous Inhibitors. *Nat. Protoc.* **2006**, *1*, 550–553.

Fenical, W. H.; Jensen, P. J. Developing a New Resource for Drug Discovery: Marine Actinomycetebacteria. *Nat. Chem. Biol.* **2006**, *2*, 666–673.

Fischer, R.; Emans, N.; Schuster, F.; Hellwig, S.; Drossard, J. Towards Molecular Farming in the Future: Using Plant-Cell-Suspension Cultures as Bioreactors. *Biotechnol. Appl. Biochem.* **1999**, *30* (2), 109–112.

Fulda, S. Betulinic Acid for Cancer Treatment and Prevention. *Int. J. Mol. Sci.* **2008**, *9* (6), 1096–1107.

Ganesan, A. The Impact of Natural Products upon Modern Drug Discovery. *Curr. Opin. Chem. Biol.* **2008**, *12*, 306–317.

Gounaris, Y. Biotechnology for the Production of Essential Oils, Flavours and Volatile Isolates. A Review. *Flavour Fragr. J.* **2010**, *25*, 367–386.

Guerriero, G.; Berni, R.; Muñoz-Sanchez, J. A.; Apone, F.; Abdel-Salam, E. M.; Qahtan, A. A.; Alatar, A. A.; Cantini, C.; Cai, G.; Hausman, J. F.; Siddiqui, K. S.; Hernández-Sotomayor, S. M. T.; Faisal, M. Production of Plant Secondary Metabolites: Examples, Tips and Suggestions for Biotechnologists. *Genes (Basel)* **2018**, *9*, 309, 1–22.

Gunasekera, S. P.; Miller, M. W.; Kwan, J. C.; Jason, C. K.; Hendrik L.; Valerie, J. P. Molassamide, a Dipeptide Serine Protease Inhibitor from the Marine Cyanobacterium *Dichothrixutahensis*. J *Nat. Prod.* **2010**, *73*, 459–462.

Gupta, B. M.; Bala, A. Alzheimer's Disease Research in India: A Scientometric Analysis of Publications output During 2002–11. *Res. Neurol. Int. J.* **2013**, *2013*, 1–11.

Gurgul, A., Litynska, A. Plant-Derived Compounds in the Treatment of Cancer. *Borgis-Postepy Fitoter.* **2017**, *3* (2), 203–208.

Gurib-Fakim, A. Medicinal Plants: Traditions of Yesterday and Drugs of Tomorrow. *Mol. Aspect. Med.* **2006**, *27*, 1–93.

Gurib-Fakim, A. Traditional Roles and Future Prospects for Medicinal Plants in Health Care. *Asian Biotechnol. Dev. Rev.* **2011**, *13* (3), 77–83.

Güthlein, F.; Burger, A. M.; Brandl, M.; Fiebig, H.; Schubert, R.; Unger, C.; Massing, U. Pharmacokinetics and Antitumor Activity of Vincristine Entrapped in Vesicular Phospholipid Gels. *Anti-Cancer Drugs.* **2002**, *13* (8), 797–805.

Ha, L. T.; Pawlicki-Jullian, N.; Pillon-Lequart, M.; Boitel-Conti, M.; Duong, H. X.; Gontier, E. Hairy Root Cultures of Panax Vietnamensis, a Promising Approach for the Production of Ocotillol-Type Ginsenosides. *Plant Cell. Tissue Organ. Cult.* **2016**, *126*, 93–103.

Hadden, M. K.; Hill, S. A.; Davenport, J.; Robert, L. M.; Brian, S. J. B. Synthesis and Evaluation of Hsp90 Inhibitors that Contain the 1,4-Naphthoquinone Scaffold. *Bioorg. Med. Chem.* **2009**, *17*, 634–640.

Huang, Y.; Mucke, L. Alzheimer Mechanisms and Therapeutic Strategies. *Cell.* **2012**, *148* (6), 1204–1222.

Hussain, M. S.; Fareed, S.; Saba Ansari, M.; Rahman, A.; Ahmad, I. Z.; Saeed, M. Current Approaches Toward Production of Secondary Plant Metabolites. *J. Pharm. Bioallied Sci.* **2012**, *4* (1), 10–20.

Hwang, E. M.; Ryu, Y. B.; Kim, H. Y.; Kim, D. -G.; Hong, S. -G.; Hwan Lee, J.; Curtis-Long, M. J.; Hun Jeong, S.; Jae-Yong, P.; Ki Hun, P. BACE1 Inhibitory Effects of Lavandulyl Flavanones from *Sophoraflavescens*. *Bioorg. Med. Chem.* **2008**, *16*, 6669–6674.

Hyun, S. K.; Sick, W. W.; Sung, L. S. Elgonica-Dimers A and B, Two Potent Alcohol Metabolism Inhibitory Constituents of *Aloe arborescens*. *J. Nat. Prod.* **1997**, *60*, 1180–1182.

Jachak, S. M.; Saklani, A. Challenges and Opportunities in Drug Discovery from Plants. *Curr. Sci.* **2007**, *92* (9), 1251–1257

Jeon, Y. H.; Heo, Y. S.; Kim, C. M. Phosphodiesterase: Overview of Protein Structures, Potential Therapeutic Applications and Recent Progress in Drug Development. *Cell. Mol. Life Sci.* **2005**, *62*, 1198–1220.

Jeong, S. H.; Ryu, Y. B.; Curtis-Long, M. J.; Hyung, W. R.; Yoon, S. B.; Jae, E. K.; Woo, S. L.; Ki, H. P. Tyrosinase Inhibitory Polyphenols from Roots of *Morusihou*. *J. Agric. Food Chem.* **2009**, *57*, 1195–1203.

Jiao, J.; Gai, Q. Y.; Wang, W.; Zang, Y. P.; Niu, L.-L.; Fu, Y. J.; Wang, X. Remarkable Enhancement of Flavonoid Production in a Co-Cultivation System of Isatis tinctoria L. Hairy Root Cultures and Immobilized Aspergillus niger. *Ind. Crops. Prod.* **2018**, *112*, 252–261.

Kang, H. S.; Kim, H. R.; Byun, D. S.; Byeng W. S.; Taek J. N.; Jae, S. C. Tyrosinase Inhibitors Isolated from the Edible Brown Alga *Ecklonia stolonifera*. *Arch. Pharmacol. Res.* **2004**, *27*, 1226–1232.

Karuppaiya, P.; Tsay, H. S. Enhanced Production of Podophyllotoxin, Kaempferol, and Quercetin from Callus Culture of Dysosma Pleiantha (Hance) Woodson: An Endangered Medicinal Plant. *Biotechnol. Appl. Biochem.* **2020**, *67*, 95–104.

Karuppusamy, S. A Review on Trends in Production of Secondary Metabolites from Higher Plants by In Vitro Tissue, Organ and Cell Culture. *J. Med. Plants Res.* **2009**, *3* (13), 1222–1239.

Khan, I.; Khan, F.; Farooqui, A.; Ansari, I. A. Andrographolide Exhibits Anticancer Potential Against Human Colon Cancer Cells by Inducing Cell Cycle Arrest and Programmed Cell Death via Augmentation of Intracellular Reactive Oxygen Species Level. *Nutr. Cancer* **2018**, *70* (5), 787–803.

Khan, T.; Khan, M. A.; Karam, K.; Ullah, N.; Mashwani, Z. U. R.; Nadhman, A. Plant In Vitro Culture Technologies; A Promise Into Factories of Secondary Metabolites Against COVID-19. *Front. Plant Sci.* **2021**, *12* (610194), 1–21.

Khazir, J.; Mir, B. A.; Pilcher, L.; Riley, D. L. Role of Plants in Anticancer Drug Discovery. *Phytochem. Lett.* **2014**, *7* (4), 173–181.

Kinghorn, A. D. Pharmacognosy in the 21st Century. *J. Pharm. Pharmacol.* **2001**, *53*: 135–148.

Klayman, D. L.; Lin, A. J.; Acton, N.; Scovill, J. P.; Hoch, J. M.; Milhous, W. K. Theoharides, A. D.; Dobek, A. S. Isolation of Artemisinin (qinghaosu) from *Artemisia annua* Growing in the United States. *J. Nat. Prod.* **1984**, *47*, 715–717.

Ko, P. C.; Higgins, J. A.; Kilduff, P. T. Evidence for Intact Selective Attention in Alzheimer's Disease Patients Using a Location Priming Task. *Neuropsychology* **2005**, *19*, 381–389.

Kolewe, M. E.; Gaurav, V.; Roberts, S. C. Pharmaceutical Active Natural Product Synthesis and Supply Viaplant Cell Culture Technology. *Mol. Pharm.* **2008**, *5*, 243–256.

Kopustinskiene, D. M.; Jakstas, V.; Savickas, A.; Bernatoniene, J. Flavonoids as Anticancer Agents. *Nutrients* **2020**, *12* (2), 457–467.

Kruczynski, A.; Hill, B. T. Vinflunine, the Latest Vinca Alkaloid in Clinical Development. A Review of Its Preclinical Anticancer Properties. *Crit. Rev. Oncol. Hematol.* **2001,** *40* (2), 159–173.

Kusakari, K.; Yokoyama, M.; Inomata, S. Enhanced Production of Saikosaponins by Root Culture of Bupleurum falcatum L. Using Two-Step Control of Sugar Concentration. *Plant Cell. Rep.* **2000,** *19,* 1115–1120.

Kwak, T. W.; Park, S. B.; Kim, H. J.; Jeong, Y. I.; Kang, D. H. Anticancer Activities of Epigallocatechin-3-Gallate Against Cholangiocarcinoma Cells. *Onco. Targets Ther.* **2016,** *10,* 214–364.

Lee, H. J.; Seong, Y. H.; Bae, K. H. β-Secretase (BACE 1) Inhibitors from *Sanguisorbae Radix. Arch. Pharm. Res.* **2005,** *28,* 799–803.

Levey, M. Babylonian Chemistry: A Study of Arabic and Second Millenium B.C. Perfumery. *Osiris.* **1956,** *12,* 376–389.

Long, H. J. Paclitaxel (Taxol): A Novel Anticancer Chemotherapeutic Drug. *Mayo. Clin. Proc.* **1994,** *69* (4), 341–345.

Mahmoudian, M.; Rahimi, P. The Anti-Cancer Activity of Noscapine: A Review. *Recent Pat. Anticancer Drug Discov.* **2009,** *4* (1), 92–97.

Malikova, J.; Swaczynova, J.; Kolar, Z.; Strnad, M. Anticancer and Antiproliferative Activity of Natural Brassinosteroids. *Phytochemistry* **2008,** *69* (2), 418–426.

Mashkovsky, M. D.; Kruglikova-Lvova, R. P. On the Pharmacology of the New Alkaloid Galantamine. *Farmacol. Toxicol. (Mosk.).* **1951,** *14,* 27–30.

Montenegro, R. C.; Araujo, A. J.; Molina, M. T.; José, D. B.; Marinho, F.; Danilo, D. R.; Lopéz-Montero, E.; Marília, O. F. G.; Bento, E. S.; Ana, P. N. N. A.; Cláudia, P.; Manoel, O. M.; Costa-Lotufo, V. L. Cytotoxic Activity of Naphthoquinones with Special Emphasis on Juglone and Its 5-O-Methyl Derivative. *Chem. Biol. Interact.* **2010,** *184,* 439–448.

Muto, N.; Inouye, K.; Inaba, A.; Nakanishi, T.; Tan, L. Inhibition of Cytochrome P-450-Linked Monooxygenase Systems by Naphthoquinones. *Biochem. Biophys. Res. Commun.* **1987,** *146,* 487–494.

Nadeem, M.; Ahmad, W.; Zahir, A.; Hano, C.; Abbasi, B. H. Salicylic Acid-Enhanced Biosynthesis of Pharmacologically Important Lignans and Neo Lignans in Cell Suspension Culture of Linum ussitatsimum L. *Eng. Life Sci.* **2019,** *19,* 168–174.

Nam, Y.; Lee, D. Ameliorating Effect of Zhizi (*FructusGardeniae*) Extract and Its Glycosides on Scopolamine-Induced Memory Impairment. *J. Tradit. Chin. Med.* **2013,** *15,* 223–227.

National Biodiversity Authority. *Annual Report 2011–2012;* National Biodiversity Authority: Chennai, 2012.

Nazir, M.; Asad Ullah, M.; Mumtaz, S.; Siddiquah, A.; Shah, M.; Drouet, S.; Hano, C.; Abbasi, B. H. Interactive Effect of Melatonin and UV-C on Phenylpropanoid Metabolite Production and Antioxidant Potential in Callus Cultures of Purple Basil (Ocimum basilicum L. var. s purpurascens). *Molecules* **2020,** *25* (1072), 1–16.

Ncube, B.; Van Staden, J. Tilting Plant Metabolism for Improved Metabolite Biosynthesis and Enhanced Human Benefit. *Molecules* **2015,** *20,* 12698–12731.

Newcomb, E. W. Flavopiridol: Pleiotropic Biological Effects Enhance Its Anti-Cancer Activity. *Anticancer Drugs* **2004,** *15* (5), 411–429.

Nina, S.; Jorg, F.; Georg, N.; Jan, F.; Dariush, B.; Bernd, K. Curcumin Administered as Micellar Solution Suppresses Intestinal Inflammation and Colorectal Carcinogenesis. *Nutr. Cancer.* **2021,** *73* (4), 686–693.

Ono, K.; Yoshiike, Y.; Takashima, A.; Kazuhiro, H.; Hironobu, N.; Masahito, Y. Potent Anti-Amyloidogenic and Fibril-Destabilizing Effects of Polyphenols In Vitro: Implications for the Prevention and Therapeutics of Alzheimer's Disease. *J. Neurochem.* **2003**, *7*, 172–181.

Osman, N. I.; Sidik, N. J.; Awal, A. Efficient Enhancement of Gallic Acid Accumulation in Cell Suspension Cultures of Barringtonia racemosa L. by Elicitation. *Plant Cell. Tissue Organ. Cult.* **2018**, *135*, 203–212.

Pandey, M. M.; Rastogi, S.; Rawat, A. K. S. Indian Traditional Ayurvedic System of Medicine and Nutritional Supplementation. *Evid. Based Complement. Alternat. Med.* **2013**, *376327*, 1–12.

Patwardhan, B. Ethnopharmacology and Drug Discovery. *J. Ethnopharmacol.* **2005**, *100*, 50–52.

Pence, V. C. Evaluating Costs for the In Vitro Propagation and Preservation of Endangered Plants. *In Vitro Cell. Dev. Biol. Plant* **2011**, *47* (1), 176–187.

Perry, E. K.; Pickering, A. T.; Wang, W. W. Medicinal Plants and Alzheimer's Disease: From Ethnobotany to Phytotherapy. *J. Pharm. Pharmacol.* **1999**, *51*, 527–533.

Petronelli, A.; Pannitteri, G.; Testa, U. Triterpenoids as New Promising Anticancer Drugs. *Anticancer Drugs* **2009**, *20* (10), 880–892.

Philomena, G. Concerns Regarding the Safety and Toxicity of Medicinal Plants – An Overview. *J. Appl. Pharm. Sci.* **2011**, *1* (6), 40–44.

Pinto, N.; Prokopec, S. D.; Ghasemi, F.; Meens, J.; Ruicci, K. M.; Khan, I. M.; Mundi, N.; Patel, K.; Han, M. W.; Yoo, J.; Fung, K.; MacNeil, D.; Mymryk, J. S.; Datti, A.; Barrett, J. W.; Boutros, P. C.; Ailles, L.; Nichols, A. C. Flavopiridol Causes Cell Cycle Inhibition and Demonstrates Anti-Cancer Activity in Anaplastic Thyroid Cancer Models. *PLoS One* **2020**, *15* (9), 1122–1138.

Pourebad, N.; Motafakkerazad, R.; Nasab, M. K.; Akhtar, N. F.; Movafeghi, A. The Influence of TDZ Concentrations on In Vitro Growth and Production of Secondary Metabolites by the Shoot and Callus Culture of Lallemantia Iberica. *Plant Cell. Tissue Organ. Cult.* **2014**, *122* (2), 1–9.

Prashant, M.; Krishnan, S.; Meesha, C.; Priyanka, D.; Kondalli, L. S.; Stephen, S.; Vinodh, N.; Anish, J.; Sandeep, N.; Francis, S. R. Cancer Statistics, 2020: Report From National Cancer Registry Programme, India. *JCO Glob. Oncol.* **2020**, *6* (1), 1063–1075.

Rajagopal, S.; Kumar, R. A.; Deevi, D. S.; Satyanarayana, C.; Rajagopalan, R. Andrographolide, a Potential Cancer Therapeutic Agent Isolated from Andrographis Paniculata. *J. Exp. Ther. Oncol.* **2003**, *3* (3), 147–158.

Ramakrishnappa, K. Impact of Cultivation and Gathering of Medicinal Plants on Biodiversity: Case Studies from India. Food and Agriculture Organization of the United Nations (FAO). Biodiversity and the Ecosystem Approach in Agriculture, Forestry and Fisheries. In *Satellite Event on the Occasion of the Ninth Regular Session of the Commission on Genetic Resources for Food and Agriculture*, Rome, Italy, Oct 12–13, 2002.

Rao, S. R.; Ravishankar, G. A. Plant Cell Cultures: Chemical Factories of Secondary Metabolites. *Biotechnol. Adv.* **2002**, *20*, 101–153.

Razavizadeh, R.; Adabavazeh, F.; Komatsu, S. Chitosan Effects on the Elevation of Essential Oils and Antioxidant Activity of Carum copticum L. Seedlings and Callus Cultures Under In Vitro Salt Stress. *J. Plant Biochem. Biotechnol.* **2020**, *29*, 473–483

Rosenthal, G. A. The Biochemical Basis for the Deleterious Effects of Lcanavanine. *Phytochemistry* **1991**, *30*, 1055–1058.

Ross, J. A.; Kasum, C. M. Dietary Flavonoids: Bioavailability, Metabolic Effects, and Safety. *Ann. Rev. Nutr.* **2002**, *22*, 19–34.

Ross, W.; Rowe, T.; Glisson, B.; Yalowich, J.; Liu, L. Role of Topoisomerase II in Mediating Epipodophyllotoxin-Induced DNA Cleavage. *Cancer Res.* **1984**, *44* (12), 5857–5860.

Rtibi, K.; Grami, D.; Selmi, S.; Amri, M.; Sebai, H.; Marzouki, L. Vinblastine, an Anticancer Drug, Causes Constipation and Oxidative Stress as Well as Others Disruptions in Intestinal Tract in Rat. *Toxicol. Rep.* **2017**, *4* (6), 221–225.

Sajjalaguddam, R. R.; Paladugu, A. Phenylalanine Enhances Quercetin Content in In Vitro Cultures of Abutilon indicum L. *J. Appl. Pharm. Sci.* **2015**, *5*, 080–084.

Samuelsson, G. *A Textbook of Pharmacognosy, Swedish Pharamaceutical Society*, 5th Ed.; Swedish Pharmaceutical Press: Stockholm, 2004a; pp 259–269.

Samuelsson, G. *Drugs of Natural Origin: A Textbook of Pharmacognosy*; 5thEd.; Swedish Pharmaceutical Press: Stockholm, 2004b.

Santoro, V. M.; Nievas, F. L.; Zygadlo, J. A.; Giordano, W. F.; Banchio, E. Effects of Growth Regulators on Biomass and the Production of Secondary Metabolites in Peppermint (Mentha pi-perita) Micropropagated In Vitro. *Am. J. Plant Sci.* **2013**, *4*, 49–55.

Sen, S.; Chakraborty, R. Toward the Integration and Advancement of Herbal Medicine: A Focus on Traditional Indian Medicine. *Bot. Target. Ther.* **2015**, *5*, 33–44.

Sen, S.; Chakraborty, R.; De, B. Challenges and Opportunities in the Advancement of Herbal Medicine: India's Position and Role in a Global Context. *J. Herb. Med.* **2011**, *1* (3–4), 67–75.

Sen. S.; Chakraborty, R. Revival, Modernization and Integration of Indian Traditional Herbal Medicine in Clinical Practice: Importance, Challenges and Future. *J. Tradit. Complement. Med.* **2017**, *7*, 234–244.

Shaikh, S.; Shriram, V.; Khare, T.; Kumar, V. Biotic Elicitors Enhance Diosgenin Production in Helicteres isora L. Suspension Cultures via Up-Regulation of CAS and HMGR Genes. *Physiol. Mol. Biol. Plants.* **2020**, *26*, 593–604.

Shankar, D.; Majumdar, B. Beyond the Biodiversity Convention: The Challenges Facing the Biocultural Heritage of India's Medicinal Plants. In *Medicinal Plants for Forest Conservationand Health Care*; Bodeker, G., Bhat, K. K. S., Burley, J., Vantomme, P., Eds.; Food and Agriculture Organization of the United Nations: Rome, 1997; pp 87–99.

Shi, L.; Wang, C.; Zhou, X.; Zhang, Y.; Liu, Y.; Ma, C. Production of Salidroside and Tyrosol in Cell Suspension Cultures of Rhodiola Crenulata. *Plant Cell. Tissue Organ. Cult.* **2013**, *114*, 295–303.

Shimmyo, Y.; Kihara, T.; Akaike, A.; Tetsuhiro, N.; Hachiro, S. Multifunction of Myricetin on Aβ: Neuroprotection via a Conformational Change of Aβ and Reduction of Aβ via the Interference of Secretases. *J. Neurosci. Res.* **2008**, *86*, 368–377.

Singh, H. Prospects and Challenges for Harnessing Opportunities in Medicinal Plants Sector in India. *Law Environ. Develop. J.* **2007**, *2* (2), 196–211.

Skovronsky, D. M.; Lee, V. M. Y.; Trojanowsky, J. Q. Neurodegenerative Diseases: New Concepts of Pathogenesis and Their Therapeutic Implications. *Annu. Rev. Pathol.* **2006**, *1*, 151–170.

Skrzypczak-Pietraszek, E.; Piska, K.; Pietraszek, J. Enhanced Production of the Pharmaceutically Important Polyphenolic Compounds in Vitex agnus castus L. Shoot Cultures by Precursor Feeding Strategy. *Eng. Life Sci.* **2018**, *18*, 287–297.

Smit, A. J. Medicinal and Pharmaceutical Uses of Seaweed Natural Products: A Review. *J. App. Phycol.* **2004**, *16*, 245–262.

Sneader, W. *Drug Discovery: A History*; Wiley, 2005.

Solomon, H. Natural Products in Alzheimer's Disease Therapy: Would Old Therapeutic Approaches Fix the Broken Promise of Modern Medicines? *Molecules* **2019**, *24* (8), 1519.

Solowey, E.; Lichtenstein, M.; Sallon, S.; Paavilainen, H.; Solowey, E.; LorberboumGalski, H. Evaluating Medicinal Plants for Anticancer Activity. *Sci. World J.* **2014**, *13* (1), 1155–1168.

Son, I. H.; Chung, I. M.; Lee, S. I.; Yang, H. D.; Moon, H. I. Pomiferin, Histone Deacetylase Inhibitor Isolated from the Fruits of Maclura pomifera. *Bioorg. Med. Chem. Lett.* **2007**, *17* (17), 4753–4755.

Speranza, G.; Data, G.; Lunazzi, L. A New Diglucosylated 6-Phenyl-2-Pyrone from Kenya Aloe. *J. Nat. Prod.* **1994**, *645*, 800–805.

Stavri, M.; Ford, C. H.; Bucar, F.; Streit, B.; Hall, M. L.; Williamson, R. T.; Mathew, K. T.; Gibbons, S. Bioactive Constituents of Artemisia Monosperma. *Phytochemistry* **2005**, *66* (2), 233–239.

Steigerova, J.; Oklestkova, J.; Levkova, M.; Rarova, L.; Kolar, Z.; Strnad, M. Brassinosteroids Cause Cell Cycle Arrest and Apoptosis of Human Breast Cancer Cells. *Chem. Biol. Interact.* **2010**, *188* (3), 487–496.

Su, X.; Jiang, X.; Meng, L.; Dong, X.; Shen, Y.; Xin, Y. Anticancer Activity of Sulforaphane: The Epigenetic Mechanisms and the Nrf2 Signaling Pathway. *Oxid. Med. Cell. Longev.* **2018**, *28* (6), 455–472.

Subhashis, D.; Runa C.; Donita D. A Review on Role of Medicinal Plants in Immune System. *Asian J. Pharm. Technol.* **2020**, *10* (4), 273–277.

Tauraite, D.; Razumas, V.; Butkus, E. Lipophilic 1,4-Naphthoquinone Derivatives: Synthesis and Redox Properties in Solution and Entrapped in the Aqueous Cubic Liquid-Crystalline Phase of Monoolein. *Chem. Phys. Lipids.* **2009**, *159*, 45–50.

Thomford, N. E.; Senthebane, D. A.; Rowe, A.; Munro, D.; Seele, P.; Maroyi, A.; Dzobo, K. Natural Products for Drug Discovery in the 21st Century: Innovations for Novel Drug Discovery. *Int. J. Mol. Sci.* **2018**, *19* (6), (1578), 1–29.

Tomeh, M. A.; Hadianamrei, R.; Zhao, X. A Review of Curcumin and Its Derivatives as Anticancer Agents. *Int. J. Mol. Sci.* **2019**, *20* (5), 1033–1049.

Traditional Medicine Programme. *Regulatory Situation of Herbal Medicines: A Worldwide Review*. World Health Organization (WHO): Geneva, 1998.

Unschuld, P. U. *Medicine in China: A History of Pharmaceutics*; University of California Press, 1986.

Vaidya, A. B. The Status and Scope of Indian Medicinal Plants Acting on the Central Nervous System. *Indian J. Pharmacol.* **1997**, *S340*, 228–234.

Veerashree, V.; Anuradha, C.; Kumar, V. Elicitor-Enhanced Production of Gymnemic Acid in Cell Suspension Cultures of Gymnema Sylvestre R. Br. *Plant Cell. Tissue. Organ. Cult.* **2012**, *108*, 27–35.

Venditto, V. J.; Simanek, E. E. Cancer Therapies Utilizing the Camptothecins: A Review of the In Vivo Literature. *Mol. Pharm.* **2010**, *7* (2), 307–349.

West, C. M.; Price, P. Combretastatin A4 Phosphate. *Anticancer Drugs* **2004**, *15* (3), 179–187.

Wilczańska-Barska, A.; Królicka, A.; Głód, D.; Majdan, M.; Kawiak, A.; Krauze-Baranowska, M. Enhanced Accumulation of Secondary Metabolites in Hairy Root Cultures of Scutellaria lateriflora Following Elicitation. *Biotechnol. Lett.* **2012**, *34*, 1757–1763.

Xu, J.; Ge, X.; Dolan, M. C. Towards High-Yield Production of Pharmaceutical Proteins with Plant Cell Suspension Cultures. *Biotechnol. Adv.* **2011**, *29*, 278–299.

Yang, R.; Hanwell, H.; Zhang, J.; Tsao, R.; Meckling, K. A. Antiproliferative Activity of Pomiferin in Normal (MCF-10A) and Transformed (MCF-7) Breast Epithelial Cells. *J. Agric. Food Chem.* **2011,** *59* (24), 13328–13336.

Yingchun, L. Antitumor Activity of Asiaticoside Against Multiple Myeloma Drug-Resistant Cancer Cells Is Mediated by Autophagy Induction, Activation of Effector Caspases, and Inhibition of Cell Migration, Invasion, and STAT-3 Signaling Pathway. *Med. Sci. Monit.* **2019,** *25* (20), 1355–1361.

Youdim, K. A.; Qaiser, M. Z.; Begley, D. J.; Rice-Evans C. A.; Abbott, N. J. Flavonoid Permeability Across an In Situ Model of the Blood-Brain Barrier. *Free Radic. Bio. Med.* **2004,** *36,* 592–604.

Yue, W.; Ming, Q. L.; Lin, B.; Rahman, K. ; Zheng, C. J. ; Han, T.; Qin, L. P. Medicinal Plant Cell Suspension Cultures: Pharmaceutical Applications and High Yielding Strategies for the Desired Secondary Metabolites. *Crit. Rev. Biotechnol.* **2016,** *36,* 215–232.

Zhang, F.; Jiang, L. Neuroinflammation in Alzheimer's Disease. *Neuropsychiatr Dis. Treat.* **2015,** *11,* 243–256.

PART 2
STRESS RESILIENCE STRATEGIES IN CROPS

CHAPTER 4

Rice Physiology and Sustainability in the Face of Increasing Carbon Dioxide Concentration

ADITYA BANERJEE and ARYADEEP ROYCHOUDHURY

Post Graduate Department of Biotechnology,
St. Xavier's College (Autonomous), Kolkata, West Bengal, India

ABSTRACT

Rice is the principal food crop chiefly in the South-East Asian countries. India and China are the largest producers as well as the consumers of rice grains. Thus, the cultivation of rice is extensively carried out in these regions and also in other parts of the world. Global warming due to the rising level of greenhouse gases is a rapidly growing problem, which is being discussed in all major global forums across the world. Carbon dioxide (CO_2) is one of the most abundant greenhouse gases and its level is rapidly rising in the atmosphere. Thus, it is essential to analyze the effects of CO_2 on the growth and quality of rice, since this aspect is directly correlated with food security of a large population that depends on rice grains as the main source of nutrition. This chapter concisely highlights the detrimental effects of increasing CO_2 concentration on the physiology and nutritional quality of rice grains. Our discussion also aims to elaborate the decline in the food value of rice grains in response to rising CO_2 and to establish the need of cultivating alternate C_4

Crop Sustainability and Intellectual Property Rights. Soumya Mukherjee, Piyali Mukherjee & Tariq Aftab (Eds)

crops instead of C_3 plants, since the former can better tolerate the effects of rising CO_2.

4.1   INTRODUCTION

Carbon dioxide (CO_2) is a greenhouse gas largely responsible for the emerging problem of global warming and air pollution. CO_2 is an agent of climate change. It imposes burden on food security and increases the vulnerability of food production and quality. Already, a steady decline in the yield of staple food crops like rice and maize ranging between 20% and 40% is expected to occur by 2100 due to elevation in the troposphere temperature in the tropical and subtropical areas (Battisti and Naylor, 2009). Such increase in surface temperature is regulated by a number of greenhouse gases among which CO_2 is the most crucial and abundant (Schlenker and Roberts, 2009). Increase in CO_2 level in the atmosphere can severely disturb the ionomic homeostasis in most plants (Taub et al., 2008).

Rice is the staple food crop of several countries and is considered as a caloric and nutritional source in low- and lower-middle income countries of Asia (Banerjee and Roychoudhury, 2018a, 2018b; Banerjee and Roychoudhury, 2019a, 2019b; Banerjee et al., 2019a). Thus, the population is highly dependent on rice as the staple food will be directly affected by the deterioration in rice nutritional quality due to increasing CO_2 level. Some experiments have already aimed at quantifying the changes in dietary components and physiology of the crop in response to increasing CO_2. Zhu et al. (2018) performed laboratory-based experiments using CO_2 treatments varying between 568 and 590 µmol mol^{-1}. The use of this concentration of CO_2 for studying the effects on rice physiology is justified by the fact that people born today would eat rice grown at CO_2 concentration of 550 µmol mol^{-1} or higher throughout their life span. The aim of this chapter is to present a concise discussion on the deleterious effects of rising CO_2 level on the overall growth and nutritional quality of rice (Fig. 4.1).

4.2   EFFECT OF INCREASING CO_2 ON RICE GROWTH

In a recent study by Maity et al. (2019), elevated CO_2 concentration actually stimulated the overall and visual growth in the aromatic indica rice cultivar, Pusa basmati 1509. It was observed that elevated concentration of CO_2 increased plant height, number of tillers, and even straw weight in the Pusa basmati 1509 cultivar, compared to the control seedlings grown under normal concentration of CO_2 (Maity et al., 2019). In another study, it was shown that exposure of indica rice cultivar, IR-30 to super-ambient concentrations of CO_2, namely, 500, 660, and 900 µmol mol^{-1} air steadily increased root: shoot biomass, lamina area, number of panicles per plant, and yield (Baker et al., 1990). Doubling of CO_2 concentration from 330 to 660 µmol mol^{-1} air increased the grain yield by 32% in the IR-30 seedlings (Baker et al., 1990).

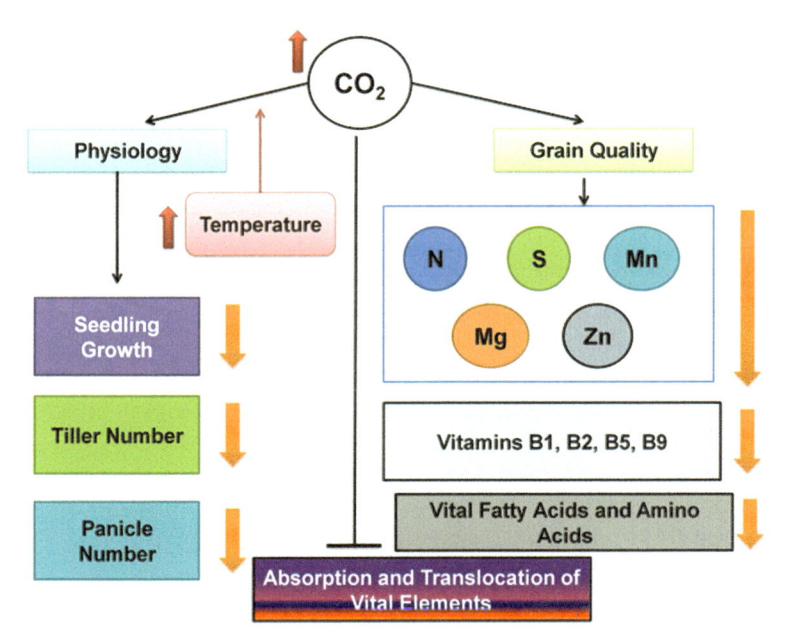

FIGURE 4.1 The effects of increasing atmospheric CO_2 level on rice growth and grain quality. Increased level of CO_2 and elevated temperature together inhibited seedling growth and reduced the number of tillers and panicles in rice plants. The grain quality was also largely affected due to the inhibition of absorption, translocation and import of beneficial elements into the cells. As a result, the accumulation of elements like nitrogen, sulphur, manganese, magnesium and zinc, along with the specified vitamins decreased within the seeds of the plants grown under high CO_2 level. The content of linoleic acid, γ-linolenic acid and essential amino acids like isoleucine, tyrosine and phenylalanine was also decreased in such grains.

In another study, it was shown that elevated CO_2 and temperature conditions triggered the accumulation of malondialdehyde in the rice cultivar Liangyou 287 (Liu et al., 2017). Malondialdehyde is a by-product of lipid peroxidation of cell membranes and is a crucial marker for membrane instability during oxidative stress within the cell (Banerjee et al., 2018; Banerjee et al., 2019b). Thus, increase in CO_2 and temperature together promoted membrane instability in Liangyou 287. Proline (Pro) is a compatible solute that is accumulated within the cell during exsiccosis. It protects the cell membrane during dehydration stress, maintains the cellular osmoticum, and is also involved in generating stress memory (Roychoudhury et al., 2015). Liu et al. (2017) observed increased Pro accumulation in the seeds during filling stage in plants exposed to elevated CO_2 and temperature. After the filling stage, the accumulation of soluble protein was observed, which indicated the inhibited nitrogen assimilatory and metabolic activity in the treated plants (Liu et al., 2017). Soluble sugar level did not vary in the rice seeds during exposure to high temperature and CO_2 level, thus verifying the inability of the plants to protect them against stress. Soluble sugar acts as a potent osmolyte in plants (Liu et al., 2017). The activity of superoxide dismutase gradually decreased during the filling stage in the Liangyou 287 seedlings exposed to both high temperature and CO_2 level. This compromised the seed development in these seedlings (Liu et al., 2017). Thus, this section clearly verifies that increased CO_2 alone might improve the visual growth characteristics of rice. However, CO_2 is a greenhouse gas and increased CO_2 concentration would always be accompanied by an elevation in temperature under realistic conditions. Thus, under such joint suboptimal situation, rice growth and grain development are compromised.

4.3   EFFECT OF INCREASING CO_2 ON RICE GRAIN QUALITY

Recent investigations have shown that rising atmospheric CO_2 negatively regulated multiple quality-associated traits in rice grains, though the biochemical mechanism behind such consequences is unknown (Jing et al., 2016). The authors showed that free-air enrichment of CO_2 in the japonica rice cultivar Wuyunjing 23 led to the partial removal of spikelets at anthesis. In addition, deterioration in the head rice percentage and concentration of protein in milled rice was reported. Overall, improvement in grain chalkiness was observed (Jing et al., 2016). Elevated level of CO_2 increased the size of large starch granules (>5 μm) within the grains, whereas the proportion of the smaller starch granules (<5 μm) was decreased (Jing et al., 2016). Field experiments with open-top chambers and elevated CO_2 concentration of 375 and 550 μmol mol^{-1} decreased the linoleic acid content in brown rice and also reduced the level of γ-linolenic acid in the husk (Goufo et al., 2014). The bioavailability of zinc and phytic acid within the grains was undisturbed.

Elevated CO_2 concentration decreased the content of vital amino acids like isoleucine, tyrosine, and phenylalanine in white rice, brown rice, and bran, respectively (Goufo et al., 2014). On the contrary, increased CO_2 level stimulated grain whiteness and starch viscosity, without milling effect (Goufo et al., 2014). In a blog published by the Scientific American (Harvey, 2018), 18 popular rice genotypes were tested for the effects of rising CO_2 in the atmosphere. It was observed that concentrations ranging between 568 and 590 parts per million reduced the iron and zinc content by 8% and 5%, respectively. The overall content of the B vitamins declined by about 30%, whereas only vitamin E level was reportedly increased in the rice grains exposed to high CO_2.

Zhu et al. (2018) reported significant reduction in the vitamin content of the grains of 18 genotypes of rice including both japonica and indica cultivars exposed to elevated concentration of CO_2. The researchers adopted multiyear and multilocation in situ free-air CO_2 enrichment studies to investigate the alteration in the vitamin content in the grains. It was observed that rising CO_2 level consistently reduced the contents of vitamins B1, B2, B5, and B9 across the seeds of the tested rice varieties (Zhu et al., 2018). In another recent report, Ujiie et al. (2019) reported significant decrease in the content of elements like nitrogen, sulfur, manganese, and zinc in both the grains and seedlings of nine rice cultivars in response to elevated CO_2 level. Further experimentation using flow analysis of elements verified low absorption and translocation of the elements in rice grains in response to high CO_2 (Ujiie et al., 2019). Microarray-dependent studies verified by gene expression revealed the downregulation ten genes associated with potassium, heavy metal transport, and most importantly import of minerals into cells (Ujiie et al., 2019). In line with these observations and deterioration of food value, the authors have suggested the cultivation of suitable C_4 crops instead of the C_3 plants in order to avoid the ensuing nutritional crisis within the human population (Ujiie et al., 2019). C_4 plants make a significant contribution to the global carbon budget and C_4 rice seems pivotal to current and future global food security (Roychoudhury et al., 2008).

4.4 CONCLUSION

CO_2 is the most abundant greenhouse gas responsible for the rising problem of global warming. At high concentration, this gas acts as a pollutant and affects the normal growth and development of plants. Rice being the staple food crop is extensively cultivated and a large population is dependent on this vital crop for receiving majority of the nutrients. Rising level of CO_2 in the atmosphere alone stimulated rice growth and tiller formation. However, during global warming, the CO_2 concentration is increased along with temperature. Under such dual suboptimal treatment, the overall rice growth and agronomic parameter was drastically decreased. Increased CO_2 level under experimental conditions steadily decreased the nutritional parameters in tested indica and japonica rice cultivars. The content of crucial trace elements along with major vitamins was observed to decrease in rice cultivars grown under high CO_2 exposure (Fig. 4.1).

4.5 FUTURE PERSPECTIVES

Rising CO_2 level negatively affects the nutritional quality of rice grains, thus posing a direct threat to food security with increasing global warming. Future perspectives should include the adoption of necessary steps to reduce and negate the risks of this gradually increasing problem. This might be accomplished through the selection of better cultivars via traditional breeding approaches or genetic engineering pursuits. Overcoming the detrimental effects of climate change on rice cultivation will be possible only if the problem and mechanism are properly and more exhaustively understood. Thus, experiments at the transcriptomic, proteomic, and metabolomic levels should be undertaken to properly illustrate the mechanism of CO_2-mediated deterioration in rice quality. This knowledge will actually help the researchers to decide where exactly to target, instead of maneuvering relentlessly in the dark.

ACKNOWLEDGMENTS

Financial assistance from Science and Engineering Research Board, Government of India through the grant (EMR/2016/004799) and Department of Higher Education, Science and Technology and Biotechnology, Government of West Bengal, through the grant [264(Sanc.)/ST/P/S&T/1G-80/2017] to Dr. Aryadeep Roychoudhury is gratefully acknowledged. Mr. Aditya Banerjee is thankful to University Grants Commission, Government of India for providing Senior Research Fellowship in course of this work.

KEYWORDS

- rice
- carbon dioxide
- growth
- nutrition
- grain quality

REFERENCES

Baker, T. J.; Allen, L. H.; Boote, K. J. Growth and Yield Response of Rice to Carbon Dioxide Concentration. *J. Agric. Sci.* **1990**, *115*, 313–320.

Banerjee, A.; Ghosh, P.; Roychoudhury, A. Salt Acclimation Differentially Regulates the Metabolites Commonly Involved in Stress Tolerance and Aroma Synthesis in Indica Rice Cultivars. *Plant Growth Regul.* **2019a**, *88*, 87–97.

Banerjee, A.; Roychoudhury, A. Seed Priming Technology in the Amelioration of Salinity Stress in Plants. In *Advances in Seed Priming*; Rakshit, A., Singh, H. B., Eds.; Springer Nature: Singapore, 2018a; pp 81–93.

Banerjee, A.; Roychoudhury, A. Role of Beneficial Trace Elements in Salt Stress Tolerance of Plants. In *Plant Nutrients and Abiotic Stress Tolerance*; Hasanuzzaman, M., Fujita, M., Oku, H., Nahar, K., Hawrylak-Nowak, B., Eds.; Springer Nature: Singapore, 2018b; pp 377–390.

Banerjee, A.; Roychoudhury, A. Rice Responses and Tolerance to Elevated Ozone. In *Advances in Rice Research for Abiotic Stress Tolerance*; Hasanuzzaman, M., Fujita, M., Nahar, K., Biswas, J. K., Eds. Woodhead Publishing, Elsevier: United Kingdom, 2019a; pp 399–412.

Banerjee, A.; Roychoudhury, A. Differential Regulation of Defence Pathways in Aromatic and Non-Aromatic Indica Rice Cultivars Towards Fluoride Toxicity. *Plant Cell Rep.* **2019b**. DOI: https://doi.org/10.1007/s00299-019-02438-6.

Banerjee, A.; Tripathi, D. K.; Roychoudhury, A. Hydrogen Sulphide Trapeze: Environmental Stress Amelioration and Phytohormone Crosstalk. *Plant Physiol. Biochem.* **2018**, *132*, 46–53.

Banerjee, A.; Tripathi, D. K.; Roychoudhury, A. The Karrikin 'Callisthenics': Can Compounds Derived from Smoke Help in Stress Tolerance? *Physiol. Plant.* **2019b**, *165*, 290–302.

Battisti, D. S.; Naylor, R. L. Historical Warnings of Future Food Insecurity with Unprecedented Seasonal Heat. *Science* **2009**, *323*, 240–244.

Goufo, P.; Falco, V.; Brites, C.; Wessel, D. F.; Kratz, S.; et al. Effect of Elevated Carbon Dioxide Concentration on Rice Quality: Nutritive Value, Color, Milling, Cooking and Eating Qualities. *Cereal Chem.* **2014**, *91*, 513–521.

Harvey, C. As CO_2 Levels Rise, Rice Becomes Less Nutritious. *Sci. Am.* **2018**. https://www.scientificamerican.com/article/as-co2-levels-rise-rice-becomes-less-nutritious/.

Jing, L. -Q.; Wu, Y. -Z.; Zhuang, S. -T.; Wang, Y. -X.; Zhu, J. -G.; et al. Effects of CO_2 Enrichment and Spikelet Removal on Rice Quality Under Open-Air Field Conditions. *J. Integr. Agric.* **2016,** *15,* 2012–2022.

Liu, S.; Waqas, M. A.; Wang, S. -H.; Xiong, X. -Y.; Wan, Y. -F. Effects of Increased Levels of Atmospheric CO_2 and High Temperatures on Rice Growth and Quality. *PLoS ONE* **2017,** *12,* e0187724.

Maity, P. P.; Chakrabarti, B.; Bhatia, A.; Purakayastha, T. J.; Saha, N. D.; Jatav, R. S.; et al. Effect of Elevated CO_2 and Temperature on Growth of Rice Crop. *Int. J. Curr. Microbiol. Appl. Sci.* **2019,** *8,* 1906–1911.

Roychoudhury, A.; Banerjee, A.; Lahiri, V. Metabolic and Molecular-Genetic Regulation of Proline Signaling and Its Cross-Talk with Major Effectors Mediates Abiotic Stress Tolerance in Plants. *Turk. J. Bot.* **2015,** *39,* 887–910.

Roychoudhury, A.; Datta, K.; Datta, S. K. C4 Plants and Abiotic Stress. In *Proceedings of The Humboldt Kolleg on Global Warming in Context to the Indian Sub-Continent*; Ghosh, S., Ed.; Humboldt Club: Calcutta, Dec 11–13, 2008; pp 86–105.

Schlenker, W.; Roberts, M. J. Nonlinear Temperature Effects Indicate Severe Damages to US Crop Yields Under Climate Change. *Proc. Natl. Acad. Sci. U.S.A.* **2009,** *106,* 15594–15598.

Taub, D. R.; Miller, B.; Allen, H. Effects of Elevated CO_2 on the Protein Concentration of Food Crops: A Meta-Analysis. *Glob. Change Biol.* **2008,** *14,* 565–575.

Ujiie, K.; Ishimaru, K.; Hirotsu, N.; Nagasaka, S.; Miyakoshi, Y.; Ota, M.; et al. How Elevated CO_2 Affects Our Nutrition in Rice, and How We Can Deal With It. *PLoS ONE* **2019,** *14,* e0212840.

Zhu, C.; Kobayashi, K.; Loladze, I.; Zhu, J.; Jiang, Q.; Xu, X.; et al. Carbon Dioxide (CO_2) Levels This Century Will Alter the Protein, Micronutrients, and Vitamin Content of Rice Grains with Potential Health Consequences for the Poorest Rice-Dependent Countries. *Sci. Adv.* **2018,** *4,* eaaq1012.

CHAPTER 5

An Analysis of the Physiological and Biochemical Attributes in Tomato Fruits Affected by Salinity Stress

MOHAMED M. EL-MOGY[1] and HANY G. ABD EL-GAWAD[2]

[1]*Vegetable Crops Department, Faculty of Agriculture, Cairo University, Giza, Egypt*

[2]*Department of Horticulture, Faculty of Agriculture, Ain Shams University, Cairo, Egypt*

ABSTRACT

Many previous works studied the influence of abiotic stress such as unfavorable environmental conditions including salinity and drought on plant growth, yield, and chemical compositions. However, very limited works investigated the effects of abiotic stress on quality, postharvest behavior, storage ability, and antioxidants of fresh vegetables. Some abiotic stress could be useful in some cases such as the enhancement of tomato sugars by moderate salinity. Thus, this chapter will discuss the influence of abiotic stress such as salinity on the quality of fresh vegetables.

5.1 INTRODCUTION

The main fruits and vegetables quality parameters are color, texture, flavor, and nutritive value. Color is related to the three main natural pigment groups,

Crop Sustainability and Intellectual Property Rights. Soumya Mukherjee, Piyali Mukherjee & Tariq Aftab (Eds)

chlorophylls, carotenoids, and anthocyanins. Flavor is consisting of taste and smell. Sugars and organic acids are mainly responsible for the taste. Volatile and nonvolatile compounds are mainly responsible for smell. Cell wall structure and its components such as pectic substances as well as turgor pressure of cell wall are the main two factors, which responsible for fruits and vegetables texture. Nutritive value is mainly related to vitamins and minerals content in fruits and vegetables.

Abiotic stress includes all adverse conditions for plant growth such as high or low temperature, radiation, salinity, drought, and all other unaffordable environmental conditions. Salinity is the main abiotic stress that causes a hug crop loss worldwide. It has been reported that 20% of total cultivated land are negatively affected by salinity stress, which is considered the first factor that causes the loss of crop production (Alzahrani et al., 2021). The effect of preharvest factor such as salinity on the postharvest behavior of vegetables is limited. There are some quality parameters affected by salinity such as:

5.2 COLOR

5.2.1 COLOR

Tomato fruits redness significantly decreased by increasing salinity level from 0.5 to 15.7 dS m^{-1} (Pascale et al., 2001). However, no significant difference was observed in L* values and a*/b* ratio values of the cucumber surface color and the internal color of melon when irrigated with saline nutrient solution compared with nonsaline nutrient solution (Colla et al., 2012, 2006). In leafy vegetables, increasing salinity caused a reduction in L*, a*, and C* of green baby lettuce (Neocleous et al., 2014) and Romaine lettuce (Kim et al., 2008).

5.2.2 CAROTENOIDS

Borghesi et al. (2011) and Pascale et al. (2001) found that carotenoids were increased in tomato fruits by increasing salinity stress. Also, Giuffrida et al. (2014) found that carotenoids content in pepper fruits increased by increasing NaCl levels in nutrient solutions. In addition, increasing salinity levels (by increasing seawater dosage) resulted in higher carotenoids in red lettuce leaves compared with the control (Sakamoto et al., 2014). The same result was observed by Neocleous et al. (2014) in red lettuce. In addition, carotenoids content was increased in romaine lettuce leaves when plants subjected to low salt level (5 mM NaCl) for long period compared with high salt level (100 mM NaCl) for short period (Kim et al., 2008). A combination between salinity treatment and CO_2 application resulted in high carotenoids content of green or red lettuce (Pérez-López et al., 2015). Contrary, Abdelgawad et al. (2019) found that carotenoids in cherry tomato fruits were decreased by increasing salinity levels. However, in pepper fruits, β-carotene content was not affected by salinity treatment (Navarro et al., 2006).

5.2.3 ANTHOCYANINS

It has been found that anthocyanins content in tomato fruits increased with increasing salt levels (Borghesi et al., 2011). In addition, Galli et al. (2016) found that mild salinity stress (40 mmol/L NaCl) increased anthocyanins content in strawberry fruits. Also, in leafy vegetables such as red lettuce, increasing salinity levels of nutrient solution resulted in higher anthocyanins content in red leaves compared with the control (Sakamoto et al., 2014). Also, it has been reported that combined treatments of salinity and CO_2 application enhanced anthocyanins content of either green or red lettuce (Pérez-López et al., 2015).

5.2.4 LYCOPENE

Salinity stress enhanced the lycopene content in tomato fruits compared with the control (Borghesi et al., 2011; Pascale et al., 2001). Also, lycopene content in pepper fruits was increased by salinity treatment (Navarro et al., 2006).

5.2.5 CHLOROPHYLL

Previously, Abdelgawad et al. (2019) found that either chlorophyll a or b content in cherry tomato fruits increased with increasing salinity levels. Also, increasing salinity levels (by sea water) resulted in higher chlorophylls in red lettuce leaves compared with the control (Sakamoto et al., 2014).

5.3   TEXTURE

Tomato fruits firmness reduced gradually with increasing salinity dosages when compared with the controls (Zahedi et al., 2020). Also, Fallik et al. (2019) reported that firmness of bell pepper fruits decreased when pepper plants irrigated with high saline water (4.5 dS m^{-1}) compared with moderate salinity (2.8 dS m^{-1}) and normal water (1.6 dS m^{-1}). However, Abdelgawad et al. (2019) recorded a rise in cherry tomato fruits firmness when treated with salinity water when compared with the controls. Also, increasing salinity level from 0 to 80 mM NaCl resulted in higher firmness of melon fruits (Colla et al., 2006). Similar results were observed previously by Botía et al. (2005). On the other hand, some previous works indicated that difference in fruit firmness was not significant when plants were irrigated with saline water compared with the control of strawberry (from 0 to 60 mM NaCl) (Garriga et al., 2015), cucumber (from 0 to 30 mM NaCl) (Navarro et al., 2010), and pepper (from 0 to 30 mM NaCl) (Colla et al., 2013).

5.4   NUTRITIVE VALUE

5.4.1   SUCROSE, GLUCOSE, FRUCTOSE, AND TOTAL SUGAR

Rivero et al. (2014) found that sucrose, glucose, and fructose increased significantly in tomato fruits (cv. Optima) under salt treatment (120 Mm NaCl) compared with the control. Also, Yan et al. (2021) found that sucrose, glucose, and fructose contents in tomato fruits increased significantly when plants treated with saline water compared with the control. Moreover, Galli et al. (2016) recorded an increase of sucrose content in strawberry fruits under mild saline stress (40 mmol/L NaCl). Also, Abdelaal et al. (2020) found that soluble sugars and sucrose in pepper fruits were increased by increasing salinity level (1500 and 3000 ppm of NaCl). Conversely, Zahedi et al. (2020) found that glucose and fructose in tomato fruits decreased linearly with increasing salinity levels when compared with the controls. Also, Navarro et al. (2002) found a reduction in glucose and fructose contents of pepper fruits by increasing salinity level (up to 8 dS m^{-1}). In addition, in leafy vegetables such as lettuce, a reduction in glucose, and fructose was observed when plants were irrigated by saline water up to 3.6 dS m^{-1} (Fallovo et al., 2009). On the other hand, no significant difference was observed in individual sugars of zucchini fruits by increasing the salinity of nutrient solutions (Rouphael and Colla, 2009).

5.4.2 STARCH

Rivero et al. (2014) found that starch content increased in tomato fruits (cv. Optima) under NaCl treatment (120 Mm) compared with the control. The same result was obtained by Abdelaal et al. (2020) who found that starch content in pepper fruits was increased by increasing salinity levels.

5.4.3 TOTAL SOLUBLE SOLIDS

At low salt concentration (2.5 dS m^{-1}), total soluble solids (TSS) content in tomato fruits was not changed compared with the control whereas under 10 dS m^{-1} NaCl level, the TSS content was doubled (Yurtseven et al., 2005). It has been reported that increasing salinity increased TSS content in tomato fruits and cherry tomatoes (Pascale et al., 2001; Magán et al., 2008; El-Mogy et al., 2018; Yan et al., 2021). Also, Liu et al. (2014) found that tomato plants irrigated with 150 mM NaCl had higher TSS in fruits than the controls. In addition, TSS content in eggplant fruits (Savvas and Lenz, 2000; Zipelevish et al., 2000), watermelon fruits (Tingwu et al., 2003), melon fruits (Botía et al., 2005; Rouphael et al., 2012), strawberry fruits (Keutgen and Pawelzik, 2007; Galli et al., 2016), bell pepper green or colored fruits (Marín et al., 2009; Hand et al., 2017; Fallik et al., 2019), squash fruits (Rouphael et al., 2017), broccoli head (Zaghdoud et al., 2012; Di Gioia et al., 2018), and cauliflower head (Giuffrida et al., 2017) was increased by increasing salinity level. Contrary, Zahedi et al. (2020) found that TSS in tomato fruits decreased linearly with increasing salinity levels when compared with the controls. Also, Navarro et al. (2002) found a reduction in TSS content of pepper fruits by increasing salinity level (up to 8 dS m^{-1}).

5.4.4 TITRATABLE ACIDITY CONTENT

Magán et al. (2008), Pascale et al. (2001), and Liu et al. (2014) found that titratable acidity (TA) in tomato fruits was increased with increasing NaCl levels. The same results have been reported for cauliflower head (Giuffrida et al., 2017). Contrary, Zahedi et al. (2020) found that TA in tomato fruits decreased with increasing salinity levels when compared with the controls. TA of broccoli heads was not influenced by salinity treatments (Di Gioia et al., 2018).

5.4.5 VITAMIN C

Yan et al. (2021) found that vitamin C content in tomato fruits increased when plants were treated with saline water compared with the control. A similar pattern of result was recorded by Marín et al. (2009) who indicated that vitamin C content in red pepper fruit increased when the plants irrigated with saline nutrient solution EC up to 30 mM NaCl, however, no significant effect was observed in green fruit. Also, Moreno et al. (2008) observed an increase of vitamin C in broccoli inflorescences by salinity treatment.

Contrary, Zahedi et al. (2020) and Abdelgawad et al. (2019) in tomato fruits and Navarro et al. (2006) in pepper fruits found that vitamin C decreased significantly with increasing salinity stress in compared with the controls. On the other hand, no significant difference was observed in vitamin C content of zucchini squash fruits and broccoli heads by increasing the salinity of nutrient solutions (Rouphael and Colla, 2009; López-Berenguer et al., 2009).

5.4.6 ANTIOXIDANT

Zahedi et al. (2020) found that total antioxidant capacity in tomato fruits decreased significantly with increasing salinity stress when compared with the controls. Also, Chisari et al. (2010) recorded a reduction in antioxidant activity of baby romaine lettuce leaves by increasing salinity levels. Contrary, Cardeñosa et al. (2015) found that salinity stress enhanced antioxidant content of strawberry fruit in the first harvest time. However, the salinity stress had no effect on antioxidant in further harvests. In the same line with previous result, Keutgen and Pawelzik (2008) reported a rise of antioxidant content in strawberry fruit exposed to salinity stress. It has been also reported that antioxidant activity in the leaves of either rocket or artichoke increased with moderate salinity level (3.5 and 6.9 dS m^{-1}, respectively) (Bonasia et al., 2017; Rezazadeh et al., 2012).

5.4.7 TOTAL PHENOLIC COMPOUNDS

Zahedi et al. (2020) in tomato fruits and Navarro et al. (2006) in pepper fruits found that total phenolic compounds decreased significantly with increasing salinity stress when compared with the controls. Also, Chisari et al. (2010) recorded a reduction in total phenolic compounds of baby romaine lettuce leaves by increasing salinity levels. In contrast to these previous works,

Cardeñosa et al. (2015) found that salinity stress enhanced total phenols of strawberry fruit in the first harvest time. However, the salinity stress had no effect on total phenols in further harvests. In addition, Galli et al. (2016) found that high salinity stress (80 mmol/L NaCl) induced phenolic compounds content in strawberry fruits compared with mild salinity stress (40 mmol/L NaCl). The same previous results were obtained by Giuffrida et al. (2014) and Hand et al. (2017) who found that total phenolic compounds in pepper fruits increased by increasing NaCl levels in nutrient solutions. Also, Rezazadeh et al. (2012) observed an increase in total phenolic of artichoke leaves when plants subjected to moderate salinity levels (6.9 dS m^{-1}).

5.4.8 GLUCOSINOLATES

Application of salinity during the first growth-phase of broccoli plants increased indolic glucosinolates (glucobrassicin and neoglucobrassicin) compared with the control plants (Di Gioia et al., 2018). Moreover, Giuffrida et al. (2017) recorded a rise of individual glucosinolates contents in cauliflower heads (cv. "Conero") when plants were irrigated by saline nutrient solution. Also, previous studies mentioned that glucosinolates content in pakchoi heads and broccoli heads was increased by increasing salinity levels (Hu and Zhu, 2010; Zaghdoud et al., 2012).

5.4.9 FLAVONOIDS CONTENT

Hand et al. (2017) found that flavonoids content in pepper fruits decreased by increasing NaCl levels (200 mM NaCl) in nutrient solutions. Contrary, Moreno et al. (2008) on broccoli inflorescences and Garrido et al. (2014) on lettuce observed an increase in total flavonoids by salinity treatment. In contrast to previous works, no significant difference was observed in flavonoids content of broccoli heads by increasing salinity level (López-Berenguer et al., 2009).

5.5 SIZE AND SHAPE

The mean of tomato fruits (cv. H2274-Oturak) height and diameter was decreased linearly with increasing salinity level from 0.26 to 9.7 (dS m^{-1}) (Yurtseven et al., 2005). Also, Magán et al. (2008) and Pascale et al. (2001)

reported that tomato fruits size was decreased by increasing salinity level. In addition, mean of tomato fruits (cv. Zhongzha105) weight decreased under saline condition (100 mM NaCl) compared with the control (Abdel Latef et al., 2011). It has been reported that average of tomato fruits weight, length, and diameter were decreased by increasing salt level (Zahedi et al., 2020). In addition, the average weight and diameter of cherry tomato fruits were decreased by increasing salinity level (Abdelgawad et al., 2019). Di Gioia et al. (2018) found a 24% reduction of broccoli head fresh weight when plants irrigated with saline water (4 dS m^{-1}) in compared with the controls. Garriga et al. (2015) found that average of weight and size of strawberry fruits cv. Camarosa was decreased when salinity level was increased from 0 to 60 mM NaCl. On the other hand, in previous study, no significant difference on strawberry fruit weight and size were observed in "Cucao" or "Bau" cultivars. Contrary to the previous works, it has been reported that fruit shape of watermelon and cucumber was not significantly affected by saline water treatment compared with nonsaline water (Navarro et al., 2010; Colla et al., 2013).

5.6 STORAGE ABILITY AND SHELF-LIFE

As far as we know, rare previous researches have investigated the effect of salinity on storability and shelf-life of vegetable crops. For example, El-Mogy et al. (2018) found that weight loss of cherry tomato fruits during storage at 20°C was not affected by salinity treatments. Contrary, Fallik et al. (2019) reported a decrease in weight loss percentage of bell pepper fruits stored at 7°C when pepper plants irrigated with saline water (2.8 dS m^{-1}) compared with normal water (1.6 dS m^{-1}). Decay percentage of bell pepper during refrigerated storage at 7°C for 14 days was not affected by salinity treatments (Fallik et al., 2019).

5.7 CONCLUSION

In summary (Table 5.1), color decreased with salinity treatment. Carotenoids, anthocyanin, lycopene, starch, TSS, flavonoids, and glucosinolates increased by salinity stress. There was no fixed trend (increased, decreased or not changed) of salinity stress on texture, total sugar, TA, vitamin C, antioxidants, and TPC. Finally, fruit quality including size and shape as well as weight loss during refrigerated storage was decreased or not-changed by salinity treatment.

TABLE 5.1 Effect of Salinity Stress on Chemical Composition and Quality of Fresh Vegetables.

Compound	Increased with salinity stress	Decreased with salinity stress	Increased, decreased, or not changed with salinity stress	Not affected with increasing salinity
Color		×		
Carotenoids	×			
Anthocyanin	×			
Lycopene	×			
Starch	×			
TSS	×			
Flavonoids	×			
Glucosinolates	×			
Texture			×	
Total sugar			×	
TA			×	
Vitamin C			×	
Antioxidants			×	
TPC			×	
Size and shape		×		
Weight loss		×		

KEYWORDS

- *Solanum lycopersicum*
- abiotic stress
- quality
- metabolite
- NaCl
- postharvest

REFERENCES

Abdel Latef, A. A. H.; Chaoxing, H. Effect of Arbuscular Mycorrhizal Fungi on Growth, Mineral Nutrition, Antioxidant Enzymes Activity and Fruit Yield of Tomato Grown Under Salinity Stress. *Sci. Hortic.* **2011,** *127* (3), 228–233.

Abdelaal, K. A. A.; Mazrou, Y. S. A.; Hafez, Y. M. Silicon Foliar Application Mitigates Salt Stress in Sweet Pepper Plants by Enhancing Water Status, Photosynthesis, Antioxidant Enzyme Activity and Fruit Yield. *Plants* **2020,** *9.* DOI: 10.3390/plants9060733.

Abdelgawad, K. F.; El-Mogy, M. M.; Mohamed, M. I. A.; Garchery, C.; Stevens, R. G. Increasing Ascorbic Acid Content and Salinity Tolerance of Cherry Tomato Plants by Suppressed Expression of the Ascorbate Oxidase Gene. *Agronomy* **2019,** *9* DOI: 10.3390/agronomy9020051.

Alzahrani, O.; Abouseadaa, H.; Abdelmoneim, T. K.; Alshehri, M. A.; El-beltagi, H. S.; El-Mogy, M. M.; Atia, M. A. M. Agronomical, Physiological and Molecular Evaluation Reveals Superior Salt-Tolerance in Bread Wheat Through Salt-Induced Priming Approach. *Not. Bot. Horti Agrobot. Cluj-Napoca* **2021,** *49,* 12310. DOI: 10.15835/nbha49212310.

Bonasia, A.; Lazzizera, C.; Elia, A.; Conversa, G. Nutritional, Biophysical and Physiological Characteristics of Wild Rocket Genotypes as Affected by Soilless Cultivation System, Salinity Level of Nutrient Solution and Growing Period. *Front. Plant Sci.* **2017,** *8,* 300. DOI: 10.3389/fpls.2017.00300.

Borghesi, E.; González-Miret, M. L.; Escudero-Gilete, M. L.; Malorgio, F.; Heredia, F. J.; Meléndez-Martínez, A. J. Effects of Salinity Stress on Carotenoids, Anthocyanins, and Color of Diverse Tomato Genotypes. *J. Agric. Food Chem.* **2011,** *59* (21), 11676–11682.

Botía, P.; Navarro, J. M.; Cerdá, A.; Martínez, V. Yield and Fruit Quality of Two Melon Cultivars Irrigated with Saline Water at Different Stages of Development. *Eur. J. Agron.* **2005,** *23,* 243–253. DOI: https://doi.org/10.1016/j.eja.2004.11.003.

Cardeñosa, V.; Medrano, E.; Lorenzo, P.; Sánchez-Guerrero, M. C.; Cuevas, F.; Pradas, I.; Moreno-Rojas, J. M. Effects of Salinity and Nitrogen Supply on the Quality and Health-Related Compounds of Strawberry Fruits (Fragaria × ananassa cv. Primoris). *J. Sci. Food Agric.* **2015,** *95,* 2924–2930. DOI: https://doi.org/10.1002/jsfa.7034.

Chisari, M.; Todaro, A.; Barbagallo, R. N.; Spagna, G. Salinity Effects on Enzymatic Browning and Antioxidant Capacity of Fresh-Cut Baby Romaine Lettuce (Lactuca sativa L. cv. Duende). *Food Chem.* **2010,** *119,* 1502–1506. DOI: https://doi.org/10.1016/j.foodchem.2009.09.033.

Chrysargyris, A.; Papakyriakou, E.; Petropoulos, S. A.; Tzortzakis, N. The Combined and Single Effect of Salinity and Copper Stress on Growth and Quality of Mentha Spicata Plants. *J. Hazard. Mater.* **2019,** *368,* 584–593.

Colla, G.; Rouphael, Y.; Cardarelli, M.; Massa, D.; Salerno, A.; Rea, E. Yield, Fruit Quality and Mineral Composition of Grafted Melon Plants Grown Under Saline Conditions. *J. Hortic. Sci. Biotechnol.* **2006,** *81,* 146–152. DOI: 10.1080/14620316.2006.11512041.

Colla, G.; Rouphael, Y.; Rea, E.; Cardarelli, M. Grafting Cucumber Plants Enhance Tolerance to Sodium Chloride and Sulfate Salinization. *Sci. Hortic.* **2012,** *135,* 177–185. DOI: https://doi.org/10.1016/j.scienta.2011.11.023.

Colla, G.; Rouphael, Y.; Jawad, R.; Kumar, P.; Rea, E.; Cardarelli, M. The Effectiveness of Grafting to Improve NaCl and CaCl2 Tolerance in Cucumber. *Sci. Hortic.* **2013,** *164,* 380–391. DOI: https://doi.org/10.1016/j.scienta.2013.09.023.

Di Gioia, F.; Rosskopf, E. N.; Leonardi, C.; Giuffrida, F. Effects of Application Timing of Saline Irrigation Water on Broccoli Production and Quality. *Agric. Water Manag.* **2018,** *203,* 97–104.

El-Mogy, M. M.; Garchery, C.; Stevens, R. Irrigation with Salt Water Affects Growth, Yield, Fruit Quality, Storability and Marker-Gene Expression in Cherry Tomato. *Acta Agric Scand. Sect. B Soil Plant Sci.* **2018,** *68,* 727–737. DOI: 10.1080/09064710.2018.1473482.

Fallik, E.; Alkalai-Tuvia, S.; Chalupowicz, D.; Zaaroor-Presman, M.; Offenbach, R.; Cohen, S.; Tripler, E. How Water Quality and Quantity Affect Pepper Yield and Postharvest Quality. *Horticulturae* **2019,** *5.* DOI: 10.3390/horticulturae5010004.

Fallovo, C.; Rouphael, Y.; Rea, E.; Battistelli, A.; Colla, G. Nutrient Solution Concentration and Growing Season Affect Yield and Quality of Lactuca sativa L. var. Acephala in Floating Raft Culture. *J. Sci. Food Agric.* **2009,** *89,* 1682–1689. DOI: https://doi.org/10.1002/jsfa.3641.

Galli, V.; da Silva Messias, R.; Perin, E. C.; Borowski, J. M.; Bamberg, A. L.; Rombaldi, C. V. Mild Salt Stress Improves Strawberry Fruit Quality. *LWT* **2016,** *73,* 693–699. DOI: https://doi.org/10.1016/j.lwt.2016.07.001.

Garrido, Y.; Tudela, J. A.; Marín, A.; Mestre, T.; Martínez, V.; Gil, M. I. Physiological, Phytochemical and Structural Changes of Multi-Leaf Lettuce Caused by Salt Stress. *J. Sci. Food Agric.* **2014,** *94,* 1592–1599. DOI: 10.1002/jsfa.6462.

Garriga, M.; Muñoz, C. A.; Caligari, P. D. S.; Retamales, J. B. Effect of Salt Stress on Genotypes of Commercial (Fragaria × ananassa) and Chilean Strawberry (F. chiloensis). *Sci. Hortic.* **2015,** *195,* 37–47. DOI: https://doi.org/10.1016/j.scienta.2015.08.036.

Giuffrida, F.; Graziani, G.; Fogliano, V.; Scuderi, D.; Romano, D.; Leonardi, C. Effects of Nutrient and NaCl Salinity on Growth, Yield, Quality and Composition of Pepper Grown in Soilless Closed System. *J. Plant Nutr.* **2014,** *37,* 1455–1474. DOI: 10.1080/01904167.2014.881874.

Giuffrida, F.; Cassaniti, C.; Malvuccio, A.; Leonardi, C. Effects of Salt Stress Imposed During Two Growth Phases on Cauliflower Production and Quality. *J. Sci. Food Agric.* **2017,** *97,* 1552–1560. DOI: 10.1002/jsfa.7900.

Hand, M. J.; Taffuo, V. D.; Nouck, A. E.; Nyemene, K. P. J.; Tonfack, B.; Meguekam, T. L.; Youmbi, E. Effects of Salt Stress on Plant Growth, Nutrient Partitioning, Chlorophyll Content, Leaf Relative Water Content, Accumulation of Osmolytes and Antioxidant Compounds in Pepper (Capsicum annuum L.) Cultivars. *Not. Bot. Horti Agrobot. Cluj-Napoca* **2017,** *45.* DOI: 10.15835/nbha45210928.

Hu, K.; Zhu, Z. Effects of Different Concentrations of Sodium Chloride on Plant Growth and Glucosinolate Content and Composition in Pakchoi. *Afr. J. Biotech.* **2010,** *9,* 4428–4433. http://dx.doi.org/10.5897/AJB10.672.

Keutgen, A.; Pawelzik, E. Modifications of Taste-Relevant Compounds in Strawberry Fruit Under NaCl Salinity. *Food Chem.* **2007,** *105,* 1487–1494. DOI: https://doi.org/10.1016/j.foodchem.2007.05.033.

Keutgen, A. J.; Pawelzik, E. Quality and Nutritional Value of Strawberry Fruit Under Long Term Salt Stress. *Food Chem.* **2008,** *107,* 1413–1420. DOI: https://doi.org/10.1016/j.foodchem.2007.09.071.

Kim, H. J.; Fonseca, J. M.; Choi, J. H.; Kubota, C.; Kwon, D. Y. Salt in Irrigation Water Affects the Nutritional and Visual Properties of Romaine Lettuce (Lactuca sativa L.). *J. Agric. Food Chem.* **2008,** *56,* 3772–3776. DOI: 10.1021/jf0733719.

Liu, F. -Y.; Li, K. -T.; Yang, W. -J. Differential Responses to Short-Term Salinity Stress of Heat-Tolerant Cherry Tomato Cultivars Grown at High Temperatures. *Hortic. Environ. Biotechnol.* **2014**, *55*, 79–90. DOI: 10.1007/s13580-014-0127-1.

López-Berenguer, C.; Martínez-Ballesta Mdel, C.; Moreno, D. A.; Carvajal, M.; García-Viguera, C. Growing Hardier Crops for Better Health: Salinity Tolerance and the Nutritional Value of Broccoli. *J. Agric. Food Chem.* **2009**, *57*, 572–578. DOI: 10.1021/jf802994p.

Magán, J. J.; Gallardo, M.; Thompson, R. B.; Lorenzo, P., Effects of Salinity on Fruit Yield and Quality of Tomato Grown in Soil-Less Culture in Greenhouses in Mediterranean Climatic Conditions. *Agric. Water Manag.* **2008**, *95* (9), 1041–1055.

Marín, A.; Rubio, J. S.; Martínez, V.; Gil, M. I. Antioxidant Compounds in Green and Red Peppers as Affected by Irrigation Frequency, Salinity and Nutrient Solution Composition. *J. Sci. Food Agric.* **2009**, *89*, 1352–1359. DOI: https://doi.org/10.1002/jsfa.3594.

Moreno, D. A.; López-Berenguer, C.; Martínez-Ballesta, M. C.; Carvajal, M.; García-Viguera, C. Basis for the New Challenges of Growing Broccoli for Health in Hydroponics. *J. Sci. Food Agric.* **2008**, *88*, 1472–1481. DOI: https://doi.org/10.1002/jsfa.3244.

Navarro, J. M.; Garrido, C.; Carvajal, M.; Martinez, V. Yield and Fruit Quality of Pepper Plants Under Sulphate and Chloride Salinity. *J. Hortic. Sci. Biotechnol.* **2002**, *77*, 52–57. DOI: 10.1080/14620316.2002.11511456.

Navarro, J. M.; Flores, P.; Garrido, C.; Martinez, V. Changes in the Contents of Antioxidant Compounds in Pepper Fruits at Different Ripening Stages, as Affected by Salinity. *Food Chem.* **2006**, *96*, 66–73. DOI: https://doi.org/10.1016/j.foodchem.2005.01.057.

Navarro, J. M.; Garrido, C.; Flores, P.; Martinez, V. The Effect of Salinity on Yield and Fruit Quality of Pepper Grown in Perlite. *Span. J. Agric. Res.* **2010**, *8*, 142–150.

Neocleous, D.; Koukounaras, A.; Siomos, A. S.; Vasilakakis, M. Changes in Photosynthesis, Yield, and Quality of Baby Lettuce Under Salinity Stress. *Mdrsjrns* **2014**, *16*, 1335–1343.

Pascale, S. D.; Maggio, A.; Fogliano, V.; Ambrosino, P.; Ritieni, A. Irrigation with Saline Water Improves Carotenoids Content and Antioxidant Activity of Tomato. *J. Hortic. Sci. Biotechnol.* **2001**, *76*, 447–453. DOI: 10.1080/14620316.2001.11511392.

Pérez-López, U.; Miranda-Apodaca, J.; Lacuesta, M.; Mena-Petite, A.; Muñoz-Rueda, A. Growth and Nutritional Quality Improvement in Two Differently Pigmented Lettuce Cultivars Grown Under Elevated CO2 and/or Salinity. *Sci. Hortic.* **2015**, *195*, 56–66. DOI: https://doi.org/10.1016/j.scienta.2015.08.034.

Rezazadeh, A.; Ghasemnezh, A.; Barani, M.; Telmadarre, T. Effect of Salinity on Phenolic Composition and Antioxidant Activity of Artichoke (*Cynara scolymus* L.) Leaves. *Res. J. Med. Plant* **2012**, *6*, 245–252.

Rivero, R. M.; Mestre, T. C.; Mittler, R. O. N.; Rubio, F.; Garcia-Sanchez, F.; Martinez, V., The Combined Effect of Salinity and Heat Reveals a Specific Physiological, Biochemical and Molecular Response in Tomato Plants. *Plant Cell Environ.* **2014**, *37* (5), 1059–1073.

Rouphael, Y.; Cardarelli, M.; Rea, E.; Colla, G. Improving Melon and Cucumber Photosynthetic Activity, Mineral Composition, and Growth Performance Under Salinity Stress by Grafting onto Cucurbita Hybrid Rootstocks. *Photosynthetica* **2012**, *50*, 180–188. DOI: 10.1007/s11099-012-0002-1.

Rouphael, Y.; Colla, G. The Influence of Drip Irrigation or Subirrigation on Zucchini Squash Grown in Closed-Loop Substrate Culture with High and Low Nutrient Solution Concentrations. *HortScience* **2009**, *44*, 306–311.

Rouphael, Y.; De Micco, V.; Arena, C.; Raimondi, G.; Colla, G.; De Pascale, S. Effect of Ecklonia Maxima Seaweed Extract on Yield, Mineral Composition, Gas Exchange, and

Leaf Anatomy of Zucchini Squash Grown Under Saline Conditions. *J. Appl. Phycol.* **2017,** *29,* 459–470. DOI: 10.1007/s10811-016-0937-x.

Sakamoto, K.; Kogi, M.; Yanagisawa, T. Effects of Salinity and Nutrients in Seawater on Hydroponic Culture of Red Leaf Lettuce. *Environ. Control Biol.* **2014,** *52,* 189–195. DOI: 10.2525/ecb.52.189.

Savvas, D.; Lenz, F. Effects of NaCl or Nutrient-Induced Salinity on Growth, Yield, and Composition of Eggplants Grown in Rockwool. *Sci. Hortic.* **2000,** *84,* 37–47. DOI: https://doi.org/10.1016/S0304-4238(99)00117-X.

Tingwu, L.; Juan, X.; Guangyong, L.; Jianhua, M.; Jianping, W.; Zhizhong, L.; Jianguo, Z. Effect of Drip Irrigation with Saline Water on Water Use Efficiency and Quality of Watermelons. *Water Resour. Manag.* **2003,** *17,* 395–408. DOI: 10.1023/B:WARM.00000 04917.16604.2c.

Yan, S.; Gao, Y.; Tian, M.; Tian, Y.; Li, J., Comprehensive Evaluation of Effects of Various Carbon-Rich Amendments on Tomato Production Under Continuous Saline Water Irrigation: Overall Soil Quality, Plant Nutrient Uptake, Crop Yields and Fruit Quality. *Agric. Water Manag.* **2021,** *255,* 106995.

Yurtseven, E.; Kesmez, G. D.; Ünlükara, A., The Effects of Water Salinity and Potassium Levels on Yield, Fruit Quality and Water Consumption of a Native Central Anatolian Tomato Species (Lycopersicon esculantum). *Agric. Water Manag.* **2005,** *78* (1), 128–135.

Zaghdoud, C.; Alcaraz-López, C.; Mota-Cadenas, C.; Martínez-Ballesta, M. D. C.; Moreno, D. A.; Ferchichi, A.; Carvajal, M. Differential Responses of Two Broccoli (Brassica oleracea L. var Italica) Cultivars to Salinity and Nutritional Quality Improvement. *Sci. World J.* **2012,** *2012,* 291435. DOI: 10.1100/2012/291435.

Zahedi, S. M.; Hosseini, M. S.; Abadía, J.; Marjani, M. Melatonin Foliar Sprays Elicit Salinity Stress Tolerance and Enhance Fruit Yield and Quality in Strawberry (Fragaria × ananassa Duch.). *Plant Physiol. Biochem.* **2020,** *149,* 313–323.

Zipelevish, E.; Grinberge, A.; Amar, S.; Gilbo, Y.; Kafkafi, U. Eggplant Dry Matter Composition Fruit Yield and Quality as Affected by Phosphate and Total Salinity Caused by Potassium Fertilizers in the Irrigation Solution. *J. Plant Nutr.* **2000,** *23,* 431–442. DOI: 10.1080/01904160009382030.

CHAPTER 6

Crop Improvement in Deserts

ADITYA BANERJEE and ARYADEEP ROYCHOUDHURY

Post Graduate Department of Biotechnology, St. Xavier's College (Autonomous), Kolkata, West Bengal, India

ABSTRACT

Desert represents a large stretch of barren land devoid of water. Such hyperarid conditions are usually unsuitable for crop production since normal glycophytes succumb to such extreme desiccation. Deserts can have both extremes of temperature, high salinity, radiation stress, and even nutrient scarcity. Thus, these stretches of land cannot be used for agricultural pursuits and hence are often considered as wastelands for the humankind. However, the immense growth of the population has urged the biotechnologists and breeders to think in an alternative way regarding the utilization of these desert lands for growing adaptable and genetically modified crop plants. Though this field of study has extremely few reports, the chapter aims to briefly state the problems of desert areas in agricultural expansion and also concisely discusses the developments in crop improvement in these lands.

6.1 INTRODUCTION

Crop improvement in desert lands is a need of the hour. Such conversion of deserts (also referred to as "the threat to biodiversity") into cultivable and

Crop Sustainability and Intellectual Property Rights. Soumya Mukherjee, Piyali Mukherjee & Tariq Aftab (Eds)

arable landscapes can be a potential solution to meet the hunger demand and food security of the ever-growing human population (Pereira and Gama, 2010; Clery, 2011). Recent advances in desert farming strategies have been observed in Egypt where the desert farmland has extended to about 40% in 2017 (Reuters, 2007). However, desert farming requires gallons of water for crop development since this strategy is solely dependent on irrigation and the availability of water resources (Barot et al., 2007; Acosta-Martinez et al., 2008). This is the major difficulty of promoting agriculture in desert lands since it might not be always possible to siphon such huge amounts of water to the remote barren stretches.

Desert presents a unique ecological condition where organisms are exposed to extremes of temperature, desiccation or drought, high soil salinity, low nutrient level, and severe UV rays (Koberl et al., 2011). The United Nations Conference on Environment and Development organized at Rio de Janeiro in 1992 cited that 'Desertification is land degradation in arid, semiarid, and dry subhumid areas resulting from various factors, including climatic variations and human activities'. Uncontrolled felling of trees, variations in climate, unplanned grazing practices, and the over-exploitation of water and land resources are the chief reasons behind desertification (Gamri, 2004). United Nations Conference on Environment and Development had adopted an integrated approach for promoting sustainable development in desert areas. The conference had also directed the UN General Assembly to advertise a convention to manage desertification (Gamri, 2004). Thus, serious approaches are being undertaken at the international scale to actively combat desertification.

Plants are sessile organisms that are invariably exposed to a multitude of suboptimal conditions. However, they remain attached to the substratum and combat these abiotic factors by exhibiting physiological plasticity and alterations in the biochemical and molecular signaling (Banerjee et al., 2016, 2018; Banerjee and Roychoudhury, 2019a, b, c; Banerjee et al., 2019a, b). The xerophytes adapted to grow in desert ecosystems have several anatomical and physiological modifications that they have gained through prolonged evolutionary processes. These adaptations or modifications actually aid the overall system to tolerate the multiple abiotic stress-like conditions witnessed on the desert soil. Crop plants are usually glycophytes, which thrive under well-watered and optimal climatic conditions. A subtle change in the climatic factors severely affect the yield and production of crop species (Banerjee and Roychoudhury, 2019d, e, f; Das et al., 2019). In this chapter, we have concisely focussed on the prospects and possibilities of crop improvement in the desert. Our discussion involves the illustration of xerophytic adaptation at the physiological, biochemical, and molecular level and potential avenues and tools to incorporate these modifications in desert-susceptible crop species.

6.2 THE CONSTRAINTS IN DESERT AGRICULTURE

Desert agriculture is a tough technology that faces several challenges. Limitations in the availability of data regarding the landscape, climatic factors, and water availability hinder the development of this technology in vast stretches of arid land (Agnew and Warren, 1996). Another retardant of agricultural development in deserts is that the general population residing in these areas is almost always inflicted with poverty and lack of food security (Soares, 2001). The paucity of moisture content due to variable precipitation along with limited soil resources are observed across the entire landscape. Usually, desert soil contains high levels of salt and exhibit unbalanced texture and shallowness (World Meteorological Organisation, WMO, 1990). Due to the barrenness of the landscape, few farmers actually own their lands and due to the prevailing poverty, few people actually can afford to spend the investment incentives to pursue agriculture. Lack of community participation is also responsible for the low expansion of desert agriculture (Gamri, 2004).

6.3 THE PROBABLE PROSPECTS OF DESERT AGRICULTURE

The abundance of sunlight and temperature regimes in desert lands has led to the proposal of the potential agriculture of crops like pearl millet, sorghum, and wheat (Sinha, 1996). Development programs of crops like date palm, pomegranate, and guava along with the pastoral crop *Cenchrus ciliaris* have been undertaken in the Thar deserts of India. Crops acting as the raw materials of pharmaceutical products, perfumes, soaps, and petroleum industry are also being cultivated in these arid areas (Sinha, 1996). The arid lands of Sudan have been cultivated with plants like *Cassia senna*, mesquite, date palm, and *Acacia senegal* known for their medicinal, pastoral, nutritive, and industrial properties, respectively (Vogt, 1995). Indigenous knowledge among the farmers is a very essential tool for extending desert agriculture (Gibbon and Pain, 1988; Reij et al., 1996).

6.4   ECOPHYSIOLOGY OF DESERT PLANTS

The ecological study dedicated to the investigation of physiotypes in particular habitats is popularly referred to as an ecophysiological study. Ecophysiology involves the culmination of function-dynamics within the habitats or ecosystems (Luttge and Scarano, 2004). This extensively increases the complexity of the study, since ecophysiology illustrates the overall conditions which surround the organism throughout its existence (Luttge and Scarano, 2004). The desert lands harbor Crassulacean Acid Metabolism plants, halophytes with high chloride concentrations in their leaf sap, cacti, and terrestrial bromeliads (Medina et al., 1989). The terrestrial bromeliads possess superficial root system, which actively absorbs the surface water during rainfall. Richardson et al. (2014) identified two blackbrush (*Coleogyne ramosissima*) shrubs found in the warm and cold deserts of Mojave and Colorado Plateau in Northern America. The warm-adapted ecotype was more abundant in the hot deserts, while the cold-adapted species became the dominant ecotype as the plateau entered higher latitudes and elevations (Richardson et al., 2014). Desert areas around the Great Salt Lake of Utah harbor a gradient of vegetation. The main communities dominating in these areas with increasing salinity are *Artemisia tridentata, Atriplex-Eurotia, Sarcobatus vermiculatus*, and *Salicornia-Allenrolfea* (Goodman and Caldwell, 1971). Chunyu et al. (2019) reported the abundance of *Phragmites australis* in the transition area where the hydrophytes were gradually replaced by the xerophytes.

6.5   PREVALENT ABIOTIC STRESSES IN THE DESERT AND THEIR EFFECTS ON GENERAL PLANT PHYSIOLOGY

Extremes of temperature, drought, salinity, and UV radiation are the most prevalent types of abiotic stress in deserts. These stresses exclusively inhibit the physiological development and reproductive processes in crops, thus severely affecting crop yield and grain production (Banerjee and Roychoudhury, 2018a, b, c). The reproductive phases linked with grain development are most vulnerable to temperature fluctuations (Sehgal et al., 2017). The temperature drastically fluctuates in the desert lands. Temperatures rise to about 50°C under the scorching sun and fall near to 15°C during the night. Prasad and Djanaguiraman (2014) showed that increase in temperature above 24°C for 2–5 days negatively regulated spikelet fertility in wheat plants. Temperature higher than 32°C impaired reproductive development and yield in pea and chickpea seedlings (Sehgal et al., 2018). Both drought as well as heat stress negatively affected the sucrose and starch metabolism in pollen grains and promoted their sterility (Sita et al., 2017, 2018). The vegetative growth is limited due to a reduction in the relative water content of leaves during drought stress in cereals and legumes (Sehgal et al., 2018; Banerjee and Roychoudhury, 2018d, e, f). The susceptible crop species experience inability to promote stomatal closure during drought or severe osmotic stress resulting from salinity stress. As a result, these species succumb to stress due to unregulated water loss via transpiration (Anjum et al., 2011). Drought stress markedly decreased the Fe, Zn, P, and N levels in common bean seeds (Ghanbari et al., 2013a). Ghanbari et al. (2013b) reported a steady decline in seed N and protein content in the seeds of white, red, and chitti bean cultivars exposed to drought. Zhang et al. (2014) performed proteomic investigations in drought-stressed wheat seeds and observed inhibited amino acid incorporation into protein chains.

Salinity is one of the most prevalent types of abiotic stress and the shallow desert soil contains toxic levels of salts, which are detrimental for the growth of crops. Salt-stressed crop plants exhibit perturbed cellular metabolism due to inhibited CO_2 assimilation (Khan et al., 2017). The metabolome of salt-stressed crop plants is represented by inhibited enzyme activity, phytohormonal dysfunction, oxidative damage, and uncontrolled necrosis (Farooq et al., 2017; Jha et al., 2019). Excessive ion toxicity due to unregulated Na^+ and Cl^- import within the cells rapidly alters the K^+/Na^+ ratio and diminishes the reducing capacity of the cells (Manchanda and Garg, 2008; Roychoudhury et al., 2008; Flowers et al., 2010; Kaashyap et al., 2017).

Intensive solar radiation can be hazardous for the normal physiological development in plants since the sunrays contain UV-A (320–400 nm), UV-B (280–320 nm), and UV-C (< 280 nm) rays. The depletion of the protective ozone layer has increased the reach of UV-B radiation to the earth's surface (Arróniz-Crespo et al., 2004). Since the desert experiences long stretch of scorching sunlight, the plants growing in these areas have to accommodate intensive UV-B radiation. Secondary growth in plants aids in the formation of lignin, which is a polyphenolic compound. Lignin protects the plants from longer wavelengths of UV rays. However, the increased reach of the shorter UV-B rays to the earth's surface has steadily increased the radiation amplification factor (Bornman, 1991). Gill et al. (2015) reported that UV-B radiation damaged the synchronization between the water-splitting complex and phytochrome 680 (P680) located on the oxidizing side of photosystem II (PSII). The chloroplast structural protein D1 was found to be degraded upon exposure to UV-B rays (Greenberg et al., 1989). UV-B irradiation destabilizes the genomic template by inducing the formation of cyclobutane pyrimidine dimers and pyrimidine (6-4) pyrimidinone dimers (6-4 PPs). This creates steric hindrance in the modified areas in the DNA and renders them inaccessible to polymerases, thus resulting in the inhibition of replication and downstream cellular metabolism (Gill et al., 2012). The desert lands also experience a severely high intensity of light, which can also be detrimental for optimal plant growth (Banerjee and Roychoudhury, 2016). High light intensity reduced the accumulation of PSI polypeptides like PsAA, PsaB, and PsaC (Jiao et al., 2004). Eickmeier et al. (1993) reported a significant decrease in the intrinsic fluorescence yield and photosynthetic efficiency in the desert resurrection plant *Selaginella lepidophylla* exposed to high proton flux density. However, the most interesting fact is that due to evolved adaptation to desert climates, these resurrection plants rapidly recovered from the high proton flux density-induced injuries once they were placed under the low intensity of light (Eickmeier et al., 1993).

6.6 PHYSIOLOGICAL AND MOLECULAR MECHANISM OF TOLERANCE IN EXTREMOPHILES

Xerophytes are extremophiles with the molecular adaptations encoded within their genomes and/ or transcriptomes (Oh et al., 2012). The advent of microarray and next-generation sequencing technologies has helped in revealing the molecular mechanism behind the tolerance of the xerophytes (Dinakar and Bartels, 2013). *Thellungiella parvula* and *Eutrema salsugineum* are robust species belonging to the Brassicaceae family and they can tolerate multiple abiotic stresses like salinity, cold, freezing, heat, and drought (Lee et al., 2012; Koch and German, 2013). It has been reported that *E. salsugineum* plants can also thrive under poor nitrogen content in the soil and are also resistant to high boron and low phosphate concentrations (Kant et al., 2008; Lamdan et al., 2012; Velasco et al., 2016). Desiccation stress triggers oxidative damage in plants due to the production of excessive reactive oxygen species (ROS) like superoxide, hydroxyl radicals, and hydrogen peroxide (H_2O_2) from the chloroplasts and the mitochondria (Banerjee and Roychoudhury, 2016, 2017, 2018g; Banerjee et al., 2019c). The resurrection plants that are highly adapted to extreme desert conditions actually can switch between a normal mode and a stress mode. During severe oxidative stress, the stress mode is activated, leading to completely suppressed photosynthesis. This significantly lowers the extent of ROS production from the chloroplasts. Thus, under situations of severe oxidative stress, when glycophytes tremendously accumulate ROS within their cells and succumb to the oxidative load, the resurrection plants endure the tough phase by going back to dormancy and controlling ROS generation (Farrant et al., 2003).

Deep transcriptome analysis of the resurrection plants like *Craterostigma plantageneum* and *Haberlea rhodopensis* revealed that these species gradually passed through different metabolic and molecular alterations on experiencing desiccation stress (Rodriguez et al., 2010; Gechev et al., 2013; Farrant et al., 2015; Giarola et al., 2017). Abscisic acid (ABA) is the universal stress hormone in plants that regulates almost every type of abiotic stress (Roychoudhury and Paul, 2012; Roychoudhury and Banerjee, 2017). It was found that after the suppression of the photosynthetic pathway during the early dehydration stage, the ABA-mediated responses to desiccation are heightened. This results in the accumulation of compatible solutes, osmolytes, and antioxidants (Roychoudhury and Chakraborty, 2013). Production of sucrose and oligosaccharides is stimulated, since these biomolecules not only act as storage food resources, but are also potent osmolytes that aid in proper maintenance of the cellular osmotic pressure (Paul et al., 2018). Interestingly as the desiccation stress is prolonged, the proteins associated with abiotic stress signaling, and classical stress adaptation like the early light-inducible proteins, late embryogenesis abundant (LEA) proteins, heat shock proteins, etc. are translated (Roychoudhury and Nayek, 2014; Banerjee and Roychoudhury, 2016; Costa et al., 2017). The signal transduction under such conditions is mediated by several groups of transcription factors like WRKYs, NACs, NF-Ys, and heat shock factors (Banerjee and Roychoudhury, 2015; Roychoudhury et al., 2015).

Soliman et al. (2019) observed the presence of shaggy-like trichomes in the *Cleome amblyocarpa* plants thriving in the arid deserts of United Arab Emirates (UAE). It was found that these plants adapted to desert conditions by releasing lipophilic metabolites from the glandular trichomes and idioblasts of the shoots and roots, respectively (Soliman et al., 2019). These metabolites mainly included hydrocarbons and terpenoids. The *C. amblyocarpa* plants also contained wax coatings over leaf surfaces in order to minimize water loss (Soliman et al., 2019). RNA-Seq study of the salt hyperaccumulating desert xerohalophyte *Zygophyllum xanthoxylum* revealed metabolic shunting towards neoglucogenesis and induction of an array of genes encoding ion transporters (Chai et al., 2019). The ion transporters actually enabled better uptake of ions and nutrients from the soil during extremely stressful conditions. Genes responsible for phyto-hormones were also distinctly up-regulated in the salt-stressed plants compared to salt-stressed *Arabidopsis* seedlings (Chai et al., 2019). Fan et al. (2018) identified novel transcripts involved in mediating tolerance to salt, osmotic shock, cold, UV, and high light stress in the extremophyte *Haloxylon ammodendron* using RNA-Seq analysis.

6.7 POTENTIAL CROP IMPROVEMENT STRATEGIES IN THE DESERT: RECENT PROSPECTS

Mining of novel genes from extremophytes can be a useful strategy to develop abiotic stress tolerance in susceptible crop species. Some strate-gies to develop desert climate-resilient crops have been illustrated (Figure 1). Muvunyi et al. (2018) identified 44 putative *LEA* genes in the genome of the desert xerophyte *Cleistogenes songorica*. In silico analyses revealed phylogenetic similarity between the identified *LEA* genes in *C. songorica* and the stress-responsive dehydrins in Arabidopsis. Thus cloning of these genes and generation of transgenic crop lines harnessing these novel *LEA* genes might lead to the development of plants tolerant to the desert climate (Muvunyi et al., 2018). Mubaiwa et al. (2018) reported that Bambara groundnut (*Vigna subterranean*) can be cultivated in the semiarid regions in sub-Saharan countries as a nutritious and value-added food crop. Ahmad et al. (2018) reported that pretreatment of maize seeds and even exogenous foliar treatment of maize seedlings with uniconazole improved the rate of seed filling and seed quality under semiarid conditions. Tolerance to the semiarid soil and climatic conditions was mainly determined by the uniconazole-mediated regulation of endogenous phytohormones like ABA, zeatin, zeatin riboside, and gibberellic acid (Ahmad et al., 2018). Zhang et al. (2018) expressed the urgent need for the domestication of sand rice (*Agriophyllum squarrosum*) in the advent of climate change. Modification of the weedy traits like the formation of dense trichomes was found to be necessary to ensure the popular acceptance of sand rice. Isolation of the *Agriophyllum squarrosum* trichomeless mutant (*astcl1*) showed that the mutant was controlled by a recessive locus. Transcriptomic analyses revealed that this locus regulated a small subset of genes and was also responsible for the formation of trichome initiation and synthesis of cuticle in sand rice (Zhang et al., 2018). Ahmed et al. (2018) performed simulation modeling to study the effects of climate change on maize cultivated in the semiarid soil of Punjab. The modeling study suggested that sowing a different variety of maize hybrids at different time points of the year ensure proper yield and production (Ahmed et al., 2018). Another field-based study suggested the suitability of the sorghum cultivar Jawar-2011 to be cultivated in the semiarid conditions of Pakistan and produce good yield and biomass (Hassan et al., 2018). Use of Landsat-8 imageries showed the spatiotemporal differences in the maize and carrot cultivation and the utilization of water by these crops in the arid climate of Saudi Arabia (Madugundu et al., 2018).

Shamaya et al. (2017) performed a genetic study regarding the basis of salt tolerance in the landraces of Afghani durum wheat. The desiccation-tolerant desert moss *Syntrichia caninervis* was screened for salt-tolerant traits (Liang et al., 2017). The authors identified and characterized a novel A-5 type *desiccation responsive element binding* factor gene, *ScDREB8*. Overexpression of this gene in *Arabidopsis* significantly improved the endogenous ability of the system to scavenge ROS and also generated tolerance to salt stress (Liang et al., 2017). This gene can be introgressed into important food crops to generate salt-resilient plants that can be grown in the desert. Analysis of nutrient utilization, growth, and yield responses of cotton seedlings in the Inner Mongolia west desert area revealed that the optimum sowing date for cotton planting was around 30th of April each year (Zhang et al., 2017). Very high deviations from this date in the case of cotton seed sowing resulted in compromised growth and yield in the desert areas (Zhang et al., 2017). Bannayan et al. (2016) reported the delay in sowing date to be advantageous for higher maize yield under adverse climatic conditions. Accumulation of salt dust during storms in the arid Ebinur Basin in Northwest China drastically affected the photosynthetic capacity and the associated physiological processes in cotton leaves (Abuduwaili et al., 2015). This led to severe ROS accumulation within the stressed cells of the seedlings, leading to necrosis and ultimately large-scale agricultural loss (Abuduwaili et al., 2015). Thus, cleansing of crops after desert sand storms is essential for the proper physiological growth of the plants and hence efficient measures have to be adopted to address the same.

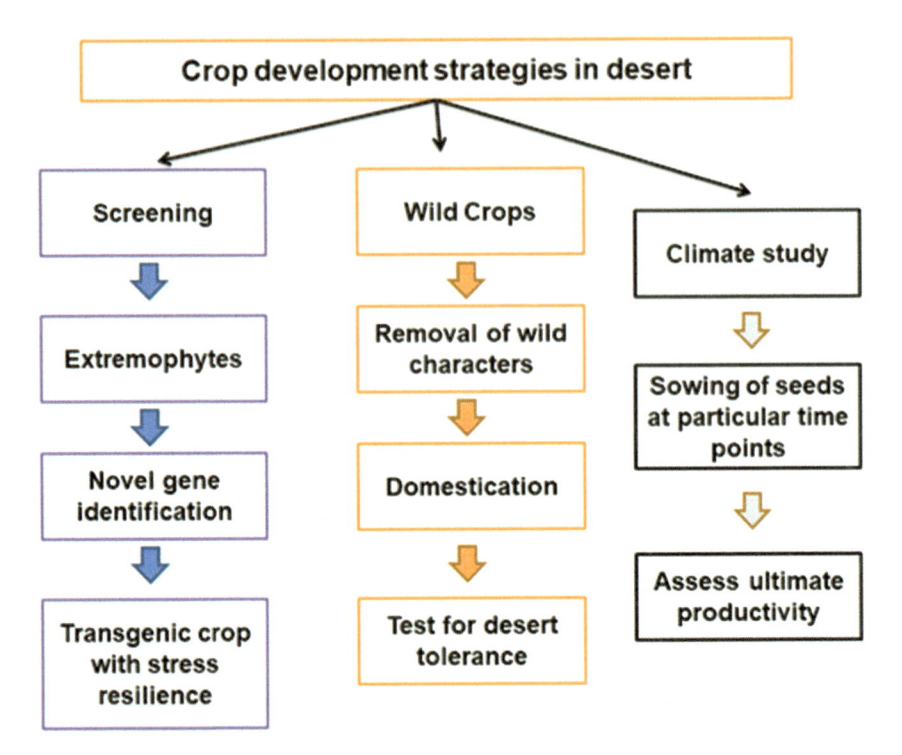

FIGURE 6.1   Strategies to accelerate crop development in desert. The first strategy involves screening of novel stress-responsive genes from xerophytes, extremophytes or resurrection plants and overexpress them in crops to promote tolerance to arid climate. The second flowchart shows the domestication of wild relatives of crop plants to be a potential approach of promoting desert cultivation. The third strategy involves agronomic knowledge and field trials to identify the particular time of the year in which seeds should be sown to receive the maximum yield under semiarid climatic conditions.

6.8   CONCLUSION

Plants acclimatized to thrive in the desert soil and arid climate are exposed to multiple abiotic stresses especially desiccation, salinity, drought, high temperature, UV-B irradiation, and high-intensity light stress. Irrigation is also a tremendous problem in desert tracts, since huge quantity of water usually cannot be traced in these remote lands. In spite of so many environmental stressors, they modulate their physiological responses and exhibit tremendous plasticity and adaptive capacity. Especially resurrection plants can remain like naive, dormant tissues for prolonged periods of time and return to their active state on exposure to few drops of water. Xerophytes and extremophytes achieve such robust physiological adaptation due to anatomical and molecular modifications that have been already discussed. Recent transcriptomic screenings have revealed that these species express novel stress-responsive genes. Introgression of these novel candidate genes into susceptible crop species can accelerate the program of crop development in the desert. Further studies have shown that the time of seed sowing is also important for a particular crop species to grow and produce an appreciable yield under semiarid conditions. Utilization of these strategies has benefitted the poor population residing in the desert areas to some extent, since these people usually lack food security.

6.9 FUTURE PERSPECTIVES

Crop improvement in the desert is a program undertaken for a novel cause of prevailing food security and social justice among the poor population residing in these areas. This field is gradually advancing and includes several potential prospects. Future perspectives involve more rapid experimentation and screening of xerophytes to identify more candidate genes, which might be able to confer multiple abiotic stress tolerance in glycophytes. The epigenomic basis of arid climate tolerance in the extremophytes is also unknown and remains to be elucidated. This would actually illustrate the actual molecular mechanism and upstream regulation of the novel genes in these native species. Crop development should also focus on the domestication of the wild relatives of the crop plants. Such wild species are physiologically more robust and might be better adapted to suboptimal conditions compared to the domesticated cultivars.

ACKNOWLEDGMENTS

Financial assistance from Science and Engineering Research Board, Government of India through the grant [EMR/2016/004799] and Department of Higher Education, Science and Technology and Biotechnology, Government of West Bengal, through the grant [264(Sanc.)/ST/P/S&T/1G-80/2017] to Dr. Aryadeep Roychoudhury is gratefully acknowledged. Mr. Aditya Banerjee is thankful to the University Grants Commission, Government of India for providing Senior Research Fellowship in the course of this work.

KEYWORDS

- desert
- arid climate
- abiotic stress
- resurrection plants
- crop improvement

REFERENCES

Abuduwaili, J.; Zhaoyong, Z.; Feng, G. J.; Dong, W. L. The Disastrous Effects of Salt Dust Deposition on Cotton Leaf Photosynthesis and the Cell Physiological Properties in the Ebinur Basin in Northwest China. *PLoS One* **2015**, *10*, e0124546.

Acosta-Martınez, V.; Dowd, S.; Sun, Y.; Allen, V. Tag-Encoded Pyrosequencing Analysis of Bacterial Diversity in a Single Soil Type as Affected by Management and Land Use. *Soil. Biol. Biochem.* **2008**, *40*, 2762–2770.

Agnew, G.; Warren, A. A Framework for Tackling Drought and Land Degradation. *J. Arid. Environ.* **1996**, *33*, 309–320.

Ahmad, I.; Kamran, M.; Ali, S.; Cai, T.; Bilegjargal, B.; Liu, T.; Han, Q. Seed Filling in Maize and Hormones Crosstalk Regulated by Exogenous Application of Uniconazole in Semiarid Regions. *Environ. Sci. Pollut. Res. Int.* **2018**, *25*, 33225–33239.

Ahmed, I.; Ur Rahman, M. H.; Ahmed, S.; Hussain, J.; Ullah, A.; Judge, J. Assessing the Impact of Climate Variability on Maize Using Simulation Modeling Under Semi-Arid Environment of Punjab, Pakistan. *Environ. Sci. Pollut. Res. Int.* **2018**, *25*, 28413–28430.

Anjum, S. A.., Xie, X. Y.; Wang, L. C.; Saleem, M. F.; Man, C.; Lei, W. Morphological, Physiological and Biochemical Responses of Plants to Drought Stress. *Afr. J. Agric. Res.* **2011**, *6*, 2026–2032.

Arróniz-Crespo, M.; Núñez-Olivera, E.; Martínez-Abaigar, J.; Tomás, R. A Survey of the Distribution of, U. V. Absorbing Compounds in Aquatic Bryophytes from a Mountain Stream. *Bryologist* **2004**, *107*, 202–208.

Banerjee, A.; Roychoudhury, A. WRKY Proteins: Signaling and Regulation of Expression During Abiotic Stress Responses. *Sci. World J.* **2015**, *2015*, 807560.

Banerjee, A.; Roychoudhury, A. Group, II Late Embryogenesis Abundant (LEA) Proteins: Structural and Functional Aspects in Plant Abiotic Stress. *Plant Growth Regul.* **2016**, *79*, 1–17.

Banerjee, A.; Roychoudhury, A. Abscisic-Acid-Dependent Basic Leucine Zipper (bZIP) Transcription Factors in Plant Abiotic Stress. *Protoplasma* **2017**, *254*, 3–16.

Banerjee, A.; Roychoudhury, A. Interactions of Brassinosteroids with Major Phytohormones: Antagonistic Effects. *J. Plant Growth Regul.* **2018a**, *37*, 1025–1032.

Banerjee, A.; Roychoudhury, A. Strigolactones: Multi-Level Regulation of Biosynthesis and Diverse Responses in Plant Abiotic Stresses. *Acta Physiol. Plant* **2018b**, *40*, 86.

Banerjee, A.; Roychoudhury, A. The Gymnastics of Epigenomics in Rice. *Plant Cell Rep.* **2018c**, *37*, 25–49.

Banerjee, A.; Roychoudhury, A. Small Heat Shock Proteins: Structural Assembly and Functional Responses Against Heat Stress in Plants. In *Plant Metabolites and Regulation Under Abiotic Stress*; Ahmad, P., Ahanger, M. A., Singh, V. P., Tripathi, D. K., Alam, P., Alyemeni, M. N., Eds.; Academic Press, Elsevier: UK and USA, 2018d; pp 367–374.

Banerjee, A.; Roychoudhury, A. Effect of Salinity Stress on Growth and Physiology of Medicinal Plants. In *Medicinal Plants and Environmental Challenges*; Ghorbanpour, M. et al., Eds.; Springer International Publishing, A. G.: Cham, Switzerland, 2018e; pp 177–188.

Banerjee, A.; Roychoudhury, A. Regulation of Photosynthesis Under Salinity and Drought Stress. In *Environment and Photosynthesis: A Future Prospect*; Singh, V. P., Singh, S., Singh, R., Prasad, S. M., Eds.; Studium Press (India) Pvt. Ltd., 2018f; pp 134–144.

Banerjee, A.; Roychoudhury, A. Abiotic Stress, Generation of Reactive Oxygen Species, and Their Consequences: An Overview. In *Revisiting the Role of Reactive Oxygen Species (ROS) in Plants: ROS Boon or Bane for Plants?*; Singh, V. P., Singh, S., Tripathi, D., Mohan Prasad, S., Chauhan, D. K., Eds.; John Wiley & Sons, Inc.: USA, 2018g; pp 23–50.

Banerjee, A.; Roychoudhury, A. Differential Regulation of Defence Pathways in Aromatic and Non-Aromatic Indica Rice Cultivars Towards Fluoride Toxicity. *Plant Cell Rep.* **2019a**. DOI: 10.1007/s00299–019–02438–6

Banerjee, A.; Roychoudhury, A. Structural Introspection of a Putative Fluoride Transporter in Plants. *3 Biotech* **2019b**, *9*, 103.

Banerjee, A.; Roychoudhury, A. Fluorine: A Biohazardous Agent for Plants and Phytoremediation Strategies for Its Removal from the Environment. *Biol Plant* **2019c**, *63*, 104–112.

Banerjee, A.; Roychoudhury, A. Rice Responses and Tolerance to Elevated Ozone. In *Advances in Rice Research for Abiotic Stress Tolerance*; Hasanuzzaman, M., Fujita, M., Nahar, K., Biswas, J. K., Eds.; Woodhead Publishing, Elsevier: UK, 2019d; pp 399–412.

Banerjee, A.; Roychoudhury, A. Role of Selenium in Plants Against Abiotic Stresses: Phenological and Molecular Aspects. In *Molecular Plant Abiotic Stress: Biology and Biotechnology*; Roychoudhury, A., Tripathi, D. K., Eds.; John Wiley & Sons, 2019e; pp. 123–133.

Banerjee, A.; Roychoudhury, A. The Regulatory Signaling of Gibberellins Metabolism and Its Crosstalk with Phytohormones in Response to Plant Abiotic Stresses. In *Plant Signaling Molecules*; Khan, M. I. R., Reddy, S. P., Ferrante, A., Khan, N. A., Eds.; Woodhead Publishers, Elsevier: UK, 2019f; pp 333–339.

Banerjee, A.; Roychoudhury, A.; Krishnamoorthi, S. Emerging Techniques to Decipher microRNAs (miRNAs) and Their Regulatory Role in Conferring Abiotic Stress Tolerance in Plants. *Plant Biotechnol. Rep.* **2016**, *10*, 185–205.

Banerjee, A.; Tripathi, D. K.; Roychoudhury, A. Hydrogen Sulphide Trapeze: Environmental Stress Amelioration and Phytohormone Crosstalk. *Plant Physiol. Biochem.* **2018**, *132*, 46–53.

Banerjee, A.; Ghosh, P.; Roychoudhury, A. Salt Acclimation Differentially Regulates the Metabolites Commonly Involved in Stress Tolerance and Aroma Synthesis in Indica Rice Cultivars. *Plant Growth Regul.* **2019a**, *88*, 87–97.

Banerjee, A.; Tripathi, D. K.; Roychoudhury, A. The Karrikin 'Callisthenics': Can Compounds Derived from Smoke Help in Stress Tolerance? *Physiol. Plant* **2019b**, *165*, 290–302.

Banerjee, A.; Roychoudhury, A.; Ghosh, P. Differential Fluoride Uptake Induces Variable Physiological Damage in a Non-Aromatic and an Aromatic Indica Rice Cultivar. *Plant Physiol. Biochem.* (Article in Press), **2019c**.

Bannayan, M.; Paymard, P.; Ashraf, B. Vulnerability of Maize Production Under Future Climate Change: Possible Adaptation Strategies. *J. Sci. Food Agric.* **2016**, *96*, 4465–4474.

Barot, S.; Blouin, M.; Fontaine, S.; Jouquet, P.; Lata, J. C. et al. A Tale of Four Stories: Soil Ecology, Theory, Evolution and the Publication System. *PLoS One* **2007**, *28*, e1248.

Bornman, J. F. UV Radiation as an Environmental Stress in Plants. *J. Photochem. Photobiol. B Biol.* **1991**, *8*, 337–342.

Chai, W. W.; Wang, W. Y.; Ma, Q.; Yin, H. J.; Hepworth, S. R.; Wang, S. M. Comparative Transcriptome Analysis Reveals Unique Genetic Adaptations Conferring Salt Tolerance in a Xerohalophyte. *Funct. Plant Biol.* **2019**. DOI: 10.1071/FP18295.

Chunyu, X.; Huang, F.; Xia, Z.; Zhang, D.; Chen, X.; Xie, Y. Assessing the Ecological Effects of Water Transport to a Lake in Arid Regions: A Case Study of Qingtu Lake in Shiyang River Basin, North China. *Int. J. Environ. Res. Public Health* **2019**, *16*, E145.

Clery, D. Environmental Technology. Greenhouse-Power Plant Hybrid Set to Make Jordan's Desert Bloom. *Science* **2011**, *331*, 136.

Costa, M. D.; Artur, M. A.S, Maia, J.; Jonkheer, E.; Derks, M. L.; Nijveen, H. et al. A Footprint of Desiccation Tolerance in the Genome of *Xerophyta viscosa*. *Nat. Plants* **2017**, *3*, 17038.

Das, P.; Banerjee, A.; Roychoudhury, A. Polyamines Ameliorate Oxidative Stress by Regulating Antioxidant Systems and Interacting with Plant Growth Regulators. In *Molecular Plant Abiotic Stress: Biology and Biotechnology*; Roychoudhury, A.; Tripathi, D. K., Eds.; John Wiley & Sons: USA, 2019; pp 135–143.

Dinakar, C.; Bartels, D. Desiccation Tolerance in Resurrection Plants: New Insights from Transcriptome, Proteome and Metabolome Analysis. *Front. Plant Sci.* **2013**, *4*, 482.

Eickmeier, W. G.; Casper, C.; Osmond, C. B. Chlorophyll Fluorescence in the Resurrection Plant *Selaginella lepidophylla* (Hook. & Grey.) Spring During High-Light and Desiccation Stress, and Evidence for Zeaxanthin-Associated Photoprotection. *Planta* **1993**, *189*, 30–38.

Fan, L.; Wang, G.; Hu, W.; Pantha, P.; Tran, K. N.; Zhang, H. et al. Transcriptomic View of Survival During Early Seedling Growth of the Extremophyte *Haloxylon ammodendron*. *Plant Physiol. Biochem.* **2018**, *132*, 475–489.

Farooq, M.; Gogoi, N.; Hussain, M.; Barthakur, S.; Paul, S.; Bharadwaj, N.; Migdadi, H. M.; Alghamdi, S. S.; Siddique, K. H.M Effects, Tolerance Mechanisms and Management of Salt Stress in Grain Legumes. *Plant Physiol. Biochem.* **2017**, *118*, 199–217.

Farrant, J. M.; Cooper, K.; Hilgart, A.; Abdalla, K. O.; Bentley, J.; Thomson, J. A. et al. A Molecular Physiological Review of Vegetative Desiccation Tolerance in the Resurrection Plant *Xerophyta viscosa* (Baker). *Planta* **2015**, *242*, 407–426.

Farrant, J. M.; Vander Willigen, C.; Loffell, D. A.; Bartsch, S.; Whittaker, A. An Investigation Into the Role of Light During Desiccation of Three Angiosperm Resurrection Plants. *Plant Cell Environ.* **2003**, *26*, 1275–1286.

Flowers, T. J.; Gaur, P. M.; Gowda, C. L.; Krishnamurthy, L.; Samineni, S.; Siddique, K. H.; Turner, N. C.; Vadez, V.; Varshney, R. K.; Colmer, T. D. Salt Sensitivity in Chickpea. *Plant Cell Environ.* **2010**, *33*, 490–509.

Gamri, T. E. Prospects and Constraints of Desert Agriculture: Lessons from West Omdurman. *Environ. Monit. Assess* **2004**, *99*, 57–73.

Gechev, T. S.; Benina, M.; Obata, T.; Tohge, T.; Sujeeth, N.; Minkov, I. et al. Molecular Mechanisms of Desiccation Tolerance in the Resurrection Glacial Relic *Haberlea rhodopensis*. *Cell Mol. Life Sci.* **2013**, *70*, 689–709.

Ghanbari, A. A.; Mousavi, S. H.; Mousapour Gorgi, A.; Rao, I. M. Effects of Water Stress on Leaves and Seeds of Bean (*Phaseolus vulgaris* L.). *Turk. J. Field Crops* **2013a**, *181*, 73–77.

Ghanbari, A. A.; Shakiba, M. R.; Toorchi, M.; Choukan, R. Nitrogen Changes in the Leaves and Accumulation of Some Minerals in the Seeds of Red, White and Chitti Beans (*Phaseolus vulgaris*) Under Water Deficit Conditions. *Aust. J. Crop Sci.* **2013b**, *7*, 706–712.

Giarola, V.; Hou, Q.; Bartels, D. Angiosperm Plant Desiccation Tolerance: Hints from Transcriptomics and Genome Sequencing. *Trends Plant Sci.* **2017**, *22*, 705–717.

Gibbon, D.; Pain, A. *Crops of the Drier Regions of the Tropics*; ELBS/Longman: UK, 1988.

Gil, M.; Pontin, M.; Berli, F.; Bottini, R.; Piccoli, P. Metabolism of Terpenes in the Response of Grape (*Vitis vinifera* L.) Leaf Tissues to UV-B Radiation. *Phytochemistry* **2012**, *77*, 89–98.

Gill, S. S.; Anjum, N. A.; Gill, R.; Jha, M.; Tuteja, N. DNA Damage and Repair in Plants Under Ultraviolet and Ionizing Radiations. *Sci. World J.* **2015**, *2015*, 250158.

Goodman, P. J.; Caldwell, M. M. Shrub Ecotypes in a Salt Desert. *Nature* **1971**, *232*, 571–572.

Greenberg, B. M.; Gaba, V.; Canaani, O.; Malkin, S.; Mattoo, A. K.; Edelman, M. Separate Photosensitizers Mediate Degradation of the 32 kDa Photosystem, I. I. Reaction Center Protein in the Visible and *W* Spectral Regions. *Proc. Natl. Acad. Sci. USA* **1989**, *86*, 6617–6620.

Hassan, M. U.; Chattha, M. U.; Mahmood, A.; Sahi, S. T. Performance of Sorghum Cultivars for Biomass Quality and Biomethane Yield Grown in Semi-Arid Area of Pakistan. *Environ. Sci. Pollut. Res. Int.* **2018**, *25*, 12800–12807.

Jha, U. C.; Bohra, A.; Jha, R.; Parida, S. K. Salinity Stress Response and 'Omics' Approaches for Improving Salinity Stress Tolerance in Major Grain Legumes. *Plant Cell Rep.* **2019**, *38*, 255–277.

Jiao, S.; Hilaire, E.; Guikema, J. A. Identification and Differential Accumulation of Two Isoforms of the, C. F.1-b Subunit Under High Light Stress in *Brassica rapa*. *Plant Physiol. Biochem.* **2004**, *42*, 883–890.

Kaashyap, M.; Ford, R.; Bohra, A.; Kuvalekar, A.; Mantri, N. Improving Salt Tolerance of Chickpea Using Modern Genomics Tools and Molecular Breeding. *Curr. Genom.* **2017**, *18*, 557–567.

Kant, S.; Bi, Y. M.; Weretilnyk, E.; Barak, S.; Rothstein, S. J. The *Arabidopsis* Halophytic Relative *Thellungiella halophila* Tolerates Nitrogen Limiting Conditions by Maintaining Growth, Nitrogen Uptake, and Assimilation. *Plant Physiol.* **2008**, *147*, 1168–1180.

Khan, H. A.; Siddique, K. H. M.; Colmer, T. D. Vegetative and Reproductive Growth of Salt-Stressed Chickpea Are Carbon-Limited: Sucrose Infusion at the Reproductive Stage Improves Salt Tolerance. *J. Expt. Bot.* **2017**, *68*, 2001–2011.

Koberl, M.; Muller, H.; Ramadan, E. M.; Berg, G. Desert Farming Benefits from Microbial Potential in Arid Soils and Promotes Diversity and Plant Health. *PLoS One* **2011**, *6*, e24452.

Koch, M. A.; German, D. A. Taxonomy and Systematics Are Key to Biological Information: *Arabidopsis*, *Eutrema* (*Thellungiella*), *Noccaea* and *Schrenkiella* (Brassicaceae) as Examples. *Front. Plant Sci.* **2013**, *4*, 267.

Lamdan, N. L.; Attia, Z.; Moran, N.; Moshelion, M. The *Arabidopsis*-Related Halophyte *Thellungiella halophila*: Boron Tolerance via Boron Complexation with Metabolites? *Plant Cell Environ.* **2012**, *35*, 735–746.

Lee, Y. P.; Babakov, A.; de Boer, B.; Zuther, E.; Hincha, D. K. Comparison of Freezing Tolerance, Compatible Solutes and Polyamines in Geographically Diverse Collections of *Thellungiella* sp. and *Arabidopsis thaliana* Accessions. *BMC Plant Biol* **2012**, *12*, 131.

Liang, Y.; Li, X.; Zhang, D.; Gao, B.; Yang, H.; Wang, Y.; Guan, K.; Wood, A. J. *ScDREB8*, a Novel A-5 Type of, *D. R.EB* Gene in the Desert Moss *Syntrichia caninervis*, Confers Salt Tolerance to *Arabidopsis*. *Plant Physiol. Biochem.* **2017**, *120*, 242–251.

Luttge, U.; Scarano, F. R. Ecophysiology. *Braz. J. Bot.* **2004**, *27*, 1–10.

Madugundu, R.; Al-Gaadi, K. A.; Tola, E.; Hassaballa, A. A.; Kayad, A. G. Utilization of Landsat-8 Data for the Estimation of Carrot and Maize Crop Water Footprint Under the Arid Climate of Saudi Arabia. *PLoS One* **2018**, *13*, e0192830.

Manchanda, G.; Garg, N. Salinity and Its Effects on the Functional Biology of Legumes. *Acta Physiol. Plant* **2008**, *30*, 595–618.

Medina, E.; Cram, W. J.; Lee, H. S.J.; Luttge, U.; Popp, M. et al. Ecophysiology of Xerophytic and Halophytic Vegetation of a Coastal Alluvial Plain in Northern Venezuela. *New Phytol.* **1989**, *111*, 233–243.

Mubaiwa, J.; Fogliano, V.; Chidewe, C.; Bakker, E. J.; Linnemann, A. R. Utilization of Bambara Groundnut (*Vigna subterranea* (L.) Verdc.) for Sustainable Food and Nutrition Security in Semi-Arid Regions of Zimbabwe. *PLoS One* **2018**, *13*, e0204817.

Muvunyi, B. P.; Yan, Q.; Wu, F.; Min, X.; Yan, Z. Z.; Kanzana, G.; Wang, Y.; Zhang, J. Mining Late Embryogenesis Abundant (LEA) Family Genes in *Cleistogenes songorica*, a Xerophytes Perennial Desert Plant. *Int. J. Mol. Sci.* **2018**, *19*, E3430.

Oh, D. H.; Dassanayake, M.; Bohnert, H. J.; Cheeseman, J. M. Life at the Extreme: Lessons from the Genome. *Genome Biol.* **2012**, *13*, 241.

Paul, S.; Banerjee, A.; Roychoudhury, A. Role of Polyamines in Mediating Antioxidant Defense and Epigenetic Regulation in Plants Exposed to Heavy Metal Toxicity. In *Plants Under Metal and Metalloid Stress*; Hasanuzzaman, M., Nahar, K., Fujita, M., Eds.; Springer Nature: Singapore Pte Ltd., 2018; pp 229–247.

Pereira, A. C.; Gama, V. F. Anthropization on the Cerrado Biome in the Brazilian Urucuí-Una Ecological Station Estimated from Orbital Images. *Braz. J. Biol.* **2010**, *70*, 969–976.

Prasad, P. V.V.; Djanaguiraman, M. Response of Floret Fertility and Individual Grain Weight of Wheat to High Temperature Stress: Sensitive Stages and Thresholds for Temperature and Duration. *Funct. Plant Biol.* **2014**, *41*, 1261–1269.

Reij, C.; Scoones, I.; Toulmin, C. *Sustaining the Soil: Indigenous Soil and Water Conservation in Africa*; Earthscan Publications Ltd.: London, 1996.

Reuters. Egypt Plan to Green Sahara Desert Stirs Controversy. Reuters website, 2007. http://www.reuters.com/article/2007/10/09/us-desertegypt-idUSL2651867020071009.

Richardson, B. A.; Kitchen, S. G.; Pendleton, R. L.; Pendleton, B. K.; Germino, M. J. et al. Adaptive Responses Reveal Contemporary and Future Ecotypes in a Desert Shrub. *Ecol. Appl.* **2014**, *24*, 413–427.

Rodriguez, M. C.; Edsgärd, D.; Hussain, S. S.; Alquezar, D.; Rasmussen, M.; Gilbert, T. et al. Transcriptomes of the Desiccation-Tolerant Resurrection Plant *Craterostigma plantagineum*. *Plant J.* **2010**, *63*, 212–228.

Roychoudhury, A.; Banerjee, A.; Lahiri, V. Metabolic and Molecular-Genetic Regulation of Proline Signaling and Its Cross-Talk with Major Effectors Mediates Abiotic Stress Tolerance in Plants. *Tuk. J. Bot.* **2015**, *39*, 887–910.

Roychoudhury, A.; Basu, S.; Sarkar, S. N.; Sengupta, D. N. Comparative Physiological and Molecular Responses of a Common Aromatic Indica Rice Cultivar to High Salinity with Non-Aromatic Indica Rice Cultivars. *Plant Cell Rep* **2008**, *27*, 1395–1410.

Roychoudhury, A.; Chakraborty, M. Biochemical and Molecular Basis of Varietal Difference in Plant Salt Tolerance. *Ann. Rev. Res. Biol.* **2013**, *3*, 422–454.

Roychoudhury, A.; Nayek, S. Structural Aspects and Functional Regulation of Late Embryogenesis Abundant (LEA) Genes and Proteins Conferring Abiotic Stress Tolerance in Plants. In *Abiotic Stress: Role in Sustainable Agriculture, Detrimental Effects and Management Strategies*; Ferro, A., Ed.; Nova Science Publishers: New York, 2014; pp 43–109.

Roychoudhury, A.; Paul, A. Abscisic Acid-Inducible Genes During Salinity and Drought Stress. In *Advances in Medicine and Biology*; Berhardt, L. V., Eds., Vol. 51; Nova Science Publishers: New York, 2012; pp 1–78.

Sehgal, A.; Sita, K.; Siddique, K. H.M, Kumar, R.; Bhogireddy, S.; Varshney, R. K.; Hanumantha Rao, B.; Nair, R. M.; Prasad, P. V.V, Nayyar, H. Drought or/and Heat-Stress Effects on Seed Filling in Food Crops: Impacts on Functional Biochemistry, Seed Yields, and Nutritional Quality. *Front. Plant Sci.* **2018,** *9*, 1705.

Shamaya, N. J.; Shavrukov, Y.; Langridge, P.; Roy, S. J.; Tester, M. Genetics of Na$^+$ Exclusion and Salinity Tolerance in Afghani Durum Wheat Landraces. *BMC Plant Biol.* **2017,** *17*, 209.

Sinha, R. K. Making Aridlands Productive: Conservation and Utilization of Thar Desert Biodiversity for Sustainable Development in India. *Desert Control Bull.* **1996,** *29*, 17–22.

Sita, K.; Sehgal, A.; Bhandari, K.; Kumar, J.; Kumar, S.; Singh, S. et al. Impact of Heat Stress During Seed Filling on Seed Quality and Seed Yield in Lentil (*Lens culinaris* Medikus) Genotypes. *J. Sci. Food Agric.* **2018,** *98*, 5134–5141.

Sita, K.; Sehgal, A.; Kumar, J.; Kumar, S.; Singh, S.; Siddique, K. H. M. et al. Identification of High-Temperature Tolerant Lentil (*Lens culinaris* Medik.) Genotypes Through Leaf and Pollen Traits. *Front. Plant Sci.* **2017,** *8*, 744.

Soares, H. S. Poverty Eradication from Dream to Reality: Hope in the Sahel. *Spore Bull.* **2001,** *96*, 16.

Soliman, S. S. M.; Abouleish, M.; Abou-Hashem, M. M. M.; Hamoda, A. M.; El-Kablawy, A. A. Lipophilic Metabolites and Anatomical Acclimatization of *Cleome amblyocarpa* in the Drought and Extra-Water Areas of the Arid Desert of UAE. *Plants (Basel)* **2019,** *8*, E132.

Velasco, V. M.; Mansbridge, J.; Bremner, S.; Carruthers, K.; Summers, P. S.; Sung, W. W. et al. Acclimation of the Crucifer *Eutrema salsugineum* to Phosphate Limitation Is Associated with Constitutively High Expression of Phosphate-Starvation Genes. *Plant Cell Environ.* **2016,** *39*, 1818–1834.

Vogt, K. *A Field Worker's Guide to the Identification, Propagation and Uses of Common Trees and Shrubs of Dryland Sudan*; SOS Sahel International: UK, 1995.

WMO. *Final Report: Working Group on Monitoring, Assessment and Combat of Desertification*; WMO, 1990.

Zhang, G. W.; Zhang, X.; Chen, H.; Zhou, Z. G. Effect of Sowing Dates on Cotton Yield, Fiber Quality, and Uptake and Utilization of Nutrients in Inner Mongolia West Desert Area, China. *Ying Yong Sheng Tai Xue Bao* **2017,** *28*, 863–870.

Zhang, J.; Zhao, P.; Zhao, J.; Chen, G. Synteny-Based Mapping of Causal Point Mutations Relevant to Sand Rice (*Agriophyllum squarrosum*) Trichomeless1 Mutant by RNA-Sequencing. *J. Plant Physiol.* **2018,** *231*, 86–95.

Zhang, Y. F.; Huang, X. W.; Wang, L. L.; Wei, L.; Wu, Z. H.; You, M. S. et al. Proteomic Analysis of Wheat Seed in Response to Drought Stress. *J. Integr. Agric.* **2014,** *13*, 919–925.

CHAPTER 7

Grafting, Seed Soaking/Priming, Soil Amendment, and Foliar Application as Tools to Increase Abiotic Stress Tolerance of Crops

S. A. SHEHATA and MOHAMED F. M. IBRAHIM

Agricultural Botany Department, Ain-Shams University, Cairo, Egypt

ABSTRACT

Climatic changes had negative impact on crop productivity and this effect was obviously detected by elevating abiotic stresses, including water scarcity [drought stress], salinization [salt stress] and raising temperature [heat stress]. In last decade, several tremendous attempts were made to increase plant tolerance to abiotic stress/es. Focusing on different application methods, grafting technique were used to solve soil salinity by grafted salt sensitive plant onto salt tolerant rootstock. It could be able to substitute root system of salt-sensitive plant with root system of salt tolerant one. Grafting could be considered as indirect methods to enhance salt tolerance of several crops. Transplanting takes also into account. Seed soaking and priming in different chemical or bio- regulators offer another alternative application to increase abiotic stress tolerance. Soil amendment by the addition humic acid, compost, biochar, and seaweed extract alleviates salinity and drought or heavy metal stresses. Foliar application of different bio regulators, growth regulators, osmolytes, and nutrient

Crop Sustainability and Intellectual Property Rights. Soumya Mukherjee, Piyali Mukherjee & Tariq Aftab (Eds)

element enhance crop tolerance to abiotic stresses. Salicylic acid, ascorbic acid, brassinosteroid, polyamines, glycine betaine, kinetin, thidiazuron, boron, $CaCl_2$, KCl, and silicate were usually used in this concern. Physiological mechanisms of abiotic stress tolerant crops were investigated in view of water status (water potential, water use efficiency, relative water content, stomatal conductance/resistance, and transpiration rate), plant pigments, osmolytes (proline, amino acid and sugars), lipid peroxidation and antioxidant enzymes.

7.1 INTRODUCTION

Looking in nature, plants are evolving several physiological mechanisms to cope with abiotic stress/es. Consequently, some plants, but not all could be acclimatized and adapted under environmental stress condition. Acclimatization and adaptation of crops under stressful abiotic condition require adopted physiological mechanism/es which in turn, maintain crop productivity. Ferrante and Mariani (2018) used the term "fitness cost" to reveal energy consumption for keeping plant fit under abiotic stress/es. This waste energy, on the other hand, reduces plant growth and productivity. However, crops were classified to abiotic stress tolerance and abiotic stress-sensitive and between them a wide scale of tolerance degree (FAO, 2017). Unfortunately, most of the agricultural crops are abiotic stress-sensitive. Plant physiological researchers have challenged to increase tolerance of abiotic stress-sensitive crops by several means. This chapter will be covering the most means, such as grafting, root/soil drench, seed soaking/priming, and foliar spraying to increase abiotic stress tolerance of crops, except plant breeding and agronomic management. The later was reviewed from an agricultural practices point of view to enhance crop tolerance to different abiotic stresses, salinity, drought, hypoxia, lodging (Ferrante and Mariani, 2017; Elkelish et al., 2021a, b), high/low temperature, and light intensity (Ferrante and Mariani, 2018).

Also, enhancing crop tolerance to abiotic stresses by traditional breeding programs, and genetic transformation was reported by (Cuartero et al., 2006). However, commercial success of producing genetically modified plant tolerance to abiotic stress/es has been very limited due to the complexity of genetically and physiologically plant tolerance to salt (Flowers, 2004; Ola et al., 2021).

7.2 ABIOTIC STRESS, DEFINITION, AND CONCEPT

Abiotic stress is defined as stress caused by nonliving factors, and is more related to ecological factors, including, hydrosphere (rains-irrigation water, lithosphere (soil type-fertility), atmosphere (air temperature and components). For optimal crop production, plants require healthy and non-polluted environmental conditions. Industrial revolution and human activities (anthropogenic) with excessive use of chemical plant fertilizers had compulsorily induced climatic changes viz., global warming, pollution, salinization, and desertification.

Ultimately, all these factors caused abiotic stresses to plant growth and productivity which negatively affect the food production worldwide. Abiotic stresses are saline in water or soil, soil infertility, heavy metal, acid rains, alkalinity, drought, heat, solar radiation, and air pollution. During plant span, plant may encounter one of the previous stress or combination of two, three or more stresses.

Saline and drought were found to be the major stresses, and heat stress ranked in the third order (Fig. 7.1). Most of the literatures deal with salinity and drought stresses. Both stresses are related together, saline stress causes physiological drought and affect plant water status. Finally, availability of free water is also a constraint due to lower osmotic and water potential.

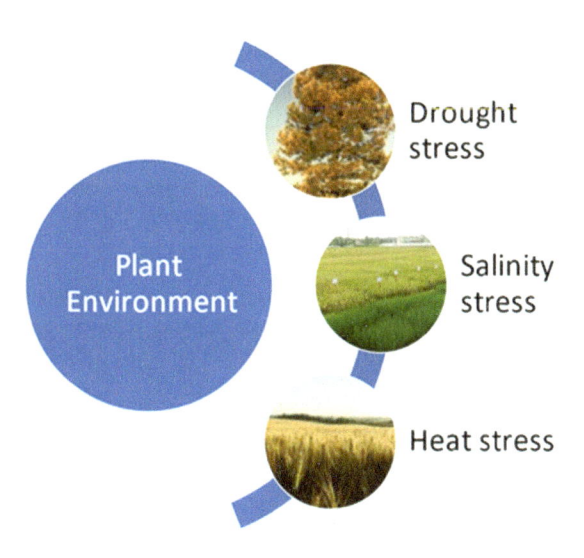

FIGURE 7.1 Drought, salinity, and heat, the most important abiotic stresses affect crops.

7.3   NEGATIVE IMPACT OF ABIOTIC STRESSES

Abiotic stressors have a detrimental influence on the development and production of plants. Salinity stresses either soil or irrigation water and negatively affected plant growth and development. A survey has classified vegetable crops into salinity stress-sensitive, moderate sensitive, moderate tolerance, and tolerance (FAO, 2017).

Soil salinity has become an immense environmental problem. Degradation of food production due to salinization is more observed in the last decade. Currently, approximately 1125 million hectares of lands are salt-affected, of which, approximately 76 million hectares are affected by human-induced salinization (Hossain, 2019). Salinity constrains plants in two ways: osmotically due to the reduced soil water potential, and ionically due to the toxicity of salty ions, and the resultant is ionic imbalance in the plants themselves (Munns and Tester, 2008; Hasan et al., 2021; Abd El-Gawad et al., 2021). Reactive oxygen species (ROS) are known to be produced when plants are exposed to salt stress. These ROS are so reactive that they seriously disrupt the normal metabolism of plants through oxidation of membrane lipids, proteins, and nucleic acids particularly, if the plant does not have any sufficient protective mechanisms (Apel and Hirt, 2004; Ibrahim et al., 2021). Plant sensitivity to salt stress manifests itself in the form of turgor loss, growth decrease, and, in extreme cases, death of the plant (Jones et al., 1989; Cheeseman, 1988; Jahan et al., 2021).

Marschner (1995) reported three physiological mechanisms by which plant suffers from salinity stress: (a) lower water potential of the root medium, (b) toxic effects of Na^+ and Cl^- ,and (c) nutrient imbalance by depression in uptake and/or shoot transport. High concentrations of salt impose both osmotic and ionic stresses on the plants which lead to several morphological and physiological changes (Munns and Tester, 2008). Many agricultural crops, including most vegetables, which are typically sensitive throughout the plant's life cycle, suffer production losses due to excessive soil salt. Several researchers strive to overcome salt stress problems, and these would have a positive impact on sustainability of agriculture production. Great efforts have been made to improve the salt tolerance of many crops by several means of applications (grafting, seed priming, soil/root drench, and foliar spray). In arid and semiarid land, salinity and drought are the major abiotic stresses that affect plant growth and productivity (Parida and Das, 2005). Generally, the negative impact of salinity stress on physiological plant parameters is profoundly varied between salt tolerance and salt-sensitive plant as postulated in Figure 7.2. This postulation could be valid to some extent for drought or heat stresses. The second major abiotic stress is drought, drought stress had negative physiological effect similar to salinity stress. However, drought represents direct problem for plant growth worldwide because of irrigation water shortage or water contamination (saline, heavy metal, and pesticide) (Ben Hassine et al., 2010). Drought stress has dramatically reduced plant growth and development due to disruption of photosynthesis (Gong et al., 2005; Abd Elhady et al., 2021), decreasing water potential and affected stomatal conductance and transpiration (Tahi et al., 2007; Özenç, 2008). Drought also inhibits nutrient translocation, enzyme activities, protein synthesis, plant pigments, and impairs cell membrane stability (Saneoka et al., 2004; Saraswathi and Paliwal, 2011).

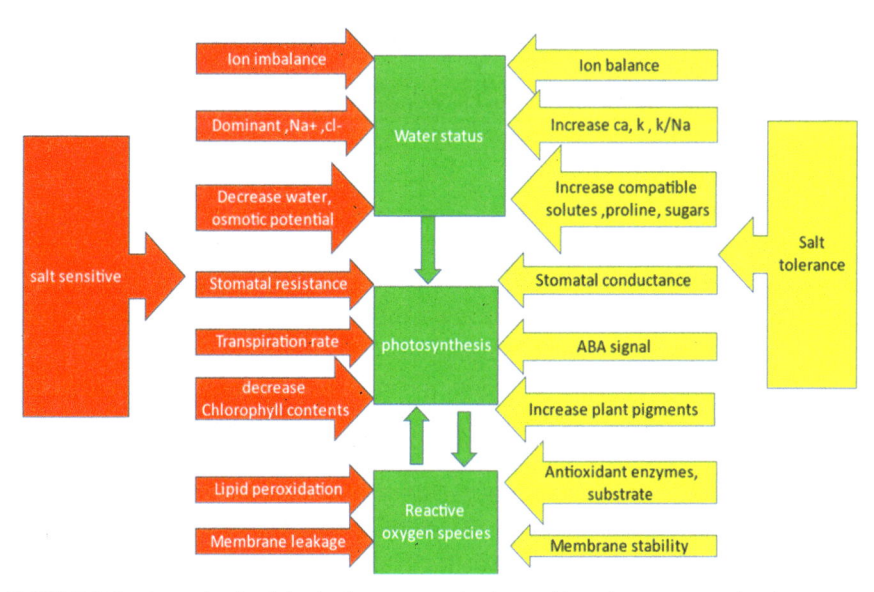

FIGURE 7.2   Reveals physiological response of salt-sensitive plant versus salt tolerance.

Ambient air and soil temperature of plant (heat/cold stress) have nega-tively affected plant growth and development, particularly during flowering and fruit setting. Heat stress was found to be more effective under drought stress, usually both stresses simultaneously affect plant growth. The most common problem in crop cultivation is the abortion of flowers due to high temperatures, where heat stress affects pollen grain viability, osmotic pres-sure, fruit setting, and fruit yield (Sato et al., 2000). At high temperature, the height of the stigma in the antheridia cone was the main factor reducing fruit-set (Charles and Harris, 1972). Heat stress as similar as salinity and drought for inducing excessive reactive oxygen species (ROS) resulting in cellular injuries by oxidation to all vital cell organelles and membranes, and even may cause the collapse of cellular organization. The inhibition of photosynthesis by heat stress has been attributed to the impairment of structural organization and decrease of photosystem II (PSII) activity (Khan et al., 2013). Heat stress causes changes in the antioxidant enzyme activities of tomato plant (Abd Elkader et al., 2016). Both heat stress and cold stress had negative impact upon plant growth. On the same context, Yadav (2010) reviewed the negative effect of cold stress on plant growth starting from seed germination to fruit setting.

Heavy metals stress induces toxicity during plant nutrition. Co, Cu, Fe, Mn, Mo, Ni, V, and Zn are required in minute quantities by organisms, excessive amounts of these elements can become harmful to plants. Heavy metals, such as Pb, Cd, Hg, and Al disrupt plant growth, they have no role for plant nutrition (Asati et al., 2016; Khan et al., 2017). Plants require solar radiation for photomorphogenesis, including building pigments, activating photoreceptors, and photosynthesis (Stapleton, 1992). Solar radiation has a wide range of spectral wavelength, and consequently, plants are exposed to all wavelengths, including the ultraviolet (UV) radiation. UV radiation is generally divided into three classes: UVA (400–320 nm), UVB (320–280 nm), and UVC (280–100 nm). The UV-C region of the UV spectrum includes wavelengths below 280 nm has highly energetic wavelengths and their intensity in atmosphere depends on ozone layer in the stratosphere (Caldwell et al., 1989). All types of UV radiation are known to damage various plant processes. Such damage can be classified into two categories: damage to DNA (which can cause heritable mutations) and damage to physiological processes. A diverse deleterious effect of UV radiation on photosynthetic characteristics of several plants was reported by Gurjar et al. 2017). Another abiotic stress is the lack of oxygen (hypoxic conditions) in plant soil due to excess water by flooding and heavy rains, particularly, in bad drainage soils (Ferrante and Mariani, 2017).

7.4 APPLICATIONS TO ENHANCE ABIOTIC STRESS TOLERANCE OF CROPS

Different methods of applications were applied to enhance crop tolerance to abiotic stress/es. Modern irrigation system allows to feed plants either via foliage parts (sprinkler -pivot) or root system (drip irrigation, hydroponics). Both methods are widely used to apply biochemical, fertilizer, or even pesticide. Therefore, root/soil drench and foliar spraying with various biochemicals to enhance plant tolerance are commonly used, with seed priming and grafting as well. Moreover, it could be possible to combine two or three applications together to obtain the maximum improvement in abiotic stress tolerance.

The most important application methods to increase abiotic stress tolerance of crops are recapitulated in Figure 7.3.

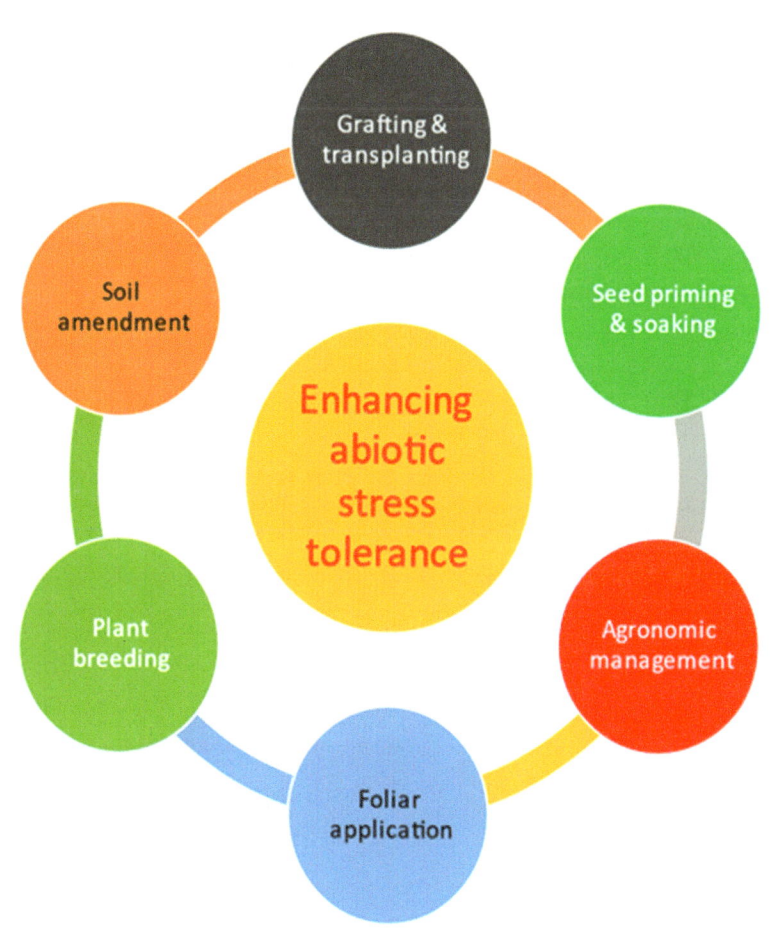

FIGURE 7.3   Reveals different methods of applications to increase abiotic stress tolerance of crops.

7.4.1 PLANT GRAFTING/TRANSPLANTING

Grafting in vegetable crops was first launched in Japan and Korea in the late of 1920. Little attention was paid to improve and increase commercial uses of grafted plant till 1980. Today, grafting was spread worldwide.

Grafting technique was made either by hand grafting or using automation (robot) (Lee et al., 2010). Grafted crops have many benefits like high yield and are environment-friendly by minimizing pesticide, biotic and abiotic stress tolerance (Lee et al 2010).

The first step for using grafting to increase abiotic stress tolerance is to find out the appropriate rootstock (abiotic stress tolerance) which is highly compatible to target scion. Therefore, survey should be made among cultivars or species to select abiotic stress tolerance of rootstock. For testing salinity stress tolerance, experiment was conducted by EL-Shraiy et al. (2011) among four cucurbitaceous cultivars: Shintosa Supreme pumpkin (Cucurbita maxima × C. moschata) "Major," Bottle gourd (Lagenaria siceraria), Gourd Black Seed (Cucurbita maxima) and pumpkin (Cucurbita moschata), under 4500 ppm NaCl salinity. Shintosa Supreme pumpkin (Cucurbita maxima x C. moschata) cultivar was found to be the highest salt tolerant as compared with other cucurbitaceous cultivars. Salt tolerant rootstock (Shintosa Supreme pumpkin) has been selected according to growth and physiological parameters as given in Table 7.1.

TABLE 7.1 Screening for Salt Tolerance Between Four Cucurbitaceous Cultivars by Measuring Different Parameters Under Salinity Stress (4500 ppm).

Parameters	Cv1	Cv2	Cv3	Cv4
Shoot fresh	40/g	71/g	121/g	51/g
Plant height	30/cm	70/cm	189/cm	40/cm
Leaf area	111/cm^2	105/cm^2	217/cm^2	121/cm^2
Root fresh	10/g	15/g	39/g	17/g
Root length	27/cm	38/cm	60/cm	40/cm
% Membrane permeability	45%	40%	27%	45%
% Salt injury index	55%	45%	32%	42%
Cl root	5 mg/g dw	5 mg/g dw	19 mg/g dw	5 mg/g dw
Cl stem	30 mg/g dw	65 mg/g dw	70 mg/g dw	65 mg/g dw
Cl leaf	80 mg/g dw	31 mg/g dw	32 mg/g dw	41 mg/g dw
Na root	8 mg/g dw	12 mg/g dw	40 mg/g dw	8 mg/g dw

TABLE 7.1 *(Continued)*

Parameters	Cv1	Cv2	Cv3	Cv4
Na stem	70 mg/g dw	72 mg/g dw	12 mg/g dw	70 mg/g dw
Na leaf	88 mg/g dw	30 mg/g dw	10 mg/g dw	40 mg/g dw
K root	5 mg/g dw	8 mg/g dw	5 mg/g dw	4 mg/g dw
K stem	55mg/g dw	65 mg/g dw	15 mg/g dw	55 mg/g dw
K leaf	35mg/g dw	25mg/g dw	55 mg/g dw	18 mg/g dw

Modified from Sheraiy et al. (2011).Cv1, Gourd black seed, Cv2, pumpkin, Cv3, shintosa supreme pumpkin, Cv4, Bottle gourd.

The highest salt tolerance should have the best growth parameters, lowest salt injury index, lowest membrane permeability, lowest Na level in stem and leaves, and highest K level in stem and leaves.

The second step was grafting of cucumber scion (salt-sensitive) onto salt tolerance rootstock (Shintosa Supreme pumpkin). As shown in Table 7.2, grafted cucumber showed significantly increase in growth parameters by 44% and 41% for fresh and dry weights, plant height by 86%, and leaf area by 57%. Lipid peroxidation was decreased as indicated by lower amount of malondialdehyde, consequently membrane stability was increased. Compatible solutes (proline, soluble protein), plant pigments and activities of antioxidant enzymes was significant enhanced in grafted cucumber as compared to nongrafted one under salinity stress (EL-Shraiy et al., 2011).

TABLE 7.2 Induction of Salt Tolerance from Salt-Sensitive Cucumber by Grafting Onto Salt Tolerance Rootstock.

Parameters at 70 days after transplanting	Nongrafted cucumber	Grafted cucumber
Fresh weight (g)	89	128
Height (cm)	79	147
Leaf area (cm^2)	98	217
Fruit yield g/plant	2679	3370
malondialdehyde (MDA (μmol g^{-1} FWt)	13	10
membrane permeability %	41	33
LRWC %	69	78
Chlorophyll	1.6	2.2
Carotenoid	0.15	0.3
Proline	5.2	6.7
Total soluble protein	26	34

TABLE 7.2   *(Continued)*

Parameters at 70 days after transplanting	Nongrafted cucumber	Grafted cucumber
Antioxidant enzymes .min^{-1} mg^{-1} protein		
PAL	14	16
GPOD	0.26	0.3
CAT	0.4	0.46
SOD	9	10
PPO	80	120
APX	80	100

Modified after EL Sheraiy et al. (2011). Shintosa Supreme Pumpkin under salinity stress, Antioxidant enzymes: Phenylalanine ammonialyase (PAL), Guaiacol peroxidase (GPOD), Catalase (CAT), Superoxide dismutase (SOD), Polyphenol oxidase (PPO), Ascorbate peroxidase (APX).

Induction of cucumber salt tolerance by grafting resulted from the following reasons:

1. In the leaves, there is a higher concentration of proline and sugar (Xu et al., 2006).
2. Leaves have a higher level of antioxidant capability (López-Gómez et al 2007).
3. Leaves with less Na$^+$ and/or Cl$^-$ deposition (Goreta et al., 2008; Zhu et al., 2008).

Generally, grafting was found to increase abiotic stress/es tolerance in plants (Table 7.3). Grafts could mitigate several of abiotic stresses in different crops viz., thermal stress of tomato plant (Rivero et al., 2003), salinity stress of cucumber (EL-Shraiy et al 2011), cold stress of tomato plant (Venema et al., 2008). Grafting improves the ability to increase nutrient uptake of watermelon (Colla et al., 2010a). Grafting promotes the synthesis of endogenous hormones (cytokinin and abscisic acid in xylem sap of cotton rootstock (Dong et al., 2008), improves water use efficiency, leaf pigments and yield of melon (Rouphael et al., 2008a), reduces uptake of persistent organic pollutants from agricultural soils in cucumber (Otani and Seike, 2006, 2007), improves alkalinity tolerance (Colla et al., 2010), increases salt and flooding tolerance (Romero et al., 1997; Colla et al., 2006; Yetisir et al., 2006), minimizes the negative effect of boron, copper, cadmium, and manganese toxicity (Edelstein et al., 2005, 2007; Rouphael et al., 2008b; Savvas et al., 2009).

TABLE 7.3 Increasing Abiotic Stress/es Tolerance of Different Crops by Grafting.

Grafted plant	Stress mitigation	References
Tomato	Thermal shock	Rivero et al. (2003)
Tomato	Cold stress	Venema et al. (2008)
Watermelon	Nutrients	Colla et al. (2010a)
Cotton	Drought	Dong et al. (2008)
Cucumber	Organic pollutants	Otani and Seike (2006, 2007)
Cucumber	Salinity	EL Sheraiy et al. (2011), Zhu et al. (2008)
Watermelon	Alkalinity	Colla et al. (2010b)
Melon	Effluent water +Boron	Edelstein (2007)
Melon	NaCl	Edelstein (2011)
Citrus	Flooding/drought	Garcia-Sanchez et al. (2007)
Grape	Drought	Satisha et al. (2007)
Cucumber	Drought	Liu et al. (2017)
Cucumber	Copper toxicity	Rouphael et al. (2008b)

Commercial production of different grafted varieties of vegetable crops was established in different countries. In Germany, "Peter Stader" company produces different types of grafting seedling reaching 100 million plant/year using robot technique, and their production covers Stuttgart, Freiburg, osteria, and Switzerland (http://www.peterstader.de). In Egypt, a grafting company produces commercially a lot of grafted plants using hand grafting technique.

Transplanting of grafted plant in the field had also another benefit to avoid seed germination in agricultural field as it is the sensitive stage to abiotic stress.

Transplanting of nongrafted crops had ameliorated effect to abiotic stress. Transplanting complete root system of sugar beet seedling has advantages to be more abiotic tolerance (Karbalaei et al., 2012) as compared with direct seeding in main field due to more susceptibility of early growth stages to adverse conditions. In addition, transplanted sugar beet with early sowing date increases sugar beet yields as compared with direct seeding in field under salinity stress (Gohari et al., 1995; Karbalaei et al., 2012; Ibrahim et al., 2017). Also, seed germination and established of sugar beet seedling in green house or nursery protect plants from negative effect of temperature (Draycott, 2006). Generally, transplanting method prolongs the growing season by earlier planting in greenhouse when direct seeding may be impossible due to inappropriate environmental condition outside the greenhouse or occupation of main field with previous crop, and this prolongation of growing season has a positive effect on the yield of crop.

7.4.2 SEED SOAKING/PRIMING

Successful agriculture depends on the seed quality and its potential to produce strong healthy seedling at the first stage of plant life cycle. Furthermore, certified seeds are considered the first key for sustainable farming. Advantages of using certified seed were recorded to be reached up to 10 benefits in addition to the benefits of priming certified seeds. One advantage of using certified seeds is that it improves abiotic stress tolerance. Alternative method for plant grafting/transplanting to increase abiotic stress tolerance of crops is seed soaking/priming, whereas not all crops are able to be grafted or transplanted. Seed soaking or seed priming have been used to reduce the seedling emergence time (Fig. 7.4), for synchronized emergence, improved emergence rate, and better seedling stand production in many crops under non-stressed or stressed conditions (Khan, 1992; Chowdhary and Baset, 1994; Jett et al., 1996; Basra et al., 2005). Pre-sowing seed soaking is an easy, low-cost and low-risk technique used to overcome abiotic stress/es (Iqbal and Ashraf, 2005). Seed priming is a controlled hydration technique in which seeds are soaked in solutions of low-osmotic potential before the actual germination takes place (Farooq et al., 2006), and that allows metabolic activities to proceed before radicle emergence (Sirritepe et al., 2003). Soaked seeds allow imbibition in appropriate solution for 2–48 h (soaking duration) according to plant variety and directly sowing or allowing it to dry being reversal to initial state with making tiny film of solute on seed surface (priming) before sowing. Soaking solution should contain nutrient elements (Eisa et al., 2012; Jisha et al., 2013), humic acid (Hartwigsen and Evans, 2000), bioregulators (plant hormones and plant growth substances) (Sivakumar and Nandhitha, 2017), biofertilizers (Mahmood et al., 2016), and pesticide. However, according to solutes in seed soaking/priming solution and their function, seed priming has been classified as hydropriming, osmo priming, chemical priming, biological priming, hormonal priming, nutrition priming (Jisha et al., 2013), and solid matrix priming (solute absorption by seeds is delayed by matrix pressures that retain water in place (Damalas et al., 2019). In this concern, Marthandan et al. (2020) evaluated the feasibility of seed priming as a means of enhancing drought tolerance in agricultural plants. Seed priming has also been shown to affect agricultural plants in terms of morphological, physiological, biochemical, and molecular responses. Figure 7.4 outlined different substances(solutes) used for seed soaking/priming, viz., hydropriming (Rodríguez et al., 2015), osmo priming (KCl, MgSO4, mannitol, KH2PO4, KNO3, $CaCl_2$, NaCl, and polyethylene glycol) (Singh et al., 2014), chemical priming (putrescine, chitosan, selenium, paclobutrazol, choline, CuSO4, and $ZnSO_4$) (Jisha et al., 2013), Hormonal priming (kinetin, SA (salicylic acid), GA3, and ascorbate) (Bakhtavar et al., 2015), spermidine and polyamines (Zheng et al., 2016).

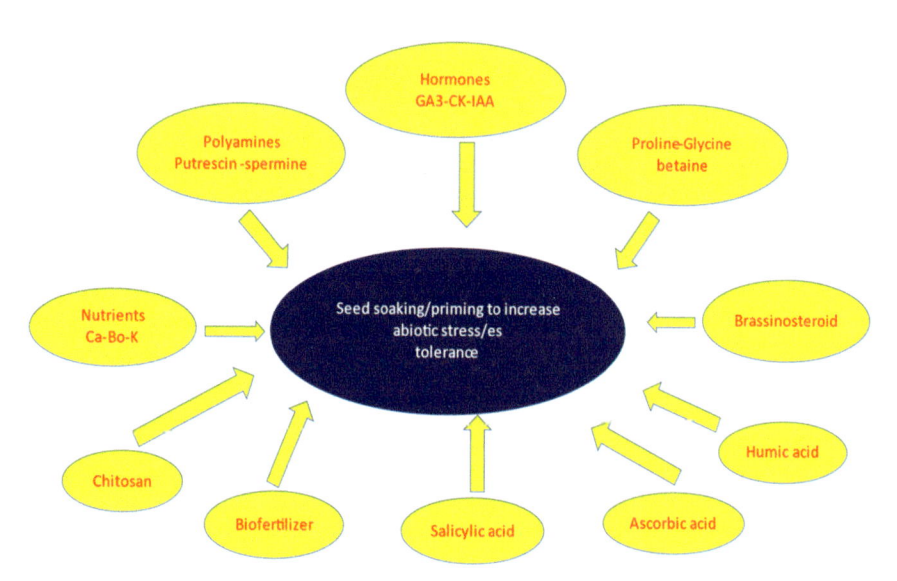

FIGURE 7.4 Reveals different substances (solutes) used in seed soaking/priming solution to increase abiotic stress/es tolerance of crops.

The above-mentioned substances were widely used in seed soaking/priming to increase the tolerance of subsequent seedling and plant against abiotic stress/es. According to abiotic stress type (drought, salinity, heat/cold, and heavy metal), nutrients, bioregulators, and biochemical agents were used as soaking/priming solution to alleviate different of abiotic stresses (Table 7.4).

TABLE 7.4 Reveals Mitigation of Abiotic Stress/es in Different Crops by Seed Soaking/Priming.

Soaking/priming solution	Seeds	Stress mitigation	References
B amino butyric acid	Cowpea	Drought	Jisha and Puthur (2016)
Selenium+ salicylic acid	Rice	Chilling	Hussain et al. (2016)
Mycorrhiza fungi	Rice	Drought	Kavitha Mary et al. (2018)
Auxin +PGPR	Wheat	Drought	Bagheri et al. (2019)
$Znso_4$	Lentil	Drought+ salinity	Ali (2005)
Hormonal Nutrient Cacl2	Chickpea	Drought	Shariatmadari et al. (2017), Kumeera et al. (2018), Shankrayya et al. (2018)
Phosphobacterium	okra	drought	Puthiyottil (2015)
Salicylic acid	Mung bean	salinity	Sivakumar and Nandhitha (2017)
Potassium humate	Sugar beet	salinity	Eisa et al. (2012)
Salicylic acid	Wheat	Drought	Al-Hakimi Hamada (2001), El-Tayeb (2005)
Salicylic acid	Barley	Salinity	
Salicylic acid	Maize	Cd	Krantev et al. (2008)
Ascorbic acid	Wheat	Salinity	Al-Hakimi Hamada (2001)
NaCl	Tomato	Salinity	Iseri et al. (2014)

Seed soaking/priming: the mode of action

Pre-sowing seed soaking had earlier germination and seedling emergence as compared with priming and unsoaked (unprimed) seeds, respectively. Secondary dormancy occurs during priming processing, and also after drying the seeds to retain initial shape. However, time-lag for breaking secondary dormancy of primed seeds is shorter than that of breaking primary dormancy of unprimed ones (Fig. 7.5). Mustafa et al. (2017) reveal that primed seed had earlier gemination and emergence than unprimed seed. Seeds of cultivar crops had eco-dormancy (quiescence seeds) rather than para-dormancy and endodormancy. Ecodormancy (external dormancy) seeds require suitable ecological conditions (humidity, temp, and oxygen) to fulfil seed germination.

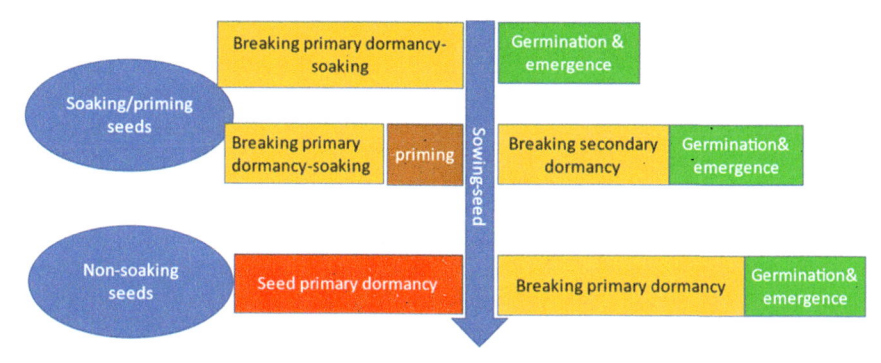

FIGURE 7.5 Revealing the effects of seed soaking, priming, and non-soaking on time- lag of germination.

Seed soaking/priming stimulates defense mechanism/s to mitigate abiotic stress/es by altering physiochemical metabolism to achieve the following benefits:

1. Accumulation of osmolytes (proline, glycine betaine, and poly-amines-sugars). Delavari et al. (2010), Hameed and Iqbal (2014), Anket et al. (2019).
2. Enhancing hydrolytic enzymes (Rodríguez et al. 2015).
3. Activation of antioxidant enzymes (Afzal et al. 2008; Jafar et al. 2012).
4. Reduce lipid peroxidation (Tani et al., 2019).
5. Increases the quantity of mitochondria and related proteins (α- and β-tubulin) that are responsible for cell division (Hussain et al., 2017).
6. Improved water use efficiency (WUE) (Koller and Hadass, 1982).
7. Increase membrane stability (Tabassum et al., 2018)
8. Regulate the water movement across the cell membrane via aquaporins. Gomes et al. (2009)

7.4.3 SOIL AMENDMENT

In addition to restoring soil fertility by adding organochemical fertilizer, biochar, compost, and seaweed extract were the most popular supplemented compounds to soil for reducing negative impact of stressed soil and increasing abiotic stress tolerance of crops.

Biochar (BC) is a soil conditioner and organic fertilizer made from activated carbon. Pyrolysis (thermal conversion in the absence of oxygen) of plant wastes under anaerobic circumstances produces it. Plant macronutrients can be slowly released by using BC in soil. The use of a variety of chemical fertilizers in agriculture has been reduced in BC (Lehmann, 2007; Laird, 2010; Zhang et al., 2012), enhance nutrients as well as water-holding capacity (Gaskin et al., 2008; Uzoma et al., 2011; Kloss et al., 2012), increase nutrient contents (total N P, K, Mg, Cu, and Zn), increase root proliferation, promote, above-ground biomass and yield (Major et al., 2010; Olmo et al., 2014; El Nahhas et al., 2021; Jabborova et al., 2021a, 2021b). Additive biochar to sandy soil increase water retention (water-holding capacity). In addition to C sequestration, soil and plant development enhancements, and energy production, biochar provides several sustainability advantages (Marris, 2006). Nutrient's composition and organic matter in sandy soil was increased after adding biochar and compost (Schulz et al., 2013).

Moreover, the addition of seaweed extract, which contains growth promoting hormones, such as indole-3-acetic acid, indole-3-butyric acid, and cytokinin, as well as trace minerals, vitamins, and amino acids, was shown to increase plant abiotic stress tolerance in stressed soil (Bhaskar and Miyashita, 2005; Khan et al., 2009). Seaweed extracts have been shown to increase the development and productivity of plants, increase their tolerance to environmental stress, increase the availability of nutrients, and increase the absorption of nutrients from the soil (Crouch et al., 1990, Rathore, 2009; Aziz et al., 2011).

Also, humus (containing humic acids, fulvic, and hymatomelonic acids), peats, and compost are widely used in mineral soils. Humic acids (HA) are high molecular weight aromatic organic acids that have a polyphenolic backbone with numerous carbohydrate and peptide side chains that give rise to a complex array of structures (Hays, 1989). The major functional groups of humic substances are carboxyl, phenolic hydroxyl, alcoholic hydroxyl, and ketone (Cacco and Agnolla, 1984). Application of humic acid was found to positively affect the plant growth parameters when grown under saline condition (Türkmen et al., 2005). Humic substances have been reported to enhance mineral uptake under salt stress by increasing the permeability of membranes of root cells (Valdrighi et al., 1996), developing strong root systems (David et al., 1994). Moreover, the positive effects of humic substance on plant growth and productivity could mainly be due to hormone-like activities of humic substances through involvement in cell respiration, photosynthesis, oxidative phosphorylation, protein synthesis, antioxidant, and various enzymatic reactions (Chen and Aviad, 1990; Zhang and Schmidt, 1999; Zhang et al., 2003).

Salicylic acid was administered as a soil drench in the same study for the purpose of increasing the abiotic stress tolerance of crops. Stevens et al. (2006) found that root soaking of tomato (*Lycopersicon esculentum*) plants with 0.1 mM SA provided protection against NaCl stress at a concentration of 200 mM. In addition to increasing the growth and photosynthetic rate of the plants, it also raised the amount of transpiration and stomatal conductance, and it decreased electrolyte leakage by 32%. Also soil drench with potassium humate (PH) (4 & 8 g/L), potassium silicate (PS) (4 & 8 g/L) increase salinity tolerance of sugar beet (Ibrahim et al., 2017).

Supplemented salt soil with SA increased survival percentage of maize plants by decreasing the Na^+ and Cl accumulation (Gunes et al., 2007).

7.4.4 FOLIAR SPRAYING

Plant foliar spraying with nutrients, bioregulators, biostimulants, biofertilizers, and hormones is a widespread application to increase the abiotic tolerance of crops and productivity as well. The following part deals with the most popular substances used to increase abiotic stress tolerance of crops when applied as foliar spray. However, most substances used in seed soaking/priming are usually applied too as foliar treatments.

7.4.4.1 EFFECT OF SALICYLIC ACID

Salicylic acid is a phenolic compound and is natural occurring in plant metabolism (Raskin, 1992). Biosynthesis of salicylic acid in plant occurs from amino acid (phenylalanine) through cinnamic acid, o-coumaric acid, and benzoic acid by shikimate-phenylpropanoid pathway (Sticher et al., 1997). Previously, salicylic acid (SA) has long been applied to acquire plant resistance against biotic stress/es (Raskin, 1992). Nowadays, SA has been applied as foliar spraying, seed soaking/priming, hydroponically, and soil drench to increase plant resistance to different of abiotic stress/es. Foliar spraying of salicylic acid increased heat resistance (acquired thermotolerance), in mustard (Dat et al., 1998), cucumber (Shi et al., 2006), potato microplants (Lopez-Delgado et al., 1998), tomato (Abd Elkader et al., 2016), and watermelon (Moustafa-Farag et al., 2020a). SA application was found to mitigate not only heat stress but also cold stress. SA also induced chilling tolerance (Mora-Herrera et al., 2005). In banana seedlings, 0.5 mM SA solution has induced chilling tolerance both when sprayed onto the leaves and/or applied in irrigation to the roots (Kang et al., 2003). Under drought stress, when wheat seedlings are treated with SA, show enhanced tolerance to drought stress by increasing moisture content, dry mass, carboxylase activity of Rubisco, superoxide dismutase activity, and total chlorophyll content (Singh and Usha, 2003).

The application of 0.05 mM SA also improved plant growth after salt stress and caused the accumulation of ABA and prolines (Shakirova et al., 2003).

SA treatment induced tolerance against copper toxicity in cucumber and tobacco (Strobel and Kuc, 1995), in sunflower plants (El-Tayeb et al., 2006). In addition, SA treatment was demonstrated to have a beneficial impact on seed germination and seedling growth in two rice plants subjected to Pb^{2+} or Hg^{2+} stressors, respectively (Mishra and Choudhuri, 1997). Cd-induced lipid peroxidation in barley seedlings was reduced by SA pretreatment, which also improved the fresh weight of the plant's shoots and roots.

Salicylic acid has hormone-like activity and is involved in signal transduction through activation of several genes expression during stress condition (Klessig et al., 2000). Recently, salicylic acid, brassinosteroids (BRs), and polyamines were added as new class of endogenous hormones. The mode of action of applied SA to increase abiotic tolerance of plant might have a process which decreases H_2O_2 and membrane electrolyte leakage, and decreased the Na^+ and Cl accumulation and caused the accretion of prolines and ABA and increases antioxidant enzyme activities. Induction of abiotic stress tolerance of crops by SA treatments was reviewed by Horvath et al. (2007).

7.4.4.2 EFFECT OF ASCORBIC ACID

Ascorbic acid (AsA) plays a crucial role in mitigating abiotic stresses, regulation of cellular redox (Foyer and Noctor, 2011; Lim et al., 2016), biosynthesis of cell wall, cell division, elongation, nutrients uptake, hormone biosynthesis (Toth et al., 2009; Sultana et al., 2015; Smirnoff, 2018). AsA possesses sturdy antioxidant potential and helps in scavenging of ROS (Müller-Moulé et al., 2003) and enzymatic reactions (Foyer and Noctor, 2011), as well it preserved photosynthetic pigments and strengthened higher relative water content (Hussain et al., 2017). Foliar application with AsA may directly influence the metabolism by regulating the key duties, such as enzyme activity, synthesis of protein, chlorophyll, and osmoregulation. This reflects on nutrients uptake (Farooq et al., 2013) which improves chlorophyll synthesis and plant growth (Alamri et al., 2018). Foliar application with AsA alleviated harmful effect of chilling and heat stresses, in addition, enhanced the photosynthetic pigments on pepino (Sivaci, et al., 2014) and tomato (Alayafi, 2020; Elkelish et al., 2020). Application of AsA mitigated the effect of salinity on common bean and increased the chlorophyll content as well as resistance to salinity (Dolatabadian and Jouneghani, 2009). Foliar spraying with AsA at 400 mg L^{-1} has mitigated the reduction of water stress on photosynthetic pigments of common bean either chlorophyll or carotenoids, activity of carbonic anhydrase, antioxidant activities, vegetative growth and seed yield (Gaafar et al., 2020).

7.4.4.3 EFFECT OF BRASSINOSTEROID

Natural steroidal plant hormones known as brassinosteroids (BRs) are a widely used class of natural steroidal plant hormones that play an important role in the regulation of plant growth, cell division and elongation, photosynthesis regulation, as well as the mitigation of biotic and abiotic stresses (Khana et al., 2019; Yu et al., 2004). Foliar application of BRs could alleviate reduction of photosynthesis and damage induced by water deficiency (Hu et al., 2013), salinity (Ali and Ashraf, 2008), chilling (Fang et al., 2019), and heat stress (Sadura and Janeczko, 2018). BRs application could also enhance genes expression of related genes induced by water deficit and produce a defense response (Kagale et al., 2007). Under water deficit conditions, BRs can maintain vegetative growth of apple and physiobiochemical processes, such as photosynthesis, stomatal conductance, water use efficiency, proline, TSS, peroxidase, catalase, rate of net assimilation, and transpiration (Kumari and Thakur, 2019). The alleviation of drought by BRs was related to enhanced activity of scavenging and elimination of ROS as reflected by high activities of antioxidant enzymes, such as catalase and peroxidase.

Foliar application of BRs at 3 µM could ameliorate the hard effect of salinity stress on tomato growth, yield, and biochemical parameters (Kutby et al., 2020). On other hand, in some cases, BRs at either high or low doses have adverse impact on the plant's salinity tolerance, When the salinity concentration is less than 150 mM, BRs inhibit and stimulate the oxidative metabolism of putrescine, resulting in a significant quantity of H_2O_2 being produced (Liu et al., 2020).

Foliar application of BRs with three different concentrations could alleviate chilling stress on chlorophyll contents of maize, which promotes photosynthesis that is the primary operation of vegetative growth and organic matter accumulation. Increasing the BRs concentration increased the chlorophyll contents of seedlings at first and then decreased (Sun et al., 2019). Foliar spraying with BRs improved chilling tolerance via maintaining photosynthesis process and increased nitrogen metabolism on pepper (Yang et al., 2019) and maize (Sun et al., 2020).

Under heat stress, foliar application with BRs could mitigate the harmful effect on leaf area, total dry weight, and yield of wheat plant (Kumari and Hemantaranjan, 2018).

7.4.4.4 EFFECT OF POLYAMINES (PUTRESCIN, SPERMIDINE, SPERMINE)

Polyamines (PAs) (including putrescine, spermidine, spermine) are considered like plant growth regulators with a broad ambit of influences, and useful factor in responding to development and stresses via regulating intracellular ions that influenced the physiological processes of plant (Pottosin and Shabala, 2014). They influence the membrane stability and permeability, thus protecting plants from drought stress (Farooq et al., 2009) as well enhancing the tolerance to salinity via protecting the plasma membrane as well as preventing the damage of cell membrane and chloroplast, and in addition, they alleviate the decrease of the tissue water content and photosynthetic pigment (Zapata et al., 2004; Radhakrishnan and Lee, 2014).

Foliar application with putrescine (Put) could enhance the activities of many antioxidant enzymes under biotic and abiotic stresses (Ibrahim et al., 2015), thus ameliorating growth, chlorophyll contents, and carotenoids. Putrescine also decreases catalase and peroxidase activities under salinity stress on guava seedlings and show scavenging of free radicals (Ghalati et al., 2020). In addition, Put decreases anatomical features of thyme plants under drought stress and accumulates phenolic compounds and enzymes activities (Abd Elbar et al., 2019), growth and wheat yield (Ahmed and Sadak, 2016). It is also improving the quality of melon under heat stress (Piñero et al., 2021).

Exogenous application with spermidine maintained activities of antioxidant enzymes, enhanced photosynthesis and decreased osmotic pressure on goosegrass under salinity stress (Deng et al., 2019), decreased oxidative damage via increasing antioxidant components, increasing redox state of ascorbate and glutathione that play important function in ameliorating physiological characteristics under drought stress in maize (Li et al., 2018) and heat stress on cauliflower (Collado-González et al., 2021).

Foliar spraying with spermine alleviates the decrease of tomato growth under salinity stress as well promoted the activity of nitrate reductase, gas exchange, proline, sugars, and glycine betaine accumulation, thus leading to better photosynthesis and tissue water content (Ahanger et al., 2019). Under high temperature and drought stresses, spermine decreased ROS, membrane lipid peroxidation, upregulated the activities of SOD, CAT, and glutathione peroxidase that were necessary for imparting ROS-induced oxidative stresses tolerance (Nahar et al., 2017).

7.4.4.5   EFFECT OF PLANT GROWTH REGULATORS

Plant growth regulators are strongly participatory in the regulation of plant developmental operations, and abiotic stress responses of plants mitigate the impacts of environmental stresses (Khan et al., 2014). Foliar application with plant growth regulators used for overcoming salinity stress (Javid et al., 2011). Kinetin, a synthetic cytokinin, when used, increased the nutritional quality of plants under salinity stress via modulating phenolic metabolism of *Vigna sinensis* and *Zea mays* (Nemat Alla et al., 2002) and scavenges free radicals. Kinetin contributes in antioxidative mechanism concerning the protection of purine breakdown (Chakrabarti and Mukherji, 2003). The favorable function of kinetin in alternating salinity stress is evident from the levels of endogenous phytohormones that provide a serious evidence for understanding the defense mechanisms of soybean against salinity (Hamayun et al., 2015).

Cytokines play a vital function in plant response to water deficit (Hai et al., 2020) via regulating cell division, photomorphogenic cell differentiation in expanding leaves and shoots (Chiang et al., 2012; Efroni et al., 2013), suppression of leaf senescence (Zwack and Rashotte, 2015).

7.4.4.6   EFFECT OF COMPATIBLE SOLUTES

Proline exerts its influence in decreasing oxidative stress and increasing bearing salt stress (Gurmani et al., 2014) via scavenging free radicals, conserves antioxidant enzymes and cells against the destructive ROS as well decreases the cellular membrane oxidation (Wutipraditkul et al., 2015).

Under salinity stress, foliar spraying with proline enhanced the growth in carrot, photosynthetic attributes, antioxidant defense system (Qirat et al., 2018; Moustafa-Farag et al., 2020b). Application of proline alleviates the harmful effect of drought stress and improves the vegetative growth, photosynthetic pigments and yield of pea and cowpea (El-Saadony et al., 2017; Merwad et al., 2018).

Glycine betaine (GB) enhances the plant growth under stress conditions and decreases oxygen-free radicals, inhibits ROS activity and conservation of cell membrane under abiotic stress. This increases photosynthetic pigments by preventing the collapse of chlorophyll that led to enhancement of photosynthesis {216} (Alasvandyari et al., 2017). Foliar application with GB increased the osmotic pressure under salinity stress, enhanced RWC due to water retention in plant {217} (Kaya et al., 2013), antioxidant preventative influence on antioxidant enzyme activities under salt stress {218} (Hoque et al., 2007 and Ibrahim et al., 2020b) as well as enhanced the activities of CAT, SOD, and APX in onion {219} (Rady et al., 2018).

7.4.4.7   EFFECT OF NITRIC OXIDE, SODIUM NITROPRUSSIDE

Nitric oxide (NO) plays a pivotal role in the prorated balance between oxidative stresses (Corpas and Palma, 2018), enzymatic antioxidative systems (Wang et al., 2013), organization of plant growth and photosynthesis (Zhang et al., 2009). NO has an essential role in mitigating abiotic stresses (Corpas and Palma, 2018) by upregulating the osmolyte metabolism (Soliman et al., 2019; Habib et al., 2020). NO influences sugars via its role in regulating photosynthesis (Faraji and Sepehri, 2020). Foliar application of NO enhances tolerance to different negative environmental conditions, such as water deficit (Elkelish et al., 2021a, b), salinity stress (Hajivar and Zare-Bavani, 2019), and heat stress (Parankusam et al., 2017). Under abiotic stresses, spraying NO promotes APX activities in tomato plants (Nasibi and Kalantari, 2009) that play a role in banning the immoderate accumulation of H_2O_2 in the plant cells via ascorbate–glutathione cycle (Pang and Wang, 2010; Alnusairi et al., 2021).

7.5   COMBINATION OF DIFFERENT METHODS OF APPLICATIONS

It is possible to combine one or more of different application methods to achieve maximum increase of abiotic stress tolerance in one crop. El-Shraiy and Mostafa (2016) used grafted cucumber plant with foliar spraying of salicylic acid (SA) at 0.5 and 1 mM, fulvic acid (FA) at 150 and 300 ppm and seaweed extract (SWE) at 2.5 and 5% in addition to osmoregulators (compatible solute), glycine betaine (GB) at 2.5 and 5 mM to improve salt tolerance in grafted cucumber plants in salty environments. Salt tolerance was improved by crossing *Cucumis sativus* L. cv. Falcon and Hybrid F1. Different research reported that salicylic acid (SA) was applied as foliar spraying, seed soaking /priming, hydroponically, and soil drench to increase the plant resistance to different of abiotic stress/es (Kang et al., 2003).

7.6 MECHANISM/S TO INCREASE ABIOTIC STRESS TOLERANCE OF CROPS

Implementation of the above-mentioned applications strives to increase abiotic stress tolerance of crops which encounter one or more of abiotic stresses. All compounds used in different methods of application exerted their function to maintain homeostasis of one or more of the subsequent processing in ascending orders: nutrition status, osmotic status, water status, photosynthesis, respiration, metabolic compounds, and growth (Fig. 7.6). There is feedback mechanism between these processing. Nutritional status of the plant was found to be negatively affected under abiotic stress/es, therefore, application of different nutrient elements improves nutrition status, subsequently, they affect the osmotic and water status successively. Osmotic status depends on the ion concentrations (nutrient elements) and osmolytes as an alternative way to ameliorate stress condition by altering endogenous plant metabolism to increase osmolytes production or by exogenous application of glycine betaine, proline, trehalose, amino acid, and other macromolecules. Furthermore, osmotic status depends also on membrane stability by decreasing lipid peroxidation and membrane permeability. Water status in plant is regulated by equilibrium between osmosis (water uptake) and transpiration (water loss). On the other hand, stomatal opening/closing (control transpiration) is regulated by osmosis and hormonal signal (ABA).

Photosynthesis is directly correlated with stomatal (stomatal conductivity, availability of CO_2) and nonstomatal factors (RBISCO, PPCASE, Chla/Chlb, carotenoid, xanthophyll). Abiotic stress/es cause an excess production of reactive oxygen species (ROS) viz., peroxide (O^{-2}_2),

superoxide (O^-_2), single oxygen ($^1O_2$), and hydroxyl (OH) due to inter-
ruption of photosystem system I and II (PSI and PSII) in chloroplast and
electron transport chain in mitochondria. all stresses induce oxidative
damage of vital molecules viz., NAD, NADP, protein, lipids. Abiotic
plant tolerance to stresses adopted defense mechanism to increase antioxidant enzymes (Cheeseman, 2007). Also, non-photochemical quenching
increases with reduction in photochemical quenching under abiotic stress.
Plant redirects metabolic processing to induce osmolytes, antioxidant
enzymes, antioxidant substrate (phenols, ascorbic acid, and carotenoids)
under abiotic stress. Energy consumption to produce these metabolite does
not involve in plant growth, which is considered as defense mechanism to
keep plant survival under abiotic stress. Therefore, increasing enzymatic
(SOD, CAT, POD, APX, GR, GSH) and nonenzymatic antioxidant (proline,
ascorbic acid, alkaloid, tocopherol, phenolic compounds, glutathione,
carotenoid) by exogenous application of ascorbic acid, poly- amines, and
brassinosteroid elevates abiotic stress tolerance of plant and ameliorate
plant growth, and productivity as well.

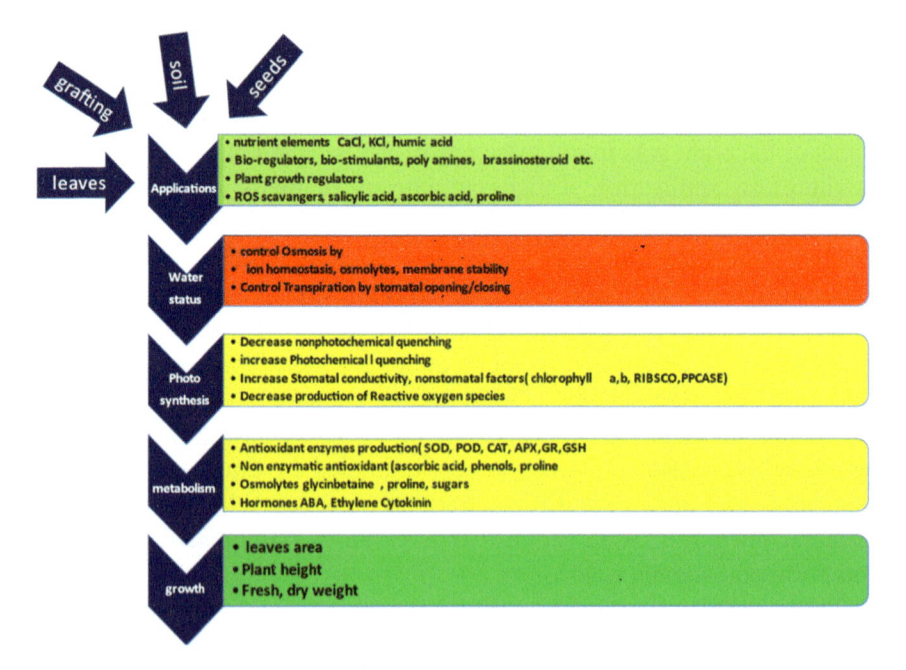

FIGURE 7.6   Reveals a cascade of reaction to increase abiotic stress tolerance of crops by
different method of applications and substances.

KEYWORDS

- **abiotic stress**
- **climate change**
- **plant productivity**
- **seed soaking**
- **signaling molecules**
- **sustainable agriculture**

REFERENCES

Abd Elbar, O. H.; Farag, R. E.; Shehata, S. A. Effect of Putrescine Application on Some Growth, Biochemical and Anatomical Characteristics of *Thymus vulgaris* L. Under Drought Stress. *Annals of Agric. Sci.* **2019,** *64,* 129–137.

Abd Elbar, O. H.; Elkelish, A.; Niedbała, G.; Farag, R.; Wojciechowski, T.; Mukherjee, S.; Abou-Hadid, A. F.; El-Hennawy, H. M.; El-Yazied, A. A.; Abd El-Gawad, H. G.; Azab, E.; Gobouri, A. A.; El-Sawy, A. M.; Bondok, A.; Ibrahim, M. F. M. Protective Effect of γ-Aminobutyric Acid Against Chilling Stress During Reproductive Stage in Tomato Plants Through Modulation of Sugar Metabolism, Chloroplast Integrity, and Antioxidative Defense Systems. *Front. Plant Sci.* **2021,** *12,* 663750. DOI: 10.3389/fpls.2021.663750.

Abd Elhady, S. A.; El-Gawad, H. G.; Ibrahim, M. F.; Mukherjee, S.; Elkelish, A. A.; Azab, E.; Gobouri, A. A.; Farag, R.; Ibrahim, H. A.; El-Azm, N. A. Hydrogen Peroxide Supplementation in Irrigation Water Alleviates Drought Stress and Boosts Growth and Productivity of Potato Plants. *Sustainability* **2021,** *13,* 899.

Abdelaal, K. A.; Mazrou, Y. A.; Hafez, Y. M. Silicon Foliar Application Mitigates Salt Stress in Sweet Pepper Plants by Enhancing Water Status, Photosynthesis, Antioxidant Enzyme Activity and Fruit Yield. *Plants* **2020,** *9,* 733.

Abd-Elkader Ali, M.; Mahmoud, M.; Shehata, S. A.; Osman, H. S.; Salama, Y. A. Induction of Thermotolerant Tomato Plants Using Salicylic Acid and Kinetin Foliar Applications. *J. Hort. Sci. Ornamental Plants* **2016,** *8* (2): 89–97. ISSN 2079–2158 © IDOSI Publications, DOI: 10.5829/idosi.jhsop.2016.8.2.1176.

Afzal, I. S.; Rauf, S. M. A.; Murtaza, G. Halopriming Improves Vigor, Metabolism of Reserves and Ionic Contents in Wheat Seedlings Under Salt Stress. *Plant Soil Environ.* **2008,** *54,* 382–388

Ahanger, M. A.; Qin C.; Maodong, Q.; Dong, X. X.; Ahmad, P.; Abd_Allah, E. F.; Zhang, L. Spermine Application Alleviates Salinity Induced Growth and Photosynthetic Inhibition in *Solanum lycopersicum* by Modulating Osmolyte and Secondary Metabolite Accumulation and Differentially Regulating Antioxidant Metabolism. *Plant Physiol. Biochem*. **2019,** *1* (144), 1–3.

Ahmed, M. M.; Sadak, M. S. Effect of Putrescine Foliar Application on Wheat Genotypes (*Triticum aestivum* L.) Under Water Stress Conditions. *Intern. J. PharmTech. Res.* **2016,** *9* (8), 94–102.

Alamri, S. A.; Siddiqui, M. H.; Al-Khaishany, M. Y.; Khan, M. N.; Ali, H. M.; Alaraidh, I. A.; Alsahli, A. A.; Al-Rabiah, H.; Mateen, M. Ascorbic Acid Improves the Tolerance of Wheat Plants to Lead Toxicity. *J. Plant Interact.* **2018**, *13*, 409–419.

Alasvandyari, F.; Mahdavi, B.; Madah, S. Glycine Betaine Affects the Antioxidant System and Ion Accumulation and Reduces Salinity-Induced Damage in Safflower Seedlings. *Arch. Biol. Sci.* **2017**, *69*, 139–147.

Alayafi, A. A. M. Exogenous Ascorbic Acid Induces Systemic Heat Stress Tolerance in Tomato Seedlings: Transcriptional Regulation Mechanism. *Environ. Sci. Pollut. Res.* **2020**, *27* (16), 19186–19199.

Al-Hakimi, A. M. A.; Hamada, A. M. Counteraction of Salinity Stress on Wheat Plants by Grain Soaking in Ascorbic Acid, Thiamin or Sodium Salicylate. *Biol. Plant* **2001**, *44*, 253–261.

Ali, M. O.; Sarkar, A.; Rahman, M. M.; Gahoonia, T. S. Improvement of Lentil Yield Through Seed Priming in Bangladesh. *J. Lentil Res.* **2005**, *2*, 54–59.

Ali, Q.; Ashraf, M. Modulation of Growth, Photosynthetic Capacity and Water Relations in Salt Stressed Wheat Plants by Exogenously Applied 24-Epibrassinolide. *Plant Growth Regul.* **2008**, *56*, 107–116.

Alnusairi, G. S.; Mazrou, Y. S.; Qari, S. H.; Elkelish, A. A.; Soliman, M. H.; Eweis, M.; Abdelaal, K. A.; El-Samad, G. A.; Ibrahim, M. F.; ElNahhas, N. Exogenous Nitric Oxide Reinforces Photosynthetic Efficiency, Osmolyte, Mineral Uptake, Antioxidant, Expression of Stress-Responsive Genes and Ameliorates the Effects of Salinity Stress in Wheat. *Plants* **2021**, *10*.

Anket, S.; Babar, S.; Vinod, K.; Sukhmeen, K. K.; Gagan, P. S. S.; Aditi, S. B.; Neha, H.; Dhriti, K.; Renu, B.; Bingsong, Z. Phytohormones Regulate Accumulation of Osmolytes under Abiotic Stress. *Biomolecules* **2019**, *9*, 285.

Apel, K.; Hirt, H. Reactive Oxygen Species: Metabolism, Oxidative Stress, and Signal Transduction. *Annu. Rev. Plant Biol.* **2004**, *55*, 373–399.

Asati, A.; Pichhode, M.; Nikhil, K. Effect of Heavy Metals on Plants: An Overview. *IJAIEM* **2016**, *5* (3). ISSN 2319. www.ijaiem.org

Aziz, N.; Mahgoub, M.; Siam, Z. Growth, Flowering and Chemical Constituents Performance of Amaranthus Tricolor Plants as Influenced by Seaweed (Ascophyllum nodosum) Extract Application Under Salt Stress Conditions. *J. Appl. Sci. Res.* **2011**, *7*, 1472–1484.

Bagheri, N.; Alizadeh, O.; Sharaf Zadeh, S.; Aref, F.; Ordookhani, K. Evaluation of Auxin Priming and Plant Growth Promoting Rhizobacteria on Yield and Yield Components of Wheat Under Drought Stress. *Eurasia J. Biosci.* **2019**, *13*, 711–716.

Bakhtavar, M. A.; Afzal, I.; Basra, S. M. A.; Ahmad, A.-U.-H.; Noor, M. A. Physiological Strategies to Improve the Performance of Spring Maize (Zea Mays L.) Planted under Early and Optimum Sowing Conditions. *PLoS One* **2015**, *10*, e0124441.

Basra, S. M. A.; Farooq, M.; Tabassum, R. Physiological and Biochemical Aspects of Seed Vigor Enhancement Treatments in Fine Rice (Oryza sativa L.) Seed. *Sci. Technol.* **2005**, *33*, 623–628.

Ben Hassine, A.; Bouzid, S.; Lutts, S. Does Habitat of *Atriplex halimus* L. Affect Plant Strategy for Osmotic Adjustment? *Acta Physiol. Plant* **2010**, *32*, 325–331. DOI: 10.1007/s11738-009-0410-4

Bhaskar, N.; Miyashita K. Lipid Composition of Padinatetratomatica (Dictyotales, Phaeophyta), a Brown Seaweed of the West Coast of India. *Indian J Fisheries* **2005**, *52*, 263–268.

Cacco, G.; Agnolla, D. Plant Growth Regulator Activity of Soluble Humic Substances. *Can. J. Soil. Sci.* **1984,** 25–28.

Caldwell, M. M.; Teramura, A. H.; and Tevini, M. The Changing Solar Ultraviolet Climate and the Ecological Consequences for Higher Plants. *Trends Ecol. Evol.* **1989,** *4,* 363–367.

Cao, B.; Ma, Q.; Xu, K. Silicon Restrains Drought-Induced ROS Accumulation by Promoting Energy Dissipation in Leaves of Tomato. *Protoplasma* **2020,** *257,* 537–547.

Chakrabarti, N.; Mukherji, S. Alleviation of NaCl Stress by Pretreatment with Phytohormones in Vigna radiata. *Biol. Plantarum* **2003,** *46,* 589–594.

Charles, W.; Harris, R. Tomato Fruit-Set at High and Low Temperatures. *Can. J. Plant Sci.* **1972,** *52* (4), 497–506.

Cheeseman, J. M. Hydrogen Peroxide and Plant Stress: A Challenging Relationship. *Plant Stress* **2007,** *1* (1): 4–15.

Cheeseman, J. Mechanisms of Salinity Tolerance in Plants. *Plant Physiol.* **1988,** *7,* 547–550.

Chen, Y.; Aviad, T. Effects of Humic Substance on Plant Growth. In *Readings;* Carthy, M. C., Calpp, P., Malcolm, C. E., Bloom, R. L., Eds.; ASA and SSSA: Madison, WI, 1990; pp 161–186.

Chiang, Y. H.; Zubo, Y. O.; Tapken, W.; Kim, H. J.; Lavanway, A. M.; Howard, L.; Schaller, G. E. Functional Characterization of the GATA Transcription Factors GNC and CGA1 Reveals Their Key Role in Chloroplast Development, Growth, and Division in Arabidopsis. *Plant Physiol.* **2012,** *160,* 332–348.

Chowdhary, A. Q.; Baset, Q. A. Effect of Soaking Period and Aerobic Condition on Germination of Wheat Seeds. *J. Chittagong Univ.* **1994,** *18,* 83–87.

Colla, G.; Rouphael, Y.; Cardarelli, M.; Massa, D.; Salerno, A.; Rea, E. Yield, Fruit Quality and Mineral Composition of Grafted Melon Plants Grown Under Saline Conditions. *J. Hortic. Sci. Biotechnol.* **2006,** *81,* 146–152.

Colla, G.; Suárez, C. M. C.; Cardarelli, M.; Rouphael, Y. Improving Nitrogen Use Efficiency in Melon by Grafting. *HortScience* **2010a,** *45,* 559–565.

Colla, G.; Rouphael, Y.; Cardarelli, M.; Salerno, A.; Rea, E. The Effectiveness of Grafting to Improve Alkalinity Tolerance in Watermelon. *Environ. Exp. Bot.* **2010b,** *68,* 283–291.

Collado-González, J.; Piñero, M. C.; Otálora, G.; López-Marín, J.; del Amor, F. M. Exogenous Spermidine Modifies Nutritional and Bioactive Constituents of Cauliflower (*Brassica oleracea* **var.** *botrytis* **L.**) Florets Under Heat Stress. *Sci Hortic.* **2021,** *277,* 109818.

Corpas, F. J.; Palma, J. M. Assessing Nitric Oxide (NO) in Higher Plants: An Outline. *Nitrogen* **2018,** *1,* 12–20.

Crouch, I. J.; Beckett, R. Van Staden, J. Effect of Seaweed Concentrate on the Growth and Mineral Nutrition of Nutrient Stressed Lettuce. *J. Appl. Phycol.* **1990,** *2,* 269–272.

Cuartero, J.; Bolarin, M. C.; Asins, M. J.; Moreno, V. Increasing Salt Tolerance in the Tomato. *J. Exp. Bot.* **2006,** *57,* 1045–1058.

Damalas, C. A.; Koutroubas, S. D.; Fotiadis, S.; Damalas, A.; Koutroubas, D. Hydro-Priming Effects on Seed Germination and Field Performance of Faba Bean in Spring Sowing. *Agriculture* **2019,** *9,* 201.

Dat, J. F.; Lopez-Delgado, H.; Foyer, C. H.; Scott, I. M. Parallel Changes in H_2O_2 and Catalase During Thermotolerance Induced by Salicylic Acid or Heat Acclimation in Mustard Seedlings. *Plant Physiol* **1998,** *116,* 1351–1357.

David, P. P.; Nelson, P. V.; Sanders, D. S. A Humic Acid Improves Growth of Tomato Seedling in Solution Culture. *J. Plant Nutr.* **1994,** *17,* 173–184.

Delavari, P. M.; Baghizadeh, A.; Enteshari, S. H.; Kalantari, K. M.; Yazdanpanah, A.; Mousavi, E. A. The Effects of Salicylic Acid on Some of Biochemical and Morphological Characteristic of Ocimum basilicum Under Salinity Stress. *Aust. J. Basic Appl. Sci.* **2010,** *4,* 4832–4845.

Deng, C; Li, J.; Liu, S.; Zhu, X.; Chen, Y.; Shen, X. Effects of Spermidine and Salinity Stress on Growth and Biochemical Response of Paraquat-Susceptibe and -Resistant Goosegrass (*Eleusine indica* **L.**). *Weed Biol. Manage* **2019,** *19,* 75–84.

Diao, M.; Ma, L.; Wang, J.; Cui, J.; Fu, A.; Liu, H.-Y. Selenium Promotes the Growth and Photosynthesis of Tomato Seedlings Under Salt Stress by Enhancing Chloroplast Antioxidant Defense System. *J. Plant Growth Regul.* **2014,** *33,* 671–682.

Dolatabadian, A.; Jouneghani, R. S. Impact of Exogenous Ascorbic Acid on Antioxidant Activity and Some Physiological Traits of Common Bean Subjected to Salinity Stress. *Not. Bot. Hort. Agrobot. Cluj.* **2009,** *37* (2), 165–172.

Dong, H. H.; Niu, Y. H.; Li, W. J.; Zhang, D. M. Effects of Cotton Rootstock on Endogenous Cytokinins and Abscisic Acid in Xylem Sap and Leaves in Relation to Leaf Senescence. *J. Exp. Bot.* **2008,** *59,* 1295–1304.

Draycott AP. *Sugar Beet,* 1st ed.; Blackwell Publishing: Oxford, 2006; 474 p.

Edelstein, M.; Ben-Hur, M.; Cohen, R.; Burger, Y.; Ravina, I. Boron and Salinity Effects on Grafted and Non-Grafted Melon Plants. *Plant Soil* **2005,** *269,* 273–284.

Edelstein, M.; Ben-Hur, M.; Plaut, Z. Grafted Melons Irrigated with Fresh or Effluent Water Tolerate Excess Boron. *J. Am. Soc. Hortic. Sci.* **2007,** *132,* 484–491.

Edelstein, M.; Plaut, Z.; Ben-Hur, M. Sodium and Chloride Exclusion and Retention by Non-Grafted and Grafted Melon and Cucurbita Plants. *J. Exp. Bot.* **2011,** *62,* 177–184.

Efroni, I.; Han, S. K.; Kim, H. J.; Wu, M. F.; Steiner, E.; Birnbaum, K. D.; Wagner, D. Regulation of Leaf Maturation by Chromatin-Mediated Modulation of Cytokinin Responses. *Dev. Cell* **2013,** *24,* 438–445.

Eisa, S. S.; Ibrahim, A. M.; Khafaga, H. S.; Shehata S. A. Alleviation of Adverse Effects of Salt Stress on Sugar Beet By Pre-Sowing Seed Treatments. *J. Appl. Sci. Res.* **2012,** *8* (2), 799–806. ISSN 1819–544X.

El Nahhas, N.; Alkahtani, M. D.; Abdelaal, K. A.; Al Husnain, L.; Algwaiz, H. I.; Hafez, Y. M.; Attia, K. A.; El-Esawi, M. A.; Ibrahim, M. F.; Elkelish, A. A. Biochar and Jasmonic Acid Application Attenuates Antioxidative Systems and Improves Growth, Physiology, Nutrient Uptake and Productivity of Faba Bean (*Vicia faba* L.) Irrigated with Saline Water. *Plant Physiol. Biochem.* **2021,** *166,* 807–817.

El-Aidy, F.; Abdalla, M.; El-sawy, M.; El kady, S.; Bayoumi, Y.; El-Ramady, H. Role of Plant Probiotics, Sucrose and Silicon in the Production of Tomato (*Solanum lycopersicum* L.) Seedlings Under Heat Stress in a Greenhouse. *Appl. Eco. Environ. Res.* **2020,** *18, 6,* 7685–7701.

Elkelish, A.; El-Mogy, M. M.; Niedbała, G.; Piekutowska, M.; Atia, M. A. M.; Hamada, M. M. A.; Shahin, M.; Mukherjee, S.; El-Yazied, A. A.; Shebl, M.; Jahan, M. S.; Osman, A.; Abd El-Gawad, H. G.; Ashour, H.; Farag, R.; Selim, S.; Ibrahim, M. F. M. Roles of Exogenous α-Lipoic Acid and Cysteine in Mitigation of Drought Stress and Restoration of Grain Quality in Wheat. *Plants* **2021a,** 10.

Elkelish, A.; Ibrahim, M.; Ashour, H.; Bondok, A.; Mukherjee, S.; Aftab, T.; Hikal, M.; Abou El-Yazied, A.; Azab, E.; Gobouri, A.; Farag, M.; Metwally, A.; Abd El-Gawad, H. Exogenous Application of Nitric Oxide Mitigates water Stress and Reduces Natural Viral

Disease Incidence of Tomato Plants Subjected to Deficit Irrigation. *Agronomy* **2021b,** *11,* 87.

Elkelish, A.; Qari, S.; Mazrou, Y.; Abdelaal, K.; Hafez, Y.; Abu-Elsaoud, A.; Batiha, G.; El-Esawi, M.; El Nahhas, N. Exogenous Ascorbic Acid Induced Chilling Tolerance in Tomato Plants Through Modulating Metabolism, Osmolytes, Antioxidants, and Transcriptional Regulation of Catalase and Heat Shock Proteins. *Plants* **2020,** *9* (431), 1–21.

El-Saadony, F.; Nawar, D.; Zyada, H. Effect of Foliar Application with Salicylic Acid, Garlic Extract and Proline on Growth, Yield and Leaf Anatomy of Pea (*Pisum sativum* L.) Grown Under Drought Stress. *Middle East J. Appl. Sci.* **2017,** *7* (3), 633–650.

El-Shraiy Amal, M.; Mostafa, M. A. Enhancing Salt Tolerance of Cucumber Using Grafting and Some Bioregulators Middle East. *J. Agric. Res.* **2016,** *5* (4), 820–840. ISSN: 2077–4605.

El-Shraiy Amal, M.; Mostafa, M. A.; Zaghlool Sanaa, A.; Shehata, S. A. M. Physiological Aspect of Nacl-Salt Stress Tolerant among Cucurbitaceous Cultivars. *Aust. J. Basic Appl. Sci.* **2011,** *5* (10), 550–559. ISSN 1991–8178.

El-Tayeb, M. A. Differential Response of Two Vicia Faba Cultivars to Drought: Growth, Pigments, Lipid Peroxidation, Organic Solutes, Catalase and Peroxidase Activity. *Acta Agron. Hung* **2006,** *54,* 25–37.

El-Tayeb, M. A. Response of Barley Grains to the Interactive Effect of Salinity and Salicylic Acid. *Plant Growth Regul.* **2005,** *45,* 215–224.

Fang, P.; Yan, M.; Chi, C.; Wang, M.; Zhou, Y.; Zhou, J.; Shi, K.; Xia, X.; Foyer, C. H.; Yu, J. Brassinosteroids Act as a Positive Regulator of Photoprotection in Response to Chilling Stress. *Plant Physiol.* **2019,** *180,* 2061–2076.

FAO. Annex 1. Crop Salt Tolerance Data. http://www.fao.org/docrep/005/y4263e/y4263e0e. htm (accessed Sept 12, 2017).

Faraji, J.; Sepehri, A. Exogenous Nitric Oxide Improves the Protective Effects of TiO_2 Nanoparticles on Growth, Antioxidant System, and Photosynthetic Performance of Wheat Seedlings Under Drought Stress. *J. Soil Sci. Plant Nutr.* **2020,** 1–12.

Farooq, M.; Irfan, M.; Aziz, T.; Ahmad, I.; Cheema, S. A. Seed Priming with Ascorbic Acid Improves Drought Resistance of Wheat. *J. Agron. Crop Sci.* **2013,** *199,* 12–22.

Farooq, M.; Shahzad, M. A.; Abdul Wahid, B. Priming of Field Sown Rice Seed Enhances Germination, Seedling Establishment, Allometry and Yield. *Plant Growth Regul.* **2006,** *49,* 285–294.

Farooq, M.; Wahid, A.; Lee, D. J. Exogenously Applied Polyamines Increase Drought Tolerance of Rice by Improving Leaf Water Status, Photosynthesis and Membrane Properties. *Acta Physiol. Plant* **2009,** *31* (5), 937–945.

Farouk, S.; Al-Amri, S. M. Exogenous Zinc Forms Counteract NaCl-Induced Damage by Regulating the Antioxidant System, Osmotic Adjustment Substances, and Ions in Canola (*Brassica napus* L. cv. Pactol) Plants. *J. Soil Sci. Plant Nutr.* **2019,** *19,* 887–899.

Farouk, S.; Elhindi, K.; Alotaibi, M. Silicon Supplementation Mitigates Salinity Stress on *Ocimum basilicum* L. via Improving Water Balance, Ion Homeostasis, and Antioxidant Defense System. *Ecotoxicol. Environ. Safet.* **2020,** *206,* 111396.

Feng, R.; Wei, C.; Tu, S. The Roles of Selenium in Protecting Plants Against Abiotic Stresses. *Environ. Exp. Bot.* **2013,** *87,* 58–68.

Ferrante, A.; Luigi, M. Agronomic Management for Enhancing Plant Tolerance to Abiotic Stresses: Drought, Salinity, Hypoxia, and Lodging. Horticulturae **2017,** *3,* 52. DOI: 10.3390/horticulturae3040052. www.mdpi.com/journal/horticulturae

Ferrante, A.; Luigi, M. Agronomic Management for Enhancing Plant Tolerance to Abiotic Stresses: High and Low Values of Temperature, Light Intensity, and Relative Humidity. *Horticulturae* **2018**, *4*, 21. DOI: 10.3390/horticulturae403. www.mdpi.com/journal/horticulturae

Flowers, T. J. Improving Crop Salt Tolerance. *J. Exp. Bot.* **2004**, *55*, 307–319.

Foyer, C. H.; Noctor, G. Ascorbate and Glutathione: The Heart of the Redox Hub. *Plant Physiol.* **2011**, *155*, 2–18.

Gaafar, A.; Ali, S.; El-Shawadfy, M.; Salama, Z.; Sekara, S.; Ulrichs, C.; Abdelhamid, M. Ascorbic Acid Induces the Increase of Secondary Metabolites, Antioxidant Activity, Growth, and Productivity of the Common Bean Under Water Stress Conditions. *Plants* **2020**, *9*, 627.

Garcia-Sanchez, F.; Syvertsen, J. P.; Gimeno, V.; Botia, P.; Perez-Perez, J. G. Responses to Flooding and Drought Stress by Two Citrus Rootstock Seedlings with Different Water-Use Efficiency. *Biol. Plant* **2007**, *130*, 532–542.

Gaskin, J. W.; Steiner, C.; Harris, K.; Das, K. C.; Bibens, B. Effect of Low-Temperature Pyrolysis Conditions on Biochar for Agricultural Use. *Trans. ASABE* **2008**, *51*, 2061–2069.

Ghalati, R.; Shamili, M.; Homaei, A. Effect of Putrescine on Biochemical and Physiological Characteristics of Guava (*Psidium guajava* L.) Seedlings Under Salt Stress. *Sci. Hortic.* **2020**, *261*, 5.

Gohari, J.; Roohi, A.; Talaee, A.; Gholi Zadeh, R. *Sugar Beet J*, **1995**, 11. (In Persian) *Biol. Res.* **2012**, *3* (7), 3474–3478.

Gomes, D.; Agasse, A.; Thiebaud, P.; Delrot, S.; Geros, H.; Chaumont, F. Aquaporins Are Multifunctional Water and Solute Transporters Highly Divergent in Living Organisms. *Biochim. Biophys. Acta* **2009**, *1788*, 1213–1228.

Gong, H. J.; Zhu, X. Y.; Chen, K. M.; Wang, S. M.; Zhang, C. L. Silicon Alleviates Oxidative Damage of Wheat Plants in Pots Under Drought. *Plant Sci.* **2005**, *169*, 313–321. DOI: 10.1016/j. plantsci.2005.02.023

Goreta, S.; Bucevic-Popovic, V.; Selak, G. V.; Pavela-Vrancic, M.; Perica, S. Vegetative Growth, Superoxide Dismutase Activity and Ion Concentration of Salt- Stressed Watermelon as in Xuenced by Rootstock. *J. Agri. Sci.* **2008**, *146*, 695–704.

Gunes, A.; Inal, A.; Alpaslan, M.; Eraslan, F.; Bagci, E. G.; Cicek, N. Salicylic Acid Induced Changes on Some Physiological Parameters Symptomatic for Oxidative Stress and Mineral Nutrition in Maize (Zea Mays L.) Grown Under Salinity. *J. Plant Physiol.* **2007**, *164*, 728–736.

Gurjar, G. N.; Swami, S.; Meena, N. K.; Lyngdoh, E. A. S. Effect of Solar Radiation in Crop Production. In *Natural Resource Management for Climate Smart Sustainable Agriculture*; Arora, S., Swami, S., Bhan, S., Eds.; Soil Conservation Society of India: New Delhi, 2017.

Gurmani, A. R.; Khan, S. U.; Mabood, F.; Ahmed, Z.; Butt, S. J.; Din, J.; Mujeeb-Kazi, A.; Smith, D. Screening and Selection of Synthetic Hexaploid Wheat Germplasm for Salinity Tolerance Based on Physiological and Biochemical Characters. *Int. J. Agric. Biol.* **2014**, *16*, 681–690.

Habib, N.; Ali, Q.; Ali, S.; Javed, M. T.; Zulqurnain Haider, M.; Perveen, R.; Shahid, M. R.; Rizwan, M.; Abdel-Daim, M. M.; Elkelish, A. A.; Bin-Jumah, M. N. Use of Nitric Oxide and Hydrogen Peroxide for Better Yield of Wheat (*Triticum aestivum* L.) Under Water Deficit Conditions: Growth, Osmoregulation, and Antioxidative Defense Mechanism. *Plants* **2020**, *9*.

Hai, N.; V, N.; Tu, N.; Kisiala, A.; Hoang, X.; Thao, N. Role and Regulation of Cytokinins in Plant Response to Drought Stress. *Plants* **2020,** *9,* 422.

Hajivar, B.; Zare-Bavani, M. R. Alleviation of Salinity Stress by Hydrogen Peroxide and Nitric Oxide in Tomato Plants. *Adv. Hortic. Sci.* **2019,** *33.*

Hamayun, M.; Hussain, A.; Khan, S. A.; Irshad, M.; Khan, A. L.; Waqas, M.; Shahzad, R.; Iqbal, A.; Ullah, N.; Rehman, G.; Kim, H. Y.; Lee, I. J. Kinetin Modulates Physio-Hormonal Attributes and Isoflavone Contents of Soybean Grown Under Salinity Stress. *Front. Plant Sci.* **2015,** *6,* 377.

Hameed, A.; Iqbal, N. Chemo-Priming with Mannose, Mannitol and H2O2 Mitigate Drought Stress in Wheat. *Cereal Res. Commun.* **2014,** *42,* 450–462.

Hany, G.; El-Gawad, A.; Mukherjee, S.; Farag, R.; Abd Elbar, O. H.; Hikal, M.; Abou El-Yazied, A.; Abd Elhady, S. A.; Helal, N.; ElKelish, A.; El Nahhas, N.; Azab, E.; Ismail, A.; Mbarki, S.; Ibrahim, M. F. M. Exogenous γ-Aminobutyric Acid (GABA)-Induced Signaling Events and Field Performance Associated with Mitigation of Drought Stress in *Phaseolus vulgaris* L. *Plant Signal. Behav.* **2021,** e1853384, 1–13. DOI: 10.1080/15592324.2020.1853384

Hartwigsen, J. A.; Evans, M. R. Humic Acid Seed and Substrate Treatments Promote Seedling Root Development. *Hortscience* **2000,** *35* (7), 1231–1233.

Hasan, M. M.; Alabdallah, N. M.; Alharbi, B. M.; Waseem, M.; Yao, G.; Liu, X.-D.; El-gawad, H. G. A.; El-yazied, A. A.; Ibrahim, M. F. M.; Jahan, M. S.; Fang, X.-W. Gaba: A Key Player in Drought Stress Resistance in Plants. *Int. J. Mol. Sci.* **2021,** *22* (18), 10136. DOI: 10.3390/ijms221810136

Hasanuzzaman, M.; Borhannuddin Bhuyan, M. H. M.; Anee, T. I.; Parvin, K.; Nahar, K.; Al Mahmud, J.; Fujita, M. Regulation of Ascorbate-Glutathione Pathway in Mitigating Oxidative Damage in Plants Under Abiotic Stress. *Antioxidants* **2019,** *8,* 384.

Hays, H. B. *Humic Substances II: In Search of Structure*; Wiley: New York, 1989.

Hoque, M. A.; Banu, M. A.; Okuma, E.; Amako, K.; Nakamura, Y.; Shimoishi, Y.; Murata, Y. Exogenous Proline and Glycine Betaine Increase NaCl-Induced Ascorbate Glutathione Cycle Enzyme Activities, and Proline Improves Salt Tolerance More Than Glycine betaine in Tobacco Bright Yellow-2 Suspension-Cultured Cells. *J. Plant Physiol.* **2007,** *164,* 1457–1468.

Horvath, E.; Szalai, G.; Janda, T. Induction of Abiotic Stress Tolerance by Salicylic Acid Signaling. *J. Plant Growth Regul.* **2007,** *26,* 290–300. DOI: 10.1007/s00344-007-9017-4

Hossain, S. Present Scenario of Global Salt Affected Soils, its Management and Importance of Salinity Research. *Int. Res. J. Biol. Sci.* **2019,** *1* (1), 1–3. ISSN: 2663–5968. eISSN: 2663–5976.

Hu, W. H.; Yan, X. H.; Xiao, Y. A.; Zeng, J. J.; Qi, H. J.; Ogweno, J. O. 24-Epibrassinosteroid Alleviate Drought-Induced Inhibition of Photosynthesis in *Capsicum annuum. Sci. Hortic.* **2013,** *150,* 232–237.

Hussain, I.; Parveen, A.; Rasheed, R.; Ashraf, M. A.; Ibrahim, M.; Riaz, S.; Afzaal, Z.; Iqbal, M. Exogenous Silicon Modulates Growth, Physio-Chemicals and Antioxidants in Barley (*Hordeum vulgare* L.) Exposed to Different Temperature Regimes. *Silicon* **2019,** *11,* 2753–2762.

Hussain, M. M.; Farooq, D. J. Evaluating the Role of Seed Priming in Improving Drought Tolerance of Pigmented and Non-Pigmented Rice. *J. Agron. Lee Crop Sci.* **2017,** *203,* 269–276.

Hussain, S.; Khan, F.; Hussain, H. A.; Nie, L. Physiological and Biochemical Mechanisms of Seed Priming-Induced Chilling Tolerance in Rice Cultivars. *Front. Plant Sci.* **2016,** *7,* 116.

Ibrahim, A. M.; Khafaga, H. S.; Abd El-Nabi, A. S.; Eisa, S. S.; Shehata, S. A.Transplanting of Sugar Beet with Soil Drench by Potassium Humate or Potassium Silicate Enhanced Plant Growth and Productivity Under Saline Soil Conditions. *Curr. Sci. Int.* **2017,** *6* (2), 303–313. ISSN 2077–4435.

Ibrahim, M. F. M.; Bondok, A. M.; Al-Senosy, N. K.; Younis, R. A. A. Stimulation Some of Defense Mechanisms in Tomato Plants Under Water Deficit and Tobacco Mosaic Virus (TMV). *World J. of Agric. Sci.* **2015,** *11,* 5, 289–302.

Ibrahim, M.; Abd El-Samad, G.; Ashour, H.; El-Sawy, A.; Hikal, M.; Elkelish, A.; Abd El-Gawad, H.; Abou El-Yazied, A.; Hozzein, W.; Farag, R. Regulation of Agronomic Traits, Nutrient Uptake, Osmolytes and Antioxidants of Maize as Influenced by Exogenous Potassium Silicate Under Deficit Irrigation and Semiarid Conditions. *Agronomy* **2020a,** *10,* 1212.

Ibrahim, M. F. M.; Abd Elbar, O. H.; Farag, R.; Hikal, M.; El-Kelish, A.; El-Yazied, A. A.; Alkahtani, J.; Abd El-Gawad, H. G. Melatonin Counteracts Drought Induced Oxidative Damage and Stimulates Growth, Productivity and Fruit Quality Properties of Tomato Plants. *Plants* **2020b,** *9* (10), 1276. DOI: 10.3390/plants9101276

Iqbal, M.; Ashraf, M. Changes in Growth, Photosynthetic Capacity and Ionic Relations in Spring Wheat (*Triticum aestivum* L.) Due to Pre-Sowing Seed Treatment with Polyamines. *Plant Growth Regul.* **2005,** *46,* 19–30.

Iseri, O. D.; Sahin, F.; Hberal, M. Sodium Chloride Priming Improves Salinity Responses of Tomato at Seedling Stage. *J. Plant Nutr.* **2014,** *37,* 374–392.

Jabborova, D.; Annapurna, K.; Paul, S.; Kumar, S.; Saad, H. A.; Desouky, S.; Ibrahim, M. F.; Elkelish, A. A. Beneficial Features of Biochar and Arbuscular Mycorrhiza for Improving Spinach Plant Growth, Root Morphological Traits, Physiological Properties, and Soil Enzymatic Activities. *J Fungi* **2021a,** *7.*

Jabborova, D.; Wirth, S.; Halwani, M.; Ibrahim, M. F.; Azab, I. H.; El-Mogy, M. M.; Elkelish, A. A. Growth Response of Ginger (*Zingiber officinale*), Its Physiological Properties and Soil Enzyme Activities after Biochar Application Under Greenhouse Conditions. *Horticulturae* **2021b.**

Jafar, M. Z.; Farooq, M.; Cheema, M. A.; Afzal, I.; Basra, S. M. A.; Wahid, M. A.; Aziz, T.; Shahid, M. Improving the Performance of Wheat by Seed Priming Under Saline Conditions. *J Agron. Crop Sci.* **2012,** *198,* 38–45.

Jahan, M. S.; Guo, S.; Sun, J.; Shu, S.; Wang, Y.; El-Yazied, A. A.; Alabdallah, N. M.; Hikal, M.; Mohamed, M. H. M.; Ibrahim, M. F. M.; Hasan, M. M. Melatonin-Mediated Photosynthetic Performance of Tomato Seedlings Under High-Temperature Stress. *Plant Physiol. Biochem.* **2021,** *167,* 309–320. DOI: 10.1016/j.plaphy.2021.08.002

Javid, M. G.; Sorooshzadeh, A.; Moradi, F.; Modarres Sanavy, S. A. M.; Allahdadi, I. The Role of Phytohormones in Alleviating Salt Stress in Crop Plants. *Aust. J. Crop Sci.* **2011,** 5,726.

Jett, L. W.; Welbaum, G. E.; Morse, R. D. Effects of Matric and Osmotic Priming Treatments on Broccoli Seed Germination. *J. Am. Soc. Hort. Sci.* **1996,** *12,* 423–429.

Jiang, C.; Zu, C.; Lu, D.; Zheng, Q.; Shen, J.; Wang, H. Effect of Exogenous Selenium Supply on Photosynthesis, Na + Accumulation and Antioxidative Capacity of Maize (*Zea mays* L.) Under Salinity Stress. *Sci. Rep.* **2017,** *7,* 42039.

Jisha, K. C.; Puthur, J. T. Seed Priming with BABA (β-Amino Butyric Acid) a Cost-Effective Method of Abiotic Stress Tolerance in Vigna Radiata (L.) Wilczek. *Protoplasma* **2016,** *253,* 277.

Jisha, K. C.; Vijayakumari, K.; Puthur, J. T. Seed Priming for Abiotic Stress Tolerance: An Overview. *Acta Physiol. Plant* **2013**, *35*, 1381–1396.

Jones, R. W.; Pike, L. M.; Yourman, L. F. Salinity Influences Cucumber Growth and Yield. *J. Am. Soc. Hort. Sci.* **1989**, *114*, 547–551.

Kagale, S.; Divi, U. K.; Krochko, J. E.; Keller, W. A.; Krishna, P. Brassinosteroid Confers Tolerance in *Arabidopsis thaliana* and *Brassica napus* to a Range of Abiotic Stresses. *Planta* **2007**, *225* (2), 353–364.

Kang, G. Z.; Wang, C. H.; Sun, G. C.; Wang, Z. X. Salicylic Acid Changes Activities of H2O2-Metabolizing Enzymes and Increases the Chilling Tolerance of Banana Seedlings. *Environ. Exp. Bot.* **2003**, *50*, 9–15.

Karbalaei, S; Mehraban, A.; Mobasser, H. R.; Bitarafan, Z. Sowing Date and Transplant Root Size Effects on Transplanted Sugar Beet in Spring Planting. 2012

Kavitha Mary, J. P.; Marimuthu, K.; Sivakumar, U. Seed Priming Effect of Arbuscular Mycorrhizal Fungi Against Induced Drought in Rice. *J. Pharmacogn. Phytochem.* **2018**, *7*, 1742–https://www.phytojournal.com/archives/2018/vol7issue2/PartY/7-2-113-703.pdf (accessed June 27, 2020).

Kaya, C.; Sönmez, O.; Aydemir, S.; Dikilitaş, M. Mitigation Effects of Glycine Betaine on Oxidative Stress and Some Key Growth Parameters of Maize Exposed to Salt Stress. *Turk. J. Agric. For.* **2013**, *37*, 188–194.

Khan, A. A. Pre-Plant Physiological Seed Conditioning. In *Reviews*; Hort, J. J., Ed., Vol. 14; Willey: NY, 1992; pp 131–181.

Khan, M. A.; Khan, S.; Khan, A.; Alam, M. Soil Contamination with Cadmium, Consequences and Remediation Using Organic Amendments. *Sci. Total Environ.* **2017**, *601–602*, 1591–1605.

Khan, M. I. R.; Asgher, M.; Khan, N. A. Alleviation of Salt-Induced Photosynthesis and Growth Inhibition by Salicylic Acid Involves Glycine Betaine and Ethylene in Mungbean (*Vigna radiata* L.). *Plant Physiol. Biochem.* **2014**, *80*, 67–74.

Khan, M. I. R.; Iqbal, N.; Masood, A.; Per, T. S.; Khan, N. A. Salicylic Acid Alleviates Adverse Effects of Heat Stress on Photosynthesis Through Changes in Proline Production and Ethylene Formation. *Plant Signal. Behav.* **2013**, *8* (11), e26374.

Khan, W.; Rayirath, U.; Subramanian, S.; Jithesh, M.; Rayorath, P.; D. Hodges, M.; Critchley, A.; Craigie, J.; Norrie, J.; Prithiviraj, B. Seaweed Extracts as Biostimulants of Plant Growth and Development. *J. Plant Growth Regul.* **2009**, *28* (4), 386–399.

Khana, T. A.; Yusufb, M.; Ahmadc, A.; Zoobia, B.; Saeede, T.; Fariduddinf, Q.; Hayatf, S.; Mocka, H.; Wu, T. Q. Proteomic and Physiological Assessment of Stress Sensitive and Tolerant Variety of Tomato Treated with Brassinosteroids and Hydrogen Peroxide Under Low Temperature Stress. *Food Chem.* **2019**, *289*, 500–511.

Klessig, D. F.; Durner, J.; Noad, R.; Navarre, D. A.; Wendehenne, D.; Kumar, D.; Zhou, J. M.; Shah, J.; Zhang, S.; Kachroo, P.; Trifa, Y.; Pontier, D.; Lam, E.; Silva, H. Nitric Oxide and Salicylic Acid Signalling in Plant Defense. *Proc. Natl. Acad. Sci. USA* **2000**, *97*, 8849–8855.

Kloss, S.; Zehetner, F.; Dellantonio, A.; Hamid, R.; Ottner, F.; Liedtke, V.; Schwanninger, M.; Gerzabek, M. H.; Soja, G. Characterization of Slow Pyrolysis Biochars: Effects of Feed Stocks and Pyrolysis Temperature on Biochar Properties. *J. Environ. Qual.* **2012**, *41* (4), 990–1000.

Koller, D.; Hadass A. *Water Relations in the Germination of Seeds*, Vol. 12; Springer: Berlin, Germany, 1982; pp 401–431.

Krantev, A.; Yordanova, R.; Janada, T.; Szalai, G.; Popova, L. Treatment with Salicylic Acid Decreases the Effect of Cadmium on Photosynthesis in Maize Plant. *J. Plant Physiol.* **2008,** *165*, 920–931.

Kumari, A. Hemantaranjan, A. Morpho-Physiological Attributes of Wheat (*Triticum aestivum* L.) Genotypes as Influenced by Brassinosteroids Under Heat Stress. *J. Pharmacogn. Phytochem.* **2018,** *7* (6), 2111–2115.

Kumari, S.; Thakur, A. The Effects of Water Stress and Brassinosteroid on Apple Varieties. *Inter. J. Econ. Plants* **2019,** *6* (1), 1–6.

Kumeera, B.; Swapnil, M.; Chaurasia, A. K.; Ramteke, P. W. Effect of Seed Priming with Inorganics on Growth, Yield and Physiological Parameters of Chickpea (*Cicer arietinum* L.) Under Drought. *Pharma Innov. J.* **2018,** *7*, 411–414.

Kutby, A.; Al-Zahrani, H.; Hakeem, K. Role of Magnetic Field and Brassinosteroids in Mitigating Salinity Stress in Tomato (*Lycopersicon Esculentum* L.). *Inter. J. Eng. Res. Tech.* **2020,** *9* (6), 306–319.

Laird, D. A. The Charcoal Vision: A Win-Win-Win Scenario for Simultaneously Producing Bioenergy, Permanently Sequestering Carbon, While Improving Soil and Water Quality. *Agron. J* **2010,** *100*, 178–181.

Lee, J. M.; Kubota, C.; Tsao, S. J.; Bie, Z.; Hevarria, P. H.; Morra, L.; Oda, M. Current Status of Vegetable Grafting: Diffusion, Grafting Techniques. *Automation Scienta Hortculturae* **2010,** 93–105.

Lehmann, J. Carbon Sequestration in Dryland Ecosystems. *J Environ. Manage.* **2007,** *33*, 528–544.

Li, L.; Gu, W.; Li, C.; Li, W.; Li, C.; Li, J.; Wei, S. Exogenous Spermidine Improves Drought Tolerance in Maize by Enhancing the Antioxidant Defence System and Regulating Endogenous Polyamine Metabolism. *Crop Pasture Sci.* **2018,** *69* **(11), 1076–1091.**

Lim, B.; Smirnoff, N.; Cobbett, C. S.; Golz, J. F. Ascorbate-Deficient VTC2 Mutants in Arabidopsis Do Not Exhibit Decreased Growth. *Front. Plant Sci.* **2016,** *7*, 1025.

Liu, J. L.; Yang, R. C.; Jian, N.; Wei, L.; Ye, L. L.; Wang, R. H.; Gao, H. L.; Zheng, Q. S. Putrescine Metabolism Modulates the Biphasic Effects of Brassinosteroids on Canola and Arabidopsis Salt Tolerance. *Plant Cell Environ.* **2020,** *43* (6), 1348–1359.

Liu, S.; Li, H.; Lv, X.; Ahammed, G. J.; Xia, X.; Zhou, J.; Zhou, Y. Grafting Cucumber Onto Luffa Improves Drought Tolerance by Increasing ABA Biosynthesis and Sensitivity. *Sci. Rep.* **2017,** *6*, 20212.

Lopez-Delgado, H.; Dat, J. F.; Foyer, C. H.; Scott, I. M. Induction of Thermotolerance in Potato Microplants by Acetylsalicylic Acid and H2O2. *J. Exp. Bot.* **1998,** *49*, 713–720.

López-Gómez, E.; San Juan, M. A.; Diaz-Vivancos, P.; Mataix Beneyto, J.; García-Legaz, M. F.; Hernández, J. A. Effect of Rootstocks Grafting and Boron on the Antioxidant Systems and Salinity Tolerance on Loquat Plants (*Eriobotrya japonica* Lindl.). *Environ. Exp. Bot.* **2007,** *60*, 151–158.

Ibrahim, M. F. M.; Ibrahim, H. A.; Abd El-Gawad, H. G. Folic Acid as a Protective Agent in Snap Bean Plants Under Water Deficit Conditions. *J. Hortic. Sci. Biotechnol.* **2021,** *96* (1), 94–109. DOI: 10.1080/14620316.2020.1793691

Mahmood, A.; Turgay, O. C.; Farooq, M.; Hayat, R. Seed Biopriming with Plant Growth Promoting Rhizobacteria: A Review. *FEMS Microbiol. Ecol.* **2016,** *92*, fiw112.

Major, J.; Rondon, M.; Molina, D.; Riha, S.J.; Lehmann, J. Maize Yield and Nutrition After 4 Years of Doing Biochar Application to a Colombian Savanna Oxisol. *Plant Soil* **2010,** *333*, 117–120.

Marris, E. Putting the Carbon Back: Black Is the New Green. *Nature* **2006,** *442* (7103), 624–626. DOI: 10.1038/442624a

Marschner, H. *Mineral Nutrition of Higher Plants*; Academic Press: London, Orlando, San Diego, New York, Austin, Boston, Sydney, Tokyo, Toronto, 1995.

Marthandan V. M.; Geetha, R.; Kumutha, K.; Gandhimeyyan, V. R.; Karthikeyan, A.; Ramalingam, J. Seed Priming: A Feasible Strategy to Enhance Drought Tolerance in Crop Plants. *Int. J. Mol. Sci.* **2020,** *21,* 8258. DOI: 10.3390/ijms21218258. www.mdpi.com/journal/ijms

Merwad, A.; Desoky, E.; Rady, M. Response of Water Deficit-Stressed *Vigna unguiculata* Performances to Silicon, Proline or Methionine Foliar Application. *Sci. Hortic.* **2018,** *228,* 132–144.

Mishra, A.; Choudhuri, M. A. Ameliorating Effects of Salicylic Acid on Lead and Mercury-Induced Inhibition of Germination and Early Seedling Growth of Two Rice Cultivars. *Seed Sci. Technol.* **1997,** *25,* 263–270.

Mora-Herrera, M. E.; Lopez-Delgado, H.; Castillo-Morales, A.; Foyer, C. H. Salicylic Acid and H2O2 Function by Independent Pathways in the Induction of Freezing Tolerance in Potato. *Physiol Plant* **2005,** *125,* 430–440.

Moustafa-Farag, M.; Mohamed, H. I.; Mahmoud, A.; Elkelish, A. A.; Misra, A. N.; Guy, K. M.; Kamran, M.; Ai, S.; Zhang, M. Salicylic Acid Stimulates Antioxidant Defense and Osmolyte Metabolism to Alleviate Oxidative Stress in Watermelons under Excess Boron. *Plants* **2020a,** *9.*

Moustafa-Farag, M.; Elkelish, A. A.; Dafea, M.; Khan, M. A.; Arnao, M. B.; Abdelhamid, M. T.; El-Ezz, A. F.; Almoneafy, A.; Mahmoud, A.; Awad, M.; Li, L.; Wang, Y.; Hasanuzzaman, M.; Ai, S. Role of Melatonin in Plant Tolerance to Soil Stressors: Salinity, pH and Heavy Metals. *Molecules* **2020b,** *25.*

Müller-Moulé, P.; Havaux, M.; Niyogi, K. K. Zeaxanthin Deficiency Enhances the High Light Sensitivity of an Ascorbate-Deficient Mutant of Arabidopsis. *Plant Physiol.* **2003,** *133,* 748–760.

Munns, R.; Tester, M. Mechanisms of Salinity Tolerance. *Annu. Rev. Plant Biol.* **2008,** *59,* 651–681.

Mustafa, H. S.; Mahmood, T.; Ullah, A.; Sharif, A.; Bhatti, A. N.; Nadeem, M.; Ali, R. Role of Seed Priming to Enhance Growth and Development of Crop Plants Against Biotic and Abiotic Stress. *Bull. Biol. All. Sci.* **2017,** *2,* 1–11.

Nahar, K.; Hasanuzzaman, M.; Alam, M. M.; Rahman, A.; Mahmud, J. A.; Suzuki, T.; Fujita, M. Insights Into Spermine-Induced Combined High Temperature and Drought Tolerance in Mung Bean: Osmoregulation and Roles of Antioxidant and Glyoxalase System. *Protoplasma.* **2017,** *254* **(1), 445–460.**

Nasibi, F.; Kalantari, K. M. Influence of Nitric Oxide in Protection of Tomato Seedling Against Oxidative Stress Induced by Osmotic Stress. *Acta Physiol. Plant.* **2009,** *31,* 1037–1044.

Nemat Alla, M. M.; Younis, M. E.; El-Shihaby, O. A.; El-Bastawisy, Z. M. Kinetin Regulation of Growth and Secondary Metabolism in Waterlogging and Salinity Treated *Vigna sinensis* and *Zea mays. Acta Physiol. Plantarum* **2002,** *24,* 19–27.

Olmo, M.; Alburquerque, J. A.; Barrón, V.; del Campillo, M. C.; Gallardo A.; Fuentes, M. Wheat Growth and Yield Responses to Biochar Addition Under Mediterranean Climate Conditions. SPECIAL ISSUE, *Biol. Fertil. Soils*; Springer-Verlag: Berlin, Heidelberg, 2014. DOI: 10.1007/s00374014–0959

Otani, T.; Seike, N. Comparative Effects of Rootstock and Scion on Dieldrin and Endrin Uptake by Grafted Cucumber (*Cucumis sativus*). *J. Pestic. Sci.* **2006**, *31*, 316–321.

Otani, T.; Seike, N. Rootstock Control of Fruit Dieldrin Concentration in Grafted Cucumber (*Cucumis sativus*). *J. Pestic. Sci.* **2007**, *32*, 235–242.

Özenç, D. B. Growth and Transpiration of Tomato Seedlings Grown in Hazelnut Husk Compost Under Water-Deficit Stress. *Compost Sci. Util.* **2008**, *16*, 125–131. DOI: 10.1080/1065657X.2008.10702367

Palai, J. B.; Jena, J.; Lenka, S. K. Growth, Yield and Nutrient Uptake of Maize as Affected by Zinc Application -A Review. *Ind. J. Pure App. Biosci.* **2020**, *8*, 332–339.

Pang, C. H.; Wang, B. S. Role of Ascorbate Peroxidase and Glutathione Reductase in Ascorbate–Glutathione Cycle and Stress Tolerance in Plants. In *Ascorbate-Glutathione Pathway and Stress Tolerance in Plants*; Springer: New York, 2010; pp 91–113.

Parankusam, S.; Adimulam, S. S.; Bhatnagar-Mathur, P.; Sharma, K. K. Nitric oxide (NO) in Plant Heat Stress Tolerance: Current Knowledge and Perspectives. *Front. Plant Sci.* **2017**, *8*, 1582.

Parida, A. K.; Das, A. B. Salt Tolerance and Salinity Effects on Plants: A Review. *Ecotoxicol. Environ. Saf.* **2005**, *60*, 324–349.

Piñero, M.; Otálora, G.; Collado, J.; Marín, J.; Amor, F. Foliar Application of Putrescine Before a Short-Term Heat Stress Improves the Quality of Melon Fruits (*Cucumis melo* L.). *J. Sci. Food Agric.* **2021**, *101*, 1428–1435.

Pottosin, I.; Shabala, S. Polyamines Control of Cation Transport Across Plant Membranes: Implications for Ion Homeostasis and Abiotic Stress Signaling. *Plant Sci.* **2014**, *5* (154), 1–16.

Puthiyottil, P. Priming of *Abelmoschus esculentus* (L.) Moench (Okra) Seeds with Liquid Phosphobacterium: An Approach to Mitigate Drought Stress. *Trop. Plant Res.* **2015**, *2*, 276–281.

Qirat, M.; Shahbaz, M.; Perveen, S. Beneficial role of foliar-applied proline on carrot (*Daucus carota* l.) Under Saline Conditions. *Pak. J. Bot.* **2018**, *50* (5), 1735–1744.

Radhakrishnan, R.; Lee, I. J. Effect of Low Dose of Spermidine on Physiological Changes in Salt-Stressed Cucumber Plants. *Russ. J. Plant Physiol.* **2014**, *61* (1), 90–96.

Rady, M.; Semida, W.; Abd El-Mageed, T.; Hemida, K. Up-Regulation of Antioxidative Defense Systems by Glycine Betaine Foliar Application in Onion Plants Confer Tolerance to Salinity Stress. *Sci. Hortic.* **2018**, *240*, 614–622.

Rady, M. M.; Belal, H. E. E.; Gadallah, F. M.; Semida, W. M. Selenium Application in Two Methods Promotes Drought Tolerance in *Solanum lycopersicum* Plant by Inducing the Antioxidant Defense System. *Sci. Hortic.* **2020**, *266*, 109290.

Raskin, I. Role of Salicylic Acid in Plants. *Annu. Rev. Plant Physiol. Plant Mol. Biol.* **1992**, *43*, 439–463.

Rathore, S. Effect of Seaweed Extract on the Growth, Yield and Nutrient Uptake of Soybean (*Glycine max*) Under Rainfed Conditions. *South Afr. J. Bot.* **2009**, *75*, 351–355

Rivero, R. M.; Ruiz, J. M.; Sanchez, E.; Romero, L. Does Grafting Provide Tomato Plants an Advantage Against H2O2 Production Under Conditions of Thermal Shock? *Physiol. Plant.* **2003**, *117*, 44–50.

Rodríguez, Z. R.; Hernández-Montiel, L. G.; Murillo-Amador, B.; Rueda-Puente, E. O.; Capistrán, L. L.; Diéguez, E. T.; Cordoba, M. Effect of Hydropriming and Biopriming on Seed Germination and Growth of Two Mexican Fir Tree Species in Danger of Extinction. *Forests* **2015**, *6*, 3109–3122.

Romero, L.; Belakbir, A.; Ragala, L.; Ruiz, J. M. Response of Plant Yield and Leaf Pigments to Saline Conditions: Effectiveness of Different Rootstocks in Melon Plants (*Cucumis melo* L.). *Soil Sci. Plant Nutr.* **1997**, *43*, 855–862.

Rouphael, Y.; Cardarelli, M.; Colla, G.; Rea, E. Yield, Mineral Composition, Water Relations, and Water Use Efficiency of Grafted Mini-Watermelon Plants Under Deficit Irrigation. *HortScience* **2008a**, *43*, 730–736.

Rouphael, Y.; Cardarelli, M.; Rea, E.; Colla, G. Grafting of Cucumber as a Means to Minimize Copper Toxicity. *Environ. Exp. Bot.* **2008b**, *63*, 49–58.

Sadura, I.; Janeczko, A. Physiological and Molecular Mechanisms of Brassinosteroid-Induced Tolerance to High and Low Temperature in Plants. *Biol. Plant* **2018**, *62* (4), 601–616.

Saneoka, H.; Moghaieb, R. E. A.; Premachandra, G. S.; Fujita, K. Nitrogen Nutrition and Water Stress Effects on Cell Membrane Stability and Leaf Water Relations in Agrostis palustris Huds. *Environ. Exp. Bot.* **2004**, *52*, 131–138. DOI: 10.1016/j.envexpbot.2004.01.011

Saraswathi, S. G.; Paliwal, K. Drought Induced Changes in Growth, Leaf Gas Exchange and Biomass Production in *Albizia lebbeck* and *Cassia siamea* Seedlings. *J. Environ Biol.* **2011**, *32*, 173–178.

Satisha, J.; Prakash, G. S.; Bhatt, R. M.; Sampath Kumar, P. Physiological Mechanisms of Water Use Efficiency in Grape Rootstocks Under Drought Conditions. *Int. J. Agric. Res.* **2007**, *2*, 159–164.

Sato, S.; Peet, M.; Thomas, J. Physiological Factors Limit Fruit Set of Tomato (*Lycopersicon esculentum* Mill.) Under Chronic, Mild Heat Stress. *Plant, Cell Environ.* **2000**, *23* (7), 719–726.

Sattar, A.; Cheema, M. A.; Abbas, T.; Sher, A.; Ijaz, M.; Hussain, M. Separate and Combined Effects of Silicon and Selenium on Salt Tolerance of Wheat Plants. *Russ. J. Plant Physiol.* **2017**, *64*, 341–348.

Savvas, D.; Papastavrou, D.; Ntatsi, G.; Ropokis, A.; Olympios, C.; Hartmann, H.; Schwarz, D. Inter Active Effects of Grafting and Mn-Supply Level on Growth, Yield and Nutrient Uptake by Tomato. *HortScience* **2009**, *44*, 1978–1982.

Schulz, H.; Dunst, G.; Glaser, B. Positive Effects of Composted Biochar on Plant Growth and Soil Fertility. *Agron. Sustain. Dev.* **2013**, *33*, 817–827. DOI: 10.1007/s13593-013-0150-0

Semida, W.; Abd El-Mageed, T.; Abdelkhalik, A.; Hemida, K.; Abdurrahman, H.; Howladar, S.; Leilah, A.; Rady, M. Selenium Modulates Antioxidant Activity, Osmoprotectants, and Photosynthetic Efficiency of Onion Under Saline Soil Conditions. *Agronomy* **2021**, *11*, 855.

Shakirova, F. M.; Sakhabutdinova, A. R.; Bezrukova, M. V.; Fatkhutdinova, R. A.; Fatkhutdinova, D. R. Changes in the Hormonal Status of Wheat Seedlings Induced by Salicylic Acid and Salinity. *Plant Sci* **2003**, *164*, 317–322.

Shalaby, T. A.; Abd-Alkarim, E.; El-Aidy, F.; Hamed, E.; Sharaf-Eldin, M.; Taha, N.; El-Ramady, H.; Bayoumi, Y. Nano-Selenium, Silicon and H2O2 Boost Growth and Productivity of Cucumber Under Combined Salinity and Heat Stress. *Ecotoxicol. Environ. Safet.* **2021**, *212*, 111962.

Shankrayya, R. G.; Teggelli, M. P. Studies on Climate Smart Intervention on Induction of Drought Tolerance by Seed Priming with CaCl$_2$ in Chickpea Growth, Yield and Quality Parameters. *Int. J. Curr. Microbiol. App. Sci.* **2018**, *7*, 3510–3514.

Shariatmadari, M. H.; Parsa, M.; Nezami, A.; Kafi, M. The Effects of Hormonal Priming on Emergence, Growth and Yield of Chickpea Under Drought Stress in Glasshouse and Field. *Biosci. Res.* **2017**, *14*, 34–41. https://www.isisnBR-14–2017/34–41–14(1)2017BR-1404. pdf (accessed Jan 15, 2020).

Shi, Q.; Bao, Z.; Zhu, Z.; Ying, Q.; Qian, Q. Effects of Different Treatments of Salicylic Acid on Heat Tolerance, Chlorophyll fluorescence, and Antioxidant Enzyme Activity in Seedlings of *Cucumis sativa* L. *Plant Growth Regul.* **2006**, *48*, 127–135.

Singh, B.; Usha, K. Salicylic Acid Induced Physiological and Biochemical Changes in Wheat Seedlings Under Water Stress. *Plant Growth Regul.* **2003**, *39*, 137–141.

Singh, A.; Dahiru, R.; Musa, M.; Haliru, B. S. Effects of Osmo-Priming Duration on Germination, Emergence and Early Growth of Cowpea (*Vigna unguiculata* (L.) Walp.) in the Sudan Savanna Nigeria. *Int. J. Agron.* **2014**, *4*, 841238.

Singh, M.; Kumar, J.; Singh, S.; Singh, V. P.; Prasad, S. M. Roles of Osmoprotectants in Improving Salinity and Drought Tolerance in Plants: A Review. *Rev. Environ. Sci. Biotechnol.* **2015**, *14*, 407–426.

Sirritepe, N.; Sivritep, H. O.; Eris, A. The Effects of NaCl Priming on Salt Tolerance in Melon Seedlings Grown Under Saline Conditions. *Sci. Hortic.* **2003**, *97*, 229–237.

Sivaci, A.; Kaya, A.; Duman, S. E_ects of Ascorbic Acid on Some Physiological Changes of Pepino (*Solanum muricatum* Ait.) Under Chilling Stress. *Acta Biol. Hung.* **2014**, *65*, 305–318.

Sivakumar, R.; Nandhitha, G. K. Impact of PGRS and Nutrients Pre-Soaking on Seed Germination and Seedling Characters of Mung Bean Under Salt Stress Legume Research **2017**, *40* (1), 125–131. Print ISSN:0250–5371 / Online ISSN:0976–0571

Smirnoff, N. Ascorbic Acid Metabolism and Functions: A Comparison of Plants and Mammals. *Free Radic. Biol. Med.* **2018**, *122*, 116–129.

Sofy, M.; Elhindi, K.; Farouk, S.; Alotaibi, M. Zinc and Paclobutrazol Mediated Regulation of Growth, Upregulating Antioxidant Aptitude and Plant Productivity of Pea Plants Under Salinity. *Plants* **2020**, *9*, 1197.

Soliman, M.; Alhaithloul, H. A.; Hakeem, K. R.; Alharbi, B. M.; El-Esawi, M.; Elkelish, A. Exogenous Nitric Oxide Mitigates Nickel-Induced Oxidative Damage in Eggplant by Upregulating Antioxidants, Osmolyte Metabolism, and Glyoxalase Systems. *Plants* **2019**, *8*, 562.

Stapleton, A. E. Ultraviolet Radiation and Plants: Burning Questions. *Plant Cell* **1992**, *4*, 1353–1358.

Stevens, J.; Senaratna, T.; Sivasithamparam, K. Salicylic Acid Induces Salinity Tolerance in Tomato (Lycopersicon esculentum cv. Roma): Associated Changes in Gas Exchange, Water Relations and Membrane Stabilisation. *Plant Growth Regul* **2006**, *49*, 77–83.

Sticher, L.; MauchMani, B.; Metraux, J. P. Systemic Acquired Resistance. *Annu. Rev. Plant Pathol.* **1997**, *35*, 235–270.

Strobel, N. E.; Kuc, A. Chemical and Biological Inducers of Systemic Acquired Resistance to Pathogens Protect Cucumber and Tobacco from Damage Caused by Paraquat and Cupric Chloride. *Phytopathology* **1995**, *85*, 1306–1310.

Sultana, N.; Florance, H. V.; Johns, A.; Smirnoff, N. Ascorbate Deficiency Influences the Leaf Cell Wall Glycoproteome in *Arabidopsis thaliana*. *Plant Cell Environ.* **2015**, *38*, 375–384.

Sun, Y.; He, Y; Irfan, A.; Liu, X.; Yu, Q.; Zhang, Q.; Yang, D. Exogenous Brassinolide Enhances the Growth and Cold Resistance of Maize (*Zea mays* L.) Seedlings Under Chilling Stress. *Agronomy* **2020**, *10*, 488.

Sun, Y. J.; Wu, Y. Ma, D. Z.; Lv, J. Y.; He, Y. H.; Gong, L.; Liu, Z.; Gao. L. D.; Li, N.; Yan, D.; Zhu, J. H.; Yang. D. G. Effects of Exogenous Brassinosteroid on Germination and Physiological Characteristics of Maize Seedlings Under Low Temperature Stress. *North China J. Agron.* **2019**, *34*, 119–128.

Tabassum, T.; Farooq, M.; Ahmad, R.; Zohaib, A.; Wahid, A.; Shahid, M. Terminal Drought and Seed Priming Improves Drought Tolerance in Wheat. Physiol. *Mol. Biol. Plants* **2018**, *24*, 845–856.

Tahi, H.; Wahbi, S.; Wakrim, R.; Aganchich, B.; Serraj, R.; Centritto, M. Water Relations, Photosynthesis, Growth and Water Use Efficiency in Tomato Plants Subjected to Partial Rootzone Drying and Regulated Deficit Irrigation. *Plant Biosyst.* **2007**, *141*, 265–274. DOI: 10.1080/11263500701401927

Tani, E.; Chronopoulou, E. G.; Labrou, N. E.; Sarri, E.; Goufa, M.; Vaharidi, X.; Tornesaki, A.; Psychogiou, M.; Bebeli, P. J.; Abraham, E. M. Growth, Physiological, Biochemical, and Transcriptional Responses to Drought Stress in Seedlings of *Medicago sativa* L., Medicago Arboreal. and Their Hybrid (Alborea). *Agronomy* **2019**, *9*, 38.

Toth, S. Z.; Puthur, J. T.; Nagy, V.; Garab, G. Experimental Evidence for Ascorbate-Dependent Electron Transport in Leaves with Inactive Oxygen-Evolving Complexes. *Plant Physiol.* **2009**, *149*, 1568–1578.

Türkmen, D.; Demir, S.; Sensoy, S.; Dursun, A. Mycorrhizal Fungus and Humic Acid on the Seedling Development and Nutrient Content of Pepper Grown Under Saline Soil Conditions. *J. Biol. Sci.* **2005**, *5* (5), 568–574.

Uzoma, K. C.; M. Inoue; H. Andry, H. Fujimaki, Z. Zahoor, Nishihara, E. Effect of Cow Manure Biochar on Maize Productivity Under Sandy Soil Condition. *Soil Use Manage.* **2011**, *27*, 205–212.

Valdrighi, M. M.; Pear, A.; Agnolucci, M.; Frassinetti, S.; Lunardi, D.; Vallini, G. Effects of Compost Derived Humic Acid on Vegetable Biomass Production and Microbial Growth Within a Plant (*Cichorium intybus*) Soil System: A Comparative Study. *Agric. Ecosyst. Environ.* **1996**, *58*, 133–144.

Venema, J. H.; Dijk, B. E.; Bax, J. M.; van Hasselt, P. R.; Elzenga, J. T. M. Grafting Tomato (*Solanum lycopersicum*) onto the Rootstock of a High-Altitude Accession of *Solanum habrochaites* Improves Suboptimal-Temperature Tolerance. *Environ. Exp. Bot.* **2008**, *63*, 359–367.

Wang, Q.; Liang, X.; Dong, Y.; Xu, L.; Zhang, X.; Hou, J.; Fan, Z. Effects of Exogenous Nitric Oxide on Cadmium Toxicity, Element Contents and Antioxidative System in Perennial Ryegrass. *Plant Growth Regul.* **2013**, *69*, 11–20.

Wutipraditkul, N.; Wongwean, P.; Buaboocha, T. Alleviation of Salt-Induced Oxidative Stress in Rice Seedlings by Proline and /or Glycine Betaine. *Biol. Plantarum.* **2015**, *59*, 1–7.

Xu, C. X.; Liu, Y. L.; Zheng, Q. S.; Liu, Z. P. Silicate Improves Growth and Ion Absorption and Distribution in *Aloe vera* under Salt Stress. *J. Plants Physiol. Mol. Biol.* **2006**, *32*, 73–78.

Yadav, S. K. Cold Stress Tolerance Mechanisms in Plants. A Review Agron. *Sustain. Dev.* **2010**, *30*, 515–527. c INRA, EDP Sciences, 2009. DOI: 10.1051/agro/2009050.

Yang, P.; Wang, Y.; Li, J.; Bian, Z. Effects of Brassinosteroids on Photosynthetic Performance and Nitrogen Metabolism in Pepper Seedlings Under Chilling Stress. *Agronomy* **2019**, *9*, 839.

Yetisir, H.; Caliskan, M. E.; Soylu, S.; Sakar, M. Some Physiological and Growth Responses of Watermelon [Citrullus lanatus (Thunb.) Matsum.and Nakai] Grafted Onto Lagenaria siceraria to Flooding. *Environ. Exp. Bot.* **2006**, *58*, 1–8.

Yin, J.; Jia, J.; Lian, Z.; Hu, Y.; Guo, J.; Huo, H.; Zhu, Y.; Gong, H. Silicon Enhances the Salt Tolerance of Cucumber Through Increasing Polyamine Accumulation and Decreasing Oxidative Damage. *Ecotoxicol. Environ. Saf.* **2019**, *169*, 8–17.

Yu, J. Q.; Huang, L. F.; Hu, W. H.; Zhou, Y. H.; Mao, W. H.; Ye, S. F.; Nogués, S. A Role for Brassinosteroids in the Regulation of Photosynthesis in *Cucumis sativus*. *J. Exp. Bot.* **2004,** *55,* 1135–1143.

Zapata, P. J.; Serrano, M.; Pretel, M. T.; Amorós, A.; Botella, M. A. Polyamines and Ethylene Changes During Germination of Different Plant Species Under Salinity. *Plant Sci.* **2004,** *167,* 781–788.

Zhang, A.; Liu, Y.; Pan, G.; Hussain, Q.; Li, L.; Zheng, J.; Zhang, X. Effect of Biochar Amendment on Maize Yield and Greenhouse Gas Emissions from a Soil Organic Carbon Poor Calcareous Loamy Soil from Central China Plain. *Plant Soil* **2012,** *351,* 263–275.

Zhang, X.; Schmidt, R. E. Antioxidant Response to Hormone Containing Product in Kentucky Bluegrass Subjected to Drought. *Crop Sci.* **1999,** *39,* 545–551.

Zhang, X.; Ervin, E. H.; Schmidt, R. E. Plant Growth Regulators Can Enhance the Recovery of Kentucky Bluegrass Sod from Heat Injury. *Crop Sci.* **2003,** *43,* 952–956.

Zhang, Y.; Han, X.; Chen, X.; Jin, H.; Cui, X. Exogenous Nitric Oxide on Antioxidative System and ATPase Activities from Tomato Seedlings Under Copper Stress. *Sci. Hortic.* **2009,** *123,* 217–223.

Zheng, M.; Tao, Y.; Hussain, S.; Jiang, Q.; Peng, S.; Huang, J.; Cui, K.; Nie, L. Seed Priming in Dry Direct-Seeded Rice: Consequences for Emergence, Seedling Growth and Associated Metabolic Events Under Drought Stress. *Plant Growth Regul.* **2016,** *78,* 167–178.

Zhu, J.; Bie, Z. L.; Huang, Y. Effects of Grafting with Different Rootstocks on the Growth, Osmotic Adjustment and Anti-Oxidant Enzyme Activities of Cucumber Seedlings Under Salt Stress. *J. Shanghai Jiaotong Univ.-Agric. Sci* **2008,** *26* (5), 393–397.

Zwack, P. J.; Rashotte, A.M. Interactions Between Cytokinin Signalling and Abiotic Stress Responses. *J. Exp. Bot.* **2015,** *66,* 4863–4871.

CHAPTER 8

Heavy Metal Stress Tolerance in Plants: Signaling Responses and Role of Plant–Microbe Association

SOMA HALDER PAUL

Associate Professor, Department of Botany, Asutosh College, Kolkata, West Bengal, India

ABSTRACT

Several different heavy metals (HMs), such as Fe, Mn. Cu, Ni, Co, Cd, Cr, Zn, Hg, Pb, and As are present in the environment through some different natural processes or due to different anthropogenic activities, that is, by release of industrial waste, sewage disposal, and by using different fungicides or pesticides during agricultural practices. Toxic HMs, such as arsenic, cadmium, chromium, lead, mercury are most vulnerable to plants as well as to various other biota. They play crucial roles in decreasing crop productivity as the plants absorb those HMs by the root system and transport them to aerial parts, where they get sequestered. Tolerance against those HMs is mediated by various kinds of molecular and biochemical mechanisms and several extracellular components, such as glutathione, cysteine, phytochelatins, metallothiones play major role to compartmentalize those metal components into vacuoles or other cell organelles. Due to the presence of HMs in plant cells, oxidative stress is pronounced and various types of stress-related proteins, hormones, anti-oxidants, and other signaling molecules get involved. Plant-associated microbes in rhizosphere, such as bacteria, fungi, and mycorrhiza play major role in bioremediation of HMs from soil by various direct or indirect ways as their natural phenomenon of cell metabolism. These microbes are being utilized to save the plants from getting absorption of HMs from soil and sustainable development in crop production. This article would like to provide an overview about how the heavy metal stress can be managed by the plant by its own defense system and also with the help of microbes in rhizosphere.

8.1 INTRODUCTION

Heavy metals (HMs) are mostly naturally occurring inorganic chemical elements which are exhibiting high density along with metallic properties (Joshi et al., 2015; Mustafa and Komatsu, 2016). Being inorganic compound, they are nonbiodegradable and largely remain bound to the soil matrix and as a result left in the environment for long time (Huang et al., 2018). According to their function, some of them, such as Mo, Mn, Fe, Co, Zn, Cu, Ni are essential in trace amount required by the plant for its several biochemical and physiological processes. The term HMs particularly used in biological sense to those metals which are potentially toxic for human as well as environment (Tchounwou et al., 2012). Although soils are natural reservoir of HMs but the amount of these toxic metals and metalloids are increasing in soil and water due to various kinds of anthropogenic activities (Bowell et al., 2014; Zhu et al., 2014, 2015). These contaminants affect agricultural productivity and ecosystem function and also creating threat to human health posing a great risk to global economic growth (Landrigan et al., 2018). These HMs are accumulated in edible parts of plants, such as cereals and vegetables which induces health risks in human.

Among the HMs arsenic, cadmium, chromium, lead, mercury and aluminum are considered as more hazardous substances (Clemens and Ma, 2016) due to their toxic behavior in plant system as well as in other organisms. When the concentration of HM increases beyond supra optimal level within the plant, crop productivity may be reduced to a great extent (Xiong et al., 2014; Pierart et al., 2015). Both underground and above ground surface of plants can absorb HMs which directly or indirectly affect plant health (Patra et al., 2004). Direct consequences are inhibition of different cytoplasmic enzymes and damage to the cell structures due to oxidative stress (Jadia and Fulekar, 2009). Whether direct or indirect, plants when are exposed to high levels of HMs, it results in the reduction or even complete cessation of different metabolic activities. Although it is known that plants possess several defense strategies by which they can avoid or tolerate HM intoxication, but beyond certain limit, these mechanisms do not work and the survival of plant becomes crucial (Clemens and Ma, 2016), (Fig. 8.1).

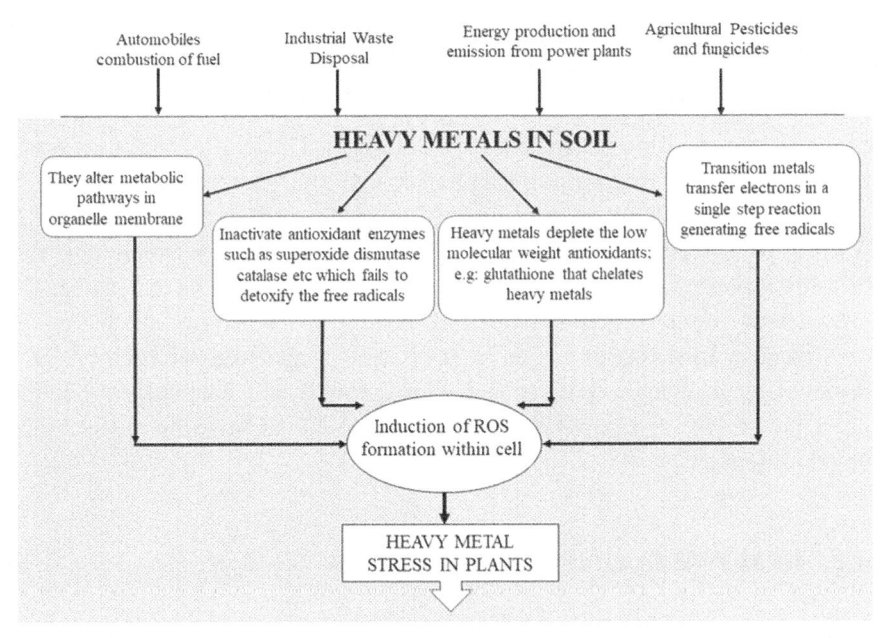

FIGURE 8.1 Heavy metals generated from different sources entering in soil creating heavy metal stress in plants.

To withstand the HM stress and metal toxicity, plants have evolved numerous defense mechanisms, viz., reduced HM uptake, sequestration of HMs into cell vacuoles, binding to some chelating compounds within the plant cell, that is, phytochelatins (PC) or metallothioneins (MTs), and activation of various antioxidants (Shahid et al., 2015). Different omics approaches, such as transcriptomics, proteomics, and metabolomics helped to understand the network involved in HM stress signaling mechanisms in plants. There are several studies which showed the application of genetic engineering in improvement of HM stress tolerance in plants (Verma et al., 2016, 2017). But this approach is more labor intensive, expensive, and time-consuming process; as well as this approach is not well accepted in agriculture due to controversy of role of GM crops plants in environment and to human health (Andersen et al., 2015).

Therefore, there is a demand for cost-effective, durable environment-friendly solutions to clean up the HMs, and it should be our priority. In this context, phytoremediation by the use of rhizospheric microbes or plant-associated microbes have emerged as important alternative strategy for sustainable agricultural practices to ensure high efficiency and better performance in HM removal. Many of the rhizospheric microbes have been reported to show abilities to protect the plant from HM stress in soil as well as help in remediation of HMs from the soil as they have the machinery to adopt and perform the metabolic activities in the presence of high concentration of various types of HMs. These microbes may form biofilm, siderophores, exopolysaccharides, and phytohormones (Tiwari et al., 2016, 2017) which help indirectly in plant growth and productivity. As this microbial role of HM remediation from the environment is not through any transgenic modification, it is totally acceptable to all; as there are various reports on this aspect, considerable investigations are going on still now. In this chapter, the role of plant-associated microbes in HM removal as well as to trigger the plant growth and productivity in the presence of HM stress will be discussed for the sustainable agricultural development.

8.2   HEAVY METAL PLANT INTERACTION

8.2.1   HEAVY METAL TOXICITY IN PLANTS: IMPACT ON PHYSIOLOGICAL PROCESSES

The toxicity of HMs that enter into plant tissues can inhibit multiple physiological processes in plants (Singh et al., 2015; López-Climent et al., 2011). At high concentration, they create damage to plants by (i) altering cell membrane permeability (Mahmood et al., 2009), (ii) inhibiting physiologically active enzymes (Mohammad et al., 2013), (iii) inactivating photosystems (Pizzeghello et al., 2013), and (iv) disturbing mineral metabolism (Gadd, 2007). The metal toxicity causes oxidative stress, disruption of different pigment function, and alteration of protein activity (Alemzadeh et al., 2014). The hypergeneration of reactive oxygen species (ROS) under HM stress may damage the cell structurally by (i) oxidation of cellular proteins and lipids, (ii) nucleic acid damage, and (iii) various enzyme inhibition and ultimately the cell death.

Seed germination and also the seedling growth are mostly affected by all toxic HMs, such as Hg, Cd, Pb, Ag, Cr, as in all these cases, the root is exposed to soil which is the reservoir of these HMs, so the root growth is maximally affected for, for example, Pb and Cr both have been reported to inhibit root growth and ultimately biomass reduction in wheat (Rees et al., 2016).

The binding of HM ions to cell wall and plasma membrane constituents, especially lipids alters the structure and functions leading to changes in other physiological functions of membranes and other cellular processes (Llamas et al., 2008). It is reported that the ATPase activity is inhibited by HM stress which is due to disruption of the cell membrane by free radicals generated within the cell (Zaidi et al., 2009; Janicka-Russak et al., 2012), which reduces the protein extrusion, and decreases the transport activity in root plasma membrane, ultimately reduces the nutrient uptake by the plant. The excessive generation of ROS under different heavy metal stress is a common feature of plants, which may react with cellular lipids, proteins, photosynthetic pigments, and other cellular organelles. It can cause lipid peroxidation of membrane lipid, which produces malondialdehyde (MDA) which can be used as a marker of oxidative stress and ultimately leading to cell death (Rozentsvet et al., 2012; Panda and Biswal, 1990). In addition to the overproduction of reactive oxygen species (ROS), ATP is exhausted, respiration rates get reduced and plant growth is affected (Singh et al., 2015; Yamamoto et al., 2002).

Another important physiological process like photosynthesis is very much affected by the HM due to (i) the deposition of HM in the foliage tissue (Gill et al., 2011), (ii) a change in the physiological activity of the chloroplast membrane and distribution of HM in leaf tissues, that is, stomata, mesophyll and bundle sheath (Romanowska et al., 2012), (iii) also the reduction in the formation of photosynthetic pigments (Shah et al., 2017; Chandra and Kang, 2016), (iv) alteration in the different cytosolic enzymes and organics (Shu et al., 2011), (v) variation in the supramolecular level action especially on PS I and PS II, membrane acyl lipids and carrier proteins of the vascular tissues (Srivastava et al., 2012), and (vi) destruction of enzyme associated with photosynthetic carbon reduction. The decrease in chlorophyll ratio under metal stress could possibly be due to the destabilization and destruction of various peripheral proteins (Shanker, 2003). Since HMs trigger the oxidative damage within the plant cell, it is very likely that these HMs could inhibit the synthesis of certain enzymes involved in chlorophyll synthesis (Fig. 8.2).

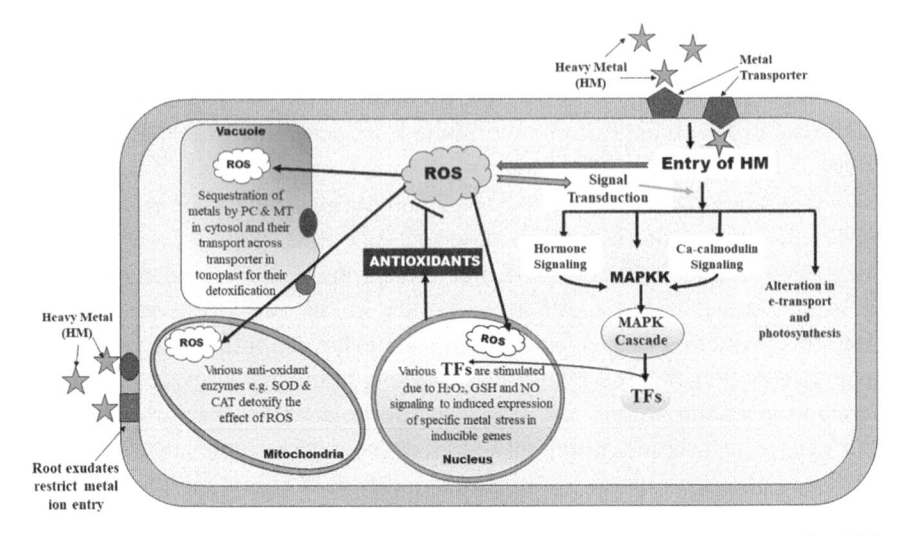

FIGURE 8.2   Interaction among different signaling system induced by heavy metals within a plant cell.

8.2.2   *SIGNALING RESPONSE UNDER HEAVY METAL STRESS*

The plants are natural bioaccumulators, that is, they uptake various HMs from soil or water, but the rate of accumulation of HMs in plant and plant's tolerance to HMs vary from species to species, in some cases those become toxic to plant system. Plants show various kinds of visible symptoms, such as stunted growth, chlorosis, root browning, and decline in growth and death (Ozturk et al., 2008, 2015). HM stress inactivates or denatures various important enzymes or other proteins, interferes with substitution reactions with other essential metal ions, which disturbs the integrity of membranes, other physiological processes, and cellular homeostasis (Hossain et al., 2012). These HMs can stimulate the production of ROS, such as superoxide radical (O_2^-), hydroxyl radical (OH-), and hydrogen peroxide (H_2O_2) along with other toxic components like methylglyoxal. These ROS may lead to lipid peroxidation of cell membrane component, damaging or causing cleavage to biomolecules and DNA which altogether is a combined activity due to oxidative stress created by high concentration of HMs in the cell.

Plants have various kinds of molecular and physiological mechanisms to overcome HM stress which are biochemical and genomic level processes, which are part of the homeostatic process and are constitutive. Plants use various types of mechanisms to avoid the exposure of HMs into the cell or some other mechanisms are activated only when the metal toxicity is encountered. All these responses are broadly classified as being avoidance or tolerant types (Krzesłowska, 2011). Plants apply first line of defense provided by the plant which helps to reduce the uptake of metals when stimulated with HM toxicity which is accomplished with the secretion of root exudates, which restricts the entry of HM ions by increasing efflux or biosorption to cell walls, that is, by avoidance mechanism. Many plants have tolerance mechanism for individual HM ions which are sequestered in different cellular compartments by chelating compounds, protecting other sensitive biomolecules present in the cell cytoplasm and metal interaction. Compounds, such as amino acids, glutathione, phytochelatins (PC), metallothioneins (MTs), and enzymes, such as superoxide dismutase and peroxidase are involved in such mechanisms (Hossain et al., 2012).

Whenever a plant is continuously exposed to HM stress, then the routine defense mechanisms may get exhausted. During then, plants activate various specialized mechanisms for the detoxification of metals during which the metals are chelated, transported, sequestered, or detoxified into plant vacuoles. Several stress-related proteins, hormones, antioxidants, and other signaling molecules are produced to resist the damage caused by those HMs.

Plants possess well-developed and complex signaling pathways involving many genes altogether. One of the most important pathways which is involved in this kind of abiotic stress is mitogen-activated protein kinase (MAPK) cascade (Jonak et al., 2002; Tena et al., 2001). This cascade is connected with phosphorylation, cell division, cell differentiation, and expression of various stress-related genes or inhibition of some. This MAPK also regulates the hormonal responses which indirectly regulate the gene expression. Concentration of trace element (heavy metal) plays a crucial role for determining whether the metal is inhibitory or having stimulatory effects on plant growth (Abolghassem et al., 2018). Generally, the toxic HMs are accumulated in the root cortex region which can impair various physiological and biochemical functions, ultimately have the impact on crop productivity. But these accumulations also can help the other parts of the plant not to get affected by the toxic effect of HMs.

8.2.2.1   HEAVY METAL UPTAKE AND TRANSLOCATION

Generally, the cationic metal ions are attracted by anionic carboxylic groups and create concentration gradient across the cell membrane which facilitates the movement of metal ions into the cells. The uptake of metal ions depends on the regulatory mechanisms, such as ATP-dependent high-affinity biphasic system, Zn-transporters help in Zn uptake, Cr and Pb enter through complexation with root exudates and through the symplast pathways or through Ca-transport channels through an active process. Chelation with the compounds, such as nicotinamine and citrate, histidine and malate present in the root exudates (induced by metal ions) prevents the uptake of unwanted metal ions. HM ions invade the xylem by the symplastic movement through electrochemical gradient across the membrane, the apoplastic pathway includes noncationic metal chelates, because the plant cell walls have relatively more cation exchange capacity.

Plants possess various types of plasma membrane transporters, such as heavy metal ATPase (HMA), copper transporter (CTR/COPT), natural resistance-associated macrophage protein (NRAMP), cation diffusion facilitators (CDF), yellow stripe-like (YSL), pleiotropic drug resistance (PDR), ABC transporters, ZTR/IRT-related protein (ZIP) for metal uptake and homeostasis (Ovečka and Takáč, 2014; Kushwaha et al., 2015). These membrane transporters belong to a particular group of proteins which are embedded into the phospholipid bilayer of plasma membrane and gets folded to form a channel which allows the ion transport. ZIP transporter family has eight transmembrane domains which help in binding of several bivalent metal ions, such as Zn, Cd, Mn, Ni, and Fe through iron-regulated transporter (IRT) (Xu et al., 2012; Milner et al., 2013). HMA transporter family utilizes the energy from ATP hydrolysis to efflux the HM ions across biological membrane, for example, Cu, Ag, Zn, Cd, and Pb (Migeon et al., 2010). Oligopeptide transporters (OPT) belonging to YSL family take part in the uptake of Zn, Cu, Fe, Ni, Cd, and Mn (Jain et al., 2019). The CDF transporter family consists of divalent metal ion transporters which are responsible in the transport of Mn, Ni, Fe, Co, Cd, and Zn from cytoplasm to the vacuole (Manara, 2012).

Metal ion translocation through the xylem vessel within the plant is regulated by transpiration or by cation exchange. After uptake, the HM that enters into the plant is regulated by transpiration or by cation exchange. After uptake through root, the HM enters into the plant through xylem loading and apoplastic transport to shoots, where these are detoxified or stored (Choppala et al., 2014). The translocation of metals may occur in complex form, that is, chelation with organic acids, amino acids or peptides, or with other types of organic chelating compound.

8.2.2.2 SEQUESTRATION AND COMPARTMENTALIZATION

Following uptake and transport, the HM gets accumulated inside the plant cells and can cause deleterious effects on the crop plants development. To deplete the toxic effects of the HM ions, many plants have developed different strategies that help in the sequestration of HM ions in different cellular compartments, such as Golgi Apparatus, vacuoles, and cell walls (Ovečka and Takáč, 2014; Saraswat and Rai, 2011) and then detoxification of those HM ions through chelation. Some of the cytosolic HMs, that is, Cd, Cu,Ag, Au, Hg, and Pb are chelated by metallothioneins (MTs), phyto-chelatins (PCs) and glutathione (GSH) to form LMW or HMW complexes and are transported into vacuoles (Joshi et al., 2015). Within the vacuoles, the HM ions are liberated and chelated with organic compounds, such as malate, oxalate, citrate and amino acids (Luo et al., 2016). In addition to these, other compartmentalization strategies include metal ion sequestration in some specialized cell types in the epidermal cells, trichomes, mesophyll cells of old leaves (Thakur et al., 2016).

8.2.2.3 SIGNAL TRANSDUCTION PATHWAYS IN PLANT UNDER HEAVY METAL STRESS

Under HM stress condition, a complex signal transduction network goes on by synthesis and activation of stress-related proteins and various signaling molecules. The signal transduction pathways usually lead to activation of various transcription factors (TFs) that induce transcription of various metal stress-responsive genes. Different signaling pathways are activated in response to different metal stresses which include MAPK cascade, ROS-signaling pathway, Ca-calmodulin pathway and hormonal signaling.

Plants utilize various strategies for heavy metal detoxification and also for the maintenance of their cellular concentration below toxicity level (Sharma and Dietz, 2009). In general, plants adopt two types of defense strategies to prevent the accumulation of HM into the plant cell cytoplasm, that is, heavy metal avoidance and tolerance to it (Thakur et al., 2016). The plant has the capacity to suppress the metal uptake using metal exclusion, complexation with various ligands, then controlling translocation within plant body, thereby avoiding the metal stress. The tolerance also depends on the types of metal ions, their concentration and moreover the plant species, organs where they are accumulated, and the developmental stage of the plant (Thakur et al., 2016). Essential metal ions, such as Mn Cu, Zn, and Fe are involved in complex metabolic processes and signal transduction mechanisms, such as Ca-Calmodulin signaling, MAPK cascade, ROS signaling and also activation of stress-related genes (Rout and Panigrahi, 2015). Transcriptional and several other types of differently produced proteins play significant role in heavy metal sequestration within cell and detoxification (Singh et al., 2016).

8.2.3   DEFENSE MECHANISM IN PLANTS UNDER HEAVY METAL STRESS

As mentioned earlier that plant possess an interrelated network of defense mechanisms to avoid or tolerate HM toxicity. There are some physical/ structural/morphological barriers, such as thick cuticle, trichomes, cell wall as well as some microbial consortium near rhizosphere which give the first line of defense when plants face HM stress (Hall, 2002; Wong et al., 2004; Harada et al., 2010).

Biosynthesis of different types cellular biomolecules is the primary way to tolerate or neutralize metal toxicity, such as low-molecular weight protein component chaperons or chelators or some cellular exudates, such as flavonoid or phenolic compounds, and amino acids, such as proline or histidine, Some hormones, such as salicylic acid, jasmonic acid, ethylene (Viehweger, 2014; Dalvi and Bhalerao, 2013; Sharma and Dietz, 2006).

When these above-mentioned strategies get failure, then increased induction of ROS takes place (Mourato et al., 2012). To mitigate the toxic effects of free HM radicles, the plant cell develops the antioxidant defense mechanism which is composed of different enzymatic antioxidants, for example, superoxide dismutase (SOD), catalase (CAT), ascorbate peroxidase (APX), glutathione reductase (GR) and many of the nonenzymatic antioxidants which act as scavengers of these free radicals (Sharma et al., 2012; Michalak, 2006; Rastgoo et al., 2011). Exploitation and upregulation of any of the mechanisms and biomolecules are plant species dependent, their level of tolerance (Solanki and Dhankhar, 2011), metal type and the stage of plant growth. Unlike constitutive mechanisms, the plants employ various adaptive mechanisms when only HM stress is encountered.

8.2.3.1 DEFENSE MECHANISM AT THE LEVEL OF ENTRY OF METAL IONS INTO THE PLANT SYSTEM

Plants system by changing the rhizospheric pH by the secretion of root exudates, which results into precipitation of metal ion, do not allow to enter the cell. It has been observed that resistant varieties of different plants have evolved more efficient secreting system and produce large amounts of specific root exudates as compared with susceptible genotypes (Ashraf et al., 2010). Various kinds of organic acids, such as malic, oxalic, citric acid play this kind of role of metal ion binding and inhibiting the entry of different HM ions such as Cd, Al by attaching them on to the cell wall, thereby restricting their movement to long distance travel within the plant body. Sometimes, the cell wall also deposits callose which does not allow the HM to pass through into the cell. The cell wall components bear many functional groups (-COOH, -SH, -OH) which can bind to bivalent or trivalent metal cations. The HGA domain of pectin is also responsible for capturing the metal ions (Krzesłowska, 2011), and in this way, this pectin layer also helps in the sequestration of metal ions by immobilization.

After entering the metal ions may get compartmentalized into the vacuole of the cell driven by the proton pump, vacuolar proton-ATPase (V-ATPase) and vacuolar proton pyrophosphatase (V-Ppase). The vacuoles contain the peptides or other molecules that can bind to metal ions in order to sequester or detoxify them. Various metal-binding chelators also help in to protect the plant cell metabolism from the toxic effect of HM ions.

8.2.3.2 ROLE OF DIFFERENT INDUCED PROTEIN MOLECULES IN DEFENSE UNDER HEAVY METAL STRESS

Phytochelatins (PCs)

These are cysteine-rich polypeptides which are found in various plant species, very well known for its cadmium binding capacity, these complexed molecules can be transported to other places of the cell by the action of ABC transporter, (Pochodylo and Aristilde, 2017; Emamverdian et al., 2015). It is well known that the biosynthesis of PCs can be regulated at the post-translational level by metalloids in many species, but its overexpression does not correlate with the tolerance level of a plant (Pomponi et al., 2006; Furini, 2012). It indicates that some other roles are played by these phytochelatins, such as involvement in essential metal ion homeostasis, antioxidant mechanism or other metabolic processes (Furini, 2012)

Metallothioneins

These are also similar kind of cysteine-rich metal-binding proteins which can immobilize, sequester and detoxify the HM ions (Capdevila and Atrian, 2011). The proposed roles of Metallothioneins (MTs) are (a) they may participate in maintaining the homeostasis of essential ions, (b) sequestration of toxic HM ions (c) can give protection against intracellular oxidative damage by HM stress (Hossain et al., 2012). Sequestration of intracellular HMs in eukaryotes is mediated by these cytosolic cysteine-rich peptides (MTs) which is also proved by the transgenic plants overexpressing MTs, which showed increased level of HM tolerance by means of metal accumulation and distribution strategies (Gu et al., 2015; Liu et al., 2015; Tomas et al., 2015). The MTs can be expressed in different plant parts, such as roots, leaf sheaths, and leaves of rice, sometimes less expressed in seeds. Recently, Irvine et al. (2017) showed an effort to develop MT-biosensor which can dramatically increase the signal associated with a metal of interest and their technology can be potentially used in monitoring the environment, specially, in the areas of HM contamination problems.

HSP Chaperones

Heavy metal ions either form complexes with functional side chain groups of different proteins or can displace essential ions from them leading to impairment of physiological functions (Tamás et al., 2014). They also interfere with the folding process, as a result nonfunctional protein or proteotoxic aggregate of deformed proteins are produced. Plants bear a set of stress genes which are induced by these toxic HMs to produce a group of proteins called HSPs (Gupta et al., 2010). In stress condition, the induced synthesis of HSP plays a significant role in the maintenance of the cellular homeostasis by assisting correct folding and preventing protein aggregation of misfolded protein (Hüttner et al., 2012; Park and Seo, 2015). HSPs function as molecular chaperons having molecular weight 10–200 KD localized into the cytosol, or in mitochondria and endoplasmic reticulum constitutively expressed in cells to assist folding of polypeptides and translations of precursor proteins (Wang et al., 2004). Recent transcriptomic studies have revealed that HSP 70 is expressed under a variety of metal stress including HM, HSP 60 is essential for many cellular functions both at normal and stress environment, it prevents the denaturation of proteins under stress condition (Sarry et al., 2006; Rodríguez-Celma et al., 2010). HSP 90 family proteins also play major roles in protein folding and regulation of signal transduction networks, cell cycle control, protein degradation, and also in protein trafficking. They are also found to be associated with the intercellular proteins, such as calmodulin, actin, tubulin, and some other receptor signaling kinases (Gupta et al., 2010; Park and Seo, 2015; Wang et al., 2004).

Misfolded Protein

Several abiotic stress, such as mutation, heat, active oxygen radicals, HM ions can cause disruption of protein folding pattern as well as misfolding of newly synthesized protein. HMs and metalloids inhibit proper folding and stimulate aggregation of several nascent proteins within living cells (Sharma et al., 2011; Jacobson et al., 2012). In various plants, the HMs, such as Cd, As, Pb, Hg, and Cr interfere with the protein folding in living cells and the potency of these misfolded and aggregated proteins differs (Tamás et al., 2014). In this situation, a coordinated adaptive program called unfolded protein response (UPR) is initiated which is a homeostatic response to alleviate HM stress. The induction of UPR has three aims: (a) to restore normal cell function by halting the production of new protein, (b) removal of misfolded protein, (c) activation of signaling pathways to increase the production of molecular chaperons involved in protein folding. If all these activities are not achieved in a cell within a certain time span, then the UPR aims at programmed cell death (PCD) (Deng et al., 2013; Liu and Howell, 2016).

In order to maintain cellular homeostasis in plant cells, the selective degradation of any unwanted misfolded or damaged protein occurs either by ubiquitin proteasome system (UPS) or by autophagosome induction (Liu and Howell, 2016). The UPS is a multistep-regulated enzymatic cascade which tightly controls the cellular homeostasis by precise degradation of unwanted proteins at a particular time. The significant role of UPS in the plant cell in the presence of HM stress has been recognized as inducer for overexpression of polyubiquitin genes and it is noted that the expression of these genes under stress condition indicates that the UPS is involved in the regulation of HM stress tolerance (Sun and Callis, 1997; Chai and Zhang, 1998). In extreme environments, overexpression of these genes involved in UPS cascade, which can enhance tolerance to multiple stresses without any kind of adverse effects on growth and development in plants (Guo et al., 2008).

On the other hand, autophagy is another biological self-destruction process by which the cells maintain their cellular homeostasis by degrading the damaged proteins or organelles into the cell vacuole during developmental transitions and also under stress condition (Liu and Bassham, 2012; Wen-Xing, 2012), these vacuoles are termed as autophagosome. There are three different kinds of autophagy which are microautophagy, macroautophagy, chaperon-mediated autophagy and organelle-mediated autophagy. The pivotal role of autophagy in HM stress and adaptive responses has been studied which showed the involvement of autophagy in plant cell towards metal tolerance and also the mechanism of adaptation (Zhang and Chen, 2010; Zheng et al., 2012). The cellular sites for ROS production and signaling are the primary targets of autophagy which can lead to either survival and death of cells (Scherz-Shouval and Elazar, 2007; Minibayeva et al., 2012). HMs are strong inducers of oxidative stress which is due to excessive accumulation of ROS that alters the cellular homeostasis. This excess accumulation of ROS can cause many types cellular injuries including damage to protein, lipid and DNA, some of which may lead to apoptosis and autophagy (Farah et al., 2016). Yang et al. (2016) demonstrated that unfolded protein accumulation in endoplasmic reticulum also triggers the process of autophagy under stress condition. Several other reports also suggest that heavy metal-induced intracellular ROS production may function in the signal transduction pathways, which lead to induction of autophagy (Zheng et al., 2012; Farah et al., 2016).

8.2.3.3   ROLE OF DIFFERENT SIGNALING PATHWAYS UNDER HEAVY METAL STRESS

In the presence of high level of HMs, the plant cells response through modulation of molecular and biochemical mechanism of cell which is evoked by important signal transduction network which are operated by several signal transduction units. The ultimate response of plant cells shown by synthesizing metal transfer protein, metal-binding protein and other modification of metabolic pathways helping the plants to counteract during HM stress

(Peng and Gong, 2014; Singh et al., 2015). In many crops, the early sign of heavy metal toxicity is shown by defects in nutrient imbalance, deficiency in photosynthesis, growth and development which is due to the interplay and convergence of many of these signaling pathways that finally result in the regulation of various transcription factors (TFs) activating several stress-responsive genes for the synthesis of metal chelators and transporters (Singh et al., 2015)

Calcium-Calmodulin Pathway

Heavy metal uptake, transport, and metabolism are regulated by calcium-calmodulin pathway as evidenced by studying under Cd, Ni, and Pb stress, which suggests that this pathway is targeted by many of the HMs. For example, Pb can bind to all four calcium-binding sites of calmodulin and stimulates its activity (Ouyang and Vogel, 1998), it has been observed that calcium can provide tolerance against chromium stress by enhancing the activity of different antioxidant enzymes (Fang et al., 2014). Most of the metals act as calcium analogues that induce calmodulin to signal transduction. Differential expression of calmodulin in response to arsenic stress indicates the role of calcium-signaling components in the stress response (Chakrabarty et al., 2009). In the presence of elevated levels of copper, the activations of intracellular calcium channels and generation of H_2O_2 was observed. There are several reports that the exogenous application of calcium can modulate many physiological and biochemical responses within cells in order to alleviate the HM stress. The antioxidant enzymes, such as ascorbate peroxidase (AP), glutathione reductase (GR), and superoxide dismutase (SD) has been shown to be enhanced upon the exogenous application of calcium (Ahmad et al., 2015). It has been evidenced that the gene expression of different antioxidant system is regulated via cross talk among the various cellular signals and levels of calcium, NO, and H_2O_2 (González et al., 2012). There are also several other reports on the role of calcium and calcium-dependent signaling pathways in imparting tolerance to HM stress in plants. Calmodulin has been reported to modulate the MAPK-signaling pathway (Tebar et al., 2002), which indicates the possibility of their interplay due to HM stress. All these findings suggest a vital role of calcium regulatory loop which is critical for the maintenance of the redox homeostasis of the cell and also ion balance in response to HM stress.

MAPK Signaling Under Heavy Metal Stress

MAPK is one of the most important signaling molecules which mediates the transmission of various stress-related signals, thus regulating large numbers of cellular processes. HM stress has profound effect on MAPK-signaling pathways, they are known to be activated by perception of several specific metal ligands and also by ROS molecules produced under the metal stress condition (Jonak et al., 2004; Smeets et al., 2013; Jalmi and Sinha, 2015). There are several reports showing activation of MAPKs in response to HMs, such as Cd, Cu, and As (Jonak et al., 2004; Smeets et al., 2013; Yeh et al., 2007; Ding et al., 2011; Rao et al., 2011) but Pb, Zn, and Fe less responsive. There are several evidences suggesting that metal ions, such as arsenic and chromium are able to induce reactive oxygen and nitrogen species within the cells, and thereby altering nitric oxide (NO)-induced cell signaling, which has been shown to modulate the activity of MAPK. Except this, MAPK cascades also exert positive feedback regulation on ROS production in cells (Asai et al., 2008). From several studies, it has been speculated that MAPK cascades might play a role in different metal stress depending on the activation induced by the ROS molecules produced (Kovtun et al., 2000).

8.2.3.4 ROLE OF PHYTOHORMONES UNDER HEAVY METAL STRESS

Phytohormones within a plant play a major role in growth and development by the activity of a single hormone regulating diverse cellular and developmental processes or by the activities of several hormones together by cross talking among each other [122]. The phytohormones act as regulators of various types of HM absorption, so used in agronomical crop management practices to reduce the metal toxicity (Piotrowska-Niczyporuk et al., 2012) which has been found to be safe for use and also gives promising result (Zhu et al., 2012, 2013; Agami and Mohamed, 2013; Masood et al., 2016). Many of the phytohormones play a significant role in signaling, in biochemical and defense pathways (Bücker-Neto et al., 2017), so the use of phytohormones to be given special attention for the management of crop production under HM stress.

Auxin is one of the most important phytohormones which directly controls the cell division in meristematic tissue but HM stress can modify the auxin metabolism and its dynamic distribution in plant (Bücker-Neto et al., 2017). Generally, HM stress induces to decrease in endogenous level of auxin, for example, As is responsible to alter IAA, NAA, and IBA level in *Brassica juncea* (Srivastava et al., 2013), Cd also effects the homeostasis in barley root tips (Zelinová et al., 2015). Exogenous application of auxin can be helpful to overcome the detrimental effect, which has been evidenced by several reports. In *Brassica*, exogenous supply of IAA could improve the plant growth under As stress, the application of L-RP, a precursor of auxin could improve the growth of root of rice seedling under Cd stress. In polluted areas, the application of auxin and some metal from outside could help in phytoremediation (Tandon et al., 2015). Co-application of selenium and auxin could improve the morphological and biochemical properties of rice seedling under As (Pandey and Gupta, 2015) and lead stress and IAA/ NAA has been found to decrease disorder of membrane organization and as a consequence, reduce HM toxicity (Hac-Wydro et al., 2016). So apart from deleterious effects of auxin in homeostasis, interaction between auxin and HM seems to be critical for survivability and also reproduction of some plants which indicates complex regulation of endogenous auxin in response to heavy metal exposure.

Cytokinins (CK) also play a regulatory role in plant development and under stress condition their endogenous concentration also gets altered (Brien and Benkova, 2013). HM stress decreases the CK production and its transport from roots to other parts of plant, participate intensively in interaction with other hormones (Ha et al., 2012). Exogenous application of CK showed increase in biomass production by increasing cell division, antioxidant capacity, and more shoot initiation (Tassi et al., 2008). CK application was also found to affect the photosynthesis rate by increasing the transpiration rate which could stimulate an increase in the plant biomass (Cassina et al., 2011). It has been observed that CK are often antagonists of abscisic acid (ABA) and change the level of plant hormones under HM stress as a result of interaction between them. ABA concentration in plant tissues is also known to increase after HM exposure which indicates its protective nature against HM toxicity. The increase in ABA level was observed in wheat seedling when germinated in the presence of metal ions, such as Hg, Cd, and Cu (Munzuro et al., 2008). In cucumber, the germination rate decreased but the level of ABA was increased in the presence of Cu and Zn (Wang et al., 2014). ABA was considered as a signaling molecule, which acts on guard cells of stomata for its regulation, decline in water potential helps in plant adaptation under unfavorable condition (Pantin et al., 2013). Exposure to toxic metals impairs the plant's water balance (Mukhopadhyay and Mondal, 2015), so it may play a role in protection against this stress. Under different HM stress in different plants, such as white bean, *Brassica*, the ABA concentration was found to increase (Rauser and Dumbroff, 1981; Poschenrieder et al., 1989; Salt et al., 1995). Exogenous application of ABA also found to affect the transport of Cd and Ni to the shoots, induces more accumulation in root (Rubio et al., 1994). As ABA is related to phloem loading, it protects the storage organs, such as fruits and seeds for accumulating HMs by growth inhibition (Vreugdenhil, 1983; Moya et al., 1995).

Brassinosteroids are also related with various plant growth and developmental activity, they are classified into different groups according to carbon number, Brassinolide (BL), 28-homobrassinolide (28-HomoBL) and 24-epibrassinolide (24-EpiBL) which are mostly bioactive under different stress condition (Vardhini et al., 2006). Exogenous application of 24-epiBL was helpful for reducing the Ni stress in *Brassica, Raphanus, Vigna*, etc. (Sharma et al., 2011; Yusuf et al., 2012; Kanwar et al., 2013), and 28-HomoBL could protect the rice. Foliar application of 28-HomoBL) and 24-EpiBL could improve the Cd tolerance in *Brassica, Phaseolus, Tomato*, and *Cicer*. Supplementation with BRs also help the plants to enhance the antioxidant enzymatic activities in response to many HMs, such as zinc, lead, and chromium (Anuradha and Rao, 2007; Arora and Bhardwaj, 2010; Choudhary et al., 2011; Rady and Osman, 2012; Ramakrishna and Rao, 2013).

Under HM stress, the plants showed the increased level of ethylene (Maksymiec, 2007; DalCorso et al., 2010; Khan et al., 2015) due to increased expression of ACC oxidase and ACC synthase. Copper has been found to induce the expression of ACS genes in potatoes. Chromium showed the increased expression of ACS in Rice, all these different findings support the view that plants show a rapid increase in ethylene production, reduced plant growth and development under heavy metal stress (Maksymiec, 2007; Schellingen et al., 2014).

Recently another class of phytohormone, Strigolactone (SL) which is carotenoid derived compound found in different plant related to plant growth and development. There are few reports related to the role of SL in HM stress like Cd and exogenous use effect on the plant growth under HM stress. It is assumed from several studies that SLs cross talk with other hormones under HM stress.

Another naturally occurring phenolic compound, salicylic acid (SA) is linked to the defense response of plants under HM stress. SA together with ABA or jasmonic acid participates in the regulation of different abiotic stresses. SA content has been found to be increased by exposure to Cd in

Barley which can alleviate Cd toxicity by detoxification rather than activation of antioxidant defense system (Metwally et al., 2013).

From various studies, it can be concluded that plants may survive better with different phytohormones priming in HM-contaminated areas. Pretreatment using exogenous phytohormones may help the plant cells toward tolerance to HM stress and also will help in reclamation of HM-contaminated soil, and thus helps to increase the quality and crop yield in such areas.

8.3 ROLE OF RHIZOSPHERIC MICROBES UNDER HEAVY METAL STRESS

Nowadays HM-polluted soil creates a negative impact in agriculture which has become a serious issue, but these HMs are not biodegradable, but they can be altered from one organic complex to another less toxic form. There are several physical approaches for remediation of HMs, such as excavation, acid leaching, thermal treatment, electro reclamation, which are not acceptable always due to high cost and safety concern. The process of bioremediation is more acceptable due to its less undesirable effects, more nature-friendly and low cost. Soil microbes and plants are being used extensively for this purpose (Ayangbenro and Babalola, 2017). The plant rhizosphere serves as a rich source of different types of microbes which play their role differently under normal and under biotic and abiotic stress condition. The microbes, such as plant growth promoting rhizospheric (PGPR) microbes, vesicular arbuscular mycorrhiza (VAM), and the endophytic bacteria play a role in protecting the plant under HM stress as well as help in their bioaccumulation from soil, as they have the metabolic capabilities which are supported by molecular machinery to adopt and perform a high concentration of HM.

Rhizospheric microbial population involved in bioremediation of HMs mostly belong to the genera *Bacillus, Pseudomonas, Arthrobacter*, etc. *Rhizobia* are also very important for promoting plant growth, they are known for their nitrogen-fixing ability symbiotically, but also are known to detoxify HMs and improving the quality of contaminated soil (Checcucci et al., 2017). Several fungi from Ascomycota and Basidiomycota are most commonly reported from the HM-contaminated soil, but due to poor nutrition condition, the arbuscular mycorrhizal population mostly colonize there. Various intracellular functions of these AM fungi and other rhizospheric microbes are driven by metal ions present in the external environment and on the cell surface or transporting them into the cell or by transforming them into different forms. Thus, they can change the external soil quality by changing the metal speciation, their toxicity, mobility in the soil, dissolution, and deterioration (Gadd, 2010). All these metabolic activities depend on the physicochemical nature of soil, type, and concentration of the heavy metal species, and types of different species of microbes present there (Fig. 8.3).

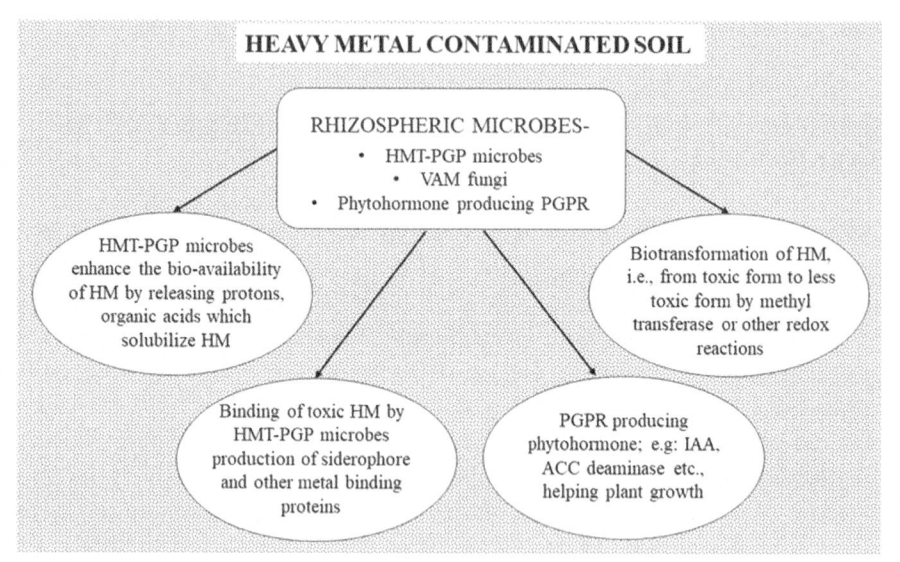

FIGURE 8.3   Rhizospheric microbes participating in heavy metal tolerance and plant growth promotion.

8.3.1   *BIOREMEDIATION OF HEAVY METAL BY RHIZOSPHERIC MICROBES*

The rhizospheric microbes play an important role in controlling the HM uptake in plants growing in polluted region through various mechanism. It has been observed that rhizospheric strains produce more extracellular polymeric substances (EPS) than non-rhizospheric strains, and this get enhanced under HM stress condition. These exopolymers bind strongly to form organometallic complexes which are less degradable naturally. The EPS production is reported as potential mechanism for Hg tolerance (Cruz, 2014), EPS producing *Azotobacter* has been found to be effective in binding Cr and Cd (Joshi and Juwarkar, 2009), likewise many of the bacterial EPS productions are involved in many heavy meal adsorption (Lau et al., 2005). The EPS have several ionizable functional groups, such as carboxyl, phosphoric, amine and hydroxyl groups which enable the EPS to sequester HMs (Liu and Fang, 2002; Seneviratne et al., 2015) as most of them are negatively charged, so easily they can form the organometallic complexes by electrostatic attraction, exchange, complexation, adsorption, and precipitation (Gutnick and Bach, 2000; Zhang et al., 2006). These complexes form the effective barrier surrounding the plant root cell (Rajkumar et al., 2010). Lead, copper, zinc were found to form strong covalent bond, whereas nickel and cadmium forms the weaker bond. The soil pH also plays a role in this ion-binding capacity.

Some of the rhizospheric microbes play a role in HM mobilization by increasing the solubility and speciation of many HM ions by secretion of various organic ligands and by decomposition of organic substances. Both bacteria and plant root cells are able to produce various organic acids which create the lower pH environment in the rhizosphere, which enhances the HM mobilization and uptake. Organic acids such as acetic and malic were found to be effective to enhance the Cd accumulation in maize root (Han et al., 2006), some other studies showed negative or no significant effect of metal uptake (Cr and Pb) in the presence of *Bacillus* in metal contaminated agricultural soil. Organic acids produced by rhizospheric bacteria act as natural chelating agent, which are also capable of solubilizing metals from soil.

Siderophores are also low-molecular weight iron chelating compound produced by many of the rhizospheric bacteria which are generally produced under low iron-containing soils and help the plants to participate in the uptake of iron. But these can also form stable complexes with other HMs (Rajkumar et al., 2010; Glick and Bashan, 1997), *Pseudomonas aeruginosa* shows the metal chelating ability of many of these HMs, such as Al, Cd, Cr, Cu, Hg, Ni, Pb (Braud et al., 2009a). The production of siderophores in rhizosphere involves in growth promoting effect of bacteria on plants (Braud et al., 2009b). These siderophore-producing bacteria play an important role in complexing toxic metals and can increase the efficiency of phytoextraction directly by enhancing the heavy metal accumulation within plant tissues (Rajkumar et al., 2010; Dimkpa et al., 2009). This phenomenon helps the plant to uptake the essential HM for plant growth under deficient condition in the soil.

Biosurfactants are microbial metabolites which also can help in metal mobilization and improve phytoremediation. These are amphiphilic molecules categorized as glycolipid, lipopeptides, phospholipids, which are either anionic or neutral in nature. The biosurfactants form complexes with HM at the soil surface, desorb metals from soil matrix, thus increasing metal solubility and bioavailability. Due to their anionic nature, low toxicity, and biodegradability and excellent surface-active properties, these can be used in heavy metal removal from the soil. *Bacillus, Pseudomonas, Torulopsis* are reported to produce the biosurfactant which could remove the HM from oil-contaminated soil also (Mulligan et al., 1999).

Metal oxidation by different rhizospheric microbes is also an interesting and an important process for phytoextraction process. Iron and sulfur-oxidizing bacteria, namely, *Thiobacillus ferrooxidans* and *T. thiooxidans* are able to leach the HMs, such as As, Cd, Co, Cu, Ni, and Zn (Gomez and Bosecker, 1999). Shi et al. (2011) reported that sulfur-oxidizing bacteria could enhance the Cu mobilization in contaminated soil and its uptake in plant tissue by reducing the soil pH and enhances the Cu availability in plant uptake. Similar activity was also found in sulfur-oxidizing bacteria which showed significant additive effect on Cd uptake and root bioaccumulation (Khorrami Vafa et al., 2012). The combined activity of iron and sulfur-oxidizing and reducing microbes on HM mobilization in contaminated soil has also been studied which showed increased mobility of Cu, Cd, Hg, and Zn by 90% (Beolchini et al., 2009).

8.3.2 ROLE OF ARBUSCULAR MYCORRHIZAL FUNGI IN HEAVY METAL REMEDIATION

Among the rhizospheric microbes, arbuscular mycorrhizal fungi are also most prominent soil microorganisms which help the plant root to facilitate the nutrient uptake (Saxena et al., 2017). Also involved in remediation of HM from the soil to protect the plant from metal toxicity. The process of remediation includes several processes: (a) HM bound to cell wall and gets deposited in the vacuoles of Arbuscular Mycorrhizal Fungi (AMF), (b) metal sequestration with the help of siderophores present in the soil or into root apoplasm, (c) metals bound to metallothioneins and phytochelatins inside the fungal or plant cells, and (d) metal transporters present at the tonoplast remove metals from cytoplasm (Jan and Parray, 2016). As this group is symbiotically associated with plant, not only the plants get benefits to avail more nutrient from soil but this association also decreases the uptake of HM by plants using those techniques described above. There is report that AMF induce tolerance by accumulating the HM into mycorrhiza-associated root cortical cells thereby transportation of HM into the plant shoot is lessened, protecting the plant from toxicity (Tiwari and Lata, 2018). In some plants,

AMF produce a glycoprotein, Glomalin, which has a metal-binding site that helps in metal accumulation.

Other than AMF, there are several filamentous rhizospheric fungi, namely, *Trichoderma, Penicillium, Aspergillus*, and *Mucor* which have been reported to have the ability to tolerate HM stress (Ezzouhri et al., 2009; Oladipo et al., 2018). Here also, the fungal cell walls also have good metal-binding capacity by presence of several functional groups (Tobin, 2001; Ong et al., 2017). Some of the fungi like *Ganoderma* has been reported to contribute in transformation of HM in less toxic forms (Kaewdoung et al., 2016). *Aspergillus* transforms Cr and helps in providing tolerance in Cr contaminated soil, Trichoderma could reduce the arsenic induced stress in rice (Tripathi et al., 2017).

8.3.3 ROLE OF HMT-PGP IN HEAVY METAL TOLERANCE AND PLANT GROWTH

Heavy metal tolerant and plant growth promoting (HMT-PGP) microbes in the rhizosphere can perform both the activities simultaneously (1) by altering the physicochemical properties of the soil to make the metal more bioavailable to microbes which trigger rapid removal of HM from the soil or (2) by detoxification and thereby helping the promotion of plant growth. There are several reports where rhizospheric PGPR (free living or symbiotic) can positively alter the plant growth and its productivity by production of different growth regulators via supplying and facilitating the nutrient uptake from soil (Nadeem et al., 2013). Also, they are potential elicitors for different abiotic stress tolerance including HM tolerance (Dary et al., 2010; Tiwari et al., 2016, 2017).

These microorganisms are considered agronomically important as they have evolved various mechanisms to avoid HM stress: (a) transport of metals across the cytoplasmic membrane, (b) biosorption and bioaccumulation of metals to the cell walls, (c) heavy metal precipitation, (d) metal entrapment in the extracellular capsule, and (e) Metal detoxification via oxidation reduction reaction (Zubair et al., 2016).

Root exudates released from plant root systems play an important role in changing the metal bioavailability to the microbes forming metal complexes. These may contain organic acids, amino acids, or phytochelatins which act as intracellular binding compounds for various HM ions. Biotransformation of some HMs by microbe-mediated redox reactions make them less or nontoxic form (Amstaetter et al., 2010). Transformation toxic form of chromium to nontoxic form of chromium by *Cellulosimicrobium cellulans* (Chatterjee et al., 2009), and toxic form of arsenic to nontoxic form of arsenic by oxidizing bacteria *Bacillus* and *Geobacillus* as reported by Majumder et al. (Majumder et al., 2013).

Many of the HMT-PGP microbes lower the soil pH and sequester soluble metal ions by producing different organic acids, these organic acids either help in HM solubilization or by chelation, they form the less toxic form, for example, from cadmium sulfate to cadmium oxalate, toxic copper sulfate to copper oxalate hydrate, lead nitrate to lead oxalate (Amstaetter et al., 2010). HM tolerant *Beauveria* sp. is reported to solubilize Cd, Cu, Pb, and Zn by producing organic acids (Gadd et al., 2014). Many PGPRs have the capacity to modulate the soil chemistry through intervening the methylation process, that is, transfer the methyl group to form the methylated metallic compounds that can be excavated easily in the soil zone (Talaat and Shawky, 2017).

Bioaccumulation is also responsible for HM uptake and further detoxification by HMT-PGP microbes, where the bioaccumulation is a combined process of passive uptake and biosorption which is metabolism-dependent and another active uptake process which occurs in living cells, requires different metabolic pathways and energy for the transport of metal ions (Gutierrez-Corona et al., 2016). The organic acids and chelating compounds produced by rhizospheric microbes take part in the biosorption process after entering into the cell, the final step of detoxification involves their sequestration or compartmentalization of HM into the different subcellular organelles.

8.3.4 PLANT GROWTH PROMOTION BY HMT-PGP MICROBES

In rhizospheric microbial communities, the HMT microbes show also the PGP traits, such as release of different extracellular enzymes, siderophores, phytohormones, solubilization of insoluble forms of minerals (phosphate, Zn, and K), also fixation of nitrogen which provide plant growth promotion and simultaneously reduce the adverse of HM on plants health. HMT-PGP microbes are reported to assist in plant growth and root development by enhanced nutrient availability and also by changing the bioavailability of HM (Wu et al., 2010; Gupta et al., 2014). Arsenic tolerant *Bacillus, Pseudomonas, Micrococcus* could help to increase the biomass of grapevine in the presence of high Arsenic. The plants growing near the mines/refineries showed good adaptability which was supported by the presence of HM tolerant (Cd, Pb, Cu) rhizospheric bacteria, for example, *Pseudomonas*, *Cupriadus*, *Bacillus*, and *Acinetobacter*. *Rhizobium* along with its nitrogen-fixing activity was found to have many other PGP traits, so it has been found to be a very good microbe which can be used for HM detoxification as well as plant growth promotion in legume plant under HM stress. Sessitsch et al. (2013) reported that HMT-PGP bacteria, such as *Bacillus*, *Pseudomonas*, *Streptomyces*, and *Methylobacterium* have the potential to improve the plant growth and production of many crops by reducing the detrimental effects. Pandey et al. (2013) reported that Cd-resistant *Ochrobactrum* and Pb- and As-resistant *Bacillus* have also the PGP traits which could help in bioremediation as well as growth promotion in rice cultivar. Several reports are also available to show that use of rhizospheric microbes along with some additives in HM-polluted soil could be more beneficial than without additives (Mishra et al., 2017), for example, addition of thiosulfate with an HMT microbe enhanced the mobilization and uptake of As and Hg in *Brassica juncea* by promoting bioavailability and phytoextraction.

The use of genetically transformed bacteria in HM bioremediation is also gaining much importance as these GM microbes are more efficient in remediation in symbiotic association (Ullah et al., 2015). These GM microbes possess one or more genes which increase the remediation capacities, such as genes for metal chelators, metal homeostasis, transporters, biodegradative enzymes, metal uptake regulators (Singh et al., 2011).

8.3.5 ROLE OF PHYTOHORMONE-PRODUCING PGPR FOR HEAVY METAL STRESS TOLERANCE

Several rhizospheric microbes have been found to produce phytohormones, including auxin, cytokinin, gibberellic acid, and ABA, though in a very small amount but play a crucial role in plant metabolic process (Ali et al., 2017; Egamberdieva et al., 2017; Mustafa et al., 2019). These phytohormone-producing microbes can alter the endogenous production of plant hormones (Sorty et al., 2016), thus changing root morphology and inducing tolerance against HM stress. HMs can affect the phytohormone production ability of the plant positively or negatively, and it has been studied in different microbes that under HM stress of Mn, Ni, Pb, As, Cu, and Cd *Escherichia,*

Serratia, Enterobacteria showed positive response, but *Klebsiella* showed negative response induced by Mn, Ni, and Cd. Egamberdieva et al. (2016) and Sorty et al. (2016) reported several genera, such as *Arthrobacter, Brevibacillus, Bacillus, Enterobacter, Mycobacterium, Pseudomonas, Pseudoxanthomonas,* and *Rhizobium* are capable of phytohormone production. Seed inoculation with these phytohormone-producing microorganisms can induce stress tolerance capacity in plants, Singh et al. (2018) presented a comprehensive study on microbial inoculation and role of phytohormones in modulating defense system in plants. Table 8.1 summarizes the role of various microbes in regulating HM tolerance in crops.

Use of PGPR microbes for improving the crop productivity under HM stress is an emerging technology. Román-ponce et al. (2017) isolated some of the rhizospheric microbes which are tolerant to high level of HMs and are also found to be positive for IAA production. Those strains could be the potential candidates for bioremediation as well as for improving plant growth by changing endogenous hormone level under HM stress. For example, *Variovorax, Pseudomonas fluorescens*, and Bacteriodetes separately or in combination helped *Brassica napus* to show tolerance against Zn and Cd. They could improve the physiological parameters as well as can reduce the accumulation of HM in roots (Dąbrowska et al., 2017). Germination and reduction in toxic effect was also observed in *Brassica* using *Bacillus* and *Pseudomonas* (Sheng and Xia, 2006; Weyens et al., 2013). Siderophore-producing strain of *Pseudomonas* has been reported to be HM tolerant which could reduce the uptake of HM in *Cucurbita* and *Brassica* under Cd stress (Pereira et al., 2015). It has been observed that these siderophore-producing HM tolerant PGPR could accelerate the bioavailability of nutrients to plant and reduce the deleterious effects of different HMs. IAA-producing HM-tolerant rhizobacteria like *Microbacterium, Achromobacter* were reported to be potential remediation agents which helped the biomass increase of *Trifolium* sp. under Zn and Cd stress (Hassan et al., 2017). IAA-producing strain of *Bacillus* could improve the root and shoot growth of *Lettuce* under Ni, Pb, and Cu stress (Seneviratne et al., 2016) and phytohormone producing, metal-tolerant strain of *Bacillus* and *Pseudomonas* were found to be effective in decreasing the toxic effects of HM on wheat under saline condition (Fatima and Ahmed, 2018; Wang et al., 2020a, 2020b). Further studies are needed to analyze the rhizospheric bacterial role in plant growth promotion by phytohormone production and as well as their induction of tolerance mechanism or giving protection by decreasing the uptake or accumulation mechanism of HM, that is, the cross talk among the procedures involved in as well as the genetic basis of expression or induction of different enzyme activities. Strategies to be made to improve the crop productivity by using these strains of rhizospheric microbes which will be beneficial for the soil as well as the environmental health (Nazli et al., 2020).

TABLE 8.1 Some of the Rhizospheric Bacteria Showing Plant Growth Promoting Traits Applied with Few Crop Plants Showed Heavy Metal Tolerance.

Bacterial strains	Plant growth promoting traits	Heavy metals tolerance	Crop plants	References
Klebsiella pneumoniae	IAA production	Cadmium	*Oryza sativa*	Pramanik et al. (2017)
	Siderophore production			
MCC 309	ACC deaminase activity			
Klebsiella pneumoniae	IAA production	Mercury	*Triticum aestivum*	Arora et al. (2016)
	Siderophore production			
HG 3	ACC deaminase activity			
	Exopolysaccharide synthesis			
	Phosphate solubilization			
Bacillus sp.	Auxin production,	Cadmium	*Zea mays*	Ahmad et al. (2016)
Klebsiella sp.	EPS production			
Enterobacter sp.	Phosphate solubilization			
	Oxidase & Catalase activity			
Staphylococcus arlettae	IAA production	Arsenic	*Brassica juncea*	Ma et al. (2008)
	Siderophore production			
	ACC deaminase activity			
Enterobacter sp.	IAA production	Cadmium	*Pisum sativum*	Naveed et al. (2020)
	Siderophore production			
Bacillus sp.	Auxin production,	Chromium	*Lens culinaris*	Khan et al. (2017)
	Phosphate solubilization			

8.3.6 *ROLE OF ENDOPHYTIC BACTERIA IN HEAVY METAL TOLERANCE*

Endophytic bacteria reside within the plant tissues without causing any damage to the host. They form small aggregates dispersed within tissues or organs of the plant where the exchange of different nutrients occurs promoting the various metabolic activity of the microbes, and in turn, helping the physiological activity of the host plant (Franco-Franklin et al., 2021). They can be nitrogen fixer or can promote the plant growth, improve stress response and also disease resistance through various mechanisms, such as mineral solubilization, phytohormone synthesis, production of siderophores (Ahemad and Khan, 2012). Thish category of bacteria, that is, plant growth promoting bacteria (PGPB) can also be proposed as a sustainable and effective alternative measure to promote crop growth and increase stress tolerance under salinity and HM stress.

Bacteria has a natural adaptive power to combat with the environment polluted with HMs showing resistance mechanism, including enzymatic oxidation and reduction of nontoxic forms (Tiwari and Lata, 2018). In addition to these, they can promote the heavy metal accumulation in plants and also enhancing the expression of stress-related genes (Srivastava et al., 2013) or by altering the levels of ACC levels which helps in metal tolerance through the manipulation of the ethylene levels in plants (Zhang et al., 2011).

Most of the studies conducted using hyperaccumulating plant and endophytic bacteria focused on the phytoremediation as well as promoting plant growth under stress condition. Now it is important to improve the ability of plant cell to accumulate metals in hyperaccumulating plant species as well as it is necessary to increase the heavy metal tolerance of non-accumulating plants of agronomic interest in order to expand the areas of cultivation from potentially arable land polluted with HM. The endophytic bacteria provide tolerance to HMs in host plants (Larimer et al., 2010; Rho et al., *2018*) and also showed positive effect on the host plant, where as Nadeem et al. (2013) showed the negative effect of these bacteria on their host as a result of overproduction of auxin and other metabolite derivatives. Some studies showed that the benefits provided by endophytic bacteria may be host-specific and also dependent on the intensity of the stress. Rho et al. (*2018*) compared the results obtained through this plant–microbe interaction under other abiotic stress condition as well as under specific HM stress condition.

Different beneficial effect of endophytic bacteria on their host plant is mediated by various mechanisms including production of auxins, siderophores, various nutrient-solubilizing compounds, increased activity of the enzymes like ACC deaminase and even changes in morphology (Tobin, 2001; Sessitsch et al., 2013). The increase of biomass has been found directly associated with an increase in phytoextraction efficiency and also stress tolerance to HMs. The HM non-accumulating plants get more benefit through this kind of plant–microbe interaction, that is, significant growth response compared with accumulating plant type (Lindblom et al., 2014; Chandra and Kumar, 2018).

8.4 FUTURE PROSPECTS

In today's world, the HM contamination in agricultural soil is a serious problem for sustainable agricultural development. We are all concerned about the use of eco-friendly as well as economical techniques to get rid out of this problem. The techniques used for HM clearance from contaminated sites include excavation, for example, physical removal of HM from contaminated sites, stabilization or in situ fixation by adding some other chelating compounds so that it is not absorbed by the plant or by reduction of HM by physical or chemical extraction method. However, all these physical processes are not so cost-effective or fruitful in application. Therefore, in recent past, much attention has been given for some alternative means, that is, biological means based on mainly to build up HM-tolerant plant species or phytoremediation by using plant species which can reduce the HM toxicity level. Through various researches and studies, it has been revealed that the signaling processes within the plant system due to the presence of high concentration of various HM uptake and tolerance depend on various factors and interactions among the various metabolic pathways which regulate the survivability of the plant and its growth in the HM-contaminated soil.

Along with this, it has also been observed through various studies that the rhizospheric microbes play a major role in this part, that is, they help the plant or protect them from the stress of HMs and induce proper growth and development of plants. These beneficial effects of microbes can be exploited in sustainable agricultural development for their phytoremediation activity along with plant growth promotion, which might have significant potential in crop improvement. HMT-PGP microbes and some associated host plants have been found to be very effective in HM remediation from polluted soil. Some reports reveal that some additive nutrients in the soil can help in removal capacity of these microbes. There are several reports that these kinds of strategies could improve the mobilization of HMs into the plants. The attention should be given that these strategies can be applied using weed plants or some non-food crops like timber yielding crops, which will result into nontransfer of HM into the food chain, also the weeds after removal from the land can be used for phytoextraction. Biostimulation is the process where the population of the rhizospheric microbes both bacteria and fungi are used for HM bioremediation to make the contaminated soil more useful for cultivation practices which is an effective approach for sustainable use of land for the development of agricultural output.

We observe that the endophytic bacteria can be considered as a good alternative for increasing HM stress tolerance, in most of the experiments, single-strain inoculum and hyperaccumulating plant types have been used in the experiments, but there are evidences that the bacterial consortia maximize the benefits compared with single strain, and non-accumulating plant types get greater advantage from this symbiotic association. So much research should go on in this direction and some more investigations are required for analyzing the role of these bacteria in gene expression and the various enzymatic function of the plant. Also, the studies should be done in natural system to get the actual impact of this plant–microbe interaction. However, further more research and studies are required to develop suitable bioformulation using different types of rhizospheric microbes (PGPR, HMT-PGP, AMF) as well as endophytic bacteria for the alleviation of HM stress and plant growth promotion in HM-contaminated soil.

More research is required in near future for the genetically improved strain of these microbes and to get more information about the cross talk between plant–microbe interaction under HM stress condition. Synergistic plan to be taken for metal mobilization, sequestration, transformation, detoxification, and ultimately the plant response to these processes and the crop or biomass yield to be monitored in the HM-contaminated field directly. At present, it is also difficult to understand the actual environmental impact of all these processes as sufficient data are not available.

KEYWORDS

- heavy metal
- tolerance
- signaling
- phytohormone
- plant-microbe interaction
- bioremediation

REFERENCES

Abolghassem, E.; Ding, Y.; Farzad, M.; Xie, Y.; Antioxidant Response of Bamboo (Indocalamus latifolius) as Affected by Heavy Metal Stress. *J. Elementol.* **2018,** *23* (1), 341–352.

Agami, R. A.; Mohamed, G. F. Exogenous Treatment with Indole3-Acetic Acid and Salicylic Acid Alleviates Cadmium Toxicity in Wheat Seedlings. *Ecotoxicol. Environ. Saf.* **2013,** *94,* 164–171.

Ahemad, M.; Khan, M. S. Alleviation of Fungicide-Induced Phytotoxicity in Green Gram [*Vigna radiata* (L.) Wilczek] Using Fungicide-Tolerant and Plant Growth Promoting Pseudomonas Strain. *Saudi J. Biol. Sci.* **2012,** *19* (4), 451–459. DOI: 10.1016/j.sjbs.2012.06.003.

Ahmad, A.; Hadi, F.; Ali, N. Effective Phytoextraction of Cadmium (Cd) with Increasing Concentration of Total Phenolics and Free Proline in Cannabis sativa (L) Plant Under Various Treatments of Fertilizers, Plant Growth Regulators and Sodium Salt. *Int. J. Phytoremed.* **2015,** *17,* 56–65. DOI: 10.1080/15226514.2013.828018.

Ahmad, I.; Akhtar, M. J.; Asghar, H. N.; Ghafoor, U.; Shahid, M. Differential Effects of Plant Growth-Promoting Rhizobacteria on Maize Growth and Cadmium Uptake. *J. Plant Growth Regul.* **2016,** *35,* 303–315.

Alemzadeh, A.; Rastgoo, L.; Tale, A.; Tazangi, S.; Eslamzadeh, T. Effects of Copper, Nickel and Zinc on Biochemical Parameters and Metal Accumulation in Gouan, Aeluropus Littoralis. *Plant Knowl. J.* **2014,** *3,* 31–38.

Ali, M. A.; Naveed, M.; Mustafa, A.; Abbas, A. The Good, the Bad, and the Ugly of Rhizosphere Microbiome. In *Probiotics and Plant Health*; Springer: Singapore, 2017; pp 253–290.

Amstaetter, K.; Borch, T.; Larese-Casanova, P.; Kappler, A. Redox Transformation of Arsenic by Fe (II)-Activated Goethite (α-FeOOH). *Environ. Sci. Technol.* **2010,** *44,* 102–108. DOI: 10.1021/es901274s.

Andersen, M. M.; Landes, X.; Xiang, W.; Anyshchenko, A.; Falhof, J.; Osterberg, J. T.; et al. Feasibility of New Breeding Techniques for Organic Farming. *Trends. Plant Sci.* **2015,** *20,* 426–434. DOI: 10.1016/j.tplants.2015.04.01.

Anuradha, S.; Rao, S. S. R. Effect of 24-Epibrassinolide on the Growth and Antioxidant Enzyme Activities in Radish Seedlings Under Lead Toxicity. *Indian J. Plant Physiol.* **2007,** *12,* 396–400.

Arora, K.; Sharma, S.; Monti, A. Bio-Remediation of Pb and Cd Polluted Soils by Switch Grass: A Case Study in India. *Int. J. Phytoremediation* **2016,** *18,* 704–709.

Arora, P.; Bhardwaj, R. 24-Epibrassinolide Induced Antioxidative Defence System of Brassica juncea L. Under Zn Metal Stress. *Physiol. Mol. Biol. Plants* **2010,** *16,* 285–293.

Asai, S.; Ohta, K.; Yoshioka, H. MAPK Signaling Regulates Nitric Oxide and NADPH Oxidase Dependent Oxidative Bursts in Nicotiana benthamiana. *Plant Cell* **2008,** *20,* 1390–1406. DOI: 10.1105/tpc.107.055855.

Ashraf, M.; Ozturk, M.; Ahmad, M. S. A.; Eds. *Plant Adaptation and Phytoremediation*; Springer: New York, 2010; p 481.

Ayangbenro, A.; Babalola, O. A New Strategy for Heavy Metal Polluted Environments: A Review of Microbial Biosorbents. *Int. J. Environ. Res. Public Health* **2017,** *14,* 94. DOI: 10.3390/ijerph14010094.

Beolchini, F.; Dell'Anno, A.; De Propris, L.; Ubaldini, S.; Cerrone, F.; Danovaro, R. Auto-and Heterotrophic Acidophilic Bacteria Enhance the Bioremediation Efficiency of Sediments Contaminated by Heavy Metals. *Chemosphere* **2009**, *74*, 1321–1326.

Bowell, R. J.; Alpers, C. N.; Jamieson, H. E.; Nordstrom, D. K.; Majzlan, J. The Environmental Geochemistry of Arsenic—An Overview. *Rev. Mineral. Geochem.* **2014**, *79*, 1–16. DOI: 10.2138/rmg.2014.79.1.

Braud, A.; Hannauer, M.; Mislin, G. L.; Schalk, I. J. The Pseudomonas Aeruginosa Pyochelin-Iron Uptake Pathway and Its Metal Specificity. *J. Bacteriol.* **2009a**, *191*, 3517–3525.

Braud, A.; Jézéquel, K.; Bazot, S.; Lebeau, T. Enhanced Phytoextraction of an Agricultural Cr-and Pb-Contaminated Soil by Bioaugmentation with Siderophore-Producing Bacteria. *Chemosphere* **2009b**, *74*, 280–286.

Brien, J. A. O.; Benkova, E. Cytokinin Cross-Talking During Biotic and Abiotic Stress Responses. *Front. Plant Sci.* **2013**, *4*, 451.

Bücker-Neto, L.; Paiva, A. L. S.; Machado, R. D.; Arenhart, R. A.; MargisPinheiro, M. Interactions Between Plant Hormones and HMs Responses. *Genet. Mol. Biol.* **2017**, *40*, 373–386.

Capdevila, M.; Atrian, S. Metallothionein Protein Evolution: A Miniassay. *J. Biol. Inorg. Chem.* **2011**, *16*, 977–989. DOI: 10.1007/s00775-011-0798-3.

Cassina, L.; Tassi, E.; Morelli, E.; Giorgetti, L.; Remorini, D.; Chaney, R. L.; Barbafieri, M. Exogenous Cytokinin Treatments of an NI Hyper-Accumulator, Alyssum Murale, Grown in a Serpentine Soil: Implications for Phytoextraction. *Int. J. Phytoremediation* **2011**, *13*, 90–101.

Chai, T.; Zhang, Y. Expression Analysis of Polyubiquitin Genes from Bean in Response to Heavy Metals. *Acta Bota. Sini.* **1998**, *41*, 1052–1057.

Chakrabarty, D.; Trivedi, P. K.; Misra, P.; Tiwari, M.; Shri, M.; Shukla, D.; et al. Comparative Transcriptome Analysis of Arsenate and Arsenite Stresses in Rice Seedlings. *Chemosphere* **2009**, *74*, 688–702. DOI: 10.1016/j.chemosphere.2008.09.082.

Chandra, R.; Kang, H. Mixed Heavy Metal Stress on Photosynthesis, Transpiration Rate, and Chlorophyll Content in Poplar Hybrids. *For. Sci. Technol.* **2016**, *12*, 55–61.

Chandra, R.; Kumar, V. Hyperaccumulator Versus Nonhyperaccumulator Plants for Environmental Waste Management. In *Phytoremediation of Environmental Pollutants;* Chandra, R., Dubey, N. K., Kumar, V., Eds., 1st ed.; CRC Press: Boca Raton, FL, 2018; pp 43–80.

Chatterjee, S.; Sau, G. B.; Mukherjee, S. K. Plant Growth Promotion by a Hexavalent Chromium Reducing Bacterial Strain, Cellulosimicrobium Cellulans KUCr3. *World J. Microbiol. Biotechnol.* **2009**, *25*, 1829–1836. DOI: 10.1007/s11274-009-0084-5.

Checcucci, A.; Bazzicalupo, M.; Mengoni, A. Exploiting Nitrogen Fixing Rhizobial Symbionts Genetic Resources for Improving Phytoremediation of Contaminated Soils. In *Enhancing Cleanup of Environmental Pollutants: Biological Approaches*; Naser, A., Anjum, A., Gill, S. S., Tuteja, N., Eds.; Vol. 1; Springer International Publishing: Cham, 2017; pp 275–288. DOI: 10.1007/978-3-319-55426-6_13.

Choppala, G.; Saifullah, B. N.; Bibi, S.; Iqbal, M.; Rengel, Z.; Kunhikrishnan, A.; Ashwath, N.; Ok, Y. S. Cellular Mechanisms in Higher Plants Governing Tolerance to Cadmium Toxicity. *Crit. Rev. Plant Sci.* **2014**, *33*, 374–391.

Choudhary, S. P.; Kanwar, M.; Bhardwaj, R.; Gupta, B. D.; Gupta, R. K. Epibrassinolide Ameliorates Cr (VI) Stress via Influencing the Levels of Indole-3-Acetic Acid, Abscisic

Acid, Polyamines and Antioxidant System of Radish Seedlings. *Chemosphere* **2011**, *84*, 592–600.

Clemens, S.; Ma, J. F. Toxic Heavy Metal and Metalloid Accumulation in Crop Plants and Foods. *Annu. Rev. Plant Biol.* **2016**, *67*, 489–512. DOI: 10.1146/annurev-arplant-043015-112301.

Cruz, K. A. *Extracellular Polysaccharides Production by Bacteria as a Mechanism of Mercury Tolerance*; Rutgers University-Graduate School; New Brunswick, 2014.

Dąbrowska, G.; Hrynkiewicz, K.; Trejgell, A.; Baum, C. The Effect of Plant Growth-Promoting Rhizobacteria on the Phytoextraction of Cd and Zn by Brassica napus L. *Int. J. Phytoremediation* **2017**, *19*, 597–604.

DalCorso, G.; Farinati, S.; Furini, A. Regulatory Networks of Cadmium Stress in Plants. *Plant Signal Behav.* **2010**, *5*, 663–667.

Dalvi, A. A.; Bhalerao, S. A. Response of Plants Towards Heavy Metal Toxicity: An Overview of Avoidance, Tolerance and Uptake Mechanism. *Ann. Plant Sci.* **2013**, *2* (9), 362–368.

Dary, M.; Chamber-Pérez, M. A.; Palomares, A. J.; Pajuelo, E. In Situ Phytostabilisation of Heavy Metal Polluted Soils Using Lupinus Luteus Inoculated with Metal Resistant Plant-Growth Promoting Rhizobacteria. *J. Hazard. Mater.* **2010**, *177*, 323–330. DOI: 10.1016/j.jhazmat.2009.12.035.

Deng, Y.; Srivastava, R.; Howell, S. H. Endoplasmic Reticulum (ER) Stress Response and Its Physiological Roles in Plants. *Int. J. Mol. Sci.* **2013**, *14*, 8188–8212. DOI: 10.3390/ijms14048188.

Dimkpa, C. O.; Merten, D.; Svatoš, A.; Büchel, G.; Kothe, E. Metal-Induced Oxidative Stress Impacting Plant Growth in Contaminated Soil is Alleviated by Microbial Siderophores. *Soil Biol. Biochem.* **2009**, *41*, 154–162.

Ding, Y.; Chen, Z.; Zhu, C. Microarray-Based Analysis of Cadmium responsive microRNAs in Rice (Oryza sativa). *J. Exp. Bot.* **2011**, *62*, 3563–3573. DOI: 10.1093/jxb/err046.

Egamberdieva, D.; Wirth, S. J.; Alqarawi, A. A.; Abd-Allah, E. F.; Hashem, A. Phytohormones and Beneficial Microbes: Essential Components for Plants to Balance Stress and Fitness. *Front. Microbiol.* **2017**, *8*, 2104.

Egamberdieva, D.; Wirth, S.; Behrendt, U.; Abd-Allah, E. F.; Berg, G. Biochar Treatment Resulted in a Combined Effect on Soybean Growth Promotion and a Shift in Plant Growth Promoting Rhizobacteria. *Front. Microbiol.* **2016**, *7*, 209.

Emamverdian, A.; Ding, Y.; Mokhberdoran, F.; Xie, Y. Heavy Metal Stress and Some Mechanisms of Plant Defense Response. Sci. World J. **2015**, *2015*, 1–18.

Ezzouhri, L.; Castro, E.; Moya, M.; Espinola, F.; Lairini, K. Heavy Metal Tolerance of Filamentous Fungi Isolated from Polluted Sites in Tangier, Morocco. *Afr. J. Microbiol. Res.* **2009**, *3*, 35–48.

Fang, H.; Jing, T.; Liu, Z.; Zhang, L.; Jin, Z.; Pei, Y. Hydrogen Sulfide Interacts with Calcium Signaling to Enhance the Chromium Tolerance in Setaria italica. *Cell Calcium* **2014**, *56*, 472–481. DOI: 10.1016/j.ceca.2014.10.004.

Farah, M. A.; Ali, M. A.; Chen, S. M.; Li, Y.; Al-Hemaid, F. M.; Abou-Tarboush, F. M.; et al. Silver Nanoparticles Synthesized from Adenium Obesum Leaf Extract Induced DNA Damage, Apoptosis and Autophagy via Generation of Reactive Oxygen Species. *Colloids Surf. B. Biointerfaces* **2016**, *141*, 158–169. DOI: 10.1016/j.colsurfb.2016.01.027.

Fatima, H. E.; Ahmed, A. Micro-Remediation of Chromium Contaminated Soils. *Peer J.* **2018**.

Franco-Franklin, V.; Moreno-Riacos, S.; Ghneim-Herrera, T. Are Endophytic Bacteria an Option for Increasing Heavy Metal Tolerance of Plants? A Meta-Analysis of the Effect Size. *Front. Environ. Sci.* **2021,** *8.* DOI: https://DOI.org/10.3389/fenvs.2020.603668.

Furini, A. *Plants and Heavy Metals*; Springer: Netherlands, 2012.

Gadd, G. M. Metals, Minerals and Microbes: Geomicrobiology and Bioremediation. *Microbiology* **2010,** *156* (Pt 3), 609–643. DOI: 10.1099/mic.0.037143-0.

Gadd, G. M.; Bahri-Esfahani, J.; Li, Q.; Rhee, Y. J.; Wei, Z.; Fomina, M.; et al. Oxalate Production by Fungi: Significance in Geomycology, Biodeterioration and Bioremediation. *Fungal Biol. Rev.* **2014,** *28,* 36–55. DOI: 10.1016/j.fbr.2014.05.001.

Gadd, G. M.; Geomycology: Biogeochemical Transformations of Rocks, Minerals, Metals and Radionuclides by Fungi, Bioweathering and Bioremediation. *Mycol. Res.* **2007,** *111,* 3–49.

Gill, S. S.; Khan, N. A.; Tuteja, N. Cadmium at High Dose Perturbs Growth, Photosynthesis and Nitrogen Metabolism While at Low Dose It Up Regulates Sulfur Assimilation and Antioxidant Machinery in Garden Cress (Lepidium sativum L.). *Plant Sci.* **2011,** *182,* 112–120.

Glick, B. R.; Bashan, Y. Genetic Manipulation of Plant Growth-Promoting Bacteria to Enhance Biocontrol of Phytopathogens. *Biotechnol. Adv.* **1997,** *15,* 353–378.

Gomez, C.; Bosecker, K. Leaching Heavy Metals from Contaminated Soil by Using Thiobacillus ferrooxidans or Thiobacillus thiooxidans. *Geomicrobiol. J.* **1999,** *16,* 233–244.

González, A.; Cabrera Mde, L.; Henríquez, M. J.; Contreras, R. A.; Morales, B.; Moenne, A. Cross Talk Among Calcium, Hydrogen Peroxide, and Nitric Oxide and Activation of Gene Expression Involving Calmodulins and Calcium dependent Protein Kinases in Ulva compressa Exposed to Copper Excess. *Plant Physiol.* **2012,** *158,* 1451–1462. DOI: 10.1104/pp.111.191759.

Gu, C. S.; Liu, L. Q.; Deng, Y. M.; Zhu, X. D.; Huang, S. Z.; Lu, X. Q. The Heterologous Expression of the Iris lactea var. chinensis Type 2 Metallothionein IlMT2b Gene Enhances Copper Tolerance in Arabidopsis thaliana. *Bull. Environ. Contam. Toxicol.* **2015,** *94,* 247–253. DOI: 10.1007/s00128-014-1444.

Guo, Q.; Zhang, J.; Gao, Q.; Xing, S.; Li, F.; Wang, W. Drought Tolerance Through Overexpression of Monoubiquitin in Transgenic Tobacco. *J. Plant Physiol.* **2008,** *165,* 1745–1755. DOI: 10.1016/j.jplph.2007.10.002.

Gupta, D. K.; Chatterjee, S.; Datta, S.; Veer, V.; Walther, C. Role of Phosphate Fertilizers in Heavy Metal Uptake and Detoxification of Toxic Metals. *Chemosphere* **2014,** *108,* 134–144. DOI: 10.1016/j.chemosphere.2014.01.030.

Gupta, S. C.; Sharma, A.; Mishra, M.; Mishra, R. K.; Chowdhuri, D. K. Heat Shock Proteins in Toxicology: How Close and How Far? *Life Sci.* **2010,** *86,* 377–384. DOI: 10.1016/j.lfs.2009.12.015.

Gutierrez-Corona, J. F.; Romo-Rodriguez, P.; Santos-Escobar, F.; Espino-Saldana, A. E.; Hernandez-Escoto, H. Microbial Interactions with Chromium: Basic Biological Processes and Applications in Environmental Biotechnology. *World J. Microbiol. Biotechnol.* **2016,** *32,* 191. DOI: 10.1007/s11274-016-2150-0.

Gutnick, D.; Bach, H. Engineering Bacterial Biopolymers for the Biosorption of Heavy Metals; New Products and Novel Formulations. *Appl. Microbiol. Biotechnol.* **2000,** *54,* 451–460.

Ha, S.; Vankova, R.; Yamaguchi-Shinozaki, K.; Shinozaki, K.; Phan Tran, L. S. Cytokinins: Metabolism and Function in Plant Adaptation to Environmental Stresses. *Trends Plant Sci.* **2012,** *17,* 172–179.

Hac-Wydro, K.; Sroka, A.; Jablo, K. The Impact of Auxins Used in Assisted Phytoextraction of Metals from the Contaminated Environment on the Alterations Caused by Lead (II) Ions in the Organization of Model Lipid Membranes. *Colloids Surf. B Biointerfaces* **2016,** *143,* 124–130.

Hall, J. L. Cellular Mechanisms for Heavy Metal Detoxification and Tolerance. *J. Exp. Bot.* **2002,** *53* (366), 1–11.

Han, F.; Shan, X.; Zhang, S.; Wen, B.; Owens, G. Enhanced Cadmium Accumulation in Maize Roots—The Impact of Organic Acids. *Plant Soil* **2006,** *289,* 355–368.

Harada, E.; Kim, J. -A.; Meyer, A. J.; Hell, R.; Clemens, S.; Choi, Y. E. Expression Profiling of Tobacco Leaf Trichomes Identifies Genes for Biotic and Abiotic Stresses. *Plant Cell Physiol.* **2010,** *51* (10), 1627–1637.

Hassan, T. U.; Bano, A.; Naz, I. Alleviation of Heavy Metals Toxicity by the Application of Plant Growth Promoting Rhizobacteria and Effects on Wheat Grown in Saline Sodic Field. *Int. J. Phytoremediation* **2017,** *19,* 522–529.

Hossain, M. A.; Piyatida, P.; Silva, J. A. T.; Fujita, M. Molecular Mechanism of Heavy Metal Toxicity and Tolerance in Plants: Central Role of Glutathione in Detoxification of Reactive Oxygen Species and Methylglyoxal and in Heavy Metal Chelation. *J. Bot.* **2012,** 1–37.

Huang, Y.; Chen, Q.; Deng, M.; Japenga, J.; Li, T.; Yang, X.; He, Z. Heavy Metal Pollution and Health Risk Assessment of Agricultural Soils in a Typical Peri-Urban Area in Southeast China. *J. Environ. Manag.* **2018,** *207,* 159–168.

Hüttner, S.; Veit, C.; Schoberer, J.; Grass, J.; Strasser, R. Unraveling the Function of Arabidopsis thaliana OS9 in the Endoplasmic Reticulum associated Degradation of Glycoproteins. *Plant Mol. Biol.* **2012,** *79,* 21–33. DOI: 10.1007/s11103-012-9891-4.

Irvine, G. W.; Tan, S. N.; Stillman, M. J. A Simple Metallothionein based Biosensor for Enhanced Detection of Arsenic and Mercury. *Biosensors* **2017,** *7,* 14. DOI: 10.3390/bios7010014.

Jacobson, T.; Navarrete, C.; Sharma, S. K.; Sideri, T. C.; Ibstedt, S.; Priya, S.; et al. Arsenite Interferes with Protein Folding and Triggers Formation of Protein Aggregates in Yeast. *J. Cell Sci.* **2012,** *125,* 5073–5083. DOI: 10.1242/jcs.107029.

Jadia, C. D.; Fulekar, M. H. Phytoremediation of Heavy Metals: Recent Techniques. *Afr. J. Biotechnol.* **2009,** *8,* 921–928.

Jain, S.; Muneer, S.; Guerriero, G.; Liu, S.; Vishwakarma, K.; Chauhan, D. K.; Dubey, N. K.; Tripathi, D. K.; Sharma, S. Tracing the Role of Plant Proteins in the Response to Metal Toxicity: A Comprehensive Review. *Plant Signal Behav.* **2019,** *13* (9), e1507401.

Jalmi, S. K.; Sinha, A. K. ROS Mediated MAPK Signaling in Abiotic and Biotic Stress-Striking Similarities and Differences. *Front. Plant Sci.* **2015,** *6,* 769. DOI: 10.3389/fpls.2015.00769.

Jan, S.; Parray, J. A.; Eds. Use of Mycorrhiza as Metal Tolerance Strategy in Plants. In *Approaches to Heavy Metal Tolerance in Plants*; Springer: Singapore, 2016; pp 57–68. DOI: 10.1007/978-981-10-1693-6_4.

Janicka-Russak, M.; Kabała, K.; Burzynski, M. Different Effect of Cadmium and Copper on H+ -ATPase Activity in Plasma Membrane Vesicles from Cucumis Sativus Roots. *J. Exp. Bot.* **2012,** *63,* 4133–4142.

Jonak, C.; Nakagami, H.; Hirt, H. Heavy metal stress. Activation of Distinct Mitogen-Activated Protein Kinase Pathways by Copper and Cadmium. *Plant Physiol.* **2004**, *136*, 3276–3283. DOI: 10.1104/pp.104.045724.

Jonak, C.; Okresz, L.; Bogre, L.; Hirt, H. Complexity, Cross Talk and Integration of Plant MAP Kinase Signalling. *Curr. Opin. Plant Biol.* **2002**, *5*, 415–424.

Joshi, P. M.; Juwarkar, A. A. In Vivo Studies to Elucidate the Role of Extracellular Polymeric Substances from Azotobacter in Immobilization of Heavy Metals. *Environ. Sci. Technol.* **2009**, *43*, 5884–5889.

Joshi, R.; Pareek, A.; Singla-Pareek, S. L. Plant Metallothioneins: Classification, Distribution, Function, and Regulation. In *Plant Metal Interaction*; Ahmad, P., Ed.; Elsevier: Amsterdam, Netherlands, 2015; pp 239–261.

Kaewdoung, B.; Sutjaritvorakul, T.; Gadd, G. M.; Whalley, A. J. S.; Sihanonth, P. Heavy Metal Tolerance and Biotransformation of Toxic Metal Compounds by New Isolates of Wood-Rotting Fungi from Thailand. *Geomicrobiol. J.* **2016**, *33*, 283–288. DOI: 10.1080/01490451.2015.1048394.

Kanwar, M. K.; Bhardwaj, R.; Chowdhary, S. P.; Arora, P.; Sharma, P.; Kumar, S. Isolation and Characterization of 24-Epibrassinolide from Brassica juncea L. and Its Effects on Growth, Ni Ion Uptake, Antioxidant Defence of Brassica Plants and in Vitro Cytotoxicity. *Acta Physiol. Plant* **2013**, *35*, 1351–1362.

Khan, M. I. R.; Nazir, F.; Asgher, M.; Per, T. S.; Khan, N. A. Selenium and Sulfur Influence Ethylene Formation and Alleviate Cadmium-Induced Oxidative Stress by Improving Proline and Glutathione Production in Wheat. *J. Plant Physiol.* **2015**, *173*, 9–18.

Khan, W. U.; Ahmad, S. R.; Yasin, N. A.; Ali, A.; Ahmad, A.; Akram, W. Application of Bacillus Megaterium MCR-8 Improved Phytoextraction and Stress Alleviation of Nickel in Vinca rosea. *Int. J. Phytoremediation* **2017**, *19*, 813–824.

Khorrami Vafa, M.; Shokri, K.; Sayyadian, K.; Rejali, F. Contribution of Microbial Associations to the Cadmium Uptake by Peppermint (Mentha piperita). *Ann. Biol. Res.* **2012**, *3*, 2325–2329.

Kovtun, Y.; Chiu, W. L.; Tena, G.; Sheen, J. Functional Analysis of Oxidative Stress-Activated Mitogen-Activated Protein Kinase Cascade in Plants. *Proc. Natl. Acad. Sci. U.S.A.* **2000**, *97*, 2940–2945. DOI: 10.1073/pnas.97.6.2940.

Krzesłowska, M. The Cell Wall in Plant Cell Response to Trace Metals: Polysaccharide Remodeling and Its Role in Defense Strategy. *Acta Physiol. Plant* **2011**, *33* (1), 35–51.

Kushwaha, A.; Rani, R.; Kumar, S.; Gautam, A. Heavy Metal Detoxification and Tolerance Mechanisms in Plants: Implications for Phytoremediation. *Environ. Rev.* **2015**, *24*, 39–51.

Landrigan, P. J.; Fuller, R.; Acosta, N. J. R.; Adeyi, O.; Arnold, R.; Basu, N.; et al. The Lancet Commission on Pollution and Health. *Lancet* **2018**, *391*, 462–512. DOI: 10.1016/S0140-6736(17)32345-0.

Larimer, A. L.; Bever, J. D.; Clay, K. The Interactive Effects of Plant Microbial Symbionts: A Review and Meta-Analysis. *Symbiosis* **2010**, *51* (2), 139–148. DOI: 10.1007/s13199-010-0083-1.

Lau, T.; Wu, X.; Chua, H.; Qian, P.; Wong, P. Effect of Exopolysaccharides on the Adsorption of Metal Ions by Pseudomonas sp. CU-1. *Water Sci. Technol.* **2005**, *52*, 63–68.

Lindblom, S. D.; Fakra, S. C.; Landon, J.; Schulz, P.; Tracy, B.; Pilon-Smits, E. A.; et al. Inoculation of Selenium Hyperaccumulator *Stanleya pinnata* and Related Non-Accumulator *Stanleya elata* with Hyperaccumulator Rhizosphere Fungi—Investigation of Effects on Se

Accumulation and Speciation. *Physiol. Plantarum* **2014,** *150* (1), 107–118. DOI: 10.1111/ppl.12094.

Liu, H.; Fang, H. H. Characterization of Electrostatic Binding Sites of Extracellular Polymers by Linear Programming Analysis of Titration Data. *Biotechnol. Bioeng.* **2002,** *80,* 806–811.

Liu, J. X.; Howell, S. H. Managing the Protein Folding Demands in the Endoplasmic Reticulum of Plants. *New Phyto.* **2016,** *211,* 418–428. DOI: 10.1111/nph.13915.

Liu, J.; Shi, X.; Qian, M.; Zheng, L.; Lian, C.; Xia, Y.; et al. Copper induced Hydrogen Peroxide Upregulation of a Metallothionein Gene, OsMT2c, from Oryza sativa L. Confers Copper Tolerance in Arabidopsis thaliana. *J. Hazard Mater.* **2015,** *294,* 99–108. DOI: 10.1016/j.jhazmat.2015.03.060.

Liu, Y.; Bassham, D. C. Autophagy: Pathways for Self-Eating in Plant Cells. *Ann. Rev. Plant Biol.* **2012,** *63,* 215–237. DOI: 10.1146/annurev-arplant-042811-105441.

Llamas, M. A.; Mooij, M. J.; Sparrius, M.; Vandenbroucke-Grauls, C. M. J. E.; Ratledge, C. Characterization of Five Novel Pseudomonas Aeruginosa Cell-Surface Signalling Systems. *Mol. Microbiol.* **2008,** *67,* 458–472.

López-Climent, M. F.; Arbona, V.; Pérez-Clemente, R. M.; Gómez-Cadenas, A. Effects of Cadmium on Gas Exchange and Phytohormone Contents in Citrus. *Biol. Plant* **2011,** *55,* 187–190.

Luo, Z. B.; He, J.; Polle, A.; Rennenberg, H. Heavy Metal Accumulation and Signal Transduction in Herbaceous and Woody Plants: Paving the Way for Enhancing Phytoremediation Efficiency. *Biotechnol. Adv.* **2016,** *34,* 1131–1148.

Ma, Y.; Rajkumar, M.; Fritas, H. Inoculation of Plant Growth Promoting Bacterium Achromobacter Xylosoxidans Strain Ax10 for the Improvement of Copper Phytoextraction by Brassica juncea. *J. Environ. Manag.* **2008,** *90,* 831–837.

Mahmood, T.; Gupta, K. J.; Kaiser, W. M. Cd Stress Stimulates Nitric Oxide Production by Wheat Roots. *Pak. J. Bot.* **2009,** *41,* 1285–1290.

Majumder, A.; Bhattacharyya, K.; Bhattacharyya, S.; Kole, S. C. Arsenic tolerant, Arsenite-Oxidising Bacterial Strains in the Contaminated Soils of West Bengal. *Indian Sci. Total Environ.* **2013,** *46,* 1006–1014. DOI: 10.1016/j.scitotenv.2013.06.068.

Maksymiec, W. Signaling Responses in Plants to Heavy Metal Stress. *Acta Physiol. Plant* **2007,** *29,* 177–187.

Manara, A. Plant Responses to Heavy Metal Toxicity. In *Plants and Heavy Metals. Springer Briefs in Molecular Science*; Furini, A., Ed.; Springer; Dordrecht, 2012; pp 27–53.

Masood, A.; Khan, M. I.; Fatma, M.; Asgher, M.; Per, T. S.; Khan, N. A. Involvement of Ethylene in Gibberellic Acid-Induced Sulfur Assimilation, Photosynthetic Responses, and Alleviation of Cadmium Stress in Mustard. *Plant Physiol. Biochem.* **2016,** *104,* 1–10.

Metwally, A.; Finkemeier, I.; Georgi, M.; Dietz, K. -J. Salicylic Acid Alleviates the Cadmium Toxicity in Barley Seedlings. *Plant Physiol.* **2013,** *132,* 272–281.

Michalak, A. Phenolic Compounds and Their Antioxidant Activity in Plants Growing Under Heavy Metal Stress. *Pol. J. Environ. Stud.* **2006,** *15* (4), 523–530.

Migeon, A.; Blaude, D.; Wilkins, O.; Montanini, B.; Campbell, M. M.; Richaud, P.; Thomine, S.; Chalot, M. Genome-Wide Analysis of Plant Metal Transporters, with an Emphasis on Poplar. *Cell Mol. Life. Sci.* **2010,** *67,* 3763–3784.

Milner, M. J.; Seamon, J.; Craft, E.; Kochian, L. V. Transport Properties of Members of the ZIP Family in Plants and Their Role in Zn and Mn Homeostasis. *J. Exp. Bot.* **2013,** *64,* 369–381.

Minibayeva, F.; Dmitrieva, S.; Ponomareva, A.; Ryabovol, V. Oxidative Stress-Induced Autophagy in Plants: The Role of Mitochondria. *Plant Physiol. Biochem.* **2012,** *59,* 11–19. DOI: 10.1016/j.plaphy.2012.02.013.

Mishra, J.; Singh, R.; Arora, N. K. Alleviation of Heavy Metal Stress in Plants and Remediation of Soil by Rhizosphere Microorganisms. *Front. Microbiol.* **2017,** *8,* 1706. DOI: 10.3389/fmicb.2017.01706.

Mohammad, J. K.; Muhammad, T.; Khalid, K. Effect of Organic and Inorganic Amendments on the Heavy Metal Content of Soil and Wheat Crop Irrigated with Wastewater. *Sarhad J. Agric.* **2013,** *29,* 145–152.

Mourato, M.; Reis, R.; Martins, L. L. Characterization of Plant Antioxidative System in Response to Abiotic Stresses: A Focus on Heavy Metal Toxicity. In *Advances in Selected Plant Physiology Aspects*; Montanaro, G., Dichio, B., Eds.; InTech: Vienna, Austria, 2012; pp 23–44. http://hdl.handle.net/10400.5/4410.h.

Moya, J. L.; Ros, R.; Picazo, I. Heavy Metal-Hormone Interactions in Rice Plants: Effects on Growth, Net Photosynthesis, and Carbohydrate Distribution. *J. Plant Growth Regul.* **1995,** *14,* 61–67.

Mukhopadhyay, M.; Mondal, T. K. Effect of Ainc and Boron on Growth and Water Relations of Camellia sinensis (L.) O. Kuntze cv. T-78. *Natl. Acad. Sci. Lett.* **2015,** *201538,* 283–286.

Mulligan, C. N.; Yong, R. N.; Gibbs, B. F. On the Use of Biosurfactants for the Removal of Heavy Metals From Oil-Contaminated Soil. *Environ. Prog.* **1999,** *18,* 50–54.

Munzuro, Ö.; Fikriye, K. Z.; Yahyagil, Z. The Abscisic Acid Levels of Wheat (Triticum aestivum L. cv. Çakmak 79) Seeds That were Germinated Under Heavy Metal (Hg++, Cd++, Cu++) Stress. *G. U. J. Sci,* **2008,** *21,* 1–7.

Mustafa, A.; Naveed, M.; Saeed, Q.; Ashraf, M. N.; Hussain, A.; Abbas, T.; Kamran, M.; Minggang, X. Application Potentials of Plant Growth Promoting Rhizobacteria and Fungi as an Alternative to Conventional Weed Control Methods. In *Crop Production*; IntechOpen: London, UK, 2019.

Mustafa, G.; Komatsu, S. Toxicity of Heavy Metals and Metal-Containing Nanoparticles on Plants. *BBA—Protein. Proteom.* **2016,** *1864,* 932–944.

Nadeem, S. M.; Ahmad, M.; Zahir, Z. A.; Javaid, A.; Ashraf, M. The Role of Mycorrhizae and Plant Growth Promoting Rhizobacteria (PGPR) in Improving Crop Productivity Under Stressful Environments. *Biotechnol. Adv.* **2013,** *32,* 429–448. DOI: 10.1016/j.biotechadv.2013.12.005.

Naveed, M.; Mustafa, A.; Majeed, S.; Naseem, Z.; Saeed, Q.; Khan, A.; Nawaz, A.; Baig, K. S.; Chen, J. T. Enhancing Cadmium Tolerance and Pea Plant Health through Enterobacter sp. MN17 Inoculation Together with Biochar and Gravel Sand. *Plants* **2020,** *9,* 530.

Nazli, F; Mustafa, A.; Ahmad, M.; Hussain. A.; Jamil, M.; Wang, X.; Shakeel, Q.; Imtiaz, M.; El-Esawi, M. A. A Review on Practical Application and Potentials of Phytohormone-Producing Plant Growth-Promoting Rhizobacteria for Inducing Heavy Metal Tolerance in Crops. *Sustainability* **2020,** *12,* 9056. DOI: 10.3390/su1221905.

Oladipo, O. G.; Awotoye, O. O.; Olayinka, A.; Bezuidenhout, C. C.; Maboeta, M. S. Heavy Metal Tolerance Traits of Filamentous Fungi Isolated from Gold and Gemstone Mining Sites. *Braz. J. Microbiol.* **2018,** *49,* 29–37. DOI: 10.1016/j. bjm.2017.06.003.

Ong, G. H.; Ho, X. H.; Shamkeeva, S.; Fernando, M. S.; Shimen, A.; Wong, L. S. Biosorption Study of Potential Fungi for Copper Remediation from Peninsular Malaysia. *Remediat. J.* **2017,** *27,* 59–63. DOI: 10.1002/rem.21531.

Ouyang, H.; Vogel, H. J. Metal Ion Binding to Calmodulin: NMR and Fluorescence Studies. *Biometals* **1998**, *11* (3), 213–222.

Ovečka, M.; Takáč, T. Managing Heavy Metal Toxicity Stress in Plants: Biological and Biotechnological Tools. *Biotechnol. Adv.* **2014**, *32* (1), 73–86.

Ozturk, M.; Ashraf, M.; Aksoy, A.; Ahmad, M. S. A.; Eds. *Plants, Pollutants & Remediation*; Springer: New York, 2015.

Ozturk, M.; Yucel, E.; Gucel, S.; Sakcali, S.; Aksoy, A. Plants as Biomonitors of Trace Elements Pollution in Soil. In *Trace Elements: Environmental Contamination, Nutritional Benefits and Health Implications*; Prasad, M. N. V., Ed.; Wiley: New York, 2008; pp 723–744.

Panda, S.; Biswal, U. C. Effect of Magnesium and Calcium Ions on Photoinduced Lipid Peroxidation and Thylakoid Breakdown of Cell-Free Chloroplasts. *Indian J. Biochem. Biophys.* **1990**, *27*, 159–163.

Pandey, C.; Gupta, M. Selenium and Auxin Mitigates Arsenic Stress in Rice (Oryza sativa L.) by Combining the Role of Stress Indicators, Modulators and Genotoxicity Assay. *J. Hazard. Mater.* **2015**, *287*, 384–391.

Pandey, S.; Ghosh, P. K.; Ghosh, S.; De, T. K.; Maiti, T. K. Role of Heavy Metal Resistant Ochrobactrum sp. and Bacillus spp. Strains in Bioremediation of a Rice Cultivar and Their PGPR Like Activities. *J. Microbiol.* **2013**, *51*, 11–17. DOI: 10.1007/s12275-013-2330-7.

Pantin, F.; Monnet, F.; Jannaud, D.; Costa, J. M.; Renaud, J.; Muller, B.; Simonneau, T.; Genty, B. The Dual Effect of Abscisic Acid on Stomata. *New Phytol.* **2013**, *197*, 65–72.

Park, C. J.; Seo, Y. S. Heat Shock Proteins: A Review of the Molecular Chaperones for Plant Immunity. *Plant Pathol. J.* **2015**, *31*, 323–333. DOI: 10.5423/PPJ.RW.08.2015.0150.

Patra, M.; Bhowmik, N.; Bandopadhyay, B.; and Sharma, A. Comparison of Mercury, Lead and Arsenic with Respect to Genotoxic Effects on Plant Systems and the Development of Genetic Tolerance. *Environ. Exp. Bot.* **2004**, *52*, 199–223. DOI: 10.1016/j.envexpbot.2004.02.009.

Peng, J. S.; Gong, J. M. Vacuolar Sequestration Capacity and Long-Distance Metal Transport in Plants. *Front. Plant Sci.* **2014**, *5*, 19. DOI: 10.3389/fpls.2014.00019.

Pereira, S. I. A.; Barbosa, L.; Castro, P. M. L. Rhizobacteria Isolated from a Metal-Polluted Area Enhance Plant Growth in Zinc and Cadmium-Contaminated Soil. *Int. J. Environ. Sci. Technol.* **2015**, *12*, 2127–2142.

Pierart, A.; Shahid, M.; Séjalon-Delmas, N.; and Dumat, C.; Antimony bioavailability: knowledge and research perspectives for sustainable agricultures. *J. Hazard. Mater.* **2015**, *289*, 219–234. DOI: 10.1016/j.jhazmat.2015.02.011.

Piotrowska-Niczyporuk, A.; Bajguz, A.; Zambrzycka, E.; GodlewskaŻyłkiewicz, B.; Phytohormones as Regulators of HM Biosorption and Toxicity in Green Alga Chlorella vulgaris (Chlorophyceae). *Plant Physiol. Biochem.* **2012**, *52*, 52–65.

Pizzeghello, D.; Francioso, O.; Ertani, A.; Muscolo, A.; Nardi, S. Isopentenyl Adenosine and Cytokinin-Like Activity of Different Humic Substances. *J. Geochem. Explor.* **2013**, *129*, 70–75.

Pochodylo, A. L.; Aristilde, L. Molecular Dynamics of Stability and Structures in Phytochelatin Complexes with Zn, Cu, Fe, Mg, and Ca: Implications for Metal Detoxification. *Environ. Chem. Lett.* **2017**, 1–6. DOI: 10.1007/s10311-017-0609-3.

Pomponi, M.; Censi, V.; Di Girolamo, V.; De Paolis, A.; Di Toppi, L. S.; Aromolo, R.; et al. Overexpression of Arabidopsis Phytochelatin Synthase in Tobacco Plants Enhances Cd^{2+} Tolerance and Accumulation but Not Translocation to the Shoot. *Planta* **2006**, *223*, 180–190. DOI: 10.1007/s00425-005-0073-3.

Poschenrieder, C.; Gunsé, B.; Barceló, J. Influence of Cadmium on Water Relations, Stomatal Resistance, and Abscisic Acid Content in Expanding Bean Leaves. *Plant Physiol.* **1989,** *90,* 1365–1371.

Pramanik, K.; Mitra, S.; Sarkar, A.; Soren, T.; Maiti, T. K. Characterization of Cadmium-Resistant Klebsiella pneumoniae MCC 3091 Promoted Rice Seedling Growth by Alleviating Phytotoxicity of Cadmium. *Env. Sci. Pollut. Res.* **2017,** *24,* 24419–24437.

Rady, M. M.; Osman, A. S. Response of Growth and Antioxidant System of Heavy Metal-Contaminated Tomato Plants to 24-Epibrassinolide. *Afr. J. Agric. Res.* **2012,** *7,* 3249–3254.

Rajkumar, M.; Ae, N.; Prasad, M. N. V.; Freitas, H. Potential of Siderophore-Producing Bacteria for Improving Heavy Metal Phytoextraction. *Trends Biotechnol.* **2010,** *28,* 142–149.

Ramakrishna, B.; Rao, S. S. R. Preliminary Studies on the Involvement of Glutathione Metabolism and Redox Status Against Zinc Toxicity in Radish Seedlings by. *Environ. Exp. Bot.* **2013,** *96,* 52–58.

Rao, K. P.; Vani, G.; Kumar, K.; Wankhede, D. P.; Misra, M.; Gupta, M.; et al. Arsenic Stress Activates MAP Kinase in Rice Roots and Leaves. *Arch. Biochem. Biophys.* **2011,** *506,* 73–82. DOI: 10.1016/j.abb.2010.11.006.

Rastgoo, L.; Alemzadeh, A.; Afsharifar, A. Isolation of Two Novel Isoforms Encoding Zinc- and Copper-Transporting P1BATPase from Gouan (Aeluropus littoralis). *Plant Omics J.* **2011,** *4* (7), 377–383.

Rauser, W. E.; Dumbroff, E. B. Effects of Excess Cobalt, Nickel and Zinc on the Water Relations of Phaseolus Vulgaris. *Environ. Exp. Bot.* **1981,** *21,* 249–255.

Rees, F.; Sterckeman, T.; Morel, J. L. Root Development of Non-Accumulating and Hyper Accumulating Plants in Metal Contaminated Soils Amended with Biochar. *Chemosphere* **2016,** *142,* 48–55.

Rho, H.; Hsieh, M.; Kandel, S. L.; Cantillo, J.; Doty, S. L.; Kim, S. H. Do Endophytes Promote Growth of Host Plants Under Stress? A Meta-Analysis on Plant Stress Mitigation by Endophytes. *Microb. Ecol.* **2018,** *75* (2), 407–418. DOI: 10.1007/s00248-017-1054-3.

Rodríguez-Celma, J.; Rellán-Álvarez, R.; Abadía, A.; Abadía, J.; LópezMillán, A. -F. Changes Induced by Two Levels of Cadmium Toxicity in the 2-DE Protein Profile of Tomato Roots. *J. Proteom.* **2010,** *73,* 1694–1706. DOI: 10.1016/j.jprot.2010.05.001.

Romanowska, E.; Wasilewska, W.; Fristedt, R.; Vener, A. V.; Zienkiewicz, M. Phosphorylation of PSII Proteins in Maize Thylakoids in the Presence of Pb Ions. *J. Plant Physiol.* **2012,** *169,* 345–352.

Román-Ponce, B.; Reza-Vázquez, D. M.; Gutiérrez-Paredes, S.; María de Jesús, D. E.; Maldonado-Hernández, J.; Bahena-Osorio, Y.; Estrada-de los Santose, P.; Wang, T.; Vásquez-Murrieta, M. S. Plant Growth-Promoting Traits in Rhizobacteria of Heavy Metal-Resistant Plants and Their Effects on Brassica Nigra Seed Germination. *Pedosphere* **2017,** *27,* 511–526. [CrossRef]

Rout, G. R.; Panigrahi, J. Analysis of Signaling Pathways During Heavy Metal Toxicity: A Functional Genomics Perspective. In *Elucidation of Abiotic Stress Signaling in Plants;* Pandey, G. K., Ed.; Springer: New York, 2015; pp 295–322.

Rozentsvet, O. A.; Nesterov, V. N.; Sinyutina, N. F. The Effect of Copper Ions on the Lipid Composition of Subcellular Membranes in Hydrilla Verticillata. *Chemosphere* **2012,** *89,* 108–113.

Rubio, M. I.; Escrig, I.; Martinez-Cortina, C.; Lopez-Benet, F. J.; Sanz, A. Cadmium and Nickel Accumulation in Rice Plants. Effects on Mineral Nutrition and Possible Interactions of Abscisic and Gibberellic Acids. *Plant Growth Regul.* **1994**, *14*, 151–157.

Salt, D. E.; Prince, R. C.; Pickering, I. J.; Raskin, I. Mechanisms of Cadmium Mobility and Accumulation in Indian Mustard. *Plant Physiol.* **1995**, *109*, 1427–1433.

Saraswat, S.; Rai, J. P. N. Complexation and Detoxification of Zn and Cd in Metal Accumulating Plants. *Rev. Environ. Sci. Biotechnol.* **2011**, *10*, 327–339.

Sarry, J. E.; Kuhn, L.; Ducruix, C.; Lafaye, A.; Junot, C.; Hugouvieux, V.; Jourdain, A.; et al. The Early Responses of Arabidopsis thaliana Cells to Cadmium Exposure Explored by Protein and Metabolite Profiling Analyses. *Proteomics* **2006**, *6*, 2180–2198. DOI: 10.1002/pmic.200500543.

Saxena, B.; Shukla, K.; Giri, B. Arbuscular Mycorrhizal Fungi and Tolerance of Salt Stress in Plants. In *Arbuscular Mycorrhizas and Stress Tolerance of Plants*; Wu, Q. S., Ed.; Springer: Singapore, 2017; pp 67–97. DOI: 10.1007/978-981-10-4115-0_4.

Schellingen, K.; Straeten, D.; Van Der Vandenbussche, F.; Prinsen, E.; Remans, T. Cadmium-Induced Ethylene Production and Responses in Arabidopsis thaliana Rely on ACS2 and ACS6 Gene Expression. *BMC Plant Biol.* **2014**, *14*, 214.

Scherz-Shouval, R.; Elazar, Z. ROS, Mitochondria and the Regulation of Autophagy. *Trends Cell Biol.* **2007**, *17*, 422–427. DOI: 10.1016/j.tcb.2007. 07.009.

Seneviratne, M.; Gunaratne, S.; Bandara, T.; Weerasundara, L.; Rajakaruna, N.; Seneviratne, G.; Vithanage, M. Plant Growh Promotion by Bradyrhizobium japonicum Under Heavy Metal Stress. *S. Afr. J. Bot.* **2016**, *105*, 19–24.

Seneviratne, M.; Seneviratne, G.; Madawala, H.; Iqbal, M.; Rajakaruna, N.; Bandara, T.; Vithanage, M. A Preliminary Study of the Role of Bacterial–Fungal Co-Inoculation on Heavy Metal Phytotoxicity in Serpentine Soil. *Aust. J. Bot.* **2015**, *63*, 261–268.

Sessitsch, A.; Kuffner, M.; Kidd, P.; Vangronsveld, J.; Wenzel, W. W.; Fallmann, K.; et al. The Role of Plant-Associated Bacteria in the Mobilization and Phytoextraction of Trace Elements in Contaminated Soils. *Soil Biol. Biochem.* **2013**, *60*, 182–194. DOI: 10.1016/j. soilbio.2013.01.012.

Shah, K.; Mankad, A. U.; Reddy, M. N. Cadmium Accumulation and Its Effects on Growth and Biochemical Parameters in Tageles erecta L. *J. Pharmacogn. Phytochem.* **2017**, *6*, 111–115.

Shahid, M.; Khalid, S.; Abbas, G.; Shahid, N.; Nadeem, M.; Sabir, M.; et al. Heavy Metal Stress and Crop Productivity. In *Crop Production and Global Environmental Issues*; Hakeem, K. R., Ed.; Springer International Publishing: Cham, 2015; pp 1–25.

Shanker, A. K. Physiological, Biochemical and Molecular Aspects of Chromium Toxicity and Tolerance in Selected Crops and Tree Species. PhD thesis, Tamil Nadu Agricultural University, Coimbatore, India, 2003.

Sharma, I.; Pati, P. K.; Bhardwaj, R. Effect of 24-Epibrassinolide on Oxidative Stress Markers Induced by Nickelion in Raphanus sativus L. *Acta Physiol. Plant* **2011**, *33*, 1723–1735.

Sharma, P.; Jha, A. B.; Dubey, R. S.; Pessarakli, M. Reactive Oxygen Species, Oxidative Damage, and Antioxidative Defense Mechanism in Plants Under Stressful Conditions. *J. Bot.* **2012**, *2012*, 26. Article ID: 217037.

Sharma, S. K.; Goloubinoff, P.; Christen, P. Non-Native Proteins as Newly-Identified Targets of Heavy Metals and Metalloids. In *Cellular Effects of Heavy Metals*; Banfalvi, G., Ed; Springer: Heidelberg, 2011; pp 263–274.

Sharma, S. S.; Dietz, K. J. The Relationship Between Metal Toxicity and Cellular Redox Imbalance. *Trends. Plant Sci.* **2009**, *14*, 43–50.

Sharma, S. S.; Dietz, K. -J. The Significance of Amino Acids and Amino Acid-Derived Molecules in Plant Responses and Adaptation to Heavy Metal Stress. *J. Exp. Bot.* **2006**, *57* (4), 711–726.

Sheng, X. F.; Xia, J. J. Improvement of Rape (Brassica napus) Plant Growth and Cadmium Uptake by Cadmium-Resistant Bacteria. *Chemosphere* **2006**, *64*, 1036–1042.

Shi, J. Y.; Lin, H. R.; Yuan, X. F.; Chen, X. C.; Shen, C. F.; Chen, Y. X. Enhancement of Copper Availability and Microbial Community Changes in Rice Rhizospheres Affected by Sulfur. *Molecules* **2011**, *16*, 1409–1417.

Shu, X.; Yin, L.; Zhang, Q.; Wang, W. Effect of Pb Toxicity on Leaf Growth, Antioxidant Enzyme Activities, and Photosynthesis in Cuttings and Seedlings of Jatropha curcas L. *Environ. Sci. Pollut. Res. Int.* **2011**, *19*, 893–902.

Singh, J. S.; Abhilash, P. C.; Singh, H. B.; Singh, R. P.; Singh, D. P. Genetically Engineered Bacteria: An Emerging Tool for Environmental Remediation and Future Research Perspectives. *Gene* **2011**, *480*, 1–9. DOI: 10.1016/j. gene.2011.03.001.

Singh, S.; Parihar, P.; Singh, R.; Singh, V. P.; Prasad, S. M. Heavy Metal Tolerance in Plants: Role of Transcriptomics, Proteomics, Metabolomics, and Ionomics. *Front. Plant Sci.* **2015**, *6*, 1143. DOI: 10.3389/fpls.2015.01143.

Singh, S.; Parihar, P.; Singh, R.; Singh, V. P.; Prasad, S. M. Heavy Metal Tolerance in Plants: Role of Transcriptomics, Proteomics, Metabolomics, and Ionomics. *Front. Plant Sci.* **2016**, *6*, 1143.

Singh, S.; Parihar, P.; Singh, R.; Singh, V. P.; Prasad, S. M. Heavy Metal Tolerance in Plants: Role of Transcriptomics, Proteomics, Metabolomics, and Ionomics. *Front. Plant Sci.* **2015**, *6*, 1143. DOI: 10.3389/fpls.2015.01143.

Singh, V.; Maharshi, A.; Singh, D. P.; Upadhyay, R. S.; Sarma, B. K.; Singh, H. B. Role of Microbial Seed Priming and Microbial Phytohormone in Modulating Growth Promotion and Defense Responses in Plants. In *Advances in Seed Priming*; Rakshit, A., Singh, H., Eds.; Springer: Singapore, 2018; pp 115–126.

Smeets, K.; Opdenakker, K.; Remans, T.; Forzani, C.; Hirt, H.; Vangronsveld, J.; et al. The Role of the Kinase OXI1 in Cadmium- and Copper-Induced Molecular Responses in Arabidopsis thaliana. *Plant Cell Environ.* **2013**, *36*, 1228–1238. DOI: 10.1111/pce.12056.

Solanki, R.; Dhankhar, R. Biochemical Changes and Adaptive Strategies of Plants Under Heavy Metal Stress. *Biologia* **2011**, *66* (2), 195–204.

Sorty, A. M.; Meena, K. K.; Choudhary, K.; Bitla, U. M.; Minhas, P. S.; Krishnani, K. K. Effect of Plant Growth Promoting Bacteria Associated with Halophytic Weed (Psoralea corylifolia L.) on Germination and Seedling Growth of Wheat Under Saline Conditions. *Appl. Biochem. Biotechnol.* **2016**, *180*, 872–882.

Srivastava, G.; Kumar, S.; Dubey, G.; Mishra, V.; Prasad, S. M. Nickel and Ultraviolet-B Stresses Induce Differential Growth and Photosynthetic Responses in Pisum sativum L. Seedlings. *Biol. Trace Elem. Res.* **2012**, *149*, 86–96.

Srivastava, S.; Chiappetta, A.; Beatrice. M. Identification and Profiling of Arsenic Stress-Induced miRNAs in Brassica juncea. *J. Exp. Bot.* **2013**, *4*, 303–315.

Srivastava, S.; Verma, P. C.; Chaudhry, V.; Singh, N.; Abhilash, P. C.; Kumar, K. V.; et al. Influence of Inoculation of Arsenic-Resistant *Staphylococcus arlettae* on Growth and Arsenic Uptake in *Brassica juncea* (L.) Czern. Var. R-46. *J. Hazard Mater.* **2013**, *262*, 1039–1047. DOI: 10.1016/j.jhazmat.2012.08.019.

Sun, C. W.; Callis, J. Independent Modulation of Arabidopsis thaliana Polyubiquitin mRNAs in Different Organs and in Response to Environmental Changes. *Plant J.* **1997**, *11*, 1017–1027. DOI: 10.1046/j.1365-313X.1997.1105 1017.x.

Talaat, N. B.; Shawky, B. T. Microbe-Mediated Induced Abiotic Stress Tolerance Responses in Plants. In *Plant-Microbe Interactions in Agro-Ecological Perspectives*; Singh, D. P., Singh, H. B., Prabha, R., Eds.; Springer: Singapore, 2017; pp 101–133.

Tamás, M. J.; Sharma, S. K.; Ibstedt, S.; Jacobson, T.; Christen, P. Heavy Metals and Metalloids as a Cause for Protein Misfolding and Aggregation. *Biomolecules* **2014**, *4*, 252–267. DOI: 10.3390/biom4010252.

Tandon, S. A.; Kumar, R.; Parsana, S. Auxin Treatment of Wetland and Non-Wetland Plant Species to Enhance Their Phytoremediation Efficiency to Treat Municipal Wastewater. *J. Sci. Ind. Res.* **2015**, *74*, 702–707.

Tassi, E.; Pouget, J.; Petruzzelli, G.; Barbafieri, M.; The Effects of Exogenous Plant Growth Regulators in the Phytoextraction of HMs. *Chemosphere* **2008**, *71* (1), 66–73.

Tchounwou, P. B.; Yedjou, C. G.; Patlolla, A. K.; Sutton, D. J. Heavy Metals Toxicity and the Environment. *EXS* **2012**, *101*, 133–164. DOI: 10.1007/978-3-7643-8340-4_6.

Tebar, F.; Lladó, A.; Enrich, C. Role of Calmodulin in the Modulation of the MAPK Signalling Pathway and the Transactivation of Epidermal Growth Factor Receptor Mediated by PKC. *FEBS Lett.* **2002**, *517*, 206–210. DOI: 10.1016/S0014-5793(02)02624-8.

Tena, G.; Asai, T.; Chiu, W. L; Sheen J. Plant Mitogen Activated Protein Kinase Signaling Cascades. *Curr. Opin. Plant Biol.* **2001**, *4*, 392–400.

Thakur, S.; Singh, L.; Ab-Wahid, Z.; Siddiqui, M. F.; Atnaw, S. M.; Din, M. F. M. Plant-Driven Removal of Heavy Metals from Soil: Uptake, Translocation, Tolerance Mechanism, Challenges, and Future Perspectives. *Environ. Monit. Assess.* **2016**, *188*, 206.

Tiwari, S.; Lata, C. Heavy Metal Stress, Signaling, and Tolerance Due to Plant-Associated Microbes: An Overview. *Front. Plant Sci.* **2018**, *9*, 452.

Tiwari, S.; Lata, C.; Chauhan, P. S.; Nautiyal, C. S. Pseudomonas Putida Attunes Morphophysiological, Biochemical and Molecular Responses in Cicer arietinum L. During Drought Stress and Recovery. *Plant Physiol. Biochem.* **2016**, *99*, 108–117. DOI: 10.1016/j.plaphy.2015.11.001.

Tiwari, S., Lata, C.; Chauhan, P. S.; Nautiyal, C. S. Pseudomonas Putida Attunes Morphophysiological, Biochemical and Molecular Responses in Cicer arietinum L. During Drought Stress and Recovery. *Plant Physiol. Biochem.* **2016**, *99*, 108–117. DOI: 10.1016/j.plaphy.2015.11.001.

Tiwari, S.; Prasad, V.; Chauhan, P. S.; Lata, C. Bacillus Amyloliquefaciens Confers Tolerance to Various Abiotic Stresses and Modulates Plant Response to Phytohormones Through Osmoprotection and Gene Expression Regulation in Rice. *Front. Plant Sci.* **2017**, *8*, 1510. DOI: 10.3389/fpls.2017.01510

Tiwari, S.; Prasad, V.; Chauhan, P. S.; Lata, C. Bacillus Amyloliquefaciens Confers Tolerance to Various Abiotic Stresses and Modulates Plant Response to Phytohormones Through Osmoprotection and Gene Expression in Rice. *Front. Plant Sci.* **2017**, *8*, 1510. DOI: 10.3389/fpls.2017.01510.

Tobin, J. M. Fungal Metal Biosorption. *Br. Mycol. Soc. Symp. Ser.* **2001**, *23*, 424–444. DOI: 10.1017/CBO9780511541780.016.

Tomas, M.; Pagani, M. A.; Andreo, C. S.; Capdevila, M.; Atrian, S.; Bofill, R. Sunflower Metallothionein Family Characterisation. Study of the Zn (II)- and Cd (II)-Binding

Abilities of the HaMT1 and HaMT2 Isoforms. *J. Iinorg. Biochem.* **2015**, *148*, 35–48. DOI: 10.1016/j.jinorgbio.2015.02.016.

Tripathi, P.; Singh, P. C.; Mishra, A.; Srivastava, S.; Chauhan, R.; Awasthi, S.; et al. Arsenic Tolerant Trichoderma sp. Reduces Arsenic Induced Stress in Chickpea (Cicer arietinum). *Environ. Pollut.* **2017**, *223*, 137–145. DOI: 10.1016/j.envpol.2016.12.073.

Ullah, A.; Heng, S.; Munis, M. F. H.; Fahad, S.; Yang, X. Phytoremediation of Heavy Metals Assisted by Plant Growth Promoting (PGP) Bacteria: A Review. *Environ. Exp. Bot.* **2015**, *117*, 28–40. DOI: 10.1016/j.envexpbot.2015. 05.001.

Vardhini, B. V.; Anuradha, S.; Rao, S. S. R. Brassinosteroids—A Great Potential to Improve Crop Productivity. *Indian J. Plant Physiol.* **2006**, *11*, 1–12.

Verma, P. K.; Meher, A. K.; Dwivedi, S.; Bansiwal, A. K.; Pande, V.; et al. A Novel Arsenic Methyltransferase Gene of Westerdykella Aurantiaca Isolated from Arsenic Contaminated Soil: Phylogenetic, Physiological, and Biochemical Studies and Its Role in Arsenic Bioremediation. *Metallomics* **2016**, *8*, 344–353. DOI: 10.1039/c5mt00277j.

Verma, S.; Verma, P. K.; Meher, A. K.; Bansiwal, A. K.; Tripathi, R. D.; Chakrabarty, D. A Novel Fungal Arsenic Methyltransferase, WaarsM Reduces Grain Arsenic Accumulation in the Transgenic Rice Plant. *J. Hazard. Mater.* **2017**, *344*, 626–634. DOI: 10.1016/j. jhazmat.2017.10.037.

Viehweger, K. How Plants Cope with Heavy Metals. *Bot. Stud.* **2014**, *55* (35), 1–12.

Vreugdenhil, D. Abscisic Acid Inhibits Phloem Loading of Sucrose. *Physiol. Plant* **1983**, *57*, 463–467.

Wang, W.; Vinocur, B.; Shoseyov, O.; Altman, A. Role of Plant Heatshock Proteins and Molecular Chaperones in the Abiotic Stress Response. *Trends Plant Sci.* **2004**, *9*, 244–252. DOI: 10.1016/j.tplants.2004.03.006.

Wang, X.; Fan, J.; Xing, Y.; Xu, G.; Wang, H.; Deng, J.; Wang, Y.; Zhang, F.; Li, P.; Li, Z. The Effects of Mulch and Nitrogen Fertilizer on the Soil Environment of Crop Plants. In *Advances in Agronomy*; Sparks, D. L., Ed.; Academic Press: Cambridge, MA, USA, 2019. *Sustainability* **2020a**, *12*, 9056 24 of 24.

Wang, X.; Wang, G.; Guo, T. T.; Xing, Y.; Mo, F.; Wang, H.; Fan, J.; Zhang, F. Effects of Plastic Mulch and Nitrogen Fertilizer on the Soil Microbial Community, Enzymatic Activity, and Yield Performance in a Dryland Maize Cropping System. *Eur. J. Soil Sci.* **2020b**, 1–13.

Wang, Y.; Wang, Y.; Kai, W.; Zhao, B.; Chen, P.; Sun, L.; Ji, K.; Li, Q.; Dai, S.; Sun, Y.; et al. Transcriptional Regulation of Abscisic Acid Signal Core Components During Cucumber Seed Germination and Under Cu2+, Zn2+, NaCl and Simulated Acid Rain Stresses. *Plant Physiol. Biochem.* **2014**, *76*, 67–76.

Wen-Xing, D. Autophagy in Toxicology: Defense Against Xenobiotics. *J. Drug. Metab. Toxicol.* **2012**, *3*, e108. DOI: 10.4172/2157-7609.1000e108.

Weyens, N.; Beckers, B.; Schellingen, K.; Ceulemans, R.; Croes, S.; Janssen, J.; Haenen, S.; Witters, N.; Vangronsveld, J. Plant-Associated Bacteria and Their Role in the Success or Failure of Metal Phytoextraction Projects: First Observations of a Field-Related Experiment. *Microb. Biotechnol.* **2013**, *6*, 288–299.

Wong, H. L.; Sakamoto, T.; Kawasaki, T.; Umemura, K.; Shimamoto, K. Down-Regulation of Metallothionein, a Reactive Oxygen Scavenger, by the Small GTPase OsRac1 in Rice. *Plant Physiol.* **2004**, *135* (3), 1447–1456.

Wu, G.; Kang, H.; Zhang, X.; Shao, H.; Chu, L.; Ruan, C. A Critical Review on the Bio-Removal of Hazardous Heavy Metals from Contaminated Soils: Issues, Progress,

Eco-Environmental Concerns and Opportunities. *J. Hazard. Mater.* **2010,** *174,* 1–8. DOI: 10.1016/j.jhazmat.2009.09.113.

Xiong, T.; Leveque, T.; Shahid, M.; Foucault, Y.; Mombo, S.; Dumat, C. Lead and Cadmium Phytoavailability and Human Bioaccessibility for Vegetables Exposed to Soil or Atmospheric Pollution by Process Ultrafine Particles. *J. Environ. Qual.* **2014,** *43,* 1593–1600. DOI: 10.2134/jeq2013.11.0469.

Xu, J.; Sun, J.; Du, L.; Liu, X.; Comparative Transcriptome Analysis of Cadmium Responses in Solanum nigrum and Solanum torvum. *New Phytol.* **2012,** *196,* 110–124.

Yamamoto, Y.; Kobayashi, Y.; Devi, S. R.; Rikiishi, S.; Matsumoto, H. Aluminum Toxicity is Associated with Mitochondrial Dysfunction and the Production of Reactive Oxygen Species in Plant Cells. *Plant Physiol.* **2002,** *128,* 63–72.

Yang, X.; Srivastava, R.; Howell, S. H.; Bassham, D. C. Activation of Autophagy by Unfolded Proteins During Endoplasmic Reticulum Stress. *Plant J.* **2016,** *85,* 83–95. DOI: 10.1111/tpj.13091.

Yeh, C. M.; Chien, P. S.; Huang, H. J. Distinct Signalling Pathways for Induction of MAP Kinase Activities by Cadmium and Copper in Rice Roots. *J. Exp. Bot.* **2007,** *58,* 659–671. DOI: 10.1093/jxb/erl240.

Yusuf, M.; Fariduddin, Q.; Ahmad, A. 24-Epibrassinolide Modulates Growth, Nodulation, Antioxidant System, and Osmolyte in Tolerant and Sensitive Varieties of Vigna Radiata Under Different Levels of Nickel: A Shotgun Approach. *Plant Physiol. Biochem.* **2012,** *57,* 143–153.

Zaidi, A.; Fernandes, D.; Bean, J. L.; Michaelis, M. L. Effects of Paraquat Induced Oxidative Stress on the Neuronal Plasma Membrane Ca (2+)-ATPase. *Free Radic. Biol. Med.* **2009,** *47,* 1507–1514.

Zelinová, V.; Alemayehu, A.; Bocová, B.; Huttová, J.; Tamás, L. Cadmium-Induced Reactive Oxygen Species Generation, Changes in Morphogenic Responses and Activity of Some Enzymes in Barley Root Tip are Regulated by Auxin. *Biologia* **2015,** *70,* 356–364.

Zhang, D.; Wang, J.; Pan, X. Cadmium Sorption by EPSs Produced by Anaerobic Sludge Under Sulfate-Reducing Conditions. *J. Hazard. Mater.* **2006,** *138,* 589–593.

Zhang, W.; Chen, W. Autophagy Induction upon Reactive Oxygen Species in Cd-Stressed Arabidopsis thaliana. In *Proceedings SPIE 7568, Imaging, Manipulation, and Analysis of Biomolecules, Cells, and Tissues VIII,* San Francisco, CA, 2010. DOI: 10.1117/12.841394.

Zhang, Y. F.; He, L. Y.; Chen, Z. J.; Zhang, W. H.; Wang, Q. Y.; Qian, M.; et al. Characterization of Lead-Resistant and ACC Deaminase-Producing Endophytic Bacteria and Their Potential in Promoting Lead Accumulation of Rape. *J. Hazard Mater.* **2011,** *186* (2–3), 1720–1725. DOI: 10.1016/j.jhazmat.2010.12.069.

Zhao, F. J.; Ma, Y.; Zhu, Y. G.; Tang, Z.; McGrath, S. P. Soil Contamination in China: Current Status and Mitigation Strategies. *Environ. Sci. Technol.* **2015,** *49,* 750–759. DOI: 10.1021/es5047099.

Zheng, L.; Peer, T.; Seybold, V.; Lütz-Meindl, U. Pb-Induced Ultrastructural Alterations and Subcellular Localization of Pb in Two Species of Lespedeza by TEM-Coupled Electron Energy Loss Spectroscopy. *Environ. Exp. Bot.* **2012,** *77,* 196–206. DOI: 10.1016/j.envexpbot.2011.11.018.

Zhu, X. F.; Jiang, T.; Wang, Z. W.; Lei, G. J.; Shi, Y. Z.; Li, G. X.; Zheng, S. J. Gibberellic Acid Alleviates Cadmium Toxicity by Reducing Nitric Oxide Accumulation and Expression of IRT1 in Arabidopsis thaliana. *J. Hazard. Mater.* **2012,** *239–240,* 302–307.

Zhu, X. F.; Wang, Z. W.; Dong, F.; Lei, G. J.; Shi, Y. Z.; Li, G. X.; Zheng, S. J. Exogenous Auxin Alleviates Cadmium Toxicity in Arabidopsis thaliana by Stimulating Synthesis of Hemicellulose 1 and Increasing the Cadmium Fixation Capacity of Root Cell Walls. *J. Hazard. Mater.* **2013,** *263,* 398–403.

Zhu, Y. -G.; Yoshinaga, M.; Zhao, F. -J.; Rosen, B. P. Earth Abides Arsenic Biotransformations. *Annu. Rev. Earth. Planet Sci.* **2014,** *42,* 443–467. DOI: 10.1146/annurev-earth-060313-054942.

Zubair, M.; Shakir, M.; Ali, Q.; Rani, N.; Fatima, N.; Farooq, S.; et al. Rhizobacteria and Phytoremediation of Heavy Metals. *Environ. Technol. Rev.* **2016,** *5,* 112–119. DOI: 10.1080/21622515.2016.1259358.

CHAPTER 9

Hydrogen Peroxide as a Signaling Molecule in Plant Abiotic Stress

ANKUR SINGH and ARYADEEP ROYCHOUDHURY

Post-Graduate Department of Biotechnology, St. Xavier's College (Autonomous), 30 Mother Teresa Sarani, Kolkata, West Bengal, India

ABSTRACT

Hydrogen peroxide (H_2O_2) is widely generated as a reactive oxygen species (ROS) during abiotic stress in all plant systems. H_2O_2 is formed in an uncontrolled way during mitochondrial respiration and photosynthesis via electron transport process. In addition, activity of various enzymes like amine oxidases, oxalate, and flavin-containing enzymes also contribute toward tight regulation of H_2O_2 formation and its level in plants via both enzymatic and nonenzymatic H_2O_2 scavengers. Salinity, drought, light, and extreme temperature like external stress can lead to excess formation of H_2O_2, which acts as cytotoxic metabolite in cells. However, in recent time, it is well established that optimal concentration of H_2O_2 acts as a signaling molecule regulating a wide range of biochemical, physiological, and molecular response in plants during stressed environment. Various reports have shown that H_2O_2 plays a major role in plant development to control physiological processes like senescence, seed germination, regulation of stomatal aperture, shoot and root development, and programmed cell death. In addition, the crosstalk of H_2O_2 with other signaling molecules likes nitric oxide, ethylene, jasmonic acid, salicylic acid, abscisic acid, and calcium plays a key role to abrogate

Crop Sustainability and Intellectual Property Rights. Soumya Mukherjee, Piyali Mukherjee & Tariq Aftab (Eds)

the effects of abiotic stresses in plants. In this chapter, we will mostly focus on the metabolism of H_2O_2 in plant species and will try to decipher the major role played by H_2O_2 and the crosstalk between other signaling molecules and H_2O_2 that helps the plants for surviving in harsh environmental conditions.

9.1 INTRODUCTION

Adverse environmental factors like salinity, drought, heavy metal toxicity, and heat hamper the normal growth and development of the plants. According to the reports of Food and Agriculture Organization (FAO, 2011), the production of the crops need to be raised by 70% to feed the growing population of the world, which is likely to cross the mark of 9.1 billion by 2050. Unfavorable environmental conditions along with limited supply of freshwater damage the quality and quantity of grain formation in agricultural crops (Hu and Xiong, 2014). On being exposed to various abiotic stresses, plants showed several symptoms of damages like chlorophyll degradation, lipid membrane damage, cellular homeostatic misbalance, and formation of several cytotoxic metabolites like malondialdehyde (MDA) and reactive oxygen species (ROS) (Roychoudhury et al., 2021).

ROS are unavoidable by-products of metabolism, which are formed in the plants on being exposed to adverse environmental condition (Waszczak et al., 2018). In spite of their harmful effects in plants, their optimal level in plants is necessary because of their important role as signaling molecule during abiotic stresses. ROS is a collective term for the oxygen radicals like peroxyl ($ROO^.$), superoxide ($O_2^{.-}$), hydroxyl ($^.OH$), etc., and other oxygen nonradicals such as singlet oxygen ($^1O_2$), hydrogen peroxide (H_2O_2), ozone (O_3), etc. (Halliwell and Gutteridge, 2007). Of all the above-mentioned radicals, four of them, that is, $^1O_2$, $O_2^{.-}$, H_2O_2, and $^.OH$ are most stable and abundant in plants; however, differing in their stability, ability to transport across membrane, and reactivity. According to Smirnoff and Arnaud (2019), H_2O_2 is the most stable compound of all the ROS and is transported actively across the membranes via aquaporin channels.

To cope with unfavorable environmental conditions, plants have evolved several physiological and metabolic responses by the activation of complex signaling pathways. Besides harmful nature of ROS, they are also considered as a signaling molecule that maintains the growth and development of the plants during both abiotic and biotic stress conditions (Mittler et al., 2004). Sensory and signaling network of ROS along with their cross-talk with other signaling metabolites help the plants to overcome the negative effects of adverse environmental conditions (Noctor et al., 2017). One such important ROS formed in plant tissues in response to stressed condition is H_2O_2. In recent year, H_2O_2 has gained much attention due to its wide role in various resistance mechanisms such as phytoalexin formation, reinforcement of plant cell wall, and inducing the resistance power of the plants against various biotic and abiotic stressed conditions (Quan et al., 2008). In addition, H_2O_2 also controls a broad range of physiological processes in plants such as photorespiration and photosynthesis (Noctor and Foyer, 1998), cell cycle (Mittler et al., 2004), stomatal movement (Bright et al., 2006), senescence (Peng et al., 2005a), and growth and development (Foreman et al., 2003). Another important process, that is, cell death also occurs in plants in response to unfavorable environmental condition due to excess formation of H_2O_2. Thus, it is obvious for maintaining H_2O_2 at an optimal level to sustain the growth and development. Plants therefore possess an elaborate and efficient system for scavenging the excess H_2O_2 formed, which comprises of protective metabolites like osmolytes and enzymatic and nonenzymatic antioxidants. According to Neill et al. (2002), H_2O_2 regulates the expression level of various stress-related genes, which encode certain proteins that help in maintaining the energy formation, metabolism, cellular biogenesis of organelles, transportation of other proteins, and cell rescue and defense (Desikan et al., 2001). In addition, protective role of H_2O_2 has been demonstrated in various crops exposed to several abiotic stress such as salinity, drought, extreme temperature, and heavy metal stress (Table 9.1). H_2O_2 also interacts with other signaling molecules like calcium, salicylic acid, jasmonic acid, ethylene, abscisic acid, and nitric oxide, thus forming a complex signaling network that altogether help the plants to overcome the damaging effects of stressed environment (Liu et al., 2004; Desikan et al., 2004; Wendehenne et al., 2004).

TABLE 9.1 Protective Role of H_2O_2 in Abrogating the Negative Effects of Abiotic Stresses (Salinity, Drought, Heat, Cold, and Heavy Metal Toxicity) in Important Crops.

Stress	Plant species	Amount of H_2O_2 used	Effect on plants	References
Salt and heat	Rice	<10 µM	Induce the photosynthetic activity in leaf, activity of enzymatic antioxidants, up regulated the expression of Δ^1-pyrroline-5-carboxylate synthetase, sucrose-phosphate synthase, and heat shock protein 26	Uchida et al. (2002)

TABLE 9.1 *(Continued)*

Stress	Plant species	Amount of H_2O_2 used	Effect on plants	References
Salt	Maize	1 mM for 2 days	Reduces the level of lipid peroxidation, increases plant growth, and regulates the activity of enzymatic antioxidants	Azevedo-Neto et al. (2005)
Salt	Wheat	1, 40, 80 and 120 mM	Reduces the leakage of electrolytes from cells, induces photosynthetic attributes and expression of heat-stable stress proteins	Wahid et al. (2007)
Salt	Barley	1 and 5 mM	Higher rates of CO_2 fixation and lower MDA and H_2O_2 contents, reduced Cl^- accumulation in leaf	Fedina et al. (2009)
Salt	Wheat	0.05 μM	Reduced the MDA content, enhanced the GSH content and increased the activities of APX, CAT, SOD, and POD	Li et al. (2011)
Salt	*Suaeda fruticosa*	Sprayed twice a week with 100 μM H_2O_2	Level of proline, GSH, and soluble sugars was increased along with the activity of enzymatic antioxidants like SOD, CAT, and GPOX, whereas the activity of APX and GR along with level of MDA and H_2O_2 was reduced	Hameed et al. (2012)
Salt	Maize	10 mM	Level of MDA was lowered followed by enhanced activity of enzymatic antioxidants such as CAT, SOD, GPOX, and APX	Gondim et al. (2012)
Salt	Maize	10 mM	Photosynthesis, stomatal conductance, and intercellular CO_2 concentration were rescued, Level of ascorbate and glutathione was induced	Gondim et al. (2013)

TABLE 9.1 *(Continued)*

Stress	Plant species	Amount of H_2O_2 used	Effect on plants	References
Salt	Wheat	0, 50 or 100 nM	Reduction in Na^+ and Cl^- content; increase in proline content and nitrogen assimilation, increased water relations, photosynthetic pigments, and growth	Ashfaque et al. (2014)
Salt	*Panax ginseng*	100 μM for 2 days	Activities of APX, CAT, and GPOX along with proline formation was upregulated, production rate of MDA, H_2O_2, and superoxide radical was reduced	Sathiyaraj et al. (2014)
Drought	Cucumber		Increased activities of the antioxidant enzymes SOD, CAT, GPOX, APX, MDHAR, DHAR, GR, and the levels of ascorbate and GSH, resulting in lower levels of MDA, H_2O_2, and superoxide radical	Jing et al. (2009)
Drought	Soybean	1 mM	Delayed foliar wilting; higher photosynthetic rate, stomatal conductance, and formation of myo inositol and galactinol due to higher mRNA levels of *d-myo-inositol 3-phosphate synthase 2* and *galactinol synthase*	Ishibashi et al. (2011)
Drought	*Phaseolus vulgaris*	2% H_2O_2	A significant increase in compatible solutes, polyamine, antioxidants and abscisic acid, upregulated formation of osmolytes such as carbohydrates, soluble sugars, amino acids, proline, and polyamines	Abass and Mohamed (2011)

TABLE 9.1 *(Continued)*

Stress	Plant species	Amount of H_2O_2 used	Effect on plants	References
Drought	Marigold	600 µM	Reduced damage to mesophyll cell ultrastructure, increased leaf chlorophyll contents, chlorophyll fluorescence parameters (Fv/Fm, ΦPS II, and qP), and hypocotyl soluble carbohydrate and protein contents, and reduced starch level	Liao et al. (2012)
Drought	Mustard	50 µM	Enhanced cell membrane stability followed by lower MDA and H_2O_2 formation, higher APX, GR, CAT, GST, and Gly II activities as well as higher GSH/GSSG ratio	Hossain and Fujita (2013)
Drought	Maize	140 mM	Reduced photosynthetic pigment degradation and lipid preoxidation and increased the activities of antioxidant enzymes (CAT, SOD, and POX) and ascorbic acid levels.	Ashraf et al. (2014)
Drought	Cucumber	1.5 mM	Increased activities of antioxidative enzymes such as SOD, GPX, CAT, GPOX, APX, GR, MDHAR, DHAR, and the antioxidants ascorbate and reduced glutathione, reduced formation of MDA, H_2O_2, and superoxide anion	Liu et al. (2010)
Drought	Maize	10 mM	Decrease MDA levels and stomatal conductance, whereas an increase in endogenous H_2O_2, leaf water potential, ABA concentration, and metabolite levels, including soluble sugars, proline, and polyamines	Terzi et al. (2014)

TABLE 9.1 *(Continued)*

Stress	Plant species	Amount of H_2O_2 used	Effect on plants	References
Drought	Pea	70 mM	Less ROS-induced damage, accelerated proline synthesis and enhanced total chlorophyll, carotenoid contents, increased photosynthetic activity, growth and higher activity of APX, GPX, and CAT along with lower formation of superoxide anions and MDA	Moussa and Mohamed (2011)
Heat	Tomato and cucumber	15 mM	Decreased electrolyte leakage, H_2O_2, superoxide anions and MDA content, increased water-soluble protein and chlorophyll content, higher activity of APX, G6PDH, HPX, PPO, GPOX, and SOD	Kang et al. (2009)
Heat	Rice	100 μM	Activities of SOD, CAT, APX, GR, and total thiol content were increased, lower membrane, protein, and lipid damages	Bhattacharjee (2012)
Heat	Ryegrass and tall fescue	10 mM	Lower oxidative damage and H_2O_2 levels and increased activities of APX, GR, GST, and GPX	Wang et al. (2014a)
Cold	Mung bean	200 mM	Increased survival rates along with lower electrolyte leakage, higher formation of GSH, whereas activity of APX and CAT was hampered	Yu et al. (2003)
Cold	Mung bean	200 mM	Lower electrolyte leakage and higher GSH formation	Hung et al. (2007)
Cold	Mascarene grass and Manila grass	10 mM	Lower MDA and electrolyte leakage levels and higher protein contents, activities of APX, GPX, and CAT in Mascarene grass and APX, GR, and POD activities in manila grass were upregulated	Wang et al. (2010a)

TABLE 9.1 *(Continued)*

Stress	Plant species	Amount of H_2O_2 used	Effect on plants	References
Cold	Tomato	1 mM H_2O_2 for 1 h	Lower MDA and H_2O_2 levels followed by higher relative water content, proline formation, and APX and CAT activity	Iseri et al. (2013)
Cd	Rice	100 µM for 1 day	Decreased the Cd concentration in the shoots and lowered the shoot: root Cd ratio, MDA, and H_2O_2, activity of enzymatic antioxidants such as APX, SOD, CAT, GST, GPX along with ascorbate and GSH were induced	Hu et al. (2009)
Cd	Rice	0.5 mM	Leaf chlorosis and MDA formation was lowered, followed by higher GSH content and GSH/GSSG ratio, activity of APX and GR was lowered	Chao et al. (2009)
Al	Wheat	0.6 mM	Reduced cell death, MDA, H_2O_2, and superoxide anion formation, level of ascorbate and total glutathione was induced, followed by higher activity of CAT, POD, SOD, GPX, GR, MDHAR, and DHAR	Xu et al. (2010)
Cd	Rice		Lowered the Cd translocation in roots along with MDA formation, level of NPT, PCs, and GSH along with GST activity was induced	Bai et al. (2011)
Cd	Rice		The activities of APX and GR increased along with their gene expression	Chou et al. (2012)

TABLE 9.1 *(Continued)*

Stress	Plant species	Amount of H_2O_2 used	Effect on plants	References
Cr	Canola	200 μM	Higher Cr accumulation in roots and aerial parts of seedlings, higher activity of SOD, CAT, POD, and APX along with higher content of thiols, expression level of *BnMP1* was downregulated	Yildiz et al. (2013)
Cu	Maize	0.5 mM	Increases in growth, water content, mineral concentration (Na^+, K^+, Ca^+, Mg^{2+}), proline, total sugar, and soluble protein contents	Guzel and Terzi (2013)

MDA, malondialdehyde; GSH, reduced glutathione; APX, ascorbate peroxidase; CAT, catalase; SOD, superoxide dismutase; POD, peroxidase; GPOX, guaiacol peroxidase; GR, glutathione reductase; MDHAR, monodehydroascorbate reductase; DHAR, dehydroascorbate reductase; GST, glutathione S-transferase; Gly II, glyoxalase II; GSSG, oxidised glutathione; G6PDH, glucose-6-phosphate dehydrogenase; HPX, hydrogen peroxidase; PPO, polyphenol oxidase; GPX, glutathione peroxidase; NPT, nonprotein thiols; PCs, phytochelatins; MP1, metalloproteins 1.

For acting as a signaling molecule in plants, the synthesis of H_2O_2 is obviously regulated. Along with this, H_2O_2 must have specific responses, precise cellular targets and must be subsequently removed after their function to avoid oxidative stress in plants. Thus, in this chapter, our main focus is on the recent work that highlights the role of H_2O_2 in signaling process along with its formation and action followed by its subsequent decomposition to maintain an optimum level in plants.

9.2 ORIGIN OF H_2O_2 IN PLANTS

H_2O_2 is continuously generated in plants from various sources in different organelles during normal metabolism. H_2O_2 is produced mainly due to the disproportionation of superoxide radicals by the catalytic activity of the superoxide dismutase (SOD; EC 1.15.1.1) (eq 9.1) or by the reduction of superoxide radicals by nonenzymatic antioxidants (X) like ascorbate, ferredoxins, thiols, etc. (eq 9.2) suggesting that generation of H_2O_2 is mainly linked with formation of superoxide radicals.

$$2O_2^- + 2H^+ \rightarrow H_2O_2 + O_2 \qquad (9.1)$$

$$2O_2^- + H + XH_2 \rightarrow H_2O_2 + XH \qquad (9.2)$$

In addition to the above-mentioned processes, substrate level oxidation by various oxidases such as glucose oxidase, amino-acid oxidase, glycolate oxidase, and sulfite oxidase also contributes to the production of H_2O_2. In recent year, various reports have also demonstrated the formation of H_2O_2 in cell organelles during major metabolic processes like photosynthesis and photorespiration, suggesting that the formation of H_2O_2 is an unavoidable by-product of normal metabolism (Dat et al., 2000; Mittler, 2002). In this section, our aim will be trying to explain the formation of H_2O_2 according to their formation in different cell organelles.

9.2.1 MITOCHONDRIA

In plants, mitochondria can be a major site of H_2O_2 formation when exposed to dark condition, as compared to that of light exposed environment or in nongreen tissues (Rhoads et al., 2006). This contradicts the earlier report of Foyer and Noctor (2003) where they suggested that total H_2O_2 formation mostly remains unaffected during both light and dark environment since the consumption of O_2 during tricarboxylic acid cycle is not hampered by the light. Superoxide radicals formed by the NAD(P)H dehydrogenases (complex I) and the cytochrome bc_1 complex (complex III) of electron transport chain mainly results in the formation of H_2O_2 (Moller, 2001). Mitochondrial-specific manganese SOD (Mn-SOD) leads to the decomposition of superoxide radical formed during electron transport chain into H_2O_2 (Rhoads et al., 2006). According to Juszczuk and Rychter (2003), formation of H_2O_2 in plant tissues can be compromised by the action of alternative oxidases that catalyzes the tetravalent reduction of O_2 by ubiquinone. Further extending this work, Apel and Hirt (2004) demonstrated that alternative oxidases clash with cytochrome bc_1 complex of electron transport chain for electrons, which in turn lowers the formation of H_2O_2.

9.2.2 CHLOROPLAST

The main source of superoxide radical in chloroplast is photosynthetic electron chain. According to the reports of Foyer and Noctor (2000) and Dat et al. (2000), the components of the photosystem I, that is, Fe-S clusters, ferredoxin, and reduced thioredoxin undergoe auto-oxidation under limiting condition of NADP, which give rise to superoxide radicals that are ultimately converted to H_2O_2 via the catalytic action of SOD containing copper/zinc or iron at the active site. Along with this, Mehler (1951) demonstrated the formation of H_2O_2 via photo-reduction of O_2, which is generally known as "Mehler reaction" and is the most powerful and primary source of H_2O_2 production in the chloroplast. Mullineaux and Karpinski (2002) and Logan et al. (2006) further showed that the photo-reduction of O_2 depends on various environmental factors. Additional work on the formation of H_2O_2 via photo-reduction was also carried out by Karpinski et al. (2003) where they have showed the Mehler reaction as an alternative sink for electrons formed during excess excitation energy state. Another minor source of H_2O_2 formation in chloroplast is the reduction of superoxide radicals by ascorbate and reduced glutathione.

9.2.3 PEROXISOMES

H_2O_2 in peroxisome is mostly generated via photorespiration, which is a light-dependent O_2 consuming process along with the release of carbon dioxide (CO_2) and H_2O_2 as by-products (Wingler et al., 2000). Similar work was also done earlier by Noctor et al. (1999) where they showed that formation of H_2O_2 in peroxisome is mainly liked with the generation of glutathione in presence of glycine via oxygenase activity of ribulose-1,5-bisphosphate carboxylase/oxygenase (RuBisCO). Cornic and Fresneau (2002) noted that the necessity of the wasteful process of photorespiration (due to drainage of energy and reducing power) is for reducing the electronic pressure on photosynthetic machinery by acting as an electronic valve. Another minor process in C_3 plants that also release H_2O_2 in peroxisome is β-oxidation of fatty acid and oxidation of other substrates. Of both the processes, on a warm sunny day, that is, at excessive energy state, the process of photorespiration is the fastest and regardied as major reason of H_2O_2 formation in C_3 plants (Foyer and Noctor, 2000).

9.2.4 OTHER H_2O_2 FORMATION SITE IN CELL

In addition to above-mentioned organelles, H_2O_2 is also produced in extra-cellular matrix, cytoplasm, and plasma membrane. NADPH oxidases present in the plasma membrane catalyzes the one-electron reduction of O_2 by NADPH and formation of superoxide radicals (Apel and Hirt, 2004; Desikan et al., 2003), which is ultimately converted to H_2O_2 either spontaneously or by decomposition via catalytic action of extracellular SOD (Bolwell et al., 2002). Extracellular matrix of higher plants is made of lignins, pectins, hemicellulose, cellulose, and protein-based fiber matrices that also contain H_2O_2 producing enzymes. Thus, along with NADPH oxidases, the catalytic activity of many enzymes of extracellular matrix like germins, pH-dependent cell wall peroxidases, amine oxidases, and germin-like oxalate oxidases are also considered as sources of H_2O_2 formation in the apoplast (Bolwell et al., 2002; Kacperska, 2004). As compared to other organelles, cytosol does not act as major site of H_2O_2 production, but instead acts as a site for H_2O_2 accumulation, which are leaked from other cell organelles. Mittler et al. (2004) showed that cytochrome P450 reductase and reduced cytochrome P450 present in the endoplasmic reticulum, involved in the hydroxylation and oxidation processes, donates electrons to oxygen, ultimately producing superoxide radicals. It was also shown that desaturation of fatty acid via the action of cytochrome b_5 and cytochrome b_5 reductase also give rise to super-oxide radicals, which are finally decomposed into H_2O_2 by the enzymatic action of cytosolic SOD.

9.3 H_2O_2-MEDIATED SIGNALING IN PLANTS

To maintain their growth and development under unfavorable conditions, plants have developed special mechanisms for using H_2O_2 as a signaling molecule, which ultimately contributes to enhance the tolerance level of plants against abiotic stresses. According to Tian et al. (2016), being a highly stable molecule, H_2O_2 can easily pass through plasma membrane by diffu-sion or by aquaporin channels, which enables them to act at both intercellular (paracrine) signals and intracellular (autocrine) signals at both cellular and organ levels (Bienert et al., 2006). In this portion, we aim to explain the role of H_2O_2 as a signaling molecule in plants at both intercellular and intracel-lular level.

9.3.1 SIGNALING MEDIATED BY H_2O_2 AT INTERCELLULAR LEVEL

H_2O_2 needs to cross the cell membrane for acting as an intercellular signaling molecule, which occurs between different cells. According to Mittler et al. (2011), long-distance transport and signal of H_2O_2 occurs in a "ROS wave cascade" after primary production of oxidative burst. The contribution of the above-mentioned model is exhibited in plants toward tolerance capacity when it is linked with other signaling components such as calcium, hydraulic waves, ABA, and electric signals (Khedia et al., 2019). Intercellular signals via H_2O_2 can be elaborated in differentiation and lignifications of xylem. According to Barcelo (2005), H_2O_2 induces the synthesis of phenylalanine ammonia lyase, which ultimately enhances the process of xylem lignification and help in polymerization of coniferyl, coumaryl, and sinapyl alcohols to lignin that result in higher programmed cell death in differentiating xylem cells.

9.3.2 SIGNALING MEDIATED BY H_2O_2 AT INTRACELLULAR LEVEL

During drought stress, ABA-mediated stomatal closure is an effective strategy adapted by the plants to escape the water-limited condition by conserving the water present within the tissues. According to Pei et al. (2000), H_2O_2-mediated stomatal closure also takes place within the cells, which further enhances the process of water conservation within the organs. The formation of H_2O_2 is induced by NADPH oxidase that in turn intensifies the above-mentioned process. In addition, H_2O_2 can also act as a second messenger, which further helps in signal amplification. Kotak et al. (2007) demonstrated that oxidative stress-responsive transcription factors such as Class A heat shock factors sense the elevated level of H_2O_2 in plants and thus regulate the downstream cascade. The H_2O_2 signaling network is connected with mitogen-activated protein kinase (MAPK) cascades during plant-pathogen interactions (Somssich, 1997), osmotic stress (Hirt, 1997), and ozone (Samuel et al., 2000) via redox network (Matern et al., 2015).

9.4 CROSSTALK BETWEEN H_2O_2 AND MITOGEN-ACTIVATED PROTEIN KINASE CASCADES

A significant character of H_2O_2 is their ability to activate the MAPK cascade, which is involved in the development and growth of plants and also enable them to respond according to the surrounding environment (Komis et al., 2018). MAPK-signaling cascade consists of three kinases, that is, MAPK kinase kinases (MAP3Ks), MAPK kinases (MAP2Ks), and MAPKs that are consecutively phosphorylated, which ultimately results in inactivation or activation of a wide range of transcription factors and proteins (Liu and He, 2017). Excess accumulation of H_2O_2 in tissues due to exogenous application or intracellular formation leads to the activation of H_2O_2 signaling pathway in plants, which is further relayed by MAPK cascade in response to abiotic stresses or pathogenic attack. Savatin et al. (2014) and Kovtun et al. (2000) reported that in plants, two kinds of MAP3Ks, that is, mitogen-activated protein kinase kinase kinase1 (MAPKKK1 or MEKK1) and arabidopsis homologs of nucleus and phraglocalized kinases (ANPs) are activated by ROS. Kovtun et al. (2000) further reported that H_2O_2 signal activates the ANP1, which in turn enhances the phosphorylation of AtMPK6 and AtMPK3. H_2O_2 signaling also enhances the sulfonylation of MPK7, MPK4, and MPK2 (Waszczak et al., 2014). BnMPK4, a homolog of MPK4 accumulates after oxidation of Cys residue present at position 232 (Zhang et al., 2015). According to Pitzschke et al. (2009), one of the H_2O_2-mediated MAPK cascade consists of MEKK1–MKK1/2–MPK4, cascading those activation leads into the activation of MPK6, MPK4, and MPK3 ultimately in a proteasome-dependent manner. Miao et al. (2007) demonstrated a unique MAPK pathway where the transcription factor WRKY53 is phosphorylated via MEKK1, which subsequently induces the DNA-binding activity that ultimately induces the expression of stress-related gene. Asai and Yoshioka (2008) reported the presence of SIPK and WIPK kinases (homolog of MPK3 and MPK6 in *Arabidopsis*, respectively) in tobacco. Cross-talk between H_2O_2 and MAPK signaling pathway is also widely reported in other abiotic stresses such as cold stress, drought, and salinity. Schmidt et al. (2013) reported that salt stress-responsive transcription factor in rice, that is, ERF1 is activated by H_2O_2, which leads to binding of MAP3K6 and MAPK5 and contributes toward capability of higher stress tolerance in plants. Reduced level of H_2O_2 in transgenic poplar and cotton plants was noted in presence of *PtMKK4* and *GhMPK17* genes, respectively, in response to drought stress (Wang et al., 2014b; Zhang et al., 2014). Complex interaction between calcium-dependent protein kinase, CPK27, H_2O_2, NO, and MPK1/2 was noted in tomato plants in response to cold stress that ultimately results in higher ABA formation and stress tolerance level of plants (Lv et al., 2018). Similar results were also demonstrated by Wang et al. (2017) where they showed that chilling stress in tomato plants was ameliorated by treating the plants with H_2O_2, which results in higher expression of *SlMAPK1/2/3* and *SlCBF1* genes, regulating the level of stress-related phytohormones. Thus, it is obvious that a complex interaction occurs between H_2O_2 and MAPK cascade in plants, which contributes toward their tolerance level against abiotic stresses and an optimum level of H_2O_2 is necessary for downstream relay of H_2O_2-mediated signaling pathway via MAPK cascade.

9.5 CROSS-TALK BETWEEN H_2O_2 AND OTHER SIGNALING MOLECULES

9.5.1 CALCIUM

Calcium (Ca^{2+}) is an important secondary messenger that is mainly stored in vacuoles (Mahajan and Tuteja, 2005). Ca controls several important processes like cell wall formation, acts as an important signaling molecule during abiotic stressed environment such as unfavorable light, temperature, and limited water supply (Mahajan and Tuteja, 2005). Pei et al. (2000) suggested that abiotic stress leads to the changes in the permeability of plasma membrane, which in turn enhances the influx of Ca and proton that ultimately upregulates the formation of H_2O_2. Several Ca^{2+}/CaM proteins involved in plant metabolism such as NAD kinases also acts as a regulator of H_2O_2 signaling pathway (Harding et al., 1997). Interaction between Ca^{2+} and H_2O_2 has been widely demonstrated in *Arabidopsis*. According to Rentel and Knight (2004), elevation of Ca^{2+} ions in the cotyledons of *Arabidopsis* followed by a second peak in roots is regulated by H_2O_2. In addition, interaction between H_2O_2 and Ca was also reported in other plants like tobacco and tomato (Lachaud et al., 2011; Chico et al., 2002).

9.5.2 ABSCISIC ACID

Abscisic acid (ABA) is a crucial stress phytohormone whose role has been extensively reported by various researcher groups in abrogating the toxic effects of adverse environmental cues like shortened day length, drought, salinity, and cold stress. ABA also controls various important processes in plants like embryo morphogenesis, stomatal closure, seed dormancy, leaf senescence, development of seed germination, storage, and synthesis of lipids and proteins (Saxena et al., 2016). Recent researches have recommended an active involvement of ABA signaling pathway in the regulation of MAPK cascade during downstream signaling of abiotic stress defense mechanisms (Zong et al., 2009). Active involvement of exogenously applied H_2O_2 in MAPK cascade is mediated by other stress-related hormones like ABA, salicylic acid (SA), and jasmonic acid (JA) (Rodriguez et al., 2010). Promising role of ABA via MAPK cascade in *Arabidopsis* has been reported earlier by Menges et al. (2008) and Wang et al. (2011). According to Montillet et al. (2013), MPK9 and MPK12 are activated by H_2O_2 and ABA that plays an active role in ABA signaling pathway in guard cells. One of the major adaptive features in plants to combat drought stress is stomatal closure, which is actively controlled by ABA. According to Miao et al. (2006) and Pei et al. (2000), in *Arabidopsis*, H_2O_2 acts as a second messenger in ABA-mediated stomatal closure. Similar result was also demonstrated by Zhang et al. (2001) where they have showed the involvement of H_2O_2 in ABA-mediated stomatal movement in *Vicia faba*. Thus, these earlier works showed a complex interaction between ABA and H_2O_2 that is necessary for mediating stress tolerance ability in plants.

9.5.3 NITRIC OXIDE

Large number of studies has demonstrated the complex interplay occurring between nitric oxide (NO) and H_2O_2 in plants for alleviating the effects of abiotic stress. NO is generated in plants via nitrate-dependent reductive pathway or via hydroxylamine or arginine-dependent oxidative pathway (Gupta et al., 2011). Delledonne et al. (2001) reported that during hypersensitive response in plants, H_2O_2 and NO are simultaneously produced that ultimately results in higher hypersensitive cell death. Earlier Delledonne et al. (1998) and Clarke et al. (2000) reported synergistic and additive effects, respectively, of H_2O_2 and NO signals in inducing programmed cell death in soybean and *Arabidopsis*, respectively, during bacterial infection. According to Saxena et al. (2016), transcription factors are directly affected by oxidation of Cys and S-nitrosylation mediated by both NO and H_2O_2. De Pinto et al. (2006) suggested the involvement of both NO and H_2O_2 in programmed cell death in tobacco BrightYellow-2 (TBY-2) cells; however, on scavenging of both the molecules, the level of cell viability was restored. According to Neill et al. (2008), ABA-mediated H_2O_2 formation may provoke the formation of NO via catalytic activity of nitrate reductase and nitrogen oxide synthase-like enzymes. He et al. (2013) reported that stomatal closure in *Arabidopsis* plants in response to UV-B is stimulated by both H_2O_2 and NO.

Thus, a strong correlation exists between H_2O_2 and NO signaling pathway, which contributes toward higher stress tolerance capacity of plants.

9.5.4 JASMONIC ACID

Jasmonic acid (JA) is contemplated to be a vital stress-related signaling molecule. According to Ishiguro et al. (2001), membrane of chloroplast acts as an important site for JA formation where alfa-linolenic acid (C18:3) and hexadecatrienoic acid (C16:3) are mostly generated from membrane phospholipids. Role of JA against biotic stress is widely reported, but in case of abiotic stress, the efficacy of JA is not so well reported till date. Hu et al. (2003) demonstrated that during the cellular response, JA and H_2O_2 serve as primary signaling molecules that are involved in saponin biosynthesis, which is mediated by oligo galacturonic acid. Role of methyl jasmonate (a derivative of JA) has been reported in inducing the level of H_2O_2 in cell suspension culture of parsley (Kauss et al. 1994). Reports demonstrating the interaction between JA and H_2O_2 during abiotic stress are very scanty and more work is required in coming future to further decipher this complex interaction that will prove to be highly effective in revoking the ill effects of abiotic stress in plants.

9.5.5 SALICYLIC ACID

Unlike above-mentioned signaling molecules, the role of salicylic acid (SA) in both biotic stress and abiotic stress is widely reported in plants. According to Herrera-Vasquez et al. (2015), SA induces both systemic and local response in plants. Environmental stress such as drought, salinity, and excess light directly induces the formation of ROS, whereas SA signaling pathway directly or indirectly interacts with glutathione and ROS in stressed plants (Herrera-Vasquez et al., 2015; Lee and Park, 2010). Liu et al. (2014) demonstrated that H_2O_2 enhances SA accumulation, programmed cell death, and sesquiterpene production in *Aquilaria sinensis* suspension cell culture. H_2O_2-dependent activation of benzoic acid 2-hydroxylase is involved in the synthesis of benzoic acid as a precursor of SA (Leon et al., 1995). According to Rao et al. (1997) and Dat et al. (1998), application of SA can enhance the formation of H_2O_2 in plant tissues and the level of SA can also be induced due to surplus production of H_2O_2 (Chamnongpol et al., 1998). Thus, SA and H_2O_2 act synergistically following a strong interrelated pathway.

9.5.6 ETHYLENE

Ethylene is one of the important stress hormones whose role has been earlier explained in both biotic stress like pathogen attack and abiotic stress such as ozone, salinity, and wounding along with plant development and growth (Wang et al., 2009). Interaction between ethylene and H_2O_2 has been demonstrated in plants, but still there is no clear picture, which can fully demonstrate the mode of interaction. Wang et al. (2010b) showed the interaction between ethylene and H_2O_2 in callus during salt stress. Similarly, Schraudner et al. (1998) showed that under ozone stress, production of H_2O_2 also enhances the formation of ethylene in tobacco plants. Under hypoxia stress, the level of both ethylene and H_2O_2 is triggered. Peng et al. (2005b) demonstrated that hyoxia stress induces the activity of ethylene biosynthetic enzymes like ACC oxidase and ACC synthase in *Arabidopsis*. Similar result was also shown by Hattori et al. (2009) where they showed the efficacy of ethylene-responsive factors in inducing submergence tolerance in rice.

9.6 SCAVENGING OF H_2O_2

H_2O_2 acts as a cytotoxic metabolite on being accumulated in cells and thus it become necessary that plants remove excess H_2O_2 produced in response to abiotic stress. In cells, H_2O_2 can be scavenged by enzymatic antioxidants or by low molecular weight compounds such as nonenzymatic antioxidants and osmolytes.

9.6.1 ENZYMATIC ANTIOXIDANTS

Enzymatic antioxidants such as catalase (CAT; EC 1.11.1.6) and peroxidases (EC 1.11.1) efficiently detoxify H_2O_2 by decomposing it into H_2O and O_2. Peroxidases such as ascorbate peroxidase (APX; EC 1.11.1.11) and glutathione peroxidase (GPX; EC 1.11.1.9) catalyze the breakdown of H_2O_2 to H_2O and O_2, using ascorbate and reduced glutathione (GSH) as cofactors, respectively. According to Chang et al. (2004), APX is considered as the second line of defense against H_2O_2 in chloroplast. Glutathione reductase (GR; EC 1.8.1.7), another vital enzyme to complete the ascorbate-glutathione cycle, regenerates GSH using NADPH as an electron donor (Apel and Hirt, 2004). Another key enzyme for decomposition of cytotoxic H_2O_2 into nontoxic H_2O is catalase. Catalase is an iron-containing enzyme that spontaneously decomposes H_2O_2. Guaiacol peroxidase (GPoX; EC 1.11.1.7) also takes part in detoxification of H_2O_2 using guaiacol as a substrate, yielding tetraguaiacol as by-product.

9.6.2   NONENZYMATIC ANTIOXIDANTS AND OSMOLYTES

To counteract the negative effects of H_2O_2, plants have developed various metabolites that efficiently scavenge the excess H_2O_2 generated. Few nonenzymatic antioxidants like GSH, ascorbate, and tocopherol efficiently degrade the toxic H_2O_2 into nontoxic molecules. Of all the above-mentioned metabolites, GSH is present in millimolar level in cells and is regarded as a key determinant of cellular redox state (Pastori and Foyer, 2002). Another important group of metabolite is osmolyte that consists of amino acids, proline, and glycine betaine. The level of osmolytes are enhanced in response to abiotic stresses, which in turn scavenge the excess H_2O_2 formed and thus protect the cell membrane from peroxidation and also maintains the osmotic balance in the cells (Singh et al., 2020).

9.7   CONCLUSION

H_2O_2 represents a key signaling molecule in plants formed in response to environmental stress, connecting multiple signaling pathways for phytohormones and chemicals, and acting as a second messenger that helps in modulating the growth and development of plants. Its dose-dependent action has forced plants to develop an efficient scavenging system comprising of enzymatic and nonenzymatic antioxidants that can keep the level of H_2O_2 at the desirable level, thereby checking the formation of oxidative stress. Significant work that focuses the importance of H_2O_2 as a signaling molecule has been done in the last two decades, which has enabled us to discern the role of H_2O_2 in communication, signal transduction between external abiotic and biotic stress, and maintaining the developmental processes. Extensive works have been done earlier to demonstrate the efficacy of H_2O_2 as a signaling molecule for controlling the defense response of plants, but there still remains a immense scope for additional researches that can be further clarify the mechanism involving these pathways. H_2O_2 interacts with other signaling molecules like ABA, NO, JA, Ca, SA, and ethylene that enhancing its significance further in plants during harsh conditions. Thus, H_2O_2 is "two-faced," being "beneficial" at lower concentration and "harmful" when formed in excess and thus plants have to maintain a deliberate and fine-tuned metabolic network to maintain safe level of H_2O_2. H_2O_2 is an unavoidable by-product of cellular metabolic process like photosynthesis and photorespiration generated in chlorophyll, mitochondria, and peroxisome. H_2O_2 accumulation is lowered in cells via efficient functioning of antioxidative machineries. Thus, the merits of H_2O_2 outweigh its demerits in plants playing an essential role in regulating response and signal transduction against abiotic and biotic stresses, development, and growth of plants.

ACKNOWLEDGMENTS

Dr. Aryadeep Roychoudhury is acknowledged for his constant guidance and support. We acknowledge the Department of Biotechnology, St. Xaviers College for instrumental facility. Department of Higher Education, Science and Technology and Biotechnology, Government of West Bengal [264(Sanc.)/ST/P/S&T/1G-80/2017] and Science and Engineering Research Board, Government of India [EMR/2016/004799] for financial support.

KEYWORDS

- **hydrogen peroxide**
- **abiotic stress**
- **antioxidative machineries**
- **nitric oxide**
- **signaling cascade**

REFERENCES

Abass, S. M.; Mohamed, H. I. Alleviation of Adverse Effects of Drought Stress on Common Bean (*Phaseolus vulgaris* L.) by Exogenous Application of Hydrogen Peroxide. *Bangladesh J. Bot.* **2011,** *41,* 75–83.

Apel, K.; Hirt H. Reactive Oxygen Species: Metabolism, Oxidative Stress, and Signal Transduction. *Annu. Rev. Plant Biol.* **2004,** *55,* 373–399.

Asai, S.; Yoshioka H. The Role of Radical Burst via MAPK Signaling in Plant Immunity. *Plant Signal Behav.* **2008,** *3,* 920–922.

Ashfaque, F.; Khan, M. I. R.; Khan, N. A. Exogenously Applied H_2O_2 Promotes Proline Accumulation, Water Relations, Photosynthetic Efficiency and Growth of Wheat (*Triticum aestivum* L.) Under Salt Stress. *Annu. Res. Rev. Biol.* **2014,** *4,* 105–120.

Ashraf, M. A.; Rasheed, R.; Hussain, I.; Iqbal, M.; Haider, M. Z.; Parveen, S.; Sajid, M. A. Hydrogen Peroxide Modulates Antioxidant System and Nutrient Relation in Maize (*Zea mays* L.) Under Water-Deficit Conditions. *Arch. Agron. Soil. Sci.* **2014,** *61,* 507–523.

Azevedo-Neto, A. D.; Prisco, J. T.; Eneas-Filho, J.; Medeiros, J. V. R.; Gomes-Filho, E. Hydrogen Peroxide Pre-Treatment Induces Salt Stress Acclimation in Maize Plants. *J. Plant Physiol.* **2005,** *162,* 1114–1122.

Bai, X. J.; Liu, L. J.; Zhang, C. H.; Ge, Y.; Cheng, W. D. Effect of H_2O_2 Pretreatment on Cd Tolerance of Different Rice Cultivars. *Rice Sci.* **2011,** *18,* 29–35.

Barcelo, A. R. Xylem Parenchyma Cells Deliver the H_2O_2 Necessary for Lignification in Differentiating Xylem Vessels. *Planta* **2005,** *220,* 747–756.

Bhattacharjee, S. An Inductive Pulse of Hydrogen Peroxide Pretreatment Restores Redox-Homeostasis and Mitigates Oxidative Membrane Damage Under Extremes of Temperature in Two Rice Cultivars (*Oryza sativa* L., Cultivars Ratna and SR26B). *Plant Growth Regul.* **2012,** *68,* 395–410.

Bienert, G. P.; Schjoerring, J. K.; Jahn, T. P. Membrane Transport of Hydrogen Peroxide. *BBA Biomembr.* **2006,** *1758,* 994–1003.

Bolwell, G. P.; Bindschedler, L. V.; Blee, K. A.; Butt, V. S.; Davies, D. R.; Gardner, S. L.; Gerrish, C.; Minibayeva, F. The Apoplastic Oxidative Burst in Response to Biotic Stress in Plants: A Three-Component System. *J. Exp. Bot.* **2002,** *53,* 1367–1376.

Bright, J.; Desikan, R.; Hancock, J. T.; Weir, I. S.; Neill, S. J. ABA Induced NO Generation and Stomatal Closure in *Arabidopsis* Are Dependent on H_2O_2 Synthesis. *Plant J.* **2006,** *45,* 113–122.

Chamnongpol, S.; Willekens, H.; Moeder, W.; Langebartels, C.; Sandermann, H.; Van Montagu, M.; Inze, D.; Van Camp, W. Defense Activation and Enhanced Pathogen Tolerance Induced by H_2O_2 in Transgenic Tobacco. *Proc. Natl. Acad. Sci. U.S.A.* **1998,** *95,* 5818–5823.

Chang, C. C. C.; Ball, L.; Fryer, M. J.; Baker, N. R.; Karpinski, S.; Mullineaux, P. M. Induction of *ascorbate peroxidase 2* Expression in Wounded *Arabidopsis* Leaves Does not Involve Known Wound-Signalling Pathways but Is Associated with Changes in Photosynthesis. *Plant J.* **2004,** *38,* 499–511.

Chao, Y. Y.; Hsu, Y. T.; Kao, C. H. Involvement of Glutathione in Heat Shock-and Hydrogen Peroxide-Induced Cadmium Tolerance of Rice (*Oryza sativa* L.) Seedlings. *Plant Soil* **2009,** *318,* 37–45.

Chico, J. M.; Raíces, M.; Téllez-Iñón, M. T.; Ulloa, R. M. A Calcium-Dependent Protein Kinase is Systemically Induced upon Wounding in Tomato Plants. *Plant Physiol.* **2002,** *128,* 256–270.

Chou, T. S.; Chao, Y. Y.; Kao, C. H. Involvement of Hydrogen Peroxide in Heat Shock- and Cadmium-Induced Expression of Ascorbate Peroxidase and Glutathione Reductase in Leaves of Rice Seedlings. *J. Plant Physiol.* **2012,** *169,* 478–486.

Clarke, A.; Desikan, R.; Hurst, R. D.; Hancock, J. T.; Neill, S. J. NO Way Back: Nitric Oxide and Programmed Cell Death in *Arabidopsis thaliana* Suspension Cultures. *Plant J.* **2000,** *24,* 667–677.

Cornic, G.; Fresneau, C. Photosynthetic Carbon Reduction and Carbon Oxidation Cycles Are the Main Electron Sinks for Photosystem II Activity During a Mild Drought. *Ann. Bot.* **2002,** 89, 887–894.

Dat, J.; Vandenabeele, S.; Vranová, E; Van Montagu, M.; Inzé, D.; Breusegem, F. Dual Action of the Active Oxygen Species During Plant Stress Responses. *CMLS Cell Mol. Life Sci.* **2000,** *57,* 779–795.

Dat, J. F.; Lopez-Delago, H.; Foyer, C. H.; Scott, I. M. Parallel Changes in H_2O_2 and Catalase During Thermotolerance Induced by Salicylic Acid or Heat Acclimation in Mustard Seedlings. *Plant Physiol.* **1998,** *116,* 1351–1357.

De Pinto, M. C.; Paradiso, A.; Leonetti, P.; De Gara, L. Hydrogen Peroxide, Nitric Oxide and Cytosolic Ascorbate Peroxidase at the Crossroad Between Defence and Cell Death. *Plant J.* **2006,** *48,* 784–795.

Delledonne, M.; Xia, Y.; Dixon, R. A.; Lamb, C. Nitric Oxide Functions as a Signal in Plant Disease Resistance. *Nature* **1998,** *394,* 585–588.

Delledonne, M.; Zeier J, Marocco, A.; Lamb, C. Signal Interactions Between Nitric Oxide and Reactive Oxygen Intermediates in the Plant Hypersensitive Disease Resistance Response. *Proc. Natl. Acad. Sci. U.S.A.* **2001,** *98,* 13454–13459.

Desikan, R.; Cheung, M. -K.; Clarke, A.; Golding, S.; Sagi, M.; Fluhr, R.; Rock, C.; Hancock, J.; Neill, S. Hydrogen Peroxide is a Common Signal for Darkness and ABA-Induced Stomatal Closure in *Pisum sativum* L. *Funct. Plant Biol.* **2004,** *31,* 913–920.

Desikan, R.; Hancock, J. T.; Neill, S. J. Oxidative Stress Signalling. *Top. Curr. Genet.* **2003,** *4,* 129–149.

Desikan, R. S.; Mackerness, A. H.; Hancock, J. T.; Neill, S. J. Regulation of the *Arabidopsis* Transcriptome by Oxidative Stress. *Plant Physiol.* **2001,** *127,* 159–172.

FAO. *The State of the World's Land and Water Resources for Food and Agriculture (SOLAW)—Managing Systems at Risk*; Food and Agriculture Organization of the United Nations: Rome and Earthscan, London, 2011.

Fedina, I. S.; Nedeva, D.; Çiçek, N. Pre-Treatment with H_2O_2 Induces Salt Tolerance in Barley Seedlings. *Biol. Plant* **2009,** *53,* 321–324.

Foreman, J.; Bothwell, J. H.; Demidchik, V.; Mylona, P.; Miedema, H.; Torres, M. A.; Linstead, P.; Costa, S.; Brownlee, C.; Jones, J. D. G.; Davies, J. M.; Dolan, L. Reactive Oxygen Species Produced by NADPH Oxidase Regulate Plant Cell Growth. *Nature* **2003,** *422,* 442–446.

Foyer, C. H.; Noctor, G. Oxygen Processing in Photosynthesis: Regulation and Signalling. *New Phytol.* **2000,** *146,* 359–388.

Foyer, C. H.; Noctor, G. Redox Sensing and Signalling Associated with Reactive Oxygen in Chloroplasts, Peroxisomes and Mitochondria. *Physiol. Plant* **2003,** *119,* 355–364.

Gondim, F. A.; Gomes-Filho, E.; Costa, J. H.; Alencar, N. L. M.; Priso, J. T. Catalase Plays a Key Role in Salt Stress Acclimation Induced by Hydrogen Peroxide Pretreatment in Maize. *J. Plant Physiol. Biochem.* **2012,** *56,* 62–71.

Gondim, F. A.; Miranda, R. S.; Gomes-Filho, E.; Prisco, J. T. Enhanced Salt Tolerance in Maize Plants Induced by H_2O_2 Leaf Spraying is Associated with Improved Gas Exchange Rather than with Non-Enzymatic Antioxidant System. *Theor. Exp. Plant Physiol.* **2013,** *25,* 251–260.

Gupta, K. J.; Fernie, A. R.; Kaiser, W. M.; van Dongen, J. T. On the Origins of Nitric Oxide. *Trend. Plant Sci.* **2011,** *16,* 160–168.

Guzel, S.; Terzi, R. Exogenous Hydrogen Peroxide Increases Dry Matter Production, Mineral Content and Level of Osmotic Solutes in Young Maize Leaves and Alleviates Deleterious Effects of Copper Stress. *Bot. Stud.* **2013,** *54,* 26.

Halliwell, B.; Gutteridge, J. M. C. *Free Radicals in Biology and Medicine*; Oxford University Press, 2007.

Hameed, A.; Hussain, T.; Gulzar, S.; Aziz, F.; Gul, B.; Khan, M. A. Salt Tolerance of a Cash Crop Halophyte *Suaeda fruticosa*: Biochemical Responses to Salt and Exogenous Chemical Treatments. *Acta Physiol. Plant* **2012,** *34,* 2331–2340.

Harding, S. A.; Oh, S. H.; Roberts, D. M. Transgenic Tobacco Expressing a Foreign Calmodulin Gene Shows an Enhanced Production of Active Oxygen Species. *EMBO J.* **1997,** *16,* 1137–1144.

Hattori, Y.; Nagai, K.; Furukawa, S.; Song, X. –J.; Kawano, R.; Sakakibara, H.; Wu, J.; Matsumoto, T.; Yoshimura, A.; Kitano, H.; Matsuoka, M.; Mori, H.; Ashikari, M. The Ethylene Response Factors SNORKEL1 and SNORKEL2 Allow Rice to Adapt to Deep Water. *Nature* **2009,** *460,* 1026–1030.

He, J. -M.; Ma, X. -G.; Zhang, Y.; Sun, T. -F.; Xu, F. -F.; Chen, Y. -P.; Liu, X.; Yue, M. Role and Interrelationship of G Protein, Hydrogen Peroxide, and Nitric Oxide in Ultraviolet B-Induced Stomatal Closure in *Arabidopsis* Leaves. *Plant Physiol.* **2013,** *161,* 1570–1583.

Herrera-Vasquez, A.; Salinas, P.; Holuigue, L. Salicylic Acid and Reactive Oxygen Species Interplay in the Transcriptional Control of Defense Genes Expression. *Front. Plant Sci.* **2015,** *6,* 171.

Hirt, H. Multiple Roles of MAP Kinases in Plant Signal Transduction. *Trends Plant Sci.* **1997,** *2,* 11–15.

Hossain, M. A.; Fujita, M. Hydrogen Peroxide Priming Stimulates Drought Tolerance in Mustard (*Brassica juncea* L.). *Plant Gene Trait* **2013,** *4,* 109–123.

Hu, H.; Xiong, L. Genetic Engineering and Breeding of Drought-Resistant Crops. *Annu. Rev. Plant Biol.* **2014,** 65, 715–741.

Hu, X.; Bidney, D. L.; Yalpani, N.; Duvick, J. P.; Crasta, O.; Folkerts, O.; Lu, G. Overexpression of a Gene Encoding Hydrogen Peroxide-Generating Oxalate Oxidase Evokes Defense Responses in Sunflower. *Plant Physiol.* **2003,** *133,* 170–181.

Hu, Y.; Ge, Y.; Zhang, C.; Ju, T.; Cheng, W. Cadmium Toxicity and Translocation in Rice Seedlings are Reduced by Hydrogen Peroxide Pretreatment. *Plant Growth Regul.* **2009,** *59,* 51–61.

Hung, S. H.; Wang, C. C.; Ivanov, S. V.; Alexieva, V.; Yu, C. W. Repetition of Hydrogen Peroxide Treatment Induces a Chilling Tolerance Comparable to Cold Acclimation in Mung Bean. *J. Am. Soc. Hort. Sci.* **2007,** *132,* 770–776.

Iseri, O. D.; Körpe, D. A.; Sahin, F. I.; Haberal, M. Hydrogen Peroxide Pretreatment of Roots Enhanced Oxidative Stress Response of Tomato Under Cold Stress. *Acta Physiol. Plant* **2013,** *35,* 1905–1913.

Ishibashi, Y.; Yamaguchi, H.; Yuasa, T.; Inwaya-Inoue, M.; Arima, S.; Zheng, S. Hydrogen Peroxide Spraying Alleviates Drought Stress in Soybean Plants. *J. Plant Physiol.* **2011,** *168,* 1562–1567.

Ishiguro, S.; Kawai-Oda, A.; Ueda, J.; Nishida, I.; Okada, K. The Defective in Anther Dehiscence Gene Encodes a Novel Phospholipase A1 Catalyzing the Initial Step of

Jasmonic Acid Biosynthesis, Which Synchronizes Pollen Maturation, Anther Dehiscence, and Flower Opening in *Arabidopsis*. *Plant Cell* **2001**, *13*, 2191–2209.

Jing, L. Z.; Kui, G. Y.; Hang, L. S.; Gang, B. J. Effects of Exogenous Hydrogen Peroxide on Ultrastructure of Chloroplasts and Activities of Antioxidant Enzymes in Greenhouse-Ecotype Cucumber Under Drought Stress. *Acta Hort. Sinica.* **2009**, *36*, 1140–1146.

Juszczuk, I. M.; Rychter, A. M. Alternative Oxidase in Higher Plants. *Acta Biochim. Pol.* **2003**, *50*, 1257–1271.

Kacperska, A. Sensor Types in Signal Transduction Pathways in Plant Cells Responding to Abiotic Stressors: Do They Depend on Stress Intensity? *Physiol. Plant* **2004**, *122*, 159–168.

Kang, N. J.; Kang, Y. I.; Kang, K. H.; Jeong, B. R. Induction of Thermotolerance and Activation of Antioxidant Enzymes in H_2O_2 Pre-Applied Leaves of Cucumber and Tomato Seedlings. *J. Jpn. Soc. Hort. Sci.* **2009**, *78*, 320–329.

Karpinski, S.; Gabryś, H.; Mateo, A.; Karpinska, B.; Mullineaux, P. M. Light Perception in Plant Disease Defence Signalling. *Curr. Opin. Plant Biol.* **2003**, *6*, 390–396.

Kauss, H.; Jeblick, W.; Ziegler, J.; Krabler, W. Pretreatment of Parsley (*Petroselinum crispum*) Suspension Cultures with Methyl Jasmonate Enhanced Elicitation of Activated Oxygen Species. *Plant Physiol.* **1994**, *105*, 89–94.

Khedia, J.; Agarwal, P.; Agarwal, P. K. Deciphering Hydrogen Peroxide-Induced Signalling Towards Stress Tolerance in Plants. *3 Biotech.* **2019**, *9*, 395.

Komis, G.; Šamajová, O.; Ovĕcka, M.; Šamaj, J. Cell and Developmental Biology of Plant Mitogen-Activated Protein Kinases. *Annu. Rev. Plant Biol.* **2018**, *69*, 237–265.

Kotak, S.; Larkindale, J.; Lee, U.; von Koskull-Döring, P.; Vierling, E.; Scharf, K. D. Complexity of the Heat Stress Response in Plants. *Curr. Opin. Plant Biol.* **2007**, *10*, 310–316.

Kovtun, Y.; Chiu, W. L.; Tena, G.; Sheen, J. Functional Analysis of Oxidative Stress-Activated Mitogen-Activated Protein Kinase Cascade in Plants. *Proc. Natl. Acad. Sci. U.S.A.* **2000**, *97*, 2940–2945.

Lachaud, C.; Da Silva, D.; Amelot, N.; Béziat, C.; Brière, C.; Cotelle, V.; Grazianaa, A.; Grata, S.; Mazarsa, C.; Thuleau, P. Dihydrosphingosine-Induced Programmed Cell Death in Tobacco BY-2 Cells is Independent of H_2O_2 Production. *Mol. Plant* **2011**, *4*, 310–318.

Lee, S.; Park, C. M. Modulation of Reactive Oxygen Species by Salicylic Acid in *Arabidopsis* Seed Germination Under High Salinity. *Plant Signal Behav.* **2010**, *5*, 1534–1536.

Leon, J.; Lawton, M. A.; Raskin, I. Hydrogen Peroxide Stimulates Salicylic Acid Biosynthesis in Tobacco. *Plant Physiol.* **1995**, *108*, 1673–1678.

Li, J. T.; Qiu, Z. B.; Zhang, X. W.; Wang, L. S. Exogenous Hydrogen Peroxide Can Enhance Tolerance of Wheat Seedlings to Salt Stress. *Acta Physiol. Plant* **2011**, *33*, 835–842.

Liao, W. B.; Huang, G. B.; Yu, J. H.; Zhang, M. L. Nitric Oxide and Hydrogen Peroxide Alleviate Drought Stress in Marigold Explants and Promote Its Adventitious Root Development. *Plant Physiol. Biochem.* **2012**, *58*, 6–15.

Liu, J.; Xu, Y.; Zheng, Z.; Wei, J. Hydrogen Peroxide Promotes Programmed Cell Death and Salicylic Acid Accumulation During the Induced Production of Sesquiterpenes in Cultured Cell Suspensions of *Aquilaria sinensis*. *Funct. Plant Biol.* **2014**, *42*, 337–346.

Liu, Q.; Yu, Z. G.; Kuang, W. C. Ethylene Signal Transduction in *Arabidopsis*. *J. Plant Physiol. Mol. Biol.* **2004**, *30*, 241–250.

Liu, Y.; He, C. A Review of Redox Signaling and the Control of MAP Kinase Pathway in Plants. *Redox Biol.* **2017**, *11*, 192–204.

Liu, Z. J.; Guo, Y. K.; Bai, J. G. Exogenous Hydrogen Peroxide Changes Antioxidant Enzyme Activity and Protects Ultrastructure in Leaves of Two Cucumber Ecotypes Under Osmotic Stress. *J. Plant Growth Regul.* **2010,** *29,* 171–183.

Logan, B. A.; Kornyeyev, D.; Hardison, J.; Holaday, A. S. The Role of Antioxidant Enzymes in Photoprotection. *Photosynth. Res.* **2006,** *88,* 119–132.

Lv, X.; Li, H.; Chen, X.; Xiang, X.; Guo, Z.; Yu, J.; Zhou, Y. The Role of Calcium-Dependent Protein Kinase in Hydrogen Peroxide, Nitric Oxide and ABA-Dependent Cold Acclimation. *J. Exp. Bot.* **2018,** *69,* 4127–4139.

Mahajan, S.; Tuteja, N. Cold, Salinity and Drought Stresses: An Overview. *Arch. Biochem. Biophys.* **2005,** 444, 139–158.

Matern, S.; Peskan-Berghoefer, T.; Gromes, R.; Kiesel, R. V.; Rausch, T. Imposed Glutathione-Mediated Redox Switch Modulates the Tobacco Wound-Induced Protein Kinase and Salicylic Acid-Induced Protein Kinase Activation State and Impacts on Defence Against *Pseudomonas syringae. J. Exp. Bot.* **2015,** *66,* 1935–1950.

Mehler, A. H. Studies on Reactions of Illuminated Chloroplasts. II Stimulation and Inhibition of the Reaction with Molecular Oxygen. *Arch. Biochem. Biophys.* **1951,** *33,* 339–351.

Menges, M.; Doczi, R.; Okresz, L.; Morandini, P.; Mizzi, L.; Soloviev, M.; Murray, J. A. H.; Bogre, L. Comprehensive Gene Expression Atlas for the *Arabidopsis* MAP Kinase Signalling Pathways. *New Phytol.* **2008,** *179,* 643–662.

Miao, Y.; Laun, T. M.; Smykowski, A.; Zentgraf, U. *Arabidopsis* MEKK1 Can Take a Short Cut: It Can Directly Interact with Senescence-Related WRKY53 Transcription Factor on the Protein Level and Can Bind to Its Promoter. *Plant Mol. Biol.* **2007,** *65,* 63–76.

Miao, Y.; Lv, D.; Wang, P.; Wang, X. -C.; Chen, J.; Miao, C.; Song, C. -P. An *Arabidopsis* Glutathione Peroxidase Functions as Both a Redox Transducer and a Scavenger in Abscisic Acid and Drought Stress Responses. *Plant Cell* **2006,** *18,* 2749–2766.

Mittler, R. Oxidative Stress, Antioxidants and Stress Tolerance. *Trends Plant Sci.* **2002,** *7,* 405–410.

Mittler, R.; Vanderauwera, S.; Suzuki, N.; Miller, G. A. D.; Tognetti, V. B.; Vandepoele, K.; Gollery, M.; Shulaev, V.; Van Breusegem, F. ROS Signaling: The New Wave? *Trends Plant Sci.* **2011,** *16,* 300–309.

Mittler, R.; Vanderauwera, S.; Gollery, M.; Van Breusegem, F. Reactive Oxygen Gene Network of Plants. *Trends Plant Sci.* **2004,** *9,* 490–498.

Moller, I. M. Plant Mitochondria and Oxidative Stress: Electron Transport, NADPH Turnover, and Metabolism of Reactive Oxygen Species. *Annu. Rev. Plant Physiol. Plant Mol. Biol.* **2001,** *52,* 561–591.

Montillet, J. L.; Leonhardt, N.; Mondy, S.; Tranchimand, S.; Rumeau, D.; Boudsocq, M.; Garcia, A. V.; Douki, T.; Bigeard, J.; Lauriere, C.; Chevalier, A.; Castresana, C.; Hirt, H. An Abscisic Acid-Independent Oxylipin Pathway Controls Stomatal Closure and Immune Defense in *Arabidopsis. PLoS Biol.* **2013,** *11,* 3.

Moussa, H. R.; Mohamed MAEFH Role of nitric acid or H_2O_2 in antioxidant defense system of *Pisum sativum* L. under drought stress. Nat Sci **2011,** 9, 211–216.

Mullineaux, P. M.; Karpinski, S. Signal Transduction in Response to Excess Light: Getting Out of the Chloroplast. *Curr. Opin. Plant Biol.* **2002,** *5,* 43–48.

Neill, S.; Barros, R.; Bright, J.; Desikan, R.; Hancock, J.; Harrison, J.; Morris, P.; Ribeiro, D.; Wilson, I. Nitric Oxide, Stomatal Closure, and Abiotic Stress. *J. Exp. Bot.* **2008,** *59,* 165–176.

Neill, S. J.; Desikan, R.; Hancock, J. Hydrogen Peroxide Signaling. *Curr. Opin. Plant Biol.* **2002**, *5*, 388–395.

Noctor, G.; Arisi, A. C. M.; Jouanin, L.; Foyer, C. H. Photorespiratory Glycine Enhances Glutathione Accumulation in Both the Chloroplastic and Cytosolic Compartments. *J. Exp. Bot.* **1999**, *50*, 1157–1167.

Noctor, G.; Foyer, C. H. Ascorbate and Glutathione: Keeping Active Oxygen Under Control. *Annu. Rev. Plant Physiol. Plant Mol. Biol.* **1998**, *49*, 249–279.

Noctor, G.; Reichheld, J.-P.; Foyer, C. H. ROS-Related Redox Regulation and Signaling in Plants. *Cell Dev. Bio.* **2017**, *80*, 3–12.

Pastori, G. M.; Foyer, C. H. Common Components, Networks, and Pathways of Cross-Tolerance to Stress. The Central Role of "Redox" and Abscisic Acid-Mediated Controls. *Plant Physiol.* **2002**, *129*, 460–468.

Pei, Z. M.; Murata, Y.; Benning, G.; Thomine, S.; Klüsener, B.; Allen, G. J.; Grill, E.; Schroeder, J. I. Calcium Channels Activated by Hydrogen Peroxide Mediate Abscisic Acid Signalling in Guard Cells. *Nature* **2000**, *406*, 731.

Peng, L. T.; Jiang, Y. M.; Yang, S. Z.; Pan, S. Y. Accelerated Senescence of Fresh-Cut Chinese Water Chestnut Tissues in Relation to Hydrogen Peroxide Accumulation. *J. Plant Physiol. Mol. Biol.* **2005a**, *31*, 527–532.

Peng, H. P.; Lin, T. Y.; Wang, N. N.; Shih, M. C. Differential Expression of Genes Encoding 1-Aminocyclopropane-1-Carboxylatesynthase in *Arabidopsis* During Hypoxia. *Plant Mol. Biol.* **2005b**, *58*, 15–25.

Pitzschke, A.; Djamei, A.; Bitton, F.; Hirt, H. A Major Role of the MEKK1-MKK1/2-MPK4 Pathway in ROS Signalling. *Mol. Plant* **2009**, *2*, 120–137.

Quan, L. J.; Zhang, B.; Shi, W. W.; Li, H. Y. Hydrogen Peroxide in Plants: A Versatile Molecule of the Reactive Oxygen Species Network. *J. Integr. Plant Bio.* **2008**, *50*, 2–18.

Rao, M. V.; Paliyath, G.; Ormrod, D. P.; Murr, D. P.; Watkins, C. B. Influence of Salicylic Acid on H_2O_2 Production, Oxidative Stress and H_2O_2-Metabolizing Enzymes. *Plant Physio.* **1997**, *115*, 137–149.

Rentel, M. C.; Knight, M. R. Oxidative Stress-Induced Calcium Signalling in *Arabidopsis*. *Plant Physiol.* **2004**, *135*, 1471–1479.

Rhoads, D. M.; Umbach, A. L.; Subbaiah, C. C.; Siedow, J. N. Mitochondrial Reactive Oxygen Species. Contribution to Oxidative Stress and Interorganellar Signalling. *Plant Physiol.* **2006**, *141*, 357–366.

Rodriguez, M. C.; Petersen, M.; Mundy, J. Mitogen-Activated Protein Kinase Signalling in Plants. *Annu. Rev. Plant Biol.* **2010**, *61*, 621–649.

Roychoudhury, A.; Singh, A.; Aftab, T.; Ghosal, P.; Banik, N. Seedling Priming with Sodium Nitroprusside Rescues *Vigna radiata* from Salinity Stress-Induced Oxidative Damages. *J. Plant Growth Regul.* **2021**. DOI: 10.1007/s00344-021-10328-z.

Samuel, M. A.; Miles, G. P.; Ellis, B. E. Ozone Treatment Rapidly Activates MAP Kinase Signalling in Plants. *Plant J.* **2000**, *22*, 367–376.

Sathiyaraj, G.; Srinivasan, S.; Kim Y-J, Lee, O. R.; Parvin, S.; Balusamy, S. R. D.; Khorolragchaa, A.; Yang, D. C. Acclimation of Hydrogen Peroxide Enhances Salt Tolerance by Activating Defense-Related Proteins in *Panax ginseng* C.A. Meyer. *Mol. Biol. Rep.* **2014**, *41*, 3761–3771.

Savatin, D. V.; Bisceglia, N. G.; Marti, L.; Fabbri, C.; Cervone, F.; De Lorenzo, G. The *Arabidopsis* NUCLEUS- AND PHRAGMOPLAST-LOCALIZED KINASE1-Related

Protein Kinases are Required for Elicitor-Induced Oxidative Burst and Immunity. *Plant Physiol.* **2014**, *165*, 1188–1202.

Saxena, I.; Srikanth, S.; Chen, Z. Cross Talk Between H$_2$O$_2$ and Interacting Signal Molecules Under Plant Stress Response. *Front. Plant Sci.* **2016**, *7*, 570.

Schmidt, R.; Mieulet, D.; Hubberten, H. M.; Obata, T.; Hoefgen, R.; Fernie, A. R.; Fisahn, J.; San Segundo, B.; Guiderdoni, E.; Schippers, J. H.; Mueller-Roeber, B. SALT-RESPONSIVE ERF1 Regulates Reactive Oxygen Species-Dependent Signaling During the Initial Response to Salt Stress in Rice. *Plant Cell* **2013**, *25*, 2115–2131.

Schraudner, M.; Moeder, W.; Wiese, C.; Camp, W. V.; Inze, D.; Langebartels, C.; Sandermann, H. Ozone-Induced Oxidative Burst in the Ozone Biomonitor Plant, Tobacco Bel W3. *Plant J.* **1998**, *16*, 235–245.

Singh, A.; Banerjee, A.; Roychoudhury, A. Seed Priming with Calcium Compounds Abrogate Fluoride-Induced Oxidative Stress by Upregulating Defence Pathways in an Indica Rice Variety. *Protoplasma* **2020**, *257*, 767–782.

Smirnoff, N.; Arnaud, D. Hydrogen Peroxide Metabolism and Functions in Plants. *New Phytol.* **2019**, *221*, 1197–1214.

Somssich, I. E. MAP Kinases and Plant Defense. *Trends Plant Sci.* **1997**, *2*, 406–408.

Terzi, R.; Kadioglu, A.; Kalaycioglu, E.; Saglam, A. Hydrogen Peroxide Pretreatment Induces Osmotic Stress Tolerance by Influencing Osmolyte and Abscisic Acid Levels in Maize Leaves. *J. Plant Interact.* **2014**, *9*, 559–565.

Tian, S.; Wang, X.; Li, P.; Wang, H.; Ji, H.; Xie, J.; Qiu, Q.; Shen, D.; Dong, H. Plant Aquaporin AtPIP1; 4 Links Apoplastic H$_2$O$_2$ Induction to Disease Immunity Pathways. *Plant Physiol.* **2016**, *171*, 1635–1650.

Uchida, A.; Jagendorf, A. T.; Hibino, T.; Takabe, T.; Takabe, T. Effects of Hydrogen Peroxide and Nitric Oxide on Both Salt and Heat Stress Tolerance in Rice. *Plant Sci.* **2002**, *163*, 515–523.

Wahid, A.; Perveen, M.; Gelani, S.; Basra, S. M. A. Pretreatment of Seed with H$_2$O$_2$ Improves Salt Tolerance of Wheat Seedlings by Alleviation of Oxidative Damage and Expression of Stress Proteins. *J. Plant Physiol.* **2007**, *164*, 283–294.

Wang, Y.; Li, J.; Wang, J.; Li, Z. Exogenous H$_2$O$_2$ Improves the Chilling Tolerance of Manila Grass and Mascarene Grass by Activating the Antioxidative System. *Plant Growth Regul.* **2010a**, *61*, 195–204.

Wang, H.; Liang, X.; Huang, J.; Zhang, D.; Lu, H.; Liu, Z.; Bi, Y. Involvement of Ethylene and Hydrogen Peroxide in Induction of Alternative Respiratory Pathway in Salt-Treated *Arabidopsis* Calluses. *Plant Cell Physiol.* **2010b**, *51*, 1754–1765.

Wang, H.; Liang, X.; Wan, Q.; Wang, X.; Bi, Y. Ethylene and Nitric Oxide Are Involved in Maintaining Ion Homeostasis in *Arabidopsis* Callus Under Salt Stress. *Planta* **2009**, *230*, 293–307.

Wang, Y.; Zhang, J.; Li, J. L.; Ma, X. R. Exogenous Hydrogen Peroxide Enhanced the Thermotolerance of *Festuca arundinacea* and *Lolium perenne* by Increasing the Antioxidative Capacity. *Acta Physiol. Plant* **2014a**, *36*, 2915–2924.

Wang, L.; Su, H.; Han, L.; Wang, C.; Sun, Y.; Liu, F. Differential Expression Profiles of Poplar MAP Kinase Kinases in Response to Abiotic Stresses and Plant Hormones, and Overexpression of PtMKK4 Improves the Drought Tolerance of Poplar. *Gene* **2014b**, *545*, 141–148.

Wang, L.; Zhao, R.; Zheng, Y.; Chen, L.; Li, R.; Ma, J.; Hong, X.; Ma, P.; Sheng, J.; Shen, L. SlMAPK1/2/3 and Antioxidant Enzymes Are Associated with H_2O_2-Induced Chilling Tolerance in Tomato Plants. *J. Agric. Food Chem.* **2017,** *65*, 6812–6820.

Wang, R.-S.; Pandey, S.; Li, S.; Gookin, T. E.; Zhao, Z.; Albert, R.; Assmann, S. M. Common and Unique Elements of the ABA-Regulated Transcriptome of *Arabidopsis* Guard Cells. *BMC Genom.* **2011,** *12*, 126.

Waszczak, C.; Akter, S.; Eeckhout, D.; Persiau, G.; Wahni, K.; Bodra, N.; Mollec, I. V.; De Smeta, B.; Vertommeng, D.; Gevaerth, K.; De Jaegera, G.; Montagua, M. V.; Messensc, J.; Breusegem, F. V. Sulfenome Mining in *Arabidopsis thaliana*. *Proc. Natl. Acad. Sci. U.S.A.* **2014,** *111*, 11545–11550.

Waszczak, C.; Carmody, M.; Kangasjärvi, J. Reactive Oxygen Species in Plant Signaling. *Annu. Rev. Plant Biol.* **2018,** *69*, 209–236.

Wendehenne, D.; Dumer, J.; Klessing, D. F. Nitric Oxide: A New Player in Plant Signaling and Defense Responses. *Curr. Opin. Plant Biol.* **2004,** *7*, 449–455.

Wingler, A.; Lea, P. J.; Quick, W. P.; Leegood, R. C. Photorespiration: Metabolic Pathways and Their Role in Stress Protection. *Phil. Trans R. Soc. Lond. B.* **2000,** *355*, 1517–1529.

Xu, F. J.; Jin, C. W.; Liu, W. J.; Zhang, Y. S.; Lin, X. Y. Pretreatment with H_2O_2 Alleviates Aluminum-Induced Oxidative Stress in Wheat Seedlings. *J. Integr. Plant Biol.* **2010,** *54*, 44–53.

Yildiz, M.; Terzi, H.; Bingül, N. Protective Role of Hydrogen Peroxide Pretreatment on Defense Systems and *BnMP1* Gene Expression in Cr(VI) Stressed Canola Seedlings. *Ecotoxicology* **2013,** *22*, 1303–1312.

Yu, C. W.; Murphy, T. M.; Lin, C. H. Hydrogen Peroxide Induces Chilling Tolerance in Mung Beans Mediated Through ABA-Independent Glutathione Accumulation. *Funct. Plant Biol.* **2003,** *30*, 955–963.

Zhang, J.; Zou, D.; Li, Y.; Sun, X.; Wang, N. N.; Gong, S. Y.; Zheng, Y.; Li, X. B. GhMPK17, a Cotton Mitogen-Activated Protein Kinase, Is Involved in Plant Response to High Salinity and Osmotic Stresses and ABA Signaling. *PLoS One* **2014,** *9*, 95642.

Zhang, T.; Zhu, M.; Song, W. Y.; Harmon, A. C.; Chen, S. Oxidation and Phosphorylation of MAP Kinase 4 Cause Protein Aggregation. *Biochem. Biophys. Acta* **2015,** *1854*, 156–165.

Zhang, X.; Zhang, L.; Dong, F.; Gao, J.; Galbraith, D. W.; Song, C. P. Hydrogen Peroxide is Involved in Abscisic Acid-Induced Stomatal Closure in *Vicia faba*. *Plant Physiol.* **2001,** *126*, 1438–1448.

Zong, X. J.; Li, D. P.; Gu, L. K.; Li, D. Q.; Liu, L. X.; Hu, X. L. Abscisic Acid and Hydrogen Peroxide Induce a Novel Maize Group CMAP Kinase Gene, ZmMPK7, Which is Responsible for the Removal of Reactive Oxygen Species. *Planta* **2009,** *229*, 485–495.

CHAPTER 10

Plant Cell During Cold Stress: Sensing, Signaling, and Regulations

MUHAMMAD A. ZAYED[1] and HALA B. KHALIL[2]

[1]*Botany and Microbiology Department, Menoufia University, Shebin El-Kom, Egypt*

[2]*Department of Genetics, Faculty of Agriculture, Ain Shams University, Cairo, Egypt*

ABSTRACT

Understanding the mechanisms of cold-stress tolerance in plants is crucial for developing stress-tolerant crops. The negative impact of cold stress is mainly due to severe cell dehydration and freezing, which lead to major crop losses. The plant cells perceive cold signals that are transduced into second messengers for mediating stress tolerance. This chapter presents a broad overview of cold-stress tolerance mechanisms in plants using examples from wheat or Arabidopsis. Here, we review the cold-stress tolerance mechanisms with a focus on cold-signal sensing and transduction, as well as the regulation via cold-responsive genes and transcription factors.

10.1 INTRODUCTION

Global food security is an essential worldwide concern. The boosting of crop yield is crucial due to the rapid growth of population and changing climatic

Crop Sustainability and Intellectual Property Rights. Soumya Mukherjee, Piyali Mukherjee & Tariq Aftab (Eds)

conditions. Abiotic stresses like drought and severe temperature changes can be induced by climatic events. Crop cultivation occupies only 9% of the world's total land area, with 91% of that area under environmental stress. So, maintaining crop productivity level is a central concern in today's agriculture. Traditional breeding of wheat plants takes a lot of time and effort, and it involves multiple gene families that control molecular and physiological pathways. Understanding signaling pathways triggered in response to abiotic stress can provide valuable knowledge for developing innovative solutions to reduce yield loss caused by abiotic stress.

Wheat is one of the most significant crops, with a production size that is roughly comparable to maize and rice (Shewry and Hey, 2015). Wheat, in general, refers to either the bread wheat (*Triticum aestivum*), which covers approximately 95% of worldwide grown wheat, or pasta wheat, (*T. durum*), about 5% of cultivated wheat (Arzani and Ashraf, 2017). Wheat accounts for a large amount of human caloric consumption, so reducing its abiotic stress is vital for the preservation of global food security. Wheat faces different climatic and seasonal fluctuations at various stages of its life cycle; nevertheless, stress experienced during the reproductive phase is more detrimental than stress experienced during the vegetative phase. Stresses during the reproductive phase have a direct impact on grain setting, size, quantity, and dry weight (Kajlaa et al., 2015).

Cold stress is an essential factor that limits wheat growth in many regions of the world. Plant species, especially wheat, have developed different degrees of tolerance for adapting cold conditions. Furthermore, individual plants that can survive under severe cold stress induce various mechanisms for improving tolerance. The domestication of wheat led to the broad diversification of ecotypes. The spring and winter wheat cultivars differ in their growth conditions, vernalization process, and cold-stress responses. At least a period of one to two weeks, winter wheat should be cold-acclimated to survive under freezing conditions. Despite the fact that cold acclimation enhances the capability of winter wheat cultivars to tolerate freezing, sub-zero temperatures always cause injury even for most winter cultivars. Genetic differences among geographically distributed wheat cultivars provide information for a strong breeding program to develop cold-stress tolerant cultivars. Investigating the differences in molecular mechanisms between spring and winter wheat cultivars is essential for understanding resistance to cold conditions.

Cold stress influences all aspects of the wheat cellular function. It is sensed by a yet unknown receptor and transduced through signal transduction pathway components. This cold-stress signal leads to the regulation of transcription factors and cold-regulated genes. Understanding this process could be of great importance for agriculture. This might be helpful in developing stress-tolerant wheat cultivars. This chapter summarizes the mechanisms of cold-signal sensing, transduction, and transcription factor regulation in cold acclimation of plants.

10.2 SENSING OF COLD SIGNALS

Plants perceive cold signals at different sensory levels instead of a single protein. These signals are decoded into a cellular message. The different potential cold sensors of plants are described as follows:

10.2.1 CELL MEMBRANE AS A TEMPERATURE SENSOR

Plant cell membranes respond to cold stress by modifying polyunsaturated fatty acids and saturated fatty acids proportion. So, membranes of cold-tolerant plants can freeze at temperatures as low as –20°C, but cold-sensitive plant membranes can freeze at temperatures of 10°C (Yadav, 2010). That is why engineered fatty-acid desaturase genes of wheat have increased tolerance to cold (Hajiahmadi et al., 2020). Members of desaturase genes play several roles in conferring cold tolerance in plants. Cold-tolerant plants use low temperatures to increase the stability of their gene-of-interest transcripts to increase the amount of fatty acid desaturase proteins in the root tips of wheat (Horiguchi et al., 2000).

During cold-signal perception, plants adjust the relative proportions of existing molecules to enhance the unsaturated phosphatidylcholine and phosphatidylethanolamine. This leads to a total increase in phospholipids and membrane proteins (Uemura et al., 2006) during cold acclimation. Similarly, cold signals stimulate phosphatidic acid (PA) biosynthesis (Yadav, 2010) via acylation of lysophosphatidic acid by lysophosphatidic acid-acyltransferase, hydrolysis of structural phospholipids by phospholipase D, or phosphorylation of diacylglycerol (DAG) by diacylglycerol kinase (DGK) (Arisz et al., 2013; Yadav, 2010). Phosphatidic acid plays a role in recruiting enzymes and dehydrins (Testerink and Munnik, 2011; Yadav, 2010). PA interacts with abscisic acid-insensitive 1 (ABI1), a protein phosphatase with a calcium-binding EF-hand motif, inhibiting its detrimental influence on ABA synthesis (Zhang et al., 2004). Accumulated membrane proteins act as cryoprotectants for protecting membranes from dehydration and improving their freezing tolerance. Examples of these are membrane-anchored lipocalins (Charron et al., 2005) and early responsive dehydration protein 14 (ERD14) (Uemura et al., 2006). The plant cell membranes respond to cold signals through accumulating different types of membrane proteins, shifting their membrane fluidity, and activating specific ionic pumps. This suggests the presence of alternative upstream temperature sensors that can send appropriate signals to initiate cell acclimatization and membrane modification.

Cold-acclimated plants have higher active plasma membrane H^+-ATPase to prevent ions' exchange and water availability from decreasing due to membrane solidification. This mechanism allows ions to travel through cell membranes or via active transport processes, causing cytoplasmic alkalinization by exporting protons out of the cytoplasm (Martz et al., 2006). H^+-ATPase activation is required for many physiological events during stress, including plasma membrane hyperpolarization, proton efflux from the cytoplasm, apoplast acidification, ions uptake, stomatal opening, solute transport, and cell growth (Kasamo, 2003). During the signaling mechanism, membrane fluidity is upstream of proton-ATPase pumps (Martz et al., 2006). Consequently, plants that can perceive the cold-acclimation signal properly are those that can accumulate polyunsaturated fatty acids in a mode that coordinates the activity of ionic-pumping channels in the plasma membrane.

10.2.2   CYTOSKELETON AS A COLD-SIGNAL RECEPTOR

Plant cell cytoskeleton responds to cold signals through the depolymerization of microfilaments and microtubules. This process is responsible for the fast increase in [Ca2+]cyt as the rearrangement of the cytoskeleton requires more time (Knight and Knight, 2012). The cytoskeleton consists of microtubules (MT) and actin filaments (AF) (Collings et al., 1998). Actin-binding proteins (ABPs) and microtubule-associated proteins (MAPs) play a role in cytoskeleton rearrangements through the interaction of regulatory proteins (Wasteneys and Yang, 2004). For instance, the stress-induced heterotrimeric G-proteins regulate cytoskeleton organization, ion channels, and phospholipases activity (Dave et al., 2009; Millner, 2001). The phosphatidylinositol 4,5-bisphosphate (PIP2) regulates small GTPases and phosphatidic acid (PA) through phosphoinositide 5-kinase (PIP5K) isoforms (Oude Weernink et al.,

2007) and activates actin regulatory proteins, ion channels, and phospholipases C and D. In addition, phospholipase D (PLD) signaling protein interacts to actin, microtubules, and the plasma membrane (Drøbak et al., 2004; Gardiner et al., 2003) where cytoskeleton components control PLD activity. Therefore, cytoskeleton-binding proteins are crucial for regulating essential signaling pathways in plants and structuring the cytoskeleton. Moreover, the actin-depolymerizing factor (ADF) is a crucial actin remodeling protein that interacted with PIP2 (Yonezawa et al., 1990) and controlled by Ca2+ ions (Drøbak et al., 2004; Viswanathan and Zhu, 2002). Wheat TaADF gene was induced after two days of cold acclimation (Ouellet et al., 2001; Viswanathan and Zhu, 2002).

10.2.3 OSMOLYTE ACCUMULATION

Cold-tolerant wheat perceives low temperature by generating osmolytes that moderate the impact of cell dehydration. During periods of cold-acclimation or cold stress, many plants including wheat gradually accumulate soluble sugar (Tarkowski and Ende, 2015; Saghfi and Eivazi, 2014; Zeng et al., 2011). This accumulation is not only correlated to dehydration but acts as a protectant to the cellular membranes (Janeczko et al., 2010). In some plants, soluble sugars are associated with reducing the freezing point of the cytoplasm (Ouellet, 2007; Sauter et al., 1996), and photosynthetic activity (Yuanyuan et al., 2010; Rosa et al., 2009; Wong et al., 2003). In addition, soluble sugars could have a role as a signaling molecule to cross communicate with phytohormones, and to regulate the expression of sugar-regulatory genes (Yunsong et al., 2021; Biswal et al., 2011). Stress-responsive plants accumulate other chemicals like organic acids, amino acids, polyamines, and lipids that have been found due to cold stress and acclimation periods (Thakur and Nayyar, 2013; Yadav, 2010; Ouellet, 2007). These generated metabolites act to reduce cell damage. The variation of accumulated osmolytes often depends on stress conditions as this accumulation occurs in a slow, gradual, and constant manner for cell protection.

10.3 COLD-SIGNAL TRANSDUCTION

Plant cells perceive cold signals and transduce this signal via messengers to regulate cold-induced genes and hormones. In this case, signal transduction acts as a bridge between plant sensors and gene expression. Lipids are also very crucial in signal transduction during cold stress. Phosphatidic acid produced by both phospholipase D and C has been proposed as a secondary messenger molecule (Meijer and Munnik, 2003). This secondary messenger molecule constitutes a minor portion of membrane lipids under normal conditions but increased significantly upon exposure to cold stress (Munnik and Meijer, 2001; Meijer and Munnik, 2003).

10.3.1   ENCODING COLD-STRESS SIGNAL BY CALCIUM

In response to cold signals, living cells uptake more Ca^{2+} ions. This phenomenon is known as "calcium signature" that varies from one stress to another but cytosolic $[Ca^{2+}]_{cyt}$ may escalate 10 to 20 folds in a moment during each spike (Bose et al., 2011). The first spike indicates calcium ion intake from the external medium, while the second spike represents calcium ion release from the vacuole; the second spike in cold-acclimated Arabidopsis has a higher peak than nonacclimated plants (Knight and Knight, 2000). Subsequently, living cells use their cytosolic buffering system to quickly restore the level of $[Ca^{2+}]_{cyt}$ following each spike (Plieth et al., 1999; Schwaller, 2009). Once $[Ca^{2+}]_{cyt}$ reaches its peak concentration inside the cell, K^+ channels will open to driving potassium ions outside the cell until it achieves equilibrium potential (E_K) in the repolarization phase (Hermann et al., 2012). Signal stress activates the plant cell action potential to rise above its threshold-guided depolarization, allowing many calcium-binding proteins to catch the high calcium ions because the majority of these proteins feature "EF" hand motif(s). The "EF" hand motif is a conserved helix-loop-helix (HLH) structure composed mostly of two alpha-helices joined by a brief loop area in which the Ca2+ ion can bind (Day et al., 2002). Known examples of "EF" hand motif-containing proteins include calcium-dependent protein kinases (CPK), calmodulin-dependent protein kinases (CaM), calmodulin-like proteins (CML), calcineurin B-like proteins (CBL), calreticulin (CRT), and small calcium-binding protein (Clo) (Mohanta et al., 2017; Xiang et al., 2015; Khalil et al., 2011). Many calcium-binding proteins secrete in a cytosolic buffering system and many of them operate as signal transducers for transferring the stress signal to a specific downstream reaction cascade. Calcium channel inhibitors not only lower $[Ca^{2+}]_{cyt}$ concentration, but also suppress the expression of *KIN1* (a cold-responsive gene marker and a member of the CBF gene family), indicating that calcium signaling is upstream of the CBF gene and an important signal for proper stress perception (Knight et al., 1996). The question of whether calcium channels are the primary receptors for the stress signal or if signaling happens upstream remains unanswered.

10.3.1.1 CALCIUM INFLUX

In the plant cells, the functions of Ca^{2+} are permitted by orchestrated transport across cell membranes, mediated by Ca^{2+}-permeable ion channels, Ca^{2+}-ATPases, and Ca^{2+}/H^+ exchangers. Ca^{2+}-permeable channels regulate the temporary increase of cytosolic calcium from extracellular and intracellular compartments. Depolarization-activated calcium channels (DACCs), hyperpolarization-activated calcium channels (HACCs), and voltage-independent calcium channels are the three main categories of calcium channels classified by voltage potential (VICCs). Plants have a less variety of Ca^{2+}-permeable channels compared to animals, with only five protein families. Cyclic nucleotide-gated channels (CNGCs), glutamate-receptor-like channels (GLRs), two-pore channels (TPCs), mechanosensitive channels (MCAs), and hyperosmolality-gated calcium-permeable channels (OSCAs) are the different groups of channels (Edel et al., 2017). Calcium enters the cytosol of plant cells when Ca^{2+}-influx channels are activated (Bose et al., 2011). Some of these channels, including MCAs, CNGCs, and GLRs, were found to be restricted to the plasma membrane. As a result, the primary function of these channels is to transport Ca^{2+} from the extracellular space into the cell. MCAs and TPCs channels have an EF-hand domain that is highly selective to Ca^{2+} ions, but CNGCs and GLRs channels are permeable to both monovalent and divalent ions (Chen et al., 2015; Guo et al., 2015). The activity of Ca^{2+}-permeable channels is controlled by the interaction with other molecules. For instance, CNGCs and GLRs have negative regulation mechanisms when bound to CaM-binding domain that overlaps with either cAMP or cGMP, whereas GLR is positively regulated through binding with glutamate, glycine, alanine, serine, aspartic acid, cysteine, or glutathione (Chen et al., 2015). Releasing calcium from vacuole to cytosol is controlled by TPC channels, which have a unique two-pore domain. TPCs are only found and expressed in the vacuole membrane (Guo et al., 2015), where they are embedded by a network of transmembrane domains. Their ability to identify calcium ions is adapted by their cytosolic EF-hand domain. In addition, they are non-selective in comparison to other monovalent ions. OSCA channels primarily function as osmosensors, and loss-of-function mutants have a weaker calcium spike under osmotic stress than wild-type plants (Yuan et al., 2014). The relatively low diversity of Ca^{2+}-influx systems in plants, compared to animals, may be because many of the components in plants are not discovered yet (Edel et al., 2017).

10.3.1.2   CALCIUM EFFLUX

The calcium efflux system's principal function is to reduce the amplitude of cellular Ca^{2+} concentrations that have formed as a result of stress-signal perception. Although calcium is an essential element that is required for plasma membrane integrity, cell wall formation, and a variety of other metabolic functions in plant cells, its extracellular concentration is often lower than its cellular concentration in normal conditions. The optimal concentration of Ca^{2+} cations in plant cells was found to be 1–10 mM in the apoplast, 0.2–10 mM in the vacuole, 0.002–0.006 mM in the chloroplast stroma, and ~1 mM in the endoplasmic reticulum, while its lowest concentration was only 0.0001–0.0002 mM, indicating that all cell compartments have higher Ca^{2+} than the cytosol by at least 20 times (Bose et al., 2011). Signals bring Ca^{2+} into the cytosol by stress either from the extracellular surroundings or intracellular compartments. Thus, cytosolic calcium levels rise to greater than 0.001 mM, triggering the first cytosolic Ca^{2+} transient or oscillation. Subsequently, the role of the calcium efflux mechanism is critical in lowering its concentration. P-type ATPases, mitochondria calcium uniporter complexes, and cation/Ca^{2+} exchangers are among the calcium efflux system components (Edel et al., 2017; Tuteja and Mahajan, 2007). These are ion channels that are found in almost all living cells. Their structure involves five domains; three are cytoplasmic that consist mainly of an actuator, nucleotide-binding, and phosphorylation domains, and the remaining two are related to membrane functioning as a transport domain and a class-specific support domain (Palmgren and Nissen, 2011).

10.3.2   MITOGEN-ACTIVATED PROTEIN KINASES SIGNALING CASCADE

Mitogen-activated protein kinases (MAPK) signaling cascade transfers the signal from the cell surface to cytosol and nucleus for initiating the transcription of several transcription factors. MAPK cascade genes code for conserved proteins including three MAPK members, MAP kinase kinase kinase (MKKK or MAP3K), MAP kinase kinase (MKK or MAP2K), and MAP kinase (MAPK or MAPI K). In wheat, a MAPK signal was induced in response to cold and other abiotic stresses (Jonak et al., 1996). MAPK Arabidopsis genes like MAPK3 and MAPK5 showed more accumulation of its transcript by drought, cold, salinity, and H_2O_2, while Arabidopsis MPK7 activation was regulated through the ABA-dependent pathway (Danquah et al., 2015; Mizoguchi et al., 1996). Teige et al. (2004) showed that MKK2 is upstream of and targets MPK4- and MPK6-enhanced cold tolerance in Arabidopsis, as revealed by mkk2 null mutant plants that were sensitive to cold and salt stresses. Furthermore, MKK2 overexpression increases MPK4 and MPK6 transcripts, which can also be activated by downstream MEKK1-mediated phosphorylation that controls the activation of MKK1, MKK2, and MPK4, which can result in the cold- and salt-tolerant plants. Cold-stress tolerance in plants was found to be controlled by MPK4 and MPK6, as they are substrates for MKK2 in Arabidopsis (Sinha et al., 2011). Cold-stress upregulated several kinases in different plants like GhMAPK, ZmMPK3, ZmMAPK5, ZmMPK5, OsMAPK5, SbMAPKK, NtNPK1, OsMEK1, MsSAMK, and OsWJUMK1 (Šamajová et al., 2013; Sinha et al., 2011).

10.4 COLD-RESPONSIVE AND TRANSCRIPTION FACTOR GENES REGULATIONS

During cold acclimation, plants reprogram their gene expression through transcriptional mechanisms. One of the most important cold-regulatory pathways is the C-repeat binding factor/dehydration responsive element-binding protein, CBF/DREB pathway (Chinnusamy et al., 2007,). The CBF or DREB proteins are members of the AP2 transcription factor superfamily that bind to DRE/CRT or LTRE cis-elements' promoters of cold-regulated genes (COR) (Akhtar et al., 2012; Gilmour et al., 2000). In addition, CBF/DREB expression was regulated via a nuclear-constitutive protein ICE (Chinnusamy et al., 2003). The ICE1–CBF transcriptional cascade plays an essential role in cold acclimation in diverse plant species. Polyubiquitination of ICE1 was elevated in normal conditions by the HOS1, a really interesting new gene (RING)-type ubiquitin E3 ligase, and identified upstream of CBF. Under cold conditions, a SUMO E3 ligase SIZ1-mediated sumoylation prevents the polyubiquitination action enhanced by HOS1 and thus increases the ICE1 stability (Dong et al., 2006). To investigate the function of ICE1, Chinnusamy et al. (2003) used *ice1* Arabidopsis mutant that enhanced CBF3 expression by interacting with the MYC-recognition sequence (CANNTG) of its promotor.

A different mechanism for CBF-gene transcription regulation has been introduced by Shi et al. (2012), where cold induces the protein accumulation of ethylene-insensitive 3 (EIN3), a CBF-negative regulator transcription factor. Moreover, cold treatment was found to indirectly induce the stability of phytochrome-interacting Factor 3 (PIF3), a CBF transcriptional repressor, by enhancing the degradation of EIN3-Binding F-Box 1(EBF1) and EBF2, which target PIF3 for 26S proteasome-mediated degradation (Jiang et al., 2017). Other CBF repressor proteins work only under certain circumstances like PHYB, PIF4, and PIF7 that repress CBF only under extended-day conditions (Lee and Thomashow, 2012).

Cold stress-regulated genes that require the ABA-dependent pathway have been previously reviewed (Gusta et al., 2005) and found that they contain either DRE/CRT or ABRE cis-acting elements in their promoter. Therefore, most of the cold-regulated genes are related to other abiotic stresses like drought and salinity. Yadav, (2010) found that the ABA-responsive element (PyACGTGGC) differs from the CBF/DREB responsive element, which is (A/GCCGAC). However, the exogenous application of ABA during the period of cold acclimation enhanced the freezing tolerance of many different plants (Bravo et al., 1998; Robertson et al., 1994). Spraying ABA on cold-sensitive rice seedlings increased freezing tolerance without prior exposure to cold (Shinkawa et al., 2013). Within reason, ABA can increase the induction of DRE/CRT-containing genes as has been reviewed by Yamaguchi-Shinozaki and Shinozaki (2005). Since DRE/CRT promoter elements may act as coupling elements for the functional activation of genes-containing ABRE elements, this confirms the communication between ABA-dependent and ABA-independent pathways. Additionally, ABA, drought, and salinity induce the transcription factors bZIP and ABA-responsive element-binding protein (AREBs) which binds with ACGT-containing ABA-responsive elements (ABRE) to induce the expression of ABRE-containing genes like RD29A, RD29B, and RD22 (Uno et al., 2000). A mutant of HOS9 (hos9-1) reveals upregulation of RD29A gene expression and HOS10 positively regulates 9-cis-epoxycarotenoid dioxygenase3 (NCED3). A gene encodes a key enzyme in abscisic acid biosynthesis, where MeJA induces its expression (Hossain et al., 2011). Consequently, HOS9 and HOS10 transcription factors regulated cold tolerance in Arabidopsis through the ABA-mediated pathway (Zhu et al., 2005).

KEYWORDS

- **cold stress**
- **wheat**
- **Arabidopsis**
- **signal transduction**
- **cold-responsive genes**
- **transcription factors**

REFERENCES

Akhtar, M.; Jaiswal, A.; Taj, G.; Jaiswal, J. P.; Qureshi, M. I.; Singh, N. K. DREB1/CBF Transcription Factors: Their Structure, Function and Role in Abiotic Stress Tolerance in Plants. *J. Genet.* 2012, *91*, 385–395. https://doi.org/10.1007/s12041-012-0201-3.

Arisz, S. A.; van Wijk, R.; Roels, W.; Zhu, J. K.; Haring, M. A.; Munnik, T. Rapid Phosphatidic Acid Accumulation in Response to Low Temperature Stress in Arabidopsis Is Generated Through Diacylglycerol Kinase. *Front. Plant Sci.* **2013,** *4,* 1–15. https://doi.org/10.3389/fpls.2013.00001.

Arzani, A.; Ashraf, M. Cultivated Ancient Wheats (Triticum spp.): A Potential Source of Health-Beneficial Food Products. *Compr. Rev. Food Sci. Food Saf.* **2017.** https://doi.org/10.1111/1541-4337.12262.

Biswal, B.; Joshi, P. N.; Raval, M. K.; Biswal, U. C. Photosynthesis, a Global Sensor of Environmental Stress in Green Plants: Stress Signalling and Adaptation. *Curr. Sci.* **2011** https://doi.org/10.2307/24077862.

Bose, J.; Pottosin, I. I.; Shabala, S. S.; Palmgren, M. G.; Shabala, S. Calcium Efflux Systems in Stress Signaling and Adaptation in Plants. *Front. Plant Sci.* **2011,** *2,* 85. https://doi.org/10.3389/fpls.2011.00085.

Bravo, L. A.; Zuniga, G. E.; Alberdi, M.; Corcuera, L. J. The Role of ABA in Freezing Tolerance and Cold Acclimation in Barley. *Physiol. Plant.* **1998,** *103,* 17–23. https://doi.org/10.1034/j.1399-3054.1998.1030103.

Charron, J.-B. F.; Ouellet, F.; Pelletier, M.; Danyluk, J.; Chauve, C.; Sarhan, F. Identification, Expression, and Evolutionary Analyses of Plant Lipocalins. *Plant Physiol.* **2005,** *139,* 2017–2028. https://doi.org/10.1104/pp.105.070466.

Chen, J.; Gutjahr, C.; Bleckmann, A.; Dresselhaus, T. Calcium Signaling During Reproduction and Biotrophic Fungal Interactions in Plants. *Mol. Plant.* **2015.** https://doi.org/10.1016/j.molp.2015.01.023.

Chinnusamy, V.; Ohta, M.; Kanrar, S.; Lee, B. ha, Hong, X.; Agarwal, M.; Zhu, J. K. ICE1: A Regulator of Cold-Induced Transcriptome and Freezing Tolerance in Arabidopsis. *Genes Dev.* **2003,** *17,* 1043–1054. https://doi.org/10.1101/gad.1077503.

Chinnusamy, V.; Zhu, J.; Zhu, J. K. Cold Stress Regulation of Gene Expression in Plants. *Trends Plant Sci.* **2007.** https://doi.org/10.1016/j.tplants.2007.07.002.

Collings, D. a.; Asada, T.; Allen, N. S.; Shibaoka, H. Plasma Membrane-Associated Actin in Bright Yellow 2 Tobacco Cells1. *Plant Physiol.* **1998,** *118,* 917–928. https://doi.org/10.1104/pp.118.3.917.

Danquah, A.; de Zélicourt, A.; Boudsocq, M.; Neubauer, J.; Frei dit Frey, N.; Leonhardt, N.; Pateyron, S.; Gwinner, F.; Tamby, J.-P.; Ortiz-Masia, D.; Marcote, M. J.; Hirt, H.; Colcombet, J. Identification and Characterization of an ABA-Activated MAP Kinase Cascade in *Arabidopsis thaliana. Plant J.* **2015,** *82,* 232–244. https://doi.org/10.1111/tpj.12808.

Dave, R. H.; Saengsawang, W.; Yu, J. Z.; Donati, R.; Rasenick, M. M. Heterotrimeric G-Proteins Interact Directly with Cytoskeletal Components to Modify Microtubule-Dependent Cellular Processes. *NeuroSignals* **2009.** https://doi.org/10.1159/000186693.

Day, I. S.; Reddy, V. S.; Shad Ali, G.; Reddy, A. S. N. Analysis of EF-Hand-Containing Proteins in Arabidopsis. *Genome Biol.* **2002,** *3,* RESEARCH0056. https://doi.org/10.1186/gb-2002-3-10-research0056.

Dong, C.-H.; Agarwal, M.; Zhang, Y.; Xie, Q.; Zhu, J.-K. The Negative Regulator of Plant Cold Responses, HOS1, Is a RING E3 Ligase That Mediates the Ubiquitination and Degradation of ICE1. *Proc. Natl. Acad. Sci.* **2006,** *103,* 8281–8286. https://doi.org/10.1073/pnas.0602874103.

Drøbak, B. K.; Franklin-Tong, V. E.; Staiger, C. J. The Role of the Actin Cytoskeleton in Plant Cell Signaling. *New Phytol.* **2004.** https://doi.org/10.1111/j.1469-8137.2004.01076.

Edel, K. H.; Marchadier, E.; Brownlee, C.; Kudla, J.; Hetherington, A. M. The Evolution of Calcium-Based Signalling in Plants. *Curr. Biol.* **2017.** https://doi.org/10.1016/j.cub.2017.05.020.

Gardiner, J.; Collings, D. A.; Harper, J. D. I.; Marc, J. The Effects of the Phospholipase D-Antagonist 1-Butanol on Seedling Development and Microtubule Organisation in Arabidopsis. *Plant Cell Physiol.* **2003,** *44,* 687–696. https://doi.org/10.1093/pcp/pcg095.

Gilmour, S. J.; Sebolt, A. M.; Salazar, M. P.; Everard, J. D.; Thomashow, M. F. Overexpression of the Arabidopsis CBF3 Transcriptional Activator Mimics Multiple Biochemical Changes Associated with Cold Acclimation. *Plant Physiol.* **2000,** *124,* 1854–1865. https://doi.org/10.1104/pp.124.4.1854.

Guo, J.; Zeng, W.; Chen, Q.; Lee, C.; Chen, L.; Yang, Y.; Cang, C.; Ren, D.; Jiang, Y. Structure of the Voltage-Gated Two-Pore Channel TPC1 from *Arabidopsis thaliana. Nature* **2015,** *531,* 196–201. https://doi.org/10.1038/nature16446.

Gusta, L. V.; Trischuk, R.; Weiser, C. J. Plant Cold Acclimation: The Role of Abscisic Acid. *J. Plant Growth Regul.* **2005.** https://doi.org/10.1007/s00344-005-0079.

Hajiahmadi, Z.; Abedi, A.; Wei, H. Identification, Evolution, Expression, and Docking Studies of Fatty Acid Desaturase Genes in Wheat (*Triticum aestivum* L.). *BMC Genomics* **2020,** *21,* 778. https://doi.org/10.1186/s12864-020-07199-1

Hermann, A.; Donato, R.; Weiger, T. M.; Chazin, W. J. S100 Calcium Binding Proteins and Ion Channels. *Front. Pharmacol.* **2012,** *3* APR. https://doi.org/10.3389/fphar.2012.00067.

Horiguchi, G.; Fuse, T.; Kawakami, N.; Kodama, H.; Iba, K. Temperature-Dependent Translational Regulation of the ER Omega-3 Fatty Acid Desaturase Gene in Wheat Root Tips. *Plant J.* **2000,** *24,* 805–813. https://doi.org/10.1111/j.1365-313X.2000.00925.

Hossain, M. A.; Munemasa, S.; Uraji, M.; Nakamura, Y.; Mori, I. C.; Murata, Y. Involvement of Endogenous Abscisic Acid in Methyl Jasmonate-Induced Stomatal Closure in Arabidopsis. *Plant Physiol.* **2011,** *156,* 430–438. https://doi.org/10.1104/pp.111.172254.

Janeczko, A.; Biesaga-Kościelniak, J.; Oklešt'ková, J.; Filek, M.; Dziurka, M.; Szarek-Łukaszewska, G. Role of 24-Epibrassinolide in Wheat Production: Physiological Effects and Uptake. *J. Agron. Crop Sci.* **2010,** *196,* 311–321. doi: 10.1111/j.1439-037X.2009.00413.

Jiang, B.; Shi, Y.; Zhang, X.; Xin, X.; Qi, L.; Guo, H.; Li, J.; Yang, S. PIF3 Is a Negative Regulator of the CBF Pathway and Freezing Tolerance in Arabidopsis. *Proc. Natl. Acad. Sci. USA.* **2017,** *114,* E6695–E6702. https://doi.org/10.1073/pnas.1706226114.

Jonak, C.; Kiegerl, S.; Ligterink, Wilc.; Barkert, P. J.; Huskissont, N. S.; by Winslow Briggs, C. R. Stress Signaling in Plants: A Mitogen-Activated Protein Kinase Pathway Is Activated by Cold and Drought. *Plant Biol.* **1996,** *93,* 11274–11279. https://doi.org/10.1073/pnas.93.20.11274.

Kajlaa M.; Yadava V. K.; Khokharc J.; Singh S.; Chhokara R. S.; Meenaa R. P.; Sharmaa R. K. Increase in Wheat Production Through Management of Abiotic Stresses: A Review. *J Appl. Nat. Sci.* **2015,** *7,* 1070–1080. https://doi.org/10.31018/jans.v7i2.733.

Kasamo, K. Regulation of Plasma Membrane H + -ATPase Activity by the Membrane Environment. *J. Plant Res.* **2003,** *116,* 517–523. https://doi.org/10.1007/s10265-003-0112-8.

Khalil H. B.; Wang Z.; Wright J. A.; Ralevski A.; Donayo A. O.; Gulick P. J. Heterotrimeric Gα Subunit from Wheat (*Triticum aestivum*), GA3, Interacts with the Calcium-Binding Protein, Clo3, and the Phosphoinositide-Specific Phospholipase C, PI-PLC1. *Plant Mol. Biol.* **2011,** *77,* 145–158. https://doi.org/10.1007/s11103-011-9801-1.

Knight, H.; Knight, M. R. Imaging Spatial and Cellular Characteristics of Low Temperature Calcium Signature After Cold Acclimation in Arabidopsis. *J. Exp. Bot.* **2000,** *51,* 1679–1686. https://doi.org/10.1093/jexbot/51.351.1679.

Knight, M. R.; Knight, H. Low-Temperature Perception Leading to Gene Expression and Cold Tolerance in Higher Plants. *New Phytol.* **2012,** *195,* 737–751. https://doi.org/10.1111/j.1469-8137.2012.04239.

Knight, H.; Trewavas, A. J.; Knight, M. R. Cold Calcium Signaling in Arabidopsis Involves Two Cellular Pools and a Change in Calcium Signature after Acclimation. *Plant Cell Online* **1996,** *8,* 489–503. https://doi.org/10.1105/tpc.8.3.489.

Lee, C.-M.; Thomashow, M. F. Photoperiodic Regulation of the C-Repeat Binding Factor (CBF) Cold Acclimation Pathway and Freezing Tolerance in *Arabidopsis thaliana. Proc. Natl. Acad. Sci.* **2012,** *109,* 15054–15059. https://doi.org/10.1073/pnas.1211295109.

Martz, F.; Sutinen, M.-L.; Kiviniemi, S.; Palta, J. P. Changes in Freezing Tolerance, Plasma Membrane H+-ATPase Activity and Fatty Acid Composition in Pinus Resinosa Needles During Cold Acclimation and De-Acclimation. *Tree Physiol.* **2006,** *26,* 783–90. https://doi.org/10.1093/treephys/26.6.783.

Meijer H. J. G.; Munnik T. Phospholipid-Based Signaling in Plants. *Annu. Rev. Plant Biol.* **2003,** *54,* 265–306. https://doi.org/10.1146/annurev.arplant.54.031902.134748.

Millner, P. A. Heterotrimeric G-Proteins in Plant Cell Signaling. In: *New Phytologist*; **2001,** pp. 165–174. https://doi.org/10.1046/j.1469-8137.2001.00172.

Mizoguchi, T.; Irie, K.; Hirayama, T.; Hayashida, N.; Yamaguchi-Shinozaki, K.; Matsumoto, K.; Shinozaki, K. A Gene Encoding a Mitogen-Activated Protein Kinase Kinase Kinase Is Induced Simultaneously with Genes for a Mitogen-Activated Protein Kinase and an S6 Ribosomal Protein Kinase by Touch, Cold, and Water Stress in *Arabidopsis thaliana. Proc. Natl. Acad. Sci. USA* **1996,** *93,* 765–769. https://doi.org/10.1073/pnas.93.2.765.

Mohanta, T. K.; Kumar, P.; Bae, H. Genomics and Evolutionary Aspect of Calcium Signaling Event in Calmodulin and Calmodulin-Like Proteins in Plants. *BMC Plant Biol.* **2017,** *17,* 38. https://doi.org/10.1186/s12870-017-0989-3.

Munnik, T.; Meijer, H. J. G. Osmotic Stress Activates Distinct Lipid and MAPK Signalling Pathways in Plants. *FEBS Lett* **2001**, *498*, 172–178. https://doi.org/10.1016/s0014-5793(01)02492-9.

Oude Weernink, P. A.; López De Jesús, M.; Schmidt, M. Phospholipase D Signaling: Orchestration by PIP2 and Small GTPases. Naunyn. Schmiedebergs. *Arch. Pharmacol.* **2007**. https://doi.org/10.1007/s00210-007-0131-4.

Ouellet, F. Cold Acclimation and Freezing Tolerance in Plants. *Life Sci.* **2007**. https://doi.org/10.1002/9780470015902.a0020093.

Ouellet, F.; Carpentier, E.; Cope, M. J.; Monroy, A F.; Sarhan, F. Regulation of a Wheat Actin-Depolymerizing Factor During Cold Acclimation. *Plant Physiol.* **2001**, *125*, 360–368. https://doi.org/10.1104/pp.125.1.360.

Plieth, C.; Hansen, U. P.; Knight, H.; Knight, M. R. Temperature Sensing by Plants: The Primary Characteristics of Signal Perception and Calcium Response. *Plant J.* **1999**, *18*, 491–497. https://doi.org/10.1046/j.1365-313X.1999.00471.

Robertson, A. J.; Weninger, A.; Wilen, R. W.; Fu, P.; Gusta, L. V. Comparison of Dehydrin Gene Expression and Freezing Tolerance in Bromus Inermis and Secale Cereale Grown in Controlled Environments, Hydroponics, and the Field. *Plant Physiol.* **1994**, *106*, 1213–1216. https://doi.org/10.1104/pp.106.3.1213.

Rosa, M.; Prado, C.; Podazza, G.; Interdonato, R.; González, J. A.; Hilal, M.; Prado, F. E. Soluble Sugars—Metabolism, Sensing and Abiotic Stress: A Complex Network in the Life of Plants. *Plant Signal. Behav.* **2009**, *4*, 388–393. https://doi.org/10.4161/psb.4.5.8294.

Saghfi, S.; Eivazi, A. R. Effects of Cold Stress on Proline and Soluble Carbohydrates in Two Chickpea Cultivars. *Int. J. Curr. Microbiol. App. Sci.* **2014**, *3*, 591–595.

Šamajová, O.; Komis, G.; Šamaj, J. Emerging Topics in the Cell Biology of Mitogen-Activated Protein Kinases. *Trends Plant Sci.* **2013**. https://doi.org/10.1016/j.tplants.2012.11.004

Sauter, J. J.; Wisniewski, M.; Witt, W. Interrelationships Between Ultrastructure, Sugar Levels, and Frost Hardiness of Ray Parenchyma Cells During Frost Acclimation and Deacclimation in Poplar (Populus × canadensis Moench ‹robusta›) *Wood. J. Plant Physiol.* **1996**, *149*, 451–461. https://doi.org/10.1016/S0176-1617(96)80148-9.

Schwaller, B. The Continuing Disappearance of "Pure" Ca2+ Buffers. *Cell. Mol. Life Sci.* **2009**. https://doi.org/10.1007/s00018-008-8564-6.

Shewry, P. R.; Hey, S. J. The Contribution of Wheat to Human Diet and Health. *Food Energy Secur.* **2015**. https://doi.org/10.1002/FES3.64.

Shi, Y.; Tian, S.; Hou, L.; Huang, X.; Zhang, X.; Guo, H.; Yang, S. Ethylene Signaling Negatively Regulates Freezing Tolerance by Repressing Expression of CBF and Type-A ARR Genes in Arabidopsis. *Plant Cell* **2012**, *24*, 2578–2595. https://doi.org/10.1105/tpc.112.098640.

Shinkawa, R.; Morishita, A.; Amikura, K.; Machida, R.; Murakawa, H.; Kuchitsu, K.; Ishikawa, M. Abscisic Acid Induced Freezing Tolerance in Chilling-Sensitive Suspension Cultures and Seedlings of Rice. *BMC Res. Notes* **2013**, *6*, 351. https://doi.org/10.1186/1756-0500-6-351

Sinha, A. K.; Jaggi, M.; Raghuram, B.; Tuteja, N. Mitogen-Activated Protein Kinase Signaling in Plants Under Abiotic Stress. *Plant Signal. Behav.* **2011**, *6*, 196–203. https://doi.org/10.4161/psb.6.2.14701.

Tarkowski, Ł. P.; Van den Ende, W. Cold Tolerance Triggered by Soluble Sugars: A Multifaceted Countermeasure. *Front. Plant Sci.* **2015**, *6*, 203. https://doi.org/10.3389/fpls.2015.00203.

Teige, M.; Scheikl, E.; Eulgem, T.; Doczi, R.; Ichimura, K.; Shinozaki, K.; Dangl, J. L.; Hirt, H. The MKK2 Pathway Mediates Cold and Salt Stress Signaling in Arabidopsis. *Mol Cell* **2004**, *15*, 141–152. https://doi.org/10.1016/j.molcel.2004.06.023.

Testerink, C.; Munnik, T. Molecular, Cellular, and Physiological Responses to Phosphatidic Acid Formation in Plants. *J. Exp. Bot.* **2011**, *62*, 2349–2361. https://doi.org/10.1093/jxb/err079.

Thakur, P.; Nayyar, H. Facing the Cold Stress by Plants in the Changing Environment: Sensing, Signaling, and Defending Mechanisms. In *Plant Acclimation to Environmental Stress*, 2013; pp. 29–69. https://doi.org/10.1007/978-1-4614-5001-6_2.

Tuteja, N.; Mahajan, S. Calcium Signaling Network in Plants: An Overview. *Plant Signal. Behav.* **2007**, *2*, 79–85. https://doi.org/10.4161/psb.2.2.4176.

Uemura, M.; Tominaga, Y.; Nakagawara, C.; Shigematsu, S.; Minami, A.; Kawamura, Y. Responses of the Plasma Membrane to Low Temperatures. *Physiol. Plant.* **2006**. https://doi.org/10.1111/j.1399-3054.2005.00594.

Uno, Y.; Furihata, T.; Abe, H.; Yoshida, R.; Shinozaki, K.; Yamaguchi-Shinozaki, K. Arabidopsis Basic Leucine Zipper Transcription Factors Involved in an Abscisic Acid-Dependent Signal Transduction Pathway Under Drought and High-Salinity Conditions. *Proc. Natl. Acad. Sci. USA* **2000**, *97*, 11632–11637. https://doi.org/10.1073/pnas.190309197

Viswanathan, C.; Zhu, J.-K. Molecular Genetic Analysis of Cold-Regulated Gene Transcription. *Philos. Trans. R. Soc. Lond. B. Biol. Sci.* **2002**, *357*, 877–886. https://doi.org/10.1098/rstb.2002.1076.

Wasteneys, G. O.; Yang, Z. The Cytoskeleton Becomes Multidisciplinary. *Plant Physiol.* **2004**, *136*, 3853–3854. https://doi.org/10.1104/pp.104.900130.

Wong, B. L.; Baggett, K. L.; Rye, A. H. Seasonal Patterns of Reserve and Soluble Carbohydrates in Mature Sugar Maple (*Acer saccharum*). *Can. J. Bot.* **2003**, *81*, 780–788. https://doi.org/10.1139/b03-079.

Xiang, Y.; Hai Lu, Y.; Song, M.; Wang, Y.; Xu, W.; Wu, L.; Wang, H.; Ma, Z. Overexpression of a Triticum Aestivum Calreticulin Gene (TaCRT1) Improves Salinity Tolerance in Tobacco. *PLoS One* **2015**, *10*. https://doi.org/10.1371/journal.pone.0140591.

Yadav, S. K. Cold Stress Tolerance Mechanisms in Plants. A Review. *Agron. Sustain. Dev.* **2010**, *30*, 515–527. https://doi.org/10.1051/agro/2009050.

Yadav, S. K. Cold Stress Tolerance Mechanisms in Plants. A Review. *Agron. Sustain. Dev.* **2010**, *30*, 515–527. https://doi.org/10.1051/agro/2009050.

Yamaguchi-Shinozaki, K.; Shinozaki, K. Organization of Cis-Acting Regulatory Elements in Osmotic- and Cold-Stress-Responsive Promoters. *Trends Plant Sci.* **2005**. https://doi.org/10.1016/j.tplants.2004.12.012.

Yonezawa, N.; Nishida, E.; Iida, K.; Yahara, I.; Sakai, H. Inhibition of the Interactions of Cofilin, Destrin, and Deoxyribonuclease I with Actin by Phosphoinositides. *J. Biol. Chem.* **1990**, *265*, 8382–8386. https://doi.org/10.1016/S0021-9258(19)38897-0.

Yuan, F.; Yang, H.; Xue, Y.; Kong, D.; Ye, R.; Li, C.; Zhang, J.; Theprungsirikul, L.; Shrift, T.; Krichilsky, B.; Johnson, D. M.; Swift, G. B.; He, Y.; Siedow, J. N.; Pei, Z.-M. OSCA1 Mediates Osmotic-Stress-Evoked Ca2+ Increases Vital for Osmosensing in Arabidopsis. *Nature* **2014**, *514*, 367–371. https://doi.org/10.1038/nature13593.

Yuanyuan, M.; Yali, Z.; Jiang, L.; Hongbo, S. Roles of Plant Soluble Sugars and Their Responses to Plant Cold Stress. *J. Biotechnol.* **2010**, *8*, 2004–2010. https://doi.org/10.4314/AJB.V8I10.60470.

Yunsong G.; Shichen H.; Lin C.; Junyi M.; Luning D.; Yaxuan L.; Yueming Y.; Xiaohui L. Expression and Regulation of Genes Involved in the Reserve Starch Biosynthesis Pathway in Hexaploid Wheat (*Triticum aestivum* L.). *Crop J.* **2021,** *9* (2), 440–455. https://doi.org/10.1016/j.cj.2020.08.002.

Zeng Y.; Yu J.; Cang J.; Liu L.; Mu Y.; Wang J.; Zhang D. Detection of Sugar Accumulation and Expression Levels of Correlative Key Enzymes in Winter Wheat (*Triticum aestivum*) at Low Temperatures. *Biosci Biotechnol Biochem***2011,** *75* (4):681–687. doi: 10.1271/bbb.100813.

Zhang, W.; Qin, C.; Zhao, J.; Wang, X. Phospholipase D 1-Derived Phosphatidic Acid Interacts with ABI1 Phosphatase 2C and Regulates Abscisic Acid Signaling. *Proc. Natl. Acad. Sci.* **2004,** *101,* 9508–9513. https://doi.org/10.1073/pnas.0402112101.

Zhu, J.; Verslues, P. E.; Zheng, X.; Lee, B.; Zhan, X.; Manabe, Y.; Sokolchik, I.; Zhu, Y.; Dong, C.-H.; Zhu, J.-K.; Hasegawa, P. M.; Bressan, R. A. HOS10 Encodes an R2R3-type MYB Transcription Factor Essential for Cold Acclimation in Plants. *Proc. Natl. Acad. Sci. USA* **2005,** *102,* 9966–9971. https://doi.org/10.1073/pnas.0503960102.

PART 3

TRADITIONAL KNOWLEDGE AND INTELLECTUAL PROPERTY RIGHTS: IMPLICATIONS IN PLANT SCIENCE

CHAPTER 11

Intellectual Property Rights vis-à-vis Food Security: A Critical Analysis

RISHAV RAY

School of Law, Christ (Deemed to be University), Bangalore, Karnataka, India

ABSTRACT

The Right to Food is undoubtedly a human right since it is one of the basic necessities without which it is impossible to sustain life. Food Security refers to the availability as well as accessibility to sufficient and quality food by all individuals. However, there persists a problem of food insecurity which is a major problem especially in the underdeveloped and, to a considerable extent, the developing countries. At an individual level, food security is limited to one's access to food but on a broader sense food security cannot be isolated from agricultural policies, crop technologies, economic and trade conditions. The intersection of crop technologies with economic factors is what links food security with Intellectual Property Rights (IPR).

Over the recent past, IPR has gained immense importance in a number of fields including agriculture. It provides the incentive for the private sector development in advancement of plant science and crop technologies which helps in ensuring food security in the long term. This study aims at discussing the issues of food security with a specific focus on the developing nations, IPR regime, and its introduction into the agriculture sector. It intends to explore the connections and linkages between IPR and food security, especially how

Crop Sustainability and Intellectual Property Rights. Soumya Mukherjee, Piyali Mukherjee & Tariq Aftab (Eds)

intellectual property can act as a medium to cover the path toward achieving global food security. The author aims to put forth the ability of IPR as a means to achieve food security by incentivising human creativity through a detailed study from an international as well as region-specific perspective.

11.1 INTRODUCTION

The Right to Food is an intrinsic human right that lays down the basic foundation of the right to life. This right is closely connected to other rights as well such as the right to health, right to wellbeing, and economic rights. The Universal Declaration of Human Rights recognizes the right to food as a part of an adequate standard of living. Article 11(2) of the International Convenant on Economic, Social and Cultural rights holds the freedom from hunger to be a fundamental right. It can be inferred that every individual has the right to food as stated by the Director General of FAO, Jacques Diuof. Thus, food security becomes an essential for every individual at the personal level as well as for every country at the international level.

Food security is an issue that remains an overwhelming concern across the globe. Many nations, especially the developing ones, are battling this adversity that is a prime obstacle to their path to a wellbeing state despite having eradicated "hunger" virtually. This intrigues one to ponder upon what exactly the term "food security" means. To understand the meaning of this term one must refer to the definition of this term which was laid down during the World Food Summit, 1974, wherein Food Security had been defined as "Availability at all times of adequate world food supplies of basic foodstuffs to sustain a steady expansion of food consumption and to offset fluctuations in production and prices" (Food and Agriculture Organization, 2003, p.27). This definition however cannot be called a wholesome definition of the term since it excludes the rights dimension and another very important aspect of food security. This definition does address the aspect of availability however; it does not take into consideration the effective access to quality food. Food insecurity may exist even when there is availability of food. A prime example is the availability of food during the 1943 Bengal Famine. Amatrya Sen put forth certain factors that were the cause of the food insecurity despite the availability of food: (i) lack of buying power, (ii) excessive price, (iii) no access to transfer. Food availability (production) and food accessibility (excessive pricing and lack of buying power) are the core reasons leading to failure of food security. It must be understood that food consumers outnumber food producers in every nation, and this concern regarding the availability of food shall keep increasing if the production of food is unable to keep pace with the population growth.

Another core issue is the lack of quality food. Undernourishment due to consumption of food lacking adequate nutrition is another primary cause of food insecurity. This is where the need for achieving food security by means of incentivizing human creativity comes into picture.

11.2 INTELLECTUAL PROPERTY RIGHTS AND FOOD SECURITY

Intellectual property rights basically refer to the practise of creating monopoly over a product of human intellect by means of rights over the benefits of the fruits of their innovation. These rights are very crucial for the growth of any sector since they act as the basic incentives that play a major role in determining investment decisions. Being exclusive rights, they often create a challenge between the interests of the creators and the interests of the society at large.

It is pertinent to understand the role of IPR in the agriculture sector before delving into the question about how IP rights affect food security. The most significant role played by IPR in general and patent rights in particular, in the field of agriculture is providing incentives to the plant breeders, thereby paving the way for the private sector investment. This shall benefit the market in two major ways: (i) economic growth of the sector, (ii) high yield in production.

However, in contrast to its beneficial nature, it has been observed that IPR in the agricultural field has witnessed considerable amount of underde-velopment. A primary reason for which is agricultural management being considered an arena of free exchange of ideas and knowledge in several nations, thereby making it tough for IPR to get established. On the contrary, IPR when introduced in this sector can provide some benefits. The agro-biotechnology industry derives investment and involvement of big private players due to the incentives guaranteed by the intellectual property rights. Research and development of new plant varieties are boosted with the involvement of the private industry. These can be varieties having the ability to absorb more photosynthetic energy, thereby enhancing the growth of the crop. These high-quality yields have a better capacity to withstand pests and viruses and thereby lead to a reduction in the rate of wastage/destruction of crops.

There are two main phases in which intellectual property rights were introduced progressively. The first phase is adoption of intellectual property rights protection for plant varieties in various developed nations. The second phase is introduction of patents over plants has served as a major incentive for the growth of agro-technology (Chaturvedi, 2002).

11.3   INTERNATIONAL LEGAL FRAMEWORK

The International Law in many instances has tried to integrate intellectual property rights and food security. The treaties dealing with food security in terms of agricultural produce and IPR treaties tremendously impact the food security development around the globe. The increase in Foreign Direct Investment has been one of the primary results of the development of IPR. To understand the corelation of intellectual property and food security in the global sphere, one must look into the various international legal instruments that have been introduced under various institutional frameworks. The FAO introduced two of the most crucial legal instruments in this regard—The International Treaty on Plant Genetic Resources for Food and Agriculture of 2001 and the International Undertaking for Plant Genetic Resources of 1983 (Barton, 1998). These instruments considered plant genetic resources as a heritage of humankind and thus should be widely available to everyone. This in a way posed an obstacle to development of personal incentives for the plant breeders. There was a need to strike a balance between the interest of private innovators and the interest of achieving food security.

Internationally, the TRIPS agreement can be said to be undoubtedly the most important intellectual property agreement for all WTO nations across the globe. Despite not having provisions directly concerned with agriculture intellectual property, the intellectual property rights standards set by the agreement undoubtedly have an impact on the aspect of agricultural science. Article 8 of the agreement provides that the implementation of TRIPS shall provide the states the opportunity to take adequate measures for protecting and promoting nutrition and public interest in other vital sectors. The agreement mandates the introduction of intellectual property protection for plant varieties by the members. Article 27(3)b that stands as an exception to Article 27 provides the protection of plant varieties by the States either by patents or by "an effective sui generis system" (Barton, 1998).

It is not only patents that are protected under the TRIPS agreement, in the context of agricultural intellectual property, Geographical Indications (GI) are provided protection as well. The TRIPS Agreement provides GI protection based upon the quality of the product, its popularity/reputations or any other characteristics that are exclusive to its geographical location.

A significant legal instrument in this regard is the International Convention for the Protection of New Varieties of Plants, also known as the UPOV convention. This treaty focuses directly upon the IPR in relation to agriculture. Protection of new plant breeds and the interests of the commercial plant breeders is the main aim of this treaty. It gives exclusive rights to the breeders to produce/reproduce protected plant varieties, condition them, offer them for sale, commercialize, import/export, or stock for any commercial practice. For a variety to be regarded as novel under UPOV, it must not have been sold/commercialized. It is pertinent to note here that prior existence of the variety is not a ground for being considered novel under this instrument unlike patent law. The later versions however did bring about certain changes in the previous versions of this convention.

11.4 INTELLECTUAL PROPERTY RIGHTS AND FOOD SECURITY— GLOBAL TRENDS

The food security legal framework across the globe has eventually evolved and adapted to growing IPR regime. Biological conventions have regarded biological resources as "common concern of humankind" (The Convention on Biodiversity, 1993, p.122). Thereby, the states have an associated duty toward making the development of these resources sustainable and publicly available. Whereas the evolution of IPR depends revolves around incentivising individual innovator's interests. This gap between the two legal regimes gradually reduced with a progressive movement toward blurring this distinction (Downes, 2000). Tracing back into history it would be found that the concept of introducing intellectual property rights to plant varieties was started in certain European nations and the United States. The introduction of plant patents in the United States began back in the year 1930 (Barton, 1998). The IP regime in Europe was slightly different since they did not introduce patents, instead they adhered to the plant breeder's rights enshrined the UPOV Convention (Greengrass, 1991).

In regard to the United States' patents regime, the landmark judgement rendered by the Hon'ble US Supreme Court in the case of Diamond v. Chakrabarty (1980) played a very significant role. In its judgment the Supreme Court put emphasis on the difference between an invention and a discovery. The Court laid more importance on the fact that bacterium was manufactured by humans by means of their intellect than on the principle of nonpatentability of natural products. The US patent regime experienced a drastic progress post this decision and by 1985 the patentability of plants was widely accepted in the States (Greengrass, 1991).

The international framework in this aspect has largely been created based on the US Model. The CGIAR is one of the prime institutions that have been greatly affected by the recent law and policy changes across the international platform. The said institution has strived to strike a balance favorable for the improvement of the developing nations in the process of adapting to the new changes in the IPR regime (Greengrass, 1991).

It was not until 1994 that the intellectual property rights regime in the agro industry started their formation in the developing countries individually. The TRIPS agreement has played a major role for the introduction of plant patents in these developing nations. An interesting aspect in this respect is that even though TRIPS imposes plant patents it also provides the nations to devise a sui generis system, that is, an alternative to patents. UPOV convention's plant breeder's rights have been recognized as a valid sui generis. The scope of sui generis however is not restricted only to the UPOV convention provisions (Greengrass, 1991). Additionally, most WTO members have international obligations in this regime as well as fundamental rights; thereby, it secures the interests of the innovators as well as the right to food of the individuals and society at large. The progression of the patent regime can be well understood with the Figure 11.1.

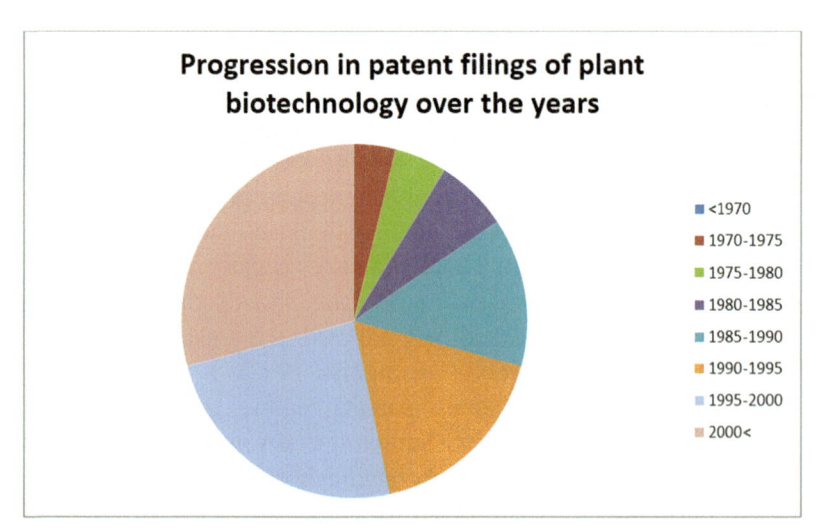

FIGURE 11.1   Progression in patent filings of plant biotechnology over the years.
Source: World Intellectual Property Report 2019.

To understand the interplay between these two regimes in the developing nations an in-depth analysis is required.

11.5 PROGRESSION IN THE DEVELOPING NATIONS

In relation to the developing countries, it cannot be denied that areas such as protection of traditional knowledge, protection of the rights of famers, and implementation of international regulations that are crucial to them have faced a huge amount of neglect over the years. There are certain factors however that have a positive effect toward accelerating the interplay between IPR and Food Security in developing nations. At the very first place, it must be understood that it is not only IPR which is responsible for the commodification taking place in the agricultural sector, but IPR was introduced in the agro-sector with an intention of fostering the over development of the economy. This is a primary reason behind the growth of the private sector in this area. Third, though the patent regime is quite developed in the IPR sector, the other areas are less explored in relation to agriculture. It is seen that the developing countries have to face a number of issues when it comes to the implementation of legal frameworks relating to IPR and food security. India can prove to be an interesting case study in this regard.

Historically, the Indian Patent regime was colonially designed and thereby had failed to estimate the innovation capability of the Indians and thereby failed to stimulate the industrial growth and commercial exploitation of new inventions in the nation. The future acts precluded the patentability of agricultural methods and innovations. Later on however, after the ratification of the TRIPS Agreement, certain progressive changes in compliance with the agreement have been brought about to the Patent Act in the form of amendments. The Biodiversity Act majorly focuses on the accessibility of resources. It directly impacts food security by linking Intellectual Property Rights and biodiversity management. It seeks to preserve India's sovereign right of its natural resources. It strengthens the growth of IPR since it mandates innovators to obtain the grant from National Biodiversity Authority before applying for IP rights. This puts a stringent restriction on exploitation of knowledge and biological resources by foreign bodies and thus condones the growth of Indian IPR regime and works toward the achievement of food security. The Seed Bill of 2004 was introduced with the objective of preventing the saving, exchanging, and reproduction of seeds by farmers. Post the amendments, the Indian law permits the patenting of seeds, plants GM's etc. This has in turn led to the monopoly control of private seed industries and has been a leading cause of farmer suicides in the nation. It can be understood that the implementation of legal framework in the context of IPR and food security in developing nations is quite a tedious task since it involves a number of restraints. Following existing models from across the globe might seem like an easy solution to this issue; however, it is to be kept in mind that every nation has a different set of problems and issues and thus there are specific needs and concerns that are to be kept in mind when tailoring such policies for a nation. The realization of the right to food, need for the protection of traditional knowledge, prevention of biopiracy, innovation of sustainable agro mechanisms, protection of the rights of the farmers, and incentives to innovators are all factors that cannot be ignored when framing policies in this regard.

11.6   CONCLUSION

The irradiation of food insecurity in the developing countries is not an easy task; it requires collective efforts on part of all the players involved. It needs to be targeted with a multi-dimensional approach. The development of IPR regime in the agro industries can be one such dimension. What impact this regime has upon its fullest implementation is yet to be ascertained since the developing nations are facing several issues in implementing these. Under the TRIPS the developing nations do not have much option to avoid the introduction of intellectual property in agriculture. The sui generis option however is open for countries to develop the IPR regime. The nations should strive to build a model of an IPR which provides incentives to the breeders as well as does not have any negative impact of the food security of the nation. The appropriation of knowledge through plant property rights will foster food security. It is seen that in the current situation, control over knowledge often creates monopoly that should not be the case. The IPR should be beneficial to all the players to foster a holistic growth. It is this ridge that needs to be filled since it is the small-scale farmers who constitute a big share in the agricultural market rather than the private players. Thereby, the development and protection of farmers' rights is necessary for ensuring food security. These rights should be extended to the fighting biopiracy and other measures to ensure sustainable agro-biodiversity management. Instead of considering them as opposites, the rights of the farmers should be considered complimentary to the existing IPR regime and the overlapping between the two in the form of property rights which would strive toward achieving food security.

KEYWORDS

- **food security**
- **developing nations**
- **intellectual property**
- **biotechnology**
- **agriculture**

BIBLIOGRAPHY

Barton, J. H. International Intellectual Property and Genetic Resource Issues Affecting Agricultural Biotechnology. In *Agricultural Biotechnology in International Development;* Ives, C. L., Bedford, B.M. Eds.; CABI: Wallingford, 1998; pp 273–274.

Chaturvedi, S. Agricultural Biotechnology and New Trends in IPRS Regime-Challenges Before Developing Countries. *Econ. Political Wkly.* **2002,** *37* (13), 1212–1222. http://ris.org.in/pdf/Chaturvedi-Sachin-2002-Agricultural-biotechnology.pdf

Coben, J. L.; Potter, C. S. Conservation of Biodiversity in Natural Habitats and the Concept of Genetic Potential. In *Perspectives on Biodiversity: Case Studies of Genetic Resource Conservation and Development;* Potter, S., et al. Eds.; AAAS: Washington, DC, 1993; pp 19–20.

de Janeiro, R. *Convention on Biological Diversity,* June 5, 1992. https://www.cbd.int/doc/publications/gbo/gbo-ch-02-en.pdf

Dhavan, R.; et al. Power Without Responsibility on Aspects of the Indian Patents Legislation. *J. Indian Law Inst.* **1991,** *33* (1), 1–2.

Diamond v. Chakrabarty, 447 U.S. 303, 1980.

Diaz-Bonilla, E.; Robinson, S. Biotechnology, Trade and Hunger. In *Biotechnology and Genetic Resource Policies;* Pardey, P. G., Koo, B., Eds.; IFPRI: Washington, DC, 2003.

Downes, D. R. How Intellectual Property Could Be a Tool to Protect Traditional Knowledge. *Columbia J. Environ. Law* **2000,** *25,* 253.

Food And Agriculture Organization of the United Nations Rome. *Trade Reforms And Food Security,* 2003. http://www.fao.org/3/y4671e/y4671e00.htm

Greengrass, B. The 1991 Act of the UPOV Convention. *Eur. Intell. Prop. Rev.* **1991,** *13,* 466.

Hamilton, N. D. Legal Issues Shaping Society's Acceptance of Biotechnology and Genetically Modified Organisms. *Drake J. Agric. Law* **2001,** *6,* 81–83.

International Convenant on Economic, Social and Cultural Rights, Dec 16, 1966. https://www.ohchr.org/EN/ProfessionalInterest/Pages/CESCR.aspx

International Convention for the Protection of New Varieties of Plants, 1991. http://vcci-ip.com/wp-content/uploads/2017/06/e_new_varieties_of_plants.pdf

International Treaty on Plant Genetic Resources for Food and Agriculture, June 29 2004 http://www.fao.org/3/i0510e/i0510e.pdf

International Undertaking for Plant Genetic Resources, 1983. https://www.google.com/search?q=International+Undertaking+for+Plant+Genetic+Resources+of+1983&rlz=1C1CHBD_enIN893IN893&oq=International+Undertaking+for+Plant+Genetic+Resources+of+1983&aqs=chrome..69i57.725j0j15&sourceid=chrome&ie=UTF-8# (accessed July 13, 2021)

Ramanna, A. Policy Implications of India's Patent Reforms—Patent Applications in the Post 1995 Era. *Econ. Political Wkly.* **2002,** *37,* 2065.

Thrupp, L.A. Linking Agricultural Biodiversity and Food Security: The Valuable Role of Agrobiodiversity for Sustainable Agriculture. *Int. Aff.* **2000,** *76,* 265.

TRIPS Agreement, Jan 1, 1995. https://www.wto.org/english/docs_e/legal_e/27-trips.pdf

Universal Declaration of Human Rights, Dec 10, 1948. https://www.ohchr.org/EN/UDHR/Documents/UDHR_Translations/eng.pdf

CHAPTER 12

Repatriation of Traditional Knowledge Through the Lens of International Legal Instruments

SARANSH CHATURVEDI[†]

Independent Consultant

ABSTRACT

The protection of Traditional Knowledge (TK) has taken a center stage at various global forums, especially at the World Intellectual Property Organization (WIPO). This move was much expected since the period saw a greater recognition of value associated with the TK. Its recognition was substantially visible through the interface of Intellectual Property (IP) in the field of TK. The primary reason for its influx on the priority agenda internationally was the sense of deriving greater benefit from it. The indigenous communities associated with the TK mostly remain unaware of any such development. This lack of awareness and more importantly the lack of active participation hinder the process of extracting sufficient economic viability from it. Talking about the interface of IP and TK, a fruitful interconnection between the international conventions becomes an important part for achieving the effective protection and conversation of TK. TRIPS, on one hand, and CBD on the other requires a reasonable interpretation in this domain. Another important perspective is the extent of vesting rights to the indigenous communities and

† Author is a practicing Advocate and deals in area of IPR and Corporate

Crop Sustainability and Intellectual Property Rights. Soumya Mukherjee, Piyali Mukherjee & Tariq Aftab (Eds)

to create a better consensus among the key stakeholders. First, this paper will touch upon the interconnection between IP with TK. In this, an important aspect will be to ponder upon the primary question of how this interconnection benefits the purpose of conversation and access to TK. Second, this paper will delve into the viability of international conventions and how this has affected various countries in framing their domestic laws in compliance with preserving TK. Third, this paper will discuss the complex consideration involving stakeholders at various levels and how this consideration affects the development of effective preservation and access to traditional knowledge.

12.1 INTRODUCTION

Intellectual Property ("IP") has seen a tremendous rise in its valuation owing to its increasing importance. Corporate and industries never fail to identify their prospective IPs to extract the maximum potential from them. The IPs, irrespective of which industry it belongs to, have a great value associated with itself (Bernieri, 2006). The philosophy behind providing an effective proprietary right for an IP to its holder can be summarized in two points. First, the need to give a commercial right to the developer to effectively recoup the investment made on IP, and second, for providing an incentive-based approach, so that a preferable environment is created for more innovations (Chaturvedi, 2020). There can be further reasons as well, but these two seem to be the basic fundamental propositions. Hence, the importance of IP cannot be negated in the present regime.

Now, moving further in our discussion, Traditional Knowledge ("TK") and IP have a very sour relation with each other—a relation that needs to be settled, but could not be done, owing to their diametric nature (Venkataraman and Swarna, 2008). There are multiple academic articles that explain the relationship between the two, but unfortunately, there has been no such improvement or rather, a consensus being reached for the betterment of the relation at the global level (Sigamany 2017). The major question, which comes for our deliberation, is with regards to why the global attitude remains lacklustre on structuring of these two aspects. For this, we will need to delve into the sour relation between the two in an anticipation to get an answer for our hypothesis. Therefore, the author will now move for discussing the core area that the article aims to address. Part I of the article will be dealing with structuring the concept of TK with the help of internationally accepted definitions. Part II will be dealing with the relation between IP and TK, and why there is a curdled relationship between the two. Part III will be dealing with International Legal Instruments and their efforts in coping up with IP and TK. Part IV will be the conclusion along with a few suggestions that the author will be putting forward.

12.2 STRUCTURING THE CONCEPT AMID MULTIPLE DEFINITIONS

Traditional knowledge, as the name suggests, means the knowledge that has been associated traditionally with a community. This knowledge or information has been developed by the community through generations and has been used within the community for years. Such knowledge holds an important place in the culture of the community because it entails in itself the experiences that the generations have added upon. Mostly, the knowledge has been used by those communities for earning their livelihood (Venkataraman and Swarna, 2008). Therefore, it can be easily argued that TK has been a key chain for the survival of various communities; hence, it needs to be preserved. On similar lines, if we talk about the Traditional Cultural Expressions ("TCEs"), this contains various activities that form an inherent part of the community's culture such as dance, arts, sculpture, and designs. We can analyze the importance of TK for those communities, but a bigger question is whether we need the protection of TK only for those communities or for the betterment of public, at large, so that they can be benefitted by it.

TK, as the author mentioned, holds a premium place in the life and survival of those communities. It needs protection from misuse and misappropriation so that it should continue to be a basis of survival for them. A particular activity can be secured for the sole purpose of communities' use. But the purpose to secure TK can also be for a larger use of information for general public, in a consolidated manner (Arewa, 2006). This aspect seems to be very confusing for various stakeholders involved in the appropriation of TK. Whether to protect TK only for the communities' purpose, or make it structured for the general use of the public at large, is something that needs interrogation. Let us see some of the internationally accepted definitions of TK.

Various organizations and scholars have defined TK in different scopes providing different structures and looks altogether. In the first place, World Intellectual Property Organization ("WIPO") defines TK as:

Traditional knowledge (TK) is knowledge, know-how, skills, and practices that are developed, sustained, and passed on from generation to generation within a community, often forming part of its cultural or spiritual identity (WIPO).

This definition provides a general framework of what constitutes TK, at large. The definition opens a lot of scope for interrogation on the varied aspects of TK. Likewise, this definition stipulates that the knowledge/know-how is passed from generation to generation, but it did not structure the ambit of knowledge/know-how, to help us retrospect the boundary of TK. Moreover, it can be said that WIPO had provided the scope of TK and had left it to the domestic countries to legislate on the same. Moving further, if we look into multiple sessions of Intergovernmental Committee on the Protection of Traditional Knowledge and Genetic Resources ("IGC"), established by the WIPO in 2000, we can find few working definitions from its sessions that can be worth mentioning here. The Ninth session (April 24–28, 2006) gives the scope of the term TK in Article 3(2) as:

[T]he term "traditional knowledge" refers to the content or substance of knowledge resulting from intellectual activity in a traditional context, and includes the know-how, skills, innovations, practices and learning that form part of traditional knowledge systems, and knowledge embodying traditional lifestyles of indigenous and local communities, or contained in codified knowledge systems passed between generations. It is not limited to any specific technical field, and may include agricultural, environmental and medicinal knowledge, and knowledge associated with genetic resources. (WIPO/GRTKF/IC/9/5)

Much later in 2018, in the 37th session of IGC, the Draft Article also tries to provide an understanding of TK as:

Traditional knowledge for the purposes of this instrument, is knowledge that is created, maintained, and developed by indigenous [peoples], local communities, [other beneficiaries], and that is linked with, or is an integral part of, the national or social identity and/or cultural heritage of indigenous [peoples], local communities; that is transmitted between or from generation to generation, whether consecutively or not; which subsists in codified, oral, or other forms; and which may be dynamic and evolving, and may take the form of know-how, skills, innovations, practices, teachings or learning. (IGC 37th Session)

Why the author provided the examples of IGC's working definition is to elaborate the contention put earlier over WIPO's definition. As the author specified, that WIPO only helped to provide a general framework; it leaves multiple scopes for varied interpretations. Now, looking at the definition by IGC, we can see a huge difference in the scope of the definition. IGC is more elaborate in conceptualizing TK to a much greater extent. The subject matter remains the same in both the draft but has been explained more lucidly. But, even though IGC has been working for possibilities for the protection of TK, still no universally accepted definition could be relied upon, where two different definitions can be seen in two different sessions of IGC.

Convention of Biological Diversity ("CBD") aims at conservation of biological diversity, with its two supplementary agreements, Cartagena Protocol and Nagoya Protocol. CBD does not provide any comprehensive definition of TK. Cartagena Protocol deals in the governance of living modified organisms; therefore, its subject matter is far away from TK. But, Nagoya Protocol has been into the fair and equitable sharing of benefits arising out of the utilization of genetic resources, where its subject matter overlaps majorly with TK. Neither CBD nor Nagoya Protocol gives any elaborate definition of TK. CBD does provide the structure of TK at few places, and has also defined genetic material and resources (United Nation), but it did not provide any definition of TK in its official document. Nevertheless, on CBD website, where it explains "Traditional Knowledge and Convention of Biological Diversity," the definition is provided as follows:

Traditional knowledge refers to the knowledge, innovations and practices of indigenous and local communities around the world. Developed from experience gained over the centuries and adapted to the local culture and environment, traditional knowledge is transmitted orally from generation to generation. It tends to be collectively owned and takes the form of stories, songs, folklore, proverbs, cultural values, beliefs, rituals, community laws, local language, and agricultural practices, including the development of plant species and animal breeds. Sometimes it is referred to as an oral traditional for it is practiced, sung, danced, painted, carved, chanted and performed down through millennia. Traditional knowledge is mainly of a practical nature, particularly in such fields as agriculture, fisheries, health, horticulture, forestry and environmental management in general. (Convention)

This definition provides an overhaul structure of TK and can be said to be one of the best definitions that can help in delving into the concept.

Interestingly, the United Nation ("UN") specialized agencies such as United Nation Educational, Scientific and Cultural Organization ("UNESCO") and the Food and Agriculture Organization ("FAO") do not provide any definition for TK, even though their work is closely related to indigenous communities.

Hence, concluding this part, we can arrive at a substantial consensus that there is no universally accepted definition that can help the stakeholders, at the international level, to frame regulations for the protection of TK. We can just take the help from the available definitions and can frame the boundary for protecting TK (Ni, 2011). These definitions do not provide any answer to the question that should TK be protected considering the indigenous communities only or should be protected in a framework where the public, at large, can use it. Let us now delve into the relational approach between TK and IPR, and why the author feels that they share a sour relation.

12.3   IPR AND TRADITIONAL KNOWLEDGE: A PATH OF UNKNOWN DESTINATION

With the previous discussion over defining TK, we saw the difficulty in arriving at a perfect definition that has also contributed in creating a bitter relationship among them. IPR's foundation lies in a perfect stature with its clearly defined boundaries and scope. With TK, having no such defined scope, applying the conventional model of IPR in its domain, does not provide any fruitful results. As specified earlier, the unanswered question concerning the claim over TK will keep haunting us, while we analyze the relation. The unanswered question of whether TK should be subjected solely to the indigenous communities or the focus must also be on public, at large, makes the job of interlinking IPR with TK difficult. Let us delve into the conventional model of IPR and analyze the position of TK therein. Multiple authors have often dealt with comparing the IPR model with TK; hence, the author will slightly touch upon this aspect and will then move into the international perspective.

In the introductory part, the author stated that the foundational principle of providing the monopoly rights over an invention is to provide a preferable environment for more innovation. It is like making the inventor feel enriched for his efforts and gets themselves "motivated" for doing more innovation (Chaturvedi, 2020). Applying this principle, can it be argued that TK protection is required to make the community "motivated" for developing more? As specified, the production of TK was never for enrichment purposes; rather, it was a part of culture and livelihood for the community (Padmavati, 2018). Therefore, the basic foundation of IPR does not fit into the TK understanding. Nevertheless, the author cannot negate the fact that there can be multiple situations where the indigenous communities, perhaps, develop something only for commercial purposes. But, the majority might not hold true for the same.

12.3.1   PATENTS

One of the most important parts of all IPs is the patents. In the majority of jurisdictions, including India, the basic conditions for the grant of patents are threefold. First, the invention needs to be new, second, it should contain inventive steps (non-obviousness), and third, it must be of industrial applicability. These prerequisites need to be proved, for the patent to be granted. Now, if we try to ascertain the requirements for a patent with that of TK, we can analyze multiple hurdles. The first and foremost requirement for a patent grant is its "newness" requirement, which does not fall in line with TK, as they have been passed from generation to generation. Similarly, the second requirement of non-obviousness fails as TK is already in the public domain and becomes much "obvious" to any person in that field. TK fits into the third requirement of industrial applicability, where it is useful in some form or the other. Multiple authors have often raised serious doubts in finding the "inventor" in the case when TK will be considered for patent granting. The indigenous community, as a whole, can be termed an inventor since finding a single inventor is close to impossible. The argument also relates to the fact that almost all TK are known, and more or less substantially, depends on the natural process, which cannot be patented owing to the newness and non-obviousness requirements. Therefore, even though TK needs to be protected, it fails to "statutorily" satisfy all the prerequisite requirements for the grant of patent. From a legal standpoint, giving patented protection to TK will result in forfeiting multiple requirements that are necessary, statutorily (Judd, 2019).

12.3.2   COPYRIGHT

Not so different from Patent Law, TK faces the same difficulty here as well. For copyright protection, the expression needs to be original and must contain some creativity. With TK, it becomes difficult to adjudge the bar of originality and creativity. Copyright does not protect the mere facts; rather it protects the unique expressions. Two frameworks can be constructed here. First, TK can be protected under the copyright, if all the prerequisites concerning originality and unique expression can be proved. Second, it is not the mere idea but the way a cultural expression is being performed is what copyright requires. Therefore, despite few difficulties, TK might pass all the requirements and can be copyrighted. Nevertheless, few scholars have pointed out two difficulties in this domain, which are authorship requirement and fixation of the work. In both, TK fell short of. In case, there are no authors for the work, the copyright can be granted to the community as a whole, if that expression is the work of any particular community. With regards to the fixation of work, this problem might change the game for TK, as most of them do not have any strong fixation of their expression, and also they are far away from technological advancement. Hence, TK will struggle to get copyright protection concerning its nonfulfilment of statutory requirements (Judd, 2019).

The author in this part tries to analyze two basic IPs and their execution over TK. The author assumes that these two IPs are more important than others, if we compare the general activity in which indigenous communities are involved in. Through these two IPs, the author wants to reiterate a similar proposition concerning the sour relation that these two engage in. After getting a glimpse of how fractured the relationship between them is, let us move onto the international instruments and get an understanding of how they have worked in this domain. Whether they have failed to get a midway solution or has it been stuck in the debatable and jammed jar, is something to ponder upon.

12.4 INTERNATIONAL INSTRUMENTS

TK has always been in discussion internationally, and at various forums. One can assume multiple reasons as to why TK has always been in the domain of regular discussion at the international level. An important one is the fact that all countries recognize its importance in terms of health, and any such nonpreferable exploitation can eventually lead to a substantial impact on the health of a considerable population (Bastida-Munoz and Patrick, 2006). Hence, it has always been on the agenda at majority of conferences. The author cannot conclude that the discussions have been unsuccessful on various fronts. There have been some successful and legitimate decisions being taken at various levels, but certainly, we do not find a strong implementation at the state level. Although the discussions were done internationally, actual results must be visible at the state level. Our first and foremost international instrument will be CBD.

12.4.1 CONVENTION OF BIOLOGICAL DIVERSITY, 1992

The CBD has been on the list of all scholars who have written over TK and its complexities. It is pertinent to note that CBD does not directly deal in TK but biodiversity. Nevertheless, the proximity of TK with biodiversity has made CBD especially focus on the indigenous right that is visible when we read its preamble. It says, "*Recognizing the close and traditional dependence of many indigenous and local communities embodying traditional lifestyles on biological resources, and the desirability of sharing equitably benefits arising from the use of traditional knowledge, innovations and practices relevant to the conservation of biological diversity and the sustainable use of its components.*" (United Nation). This might be the only international convention that gives such a clear agenda on the protection of the indigenous community. Article 8(j) of CBD specifically draws core attention to the protection of TK which says:

> *Subject to its national legislation, respect, preserve and maintain knowledge, innovations and practices of indigenous and local communities embodying traditional lifestyles relevant for the conservation and sustainable use of biological diversity and promote their wider application with the approval and involvement of the holders of such knowledge, innovations and practices and encourage the equitable sharing of the benefits arising from the utilization of such knowledge, innovations, and practices. (Convention)*

An International Convention certainly cannot comprehend the fathom of intricacies at the local level; hence, this Article specifically starts with "Subject to its national legislation." This provides a great room for different countries to formulate regulations as per the prevailing situation. Moreover, it can be argued that there is no such stretched rope for the states to formulate any plan for TK; hence, it can be said to be an invisible rope. Nevertheless, this too seems to be an important phrase as one cannot comprehend the local issues. But, on the other hand, irrespective of any prevalent situations, the tryst for the grant of protection is apparent which must be respected. Hence, providing a special "privilege" in terms of subjecting it to the national legislation can be construed to be providing more freedom for the state that can result in lack of stronger sanction internationally (Ni, 2011). On similar lines, Article 15 of CBD, too, provides for access to genetic resources but in a similar context of keeping sovereign rights of the states on the above pedestal. Article 18 deals in the promotion of technical and scientific cooperation through institutions. Therefore, Article 15 and Article 18 can be read together to give a picture of how substantially the structure of benefit sharing (Gebru, 2017).

Moreover, the Conference of Parties ("COP") which is the supreme decision-making body of CBD, at its various annual meetings, has often reiterated the framing of effective rules for the implementation of Article 8(j). In 1998, COP in its fourth annual meeting established an ad hoc open-ended Inter-sessional Working Group for implementation of Article 8(j), which drafted the program of work, helping parties to fulfil their commitments under Article 8(j). In further years, this working group worked majorly in the field of TK where it also initiated the work of documentation of TK. India's major initiative of Traditional Knowledge Digital Library ("TKDL") was one of the major programs for the documentation of TK, mainly those which have medicinal value. Even to the extent, CBD comes up with the "Elements of code of Ethical Conduct," helping parties for the preservation of TK, which was adopted at the COP-10 in 2010.

Another important point to note, before we move further into some of the guidelines, is the Strategic Plan for Biodiversity 2011–2020 (Strategic Plan for Biodiversity), where Target 16 and Target 18 relate with TK. Target 16 was:

> *By 2015, the Nagoya Protocol on Access to Genetic Resources and the Fair and Equitable Sharing of Benefits Arising from their Utilization is in force and operational, consistent with national legislation.*

Target 18 was:

> *By 2020, the traditional knowledge, innovations, and practices of indigenous and local communities relevant for the conservation and sustainable use of biodiversity, and their customary use of biological resources, are respected, subject to national legislation and relevant international obligations, and fully integrated and reflected in the implementation of the Convention with the full and effective participation of indigenous and local communities, at all relevant levels*

For the achievement of these targets, the Capacity Development Programme at CBD has worked to help initiate practices for protecting TK. The program has been active at the national and local levels, providing all kinds of support for the protection of TK. Traditional Knowledge Information Portal ("TKIP") at CBD is an important step for enhanced access and information of TK. (Traditional Knowledge Information Portal)

12.4.1.1   BONN GUIDELINES

COP at its sixth session in 2002 adopted, "Bonn Guidelines on Access to Genetic Resources and Fair and Equitable Sharing of the Benefits Arising out of their Utilization" ("Bonn Guidelines"). As earlier stated, regarding Article 15 of CBD where access to genetic resources was discussed, CBD initiated another Ad Hoc Open-ended Working Group on Access and Benefit Sharing ("ABS") for the primary aim to implement Article 15. Bonn Guidelines is the brainchild of this working group. The discussion of the Bonn Guidelines was important here so that the reader must understand that access and benefit-sharing of TK is also an important part of the preservation of TK (Secretariat of the Convention on Biological Diversity, Bonn, 2002). A very important part of Article 15 is its clause 5 where it is mentioned that the prior informed consent of the party is important, thereby giving more impetus to the actual owner of TK, similar to the case where the IPR owner enjoys monopoly and any use of IP requires the permission of its owner (Ni, 2011). Multiple principles are mentioned in Bonn Guidelines that work for creating effective coordination between the indigenous communities and the user, at large. The creation of a mechanism for participation, respecting custom and tradition are some examples. Nevertheless, if we read the General Provision of this guideline, it again gives us a clear indication that these all principles do not, in any case, create any obligation and keep the sovereign rights of States at a higher pedestal. (Daum, 2014)

12.4.1.2   NAGOYA PROTOCOL

Nagoya Protocol is generally considered one of the most important agreements that deal in the Access and Benefit Sharing ("ABS") of TK. Nagoya is the result of 6 years of negotiations of Working Group on Access and Benefit

Sharing for creating the ABS regime. There was an utmost need for the creation of strong ABS requirements than was in the already developed Bonn Guidelines, and therefore COP worked on examining relevant issues about the genetic resources and associated TK with the help of Technical experts (Ni, 2011). With multiple deliberations, Nagoya Protocol was adopted at the tenth meeting in 2011, where it aims to provide a broader overview of ABS. Article 3 confirms the scope of Nagoya stating that the scope of this Protocol will work within the scope of Article 15 of CBD. But it suffers similar issues with respect to creating a strong obligation over parties. Like in Article 5, where Protocol defines the Fair and Equitable Benefit-Sharing, in clauses 2 and 3, it mentions the phrase "as appropriate." This certainly opens up the forum for multiple interpretations and understanding, thereby creating a weak juncture for the ABS regime. Similarly, Article 6, which talks about the access to genetic resources, uses the phrase "[I]n accordance with the domestic law" in clause 2, and "as appropriate" in clause 3. Article 7 speaks about Access to Traditional Knowledge Associated with Genetic Resources where it starts with the phrase, "In accordance with domestic law, each Party shall take measures, as appropriate…" (Secretariat of the Convention on Biological Diversity, Nagoya, 2011) leaving a major room open for States to formulate unfavorable legislation. This stipulates the factor of unwillingness of States to take over any obligatory measures, showing deep uncertainty and a clouded atmosphere over TK, when it comes to benefit sharing.

12.4.2 *WIPO AND TRIPS*

The discussion started way back in the 1960s and the 1970s regarding the protection for traditional cultural expression where we see the introduction of WIPO-UNESCO Model Provisions for National Laws on the Protection of Expressions of Folklore against Illicit Exploitation and other Forms of Prejudicial Action in 1982. This came up after major demand from countries for recognizing their contribution in this field. Even in the 1996 WIPO Phonograms and Performance Treaty, the definition of Performers in Article 2(a) mentions "expressions of folklore," thereby giving impetus to the indigenous people over their work (World Intellectual Property Organization, 1996). Getting into a brief history of prioritizing TK and its protection at the International level, WIPO undertook nine fact-finding missions in 1998 and 1999. The findings tackled the need to have full-fledged provisions of TK in terms of Intellectual Property, which finally established the Intergovernmental Committee on Intellectual Property and Genetic Resources, Traditional Knowledge and Folklore ("IGC") in October 2000 (document WO/GA/26/6). The committee focused on the importance of TK and treated it as an intellectual asset for humanity. This can be considered one of the strongest international forums that initiated the deliberation over IPR for TK protection (Bastida-Munoz and Patrick, 2006). But even after two decades of negotiations at IGC, why there has been slow progress at IGC?

Among several reasons, the author feels that one of the prominent reasons for the lack of effective consensus is different approaches of different countries. Each country has a diverse atmosphere concerning TK and they have their diverse approaches for protecting TK which results in nonconsensus at various levels. This situation cannot be negated owing to the history of lack of consensus in various international discussions. At this juncture, it also becomes important for the reader to understand the coordination between Trade-Related Aspects of Intellectual Property Rights ("TRIPS") and TK and whether TRIPS has helped for the repatriation of TK.

TRIPS is a prominent document developed under the aegis of the World Trade Organisation ("WTO") which gives a solid structure for protecting Intellectual Property at the domestic level. To be precise, TRIPS does not deal in and recognize TK in any of its provisions (World Trade Organization, 1995). No such provision deals with indigenous communities' work and its protection in the IPR framework. Some scholars have argued that having any provision for the protection of TK substantially makes it a TRIPS-plus provision, meaning that it will constitute to be more than what is already mentioned in TRIPS. The author will reiterate the similar proposition that was mentioned earlier concerning the requirements for an invention to qualify for a patent. TRIPS, more or less, deal with the similar proposition of statutory requirements, and getting TK into these fundamental principles will again require substantial change in the IP requirements which will undermine the foundation of IP. Therefore, scholars and stakeholders have often found it difficult to take TK under the TRIPS regime. The 3rd WTO ministerial meeting witnessed much resistance over the framework of TRIPS where a group of people published the "Seattle Declaration" stating that TRIPS have been for western knowledge and ignore the TK aspects. But that even failed to put any strong provision of TK in TRIPS. (Gebru, 2017)

There have been various discussions at the global level determining the relation between CBD and TRIPS. The first one was the Doha Ministerial meeting in 2001 where the discussion was initiated concerning TRIPS

Article 27(3) (b) which deals in the context of patenting life forms. The discussion mandated Committee on Trade and Environment for examining the relationship between CBD and TRIPS. Majorly, the stakeholders were in disagreement with various provisions. There has been disagreement between developed and developing countries for requirements such as "disclosure of origin." The disclosure of origin requirement is one of the primary objectives of CBD where it states that the origin of TK must be at the primary place. Developed countries do not agree with the origin requirement stating that it will hinder innovation, While developing countries, such as India, have often demanded an amendment to TRIPS for including origin requirement, it had received severe opposition from developed countries. The US has never been in favor of having the origin requirement under TRIPS but instead has favored national measure and contractual commitments between interested parties rather than making a general international obligation. Similarly, Japan too negated this proposition (Gebru, 2017). Due to various such oppositions, TK protection and TRIPS have never seen an undisputed light of the horizon where the stakeholders could find a stable platform for unwavering respect and protection of TK in the quintessential world of IPR.

12.5 CONCLUSION AND SUGGESTIONS

Various scholars have shared their perspective of how international instruments have played for the protection of TK. After getting a glimpse of multiple international instruments, it is clear that none of the instruments have been majorly successful in providing stronger regulations to protect TK. But it is clear that almost all of the instruments have given their best to incorporate stronger protection. So, where it lacked? The common reason that seems to be more applicable is the lack of a consensus-based approach. The diverse culture with diverse nature of legislation has made it difficult for the instruments to provide anchored regulations. We saw that in multiple instances, these international agreements have given more impetus to the sovereign states, in terms of framing rules in accordance with their local laws. The author supports this contention because each country has its complexities that might not be generalized at the international level. Nevertheless, as the author has already stated that there can be some issues that have to be generalized. Giving due respect is altogether fine, but there can be some issues that can be generalized, irrespective of the prevailing local situation.

As a part of suggestions, the author will try to put few suggestions that can help create a protective layer around TK

12.5.1 DISCLOSURE

This is one of the suggestions that have already been negated at the international level, as stated above, but its incorporation can turn the table for the protection of TK. Why it seems to be unacceptable to countries, especially developed, is the fact that this requirement will largely affect the patent requirements, as discussed above. Nevertheless, this will give stronger protection to TK. More importantly, as has been the case in India, regarding the development of Traditional Knowledge Digital Library, the resource-rich countries must develop such database so that any patent office, while examining for a patent, can check these databases for novelty purposes. India has also shared the database with the United States Patent and Trademark Office ("USPTO"), European Patent Office ("EPO"), and Japan Patent Office ("JPO") to name a few. As per the website of TKDL, 132 cases from EPO, 26 cases from USPTO, and even 36 cases from India itself were identified as a reflection of Indian Systems of Medicine (Traditional Knowledge Digital Library). Those applications were either withdrawn, cancelled, or terminated. Therefore, the development of such database can be very useful in terms of protecting the TK of one's country, when there seems to be no substantial consensus at the international level. Still, some scholars have negated the use of database for the protection of TK and have termed it "illusory" in protecting the secrecy associated with TK (Bagley, 2019). But in a situation, where there seems to have no stronger regulations, creating a database is one of the best options.

12.5.2 TRADE SECRET

There have been a lot of scholarly articles explaining the need to take up Trade Secret as a part of protecting TK. Trade Secret is also a form of IPR which can provide strong protection to TK, where the conventional IPR does not fit. As the name suggests, trade secret holds its economic value over information, if it is kept secret, that is, if it is not in the public domain. The knowledge that is held in the community can be considered a trade secret. Giving TK, the protection of trade secrets can be a good move, as it does not require the abovementioned criteria of novelty, non-obviousness, and industrial applicability. Therefore, this method can be provided to have substantial protection of TK (Okediji, 2019). Nevertheless, this comes up with a lot of limitations as well. For those countries, like India, where there is no Trade Secret Law, it can have a serious blow to this protection. Moreover, this protection comes with the mandate of creation of contractual engagements between parties. In most cases, there is no such awareness among communities to come under the contractual arrangement with the third party, which results in the lack of protection of their TK. Despite various shortcomings of trade secret as a part of protecting TK, this could be the best-suited IP form for protecting TK, if a stronger and substantial regulation could be made at the national level in the respective States.

12.5.3 DEFINING TK AND ENFORCEABILITY OF INTERNATIONAL INSTRUMENT

The problem starts when we do not see any solid definition that can be treated as the baseline for the work to be done for TK. Undoubtedly, the definitions provided give a holistic view of TK, but the difference in approach and lack of consensus over terms used in definition shows a lack of generalized baseline for protection. The author believes that it is difficult to agree upon any single definition, owing to its nature. But it is important to adopt a definition that gives an overhaul view of TK along with the Traditional Culture Folklore and the associated Genetic Resources, as these three should not be treated differently.

Talking about enforceability, the author understands that the international legal instruments are not legally binding over member states. The author further believes that having a nonbinding nature of agreements is the reason for major obstacles in the pursuit for protecting TK. Each country will have their interest and their unique arguments. Bringing all countries under a single umbrella is close to impossible. Probably, the content that is required is sufficient in agreements, but the lack of binding nature seems to affect the practicality of agreements. The author believes that there must be at least some form of "minimum" binding provisions which should be legally enforceable to give a more stringent approach toward TK protection (Gebru, 2017).

TK is not only about the property but also about the emotions of communities. We have come a far way in protecting TK but still, a lot needs to be done. The world needs to come and agree upon a strong binding framework that can lead the path in the future for the efficient protection of TK.

KEYWORDS

- **traditional knowledge**
- **intellectual property**
- **TRIPS**
- **convention of biological diversity**
- **WIPO**

REFERENCES

Arewa, O. B. TRIPS and Traditional Knowledge: Local Communities, Local Knowledge, and Global Intellectual Property Frameworks. Marquette Intellect. *Prop. Law Rev.* **2006,** *10* (2), 155–180. HeinOnline.

Bagley, M. A. The Fallacy of Defensive Protection for Traditional Knowledge. *Washburn Law J.* **2019,** *58,* 323. HeinOnline.

Bastida-Munoz, M. C.; Patrick, G. A. Traditional Knowledge and Intellectual Property Rights: beyond Trips Agreements and Intellectual Property Chapters of FTAs. *Mich. State J. Int. Law* **2006,** *14* (2), 259–290. HeinOnline.

Bernieri, R. C. Intellectual Property Rights in Bilateral Investment Treaties and Access to Medicines: The Case of Latin America. J. World Intell. Prop. **2006,** *9* (5), 548–572. HeinOnline.

Chaturvedi, S. Understanding IP: Jurisprudential foundation of Intellectual Property. *Sunday Guard. Live* **2020.** https://www.sundayguardianlive.com/legally-speaking/understanding-ip-jurisprudential-foundation-intellectual-property (accessed May 10, 2021).

Convention of Biological Diversity. Traditional Knowledge, Innovations and Practices. *CBD* **2011.** https://www.cbd.int/traditional/intro.shtml (accessed May 18, 2021).

Daum, I. Legal Conflicts in the Protection of Traditional Knowledge and Intellectual Property in International Law. *Ger. Yearb. Int. Law* **2014,** *57,* 411–442. HeinOnline.

Gebru, A. The Global Protection of Traditional Knowledge: Searching for the Minimum Consensus. *John Marshall Rev. Intell. Prop. Law* **2017,** *42* (17), 42–91. HeinOnline.

IGC 37th Session. The Protection of Traditional Knowledge: Draft Articles, Facilitators' Rev. 2. *WIPO.* **2018.** https://www.wipo.int/edocs/mdocs/tk/en/wipo_grtkf_ic_37/wipo_grtkf_ic_37_facilitators_text_tk_rev_2.pdf (May 15, 2021).

Judd, P. L. The Difficulties in Harmonizing Legal Protections for Traditional Knowledge and Intellectual Property. *Washburn Law J.* **2019,** *58* (2), 249–270. HeinOnline.

Ni, K. -J. Traditional Knowledge and Global Lawmaking. *Northwest. Univ. J. Int. Hum. Rights* **2011,** *10* (2), 85–118. HeinOnline.

Okediji, R. L. A Tiered Approach to Rights in Traditional Knowledge. *Washburn Law J.* **2019,** *58,* 271. HeinOnline.

Padmavati, M. Ensuring Longevity of Traditional Knowledge Associated with Biodiversity to Address Climate Change. *J. Intell. Prop. Rights* **2018,** *23* (1), 35–43. Online.

Secretariat of the Convention on Biological Diversity, Bonn. Bonn Guidelines on Access to Genetic Resources and Fair and Equitable Sharing of the Benefits Arising out of their Utilization. *Conv. Biol. Divers.* **2002.** https://www.cbd.int/doc/publications/cbd-bonn-gdls-en.pdf (accessed May 30. 2021).

Secretariat of the Convention on Biological Diversity, Nagoya. Nagoya Protocol on Access to Genetic Resources and the Fair and Equitable Sharing of Benefits Arising From Their Utilization to the Convention on Biological Diversity. *Conv. Biol. Divers.* **2011.** https://www.cbd.int/abs/doc/protocol/nagoya-protocol-en.pdf (accessed June 10, 2021).

Sigamany, I. Land Rights and Neoliberalism: An Irreconcilable Conflict for Indigenous Peoples in India? *Int. J. Law Context* **2017,** *13* (3), 369–387. HeinOnline.

Strategic Plan for Biodiversity. Strategic Plan for Biodiversity 2011–2020 and the Aichi Targets. *Conv. Biol. Divers.* **2011.** https://www.cbd.int/doc/strategic-plan/2011-2020/Aichi-Targets-EN.pdf (accessed May 30, 2021).

Traditional Knowledge Digital Library. *TKDL Outcomes against Bio-Piracy*; Ministry of Ayush, 2006. http://www.tkdl.res.in/tkdl/langdefault/common/outcomemain.asp?GL=Eng (accessed June 16, 2021).

Traditional Knowledge Information Portal. About Traditional Knowledge Information Portal (TKIP). *Conv. Biol. Divers.* **2015.** https://www.cbd.int/tk/about.shtml (accessed May 30, 2021).

United Nation. Convention of Biological Diversity; United Nation, 1992. https://www.cbd.int/doc/legal/cbd-en.pdf (accessed May 18, 2021).

Venkataraman, K.; Swarna, L. S. Intellectual Property Rights, Traditional Knowledge and Biodiversity of India. *J. Intell. Prop. Rights* **2008,** *13* (4), 326–335. Online.

WIPO/GRTKF/IC/9/5, 9th Session. *Intergovernmental Committee on Intellectual Property and Genetic Resources, Traditional Knowledge and Folklore Ninth Session Geneva*, April 24–28, 2006; WIPO, 2006. https://www.wipo.int/edocs/mdocs/tk/en/wipo_grtkf_ic_9/wipo_grtkf_ic_9_5.pdf (accessed May 15, 2021).

World Intellectual Property Organization (WIPO). Traditional Knowledge. WIPO. https://www.wipo.int/tk/en/tk/ (accessed May 14, 2021).

World Intellectual Property Organization. WIPO Performances and Phonograms Treaty (WPPT); WIPO, 1996. https://www.wipo.int/edocs/pubdocs/en/wipo_pub_227.pdf (accessed June 12, 2021).

World Trade Organization. Agreement on Trade-Related Aspects of Intellectual Property Rights; WTO, 1995. https://www.wto.org/english/docs_e/legal_e/27-trips.pdf (accessed June 14, 2021).

ADDITIONAL READINGS

Andanda, P. Striking a Balance between Intellectual Property Protection of Traditional Knowledge, Cultural Preservation and Access to Knowledge. *J. Intell. Prop. Rights* **2012,** *17* (6), 547–558. Online.

Chakrabarti, G. Biological Diversity Act: A Concern for Conservation of Genetic Resource and Associated Traditional Knowledge in India. *J. Intell. Prop. Rights* **2019,** *24* (3–4), 53–61. Online.

Cross, J. T. Property Rights and Traditional Knowledge. *Potchefstroom Electron. Law J.* **2010,** *13* (4), 12–48. HeinOnline.

Dutfield, G. TRIPS-Related Aspects of Traditional Knowledge. *Case West. Reserve J. Int. Law* **2001,** *33* (2), 233–276. HeinOnline.

Gervais, D. TRIPS, Doha and Traditional Knowledge. *J. World Intell. Prop.* **2003,** *6* (3), 403–420. HeinOnline.

Kariuki, F. Notion of 'Ownership' in IP: Protection of Traditional Ecological Knowledge vis-a-vis Protection of T K and Cultural Expressions Act, 2016 of Kenya. *J. Intell. Prop. Rights* **2019,** *24* (3–4), 89–102. Online.

Kariyawasam, K. Protecting Biodiversity, Traditional Knowledge and Intellectual Property in the Pacific: Issues and Challenges. *Asia Pac. Law Rev.* **2008,** *16* (1), 73–90. HeinOnline.

Rahaman, M. R. Protection of Traditional Knowledge and Traditional Cultural Expressions in Bangladesh. *J. Intell. Prop. Rights* **2015,** *20* (3), 164–171. Online.

Sreedharan, S. K. Bridging the Time and Tide—Traditional Knowledge in the 21st Century. *J. Intell. Prop. Rights* **2010,** *15* (2), 146–150. Online.

CHAPTER 13

International Convention for Protection of Geographical Indication and Its Application in Agriculture: A Legal Perspectives

PIYALI MUKHERJEE

Department of Law, Brainware University, Barasat, Kolkata, West Bengal, India

ABSTRACT

Every territory has an identity and a reputation. Geographical indications (GIs) refer to the qualities and attributes of particular commodities that can be traced back to specific geographical regions and are marketed as "products of a specific region." GIs is a new intellectual property trend. GI has emerged as a major intellectual property issue in India. It grants the right to use the product's indication to makers and producers in that region. It also implies that they have the legal ability to prohibit the use of a sign or name that does not contain the exact qualities and characteristics stipulated by the product's GI. A geographical indicator (GI) is a technique that allows manufacturers to distinguish their products from rival items on the market and to develop a reputation and goodwill around their products that will allow them to command a higher price. GI registration has been shown to have potential economic benefits in a number of studies. This article provides an outline of

Crop Sustainability and Intellectual Property Rights. Soumya Mukherjee, Piyali Mukherjee & Tariq Aftab (Eds)
© 2023 Apple Academic Press, Inc. Co-published with CRC Press (Taylor & Francis)

current situation with protection of GIs, registration procedure, and its legal status in Indian scenario.

13.1 INTRODUCTION

Intellectual property means the creation of the human imagination and intellect, and is thus referred to as "intellectual property." Intellectual property is established by combining knowledge intangible items that can multiply an infinite number of possibilities in different places in the world. Geographical indication (GI) is one of the most argumentum matter in contemporary global agricultural aspects. It is used to identify a product's origin when its characteristics or quality are a result of geographical origin, which includes agricultural product. All agricultural production relies on seed or planting material. Seed is the cheapest component of the entire cost of crop production, but it has the greatest impact. Farmers quickly realized the value of good seeds of new and better varieties of crops after reaping the benefits of green revolution varieties' seeds. Farmers were even more willing to pay a greater premium for such excellent seeds. This presented an opportunity for seed businesses and technology developers to turn plant varieties and critical plant genes into profit-generating goods. To control the market, global strategy, pesticides, and seed businesses united to centralize capital and technology. The need to conserve biodiversity, farm level variation, crediting farmers for their traditional crop varieties, folk varieties, farmers varieties, access to benefit sharing, extending consumer assurance through geographic indications, appellations of origin, traditional knowledge, and so on were all attempted to be protected in various countries. Several new challenges are now dominating global commodities trade, which are now understood and applied in India. Plant-based goods or by-products are another component of GI in agriculture. Plant-based goods could be used as a raw material for manufacturing, processing, or preparation (Chaudhary et al., 2017).

An agricultural product or any item identify the Geographical name and if the product with a GI is used by someone from a different location, it is likely to cause confusion among customers. A GI sign should identify a product as being from a specific location, and the qualities or properties must be similar to that location. Basically, GI is conferred to agricultural, manufactured, or natural origin items, as well as handicraft gods with a known location of origin. In India, the GIs of Goods (Registration and Protection) Act of 1999 safeguards GI. GI registration is not mandatory in India. If it is registered, it will provide better legal protection, that making an action for infringement much easier.

Over the years, one of the most contentious Intellectual Property Rights problems in the The World Trade Organization's (WTO's) Agreement on Trade Related Aspects of Intellectual Property Rights (TRIPS) has been the protection of GIs. TRIPS defines GI as "any indicator that identifies a product as being from a specific location, if the product's quality, reputation, or other features are substantially due to its geographic origin." A geographical indicator grants a location or place the exclusive right to use a name for any agricultural product that has certain qualities that are unique to that location. After the GI took effective on September 15, 2003, Darjeeling Tea became the first GI-tagged product in India in 2004. Following that milestone, India added a slew of GI-labeled agricultural items. This article provides an outline of the current situation with protection of GIs, registration procedure, and its legal status in Indian scenario.

13.2 HISTORICAL BACKGROUND OF GI

Governments have protected trade names and trademarks of food products associated with a specific region since at least the end of the nineteenth century. In such circumstances, governments justify the restriction on competitive freedoms resulting from the grant of a monopoly of usage over a geographical designation by citing consumer or producer protection benefits. Items that fulfil geographical origin and quality criteria may be stamped with a government-issued stamp that serves as formal proof of the product's origins and standards. Gruyère cheese (from Switzerland) and several French wines are examples of items with such "appellations of origin." GIs have always been linked to the concept of terroir and to Europe as a whole, where there is a historical tradition of connecting different foods with specific locations. The protected designation of origin framework, which came into effect in 1992 under European Union law, governs the following systems of GIs: "Protected designation of origin", "protected GI," and "Traditional Specialities Guaranteed" (Tosato, 2013).

13.2.1 HISTORY OF TRIPS PROVISION ON GI

The basic principle of GI has developed significantly since its inception in nineteenth-century Europe. The present international framework is outlined in Article 22 of the TRIPS Agreement, which requires members to establish "legal means for interested parties" to obtain protection of their GIs.

The Uruguay Round of GATT negotiations began in 1986, at a critical juncture in India's development policymaking process. The Uruguay Round negotiations were well underway when India launched its enormous economic reform package in 1991, signaling a paradigm shift in its policies. This paved the way for Marrakesh in 1994 and the foundation of the WTO. Given its lengthy history of inward-looking development strategy and protectionist trade policy regime, India remained a cautious and rather inactive player during the early years of the Uruguay Round negotiations (The Protection of GI in India, 2009).

India, on the other hand, intended to broaden the scope of "GI" protection beyond wine and spirits to include other items in Doha. A number of countries wished to negotiate extending this greater degree of protection to other items because they saw it as a method to boost marketing by better differentiating their products from their competitors, and they object to other countries "usurping" their conditions. Others were against the initiative, and the debate centered on whether the Doha Declaration serves as a mandate for negotiations (GI in the WTO & Doha Negotiations, 2007).

Those who oppose the extension claim that the current level of protection (Article 22) is sufficient. They warn that enhancing protection would be inconvenient and would disrupt current acceptable marketing tactics. India, along with a slew of other like-minded countries, pushed for an "extension" of Article 23's scope to include all types of commodities. Countries like the United States, Australia, New Zealand, Canada, Argentina, Chile, Guatemala, and Uruguay, on the other hand, are adamantly opposed to any "extension." The question of "extension" was a key component of the Doha Work Programme (2001). Moreover, because to large divergences of opinion among WTO members, no progress has been made in the negotiations, and it remains an "outstanding implementation issue."

13.3 PROTECTION OF GI IN INDIAN CONTEXT

India has built a one-of-a-kind system of GI protection with the implementation of a law expressly dealing with the protection of GIs. The "GIs of Goods

(Registration and Protection) Act, 1999" (GI Act) and the "GIs of Goods (Registration and Protection) Rules, 2002" (GI Rules) govern GI protection in India (GI Rules). In order to execute national intellectual property laws in compliance with India's TRIPS obligations, India established its GI legislations. The central government has established a GIs Registry in Chennai with all-India jurisdiction, where right-holders can register their GI under the GI Act and the GI Rules, which went into effect on September 15, 2003. (The Protection of GI in India, 2009).

India is required to preserve GIs as a signatory to the TRIPS Agreement, and the GIs of Goods (Registration and Protection) Act, 1999 was passed to meet this need. It should also be highlighted that India believed that several of its products had a great potential for benefiting from GI registration, and that comprehensive laws for registration and proper protection for GI was required. Because unless a GI is protected in the nation of origin, the TRIPS Agreement does not obligate other countries to provide reciprocal protection. The following are the primary advantages of registering under the Act:

- Provides GI legal protection in India;
- Prevents others from using a registered geographical indicator without permission;
- Allows individuals to seek legal protection in other WTO members.

One of the best characteristics of the Indian Act from the standpoint of a developing country is the extensive definition of GI, which includes agricultural, natural, and manufactured items. This is particularly significant in the Indian context, given the large range of items that deserve protection, from agricultural products like Basmati and Darjeeling tea. "Any association of persons, producers, organization, or authority established by or under the law can request for registration of a GI, according to Section 11 of the Act". Another essential feature of the Act is the ability to keep a GI protected indefinitely by renewing its registration when it expires after 10 years. The Indian Act has attempted to extend the TRIPS-mandated additional protection for wines and spirits to commodities of national interest on a case-by-case basis. The Central Government has the jurisdiction under Section 22.2 of the Act to provide further protection to particular commodities or classes of goods (The Protection of GIs in India: Issues and Challenges, 2013).

By prohibiting the registration of a GI as a trademark, Section 25 of the Act aims to prevent the appropriation of a public property in the form of a geographical indicator as a trademark by an individual, resulting in market confusion. A GI cannot also be assigned or transmitted, pursuant to Section

24 of the Act. The Act recognizes that a GI is a form of public property owned by the manufacturers of the items in question, and that it cannot be assigned, transmitted, licensed, pledged, mortgaged, or the subject of any arrangement for transferring title or possession.

The GI Act was passed as part of India's endeavor to put in place national intellectual property legislation that complied with the TRIPS Agreement's responsibilities. Under this Act, which went into force on September 15, 2003, the Central Government established the 'GIs Registry' in Chennai with all India jurisdictions, where right holders can register their respective GIs. Any individual claiming to be the producer of the good identified by the registered GI can make an application for registration as an approved user once the GI is registered. The Registrar of GIs will manage the GI Act on behalf of the Controller General of Patents, Designs, and Trade Marks (GI: Indian Scenario, 2006).

13.4 PROCEDURE AND IMPACTS OF GI REGISTRATION IN INDIA

Any individual, producer, organization, group, or body representing the interests of the producers of the concerned commodities can apply to the Registrar of GIs for GI registration. Every application must be submitted in the required form GI – 1A to ID, signed in triplicate by the applicant and accompanied by three copies of the Statement of Case and the necessary fee. The applicant must state the producers' stake in the concern goods to be registered.

Before applying for a GI Tag, make sure the product is covered by Section 2 of the Act. After it has been validated, the GI Tag Application can be filed under Section 11 of the Act. Three copies of the application must be made, each containing the following information:

- "statement of the case for GI Tag
- class of good to which GI is applicable
- three certified copies of the geographical map of the region to which the GI belongs
- description of the GI
- details of the applicants with their addresses"

Step 2: The Registrar will examine the application to see if there are any deficiencies that must be addressed within 1 month of receiving the notification. The experts will next review the statement of the case, and an Examination Report will be issued as a result.

Step 3: If the Registrar has any objections to the Application, they will be informed to the Application, and the Application will have 2 months to respond or request a hearing on the matter. If the Applicant so desires, he or she will have 2 months to file an appeal against the decision.

Step 4: The Application must be published in the GIs Journal within 3 months of its acceptance.

Step 5: Anyone opposing the GI Application has 3 months to file a notice of opposition. The Applicant will receive a copy of the Notice and will have 2 months to respond with a counter-statement. Following that, both sides will present evidence along with supporting papers, and a date for hearing the case will be set.

If the Applicant does not send a rebuttal statement, it will be assumed that the Applicant has abandoned his or her application.

Step 6: After the hearing is completed and the Application is approved, the Registrar will register the GI. Following that, the Applicant will receive a certificate of registration with the GI Registry's seal. The date on which the Application is filed is the date on which it is registered.

Step 7: For a period of 10 years, the GI is valid and must be renewed by paying a renewal cost (Sharma, 2021)

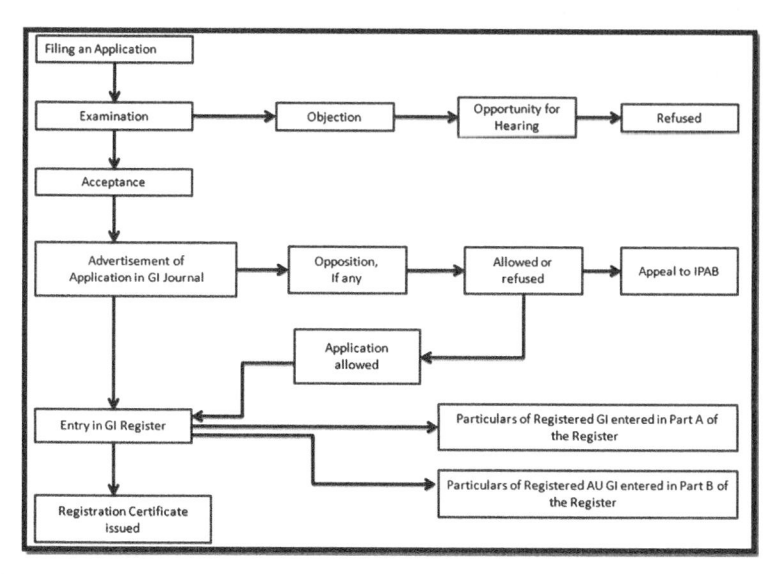

FIGURE 13.1 Flowchart on registration procedure of GI in India.

Source: Reprinted from Sharma, 2021.

According to some observers, GI may be more adaptable to the specific circumstances of developing countries than other types of intellectual property rights. GIs could aid in the protection of rural and indigenous communities' collective rights to their cultural knowledge. It ensures that the entire community benefits from the information that has been kept and passed down through generations with incremental refinement, rather than being locked up as the private property of a single individual (Sahai and Barpujari, 2007).

The scope of protection is limited to limiting the type of person who can utilize the protected indicator and/or their location. Among other benefits of GIs, the rights can theoretically be preserved in perpetuity as long as the product-place relationship is maintained (Commission on Intellectual Property Rights, 2004). Furthermore, proprietors of a GI do not have the power to assign the indication, prohibiting it from being transferred to non-local manufacturers. Although empirical data indicates that GIs have substantial repercussions for producers in both rich and developing nations (Jena and Grote, 2007). There is a scarcity of data on the socioeconomic effects of GIs in India. Surprisingly, the collective nature of GIs also highlights substantial collective action-related issues at various phases of organization and administration (Das, 2009).

13.5 GI TAG AND ITS IMPORTANCE

A GI is a term or sign given to goods that are associated with a certain geographical area or origin, such as a region, town, or country.

Utilizing of GIs as GI tag can be viewed as a certification that a product is made using traditional methods, has special attributes, or has a specific reputation due to its geographical origin. Wine and spirit drinks, cuisine, agricultural products, handicrafts, and industrial products are all examples of geographical indicators.

GI Tag ensures that the popular product name can only be used by individuals who have been registered as authorized users. A sign must identify a product as coming from a specific location in order to operate as a GI.

The Paris Convention for the Protection of Industrial Property includes GIs as a component of intellectual property rights. The WTO Agreement on Trade-Related Aspects of Intellectual Property Rights governs GI on a global scale (TRIPS). The GIs of Goods (Registration and Protection) Act,

1999 governs GIs registration in India, and it took effect in September 2003. In 2004–2005, Darjeeling tea was the first product in India to be given a GI label (List of GIs or GI Tags in India, 2021). If the origin of a product cannot be reliably determined, either both states should have control of the lands, or none of the places should have a GI tag. States and communities must shift their focus away from merely certifying products for the sake of regionalism and instead devote all resources to actively promoting the product and its industry.

13.5.1 BENEFITS OF GI TAGS

The protection of GIs improves the economic success of manufacturers and producers. Moreover, the marketing and promotion of GI-labeled products increases secondary business development in that region, which boosts growth of the economy. The following benefits can be obtained with the registration of a GI:

- Each and every Products have their legal protection.
- It makes it impossible for people to use GI tag things without permission.
- It enables customers to obtain high-quality products with required characteristics while also ensuring their authenticity.
- Enhances the demand for GI tag goods on national and the economic prosperity of GI tag makers is boosted.

GI tags provide a number of advantages, but they also have a number of drawbacks. Recently, there has been an increase in disagreements concerning the product's point of origin. The lack of unambiguous historical evidence aggravates the situation.

Each state hopes to push its own cultural and regional jingoism over the other by 'winning' a GI tag (List of GIs or GI Tags in India, 2021).

This type of unhealthy competitiveness, as a matter of discussion, tends to polarize the country along regional, cultural, and linguistic lines. In their drive to capture as many GI tags as possible, most governments haven't thought about how to boost the value of products that currently have a GI label.

13.5.2 SIGNIFICANCE OF GI TAGS AND GI LISTING IN INDIAN PERSPECTIVES

Individuals have the right to use a geographical indicator who have the right to do so to prevent it from being used by a third party whose product fails to meet the requirements the required standards.

A protected GI, on the other hand, does not give the holder the right to prohibit someone from creating a product using the same methods as those outlined in the criteria for that indication. A GI tag is usually secured by obtaining a right over the sign that serves as the indication.

Prevents others from using a Registered GI without permission. It gives Indian GIs legal protection, which boosts exports. It supports the economic well-being of producers of items produced in a certain geographical area (Intellectual Property India, 2021).

TABLE 13.1 Agricultural GI Tags Registered in India in 2021.

Products	Categories	States
Kashmir Saffron	Agriculture	Jammu & Kashmir
Manipuri Black Rice	Food Stuff	Manipur
Kandhamal Haladi	Agricultural	Odisha
Kodaikanal Malai Poondu	Agricultural	Tamil Nadu
Gulbarga Tur Dal	Agricultural	Karnataka
Tirur Betel Leaf (Tirur Vettila)	Agricultural	Kerala
Khola Chilli	Agricultural	Goa
Kaji Nemu	Agricultural	Assam

In this Table 13.1, this tag has a 10-year validity span and can be renewed. Agricultural items, meals, wine and spirit drinks, handicrafts, and industrial products are all examples of geographical indicators. It's essentially a guarantee that the goods originated in that region. Karnataka has the most 47 GI tags in India.

13.6 CASE STUDIES ON DARJEELING TEA GI TAGS

GI is a technique that allows manufacturers to distinguish their products from their competitors, develop a reputation and brand around their products, and command the premium price they deserve. Darjeeling Tea was given

GI classification in October 2004. It is the first application in India to be registered as a GI. The logo is a roundel with the word "DARJEELING" and an image of an Indian woman clutching tea leaves. Under the Tea Board's license and authority, this logo has been widely utilized by all Darjeeling tea manufacturers.

Darjeeling tea needed such protection since tea from Kenya, Sri Lanka, and even Nepal had previously been mislabeled as "Darjeeling tea" on the international market. The design of the emblem as a certified trademark began in 1983, the procedure of obtaining special protection for Darjeeling tea (Chaudhary et al., 2017).

13.6.1 HISTORY BEHIND THE DARJEELING TEA

Darjeeling tea's history dates back to the 1830s, when Captain Lloyd, a young British soldier, realized the potential of converting Darjeeling, a little village in northern West Bengal situated against the backdrop of the snow-capped Himalayas, into a hill station. Darjeeling was thereafter entrusted to a civil surgeon, who established the first tea nursery with seeds from the China kind of tea grown in the Kumaon highlands of northern India. The British were inspired by the success of this tea nursery to establish tea gardens. There are currently 87 such tea plantations spread across 17,800 hectares, employing over a lakh employees (more than half of them are women), and producing around 8–9 million Kg of tea every year. Due to a complex combination of agro-climatic circumstances, Darjeeling Tea has a distinct flavor, and it's no surprise that 70% of it is exported. And when a product gains such a high profile in the worldwide market, it's only natural for knockoffs to emerge, claiming to be "Darjeeling." These tainted or low-quality teas began to tarnish Darjeeling's reputation. Darjeeling tea, as a result, required protection.

13.6.2 CERTIFIED GI TAGS

The government has made it mandatory to certify the authenticity of all Darjeeling sold on the domestic and international market. All Darjeeling tea merchants must sign a license agreement with the Tea Board of India and pay an annual license fee as a result of this. The Tea Board can calculate and assemble the entire volume of Darjeeling tea produced and sold in a given period in this manner.

Now that the Tea Board has secured protection and logo certification, it has recruited Compumark, a World Wide Watch agency, whose purpose it is to monitor and report to the Tea Board any cases of unauthorized usage or any other attempt to register other items under the brand name "Darjeeling." This is how the Tea Board maintains this protection by attempting to prevent the misuse of this brand name, which has been brought to its attention by many.

Negotiations were used to settle several of these situations. The acquisition of this GI resulted in a beneficial outcome. It sells for a high price, and the benefits trickle down to the workers. Because of the protection, Darjeeling tea has been able to retain a consistent curve even when the tea industry as a whole was in a slump. It also resulted in fresh investments and consolidation of ownership in the Darjeeling tea industry.

In India, the notion of GI is still developing, despite the fact that GI-registered agricultural products attract a price premium of 10–15%, according to a United Nations Conference on Trade and Development study. Tea and other items on the global market still lack TRIPS protection, which is provided by an Act. Progress is being made, however, thanks to increased consumer awareness and government initiatives. Darjeeling tea, with its distinct color, flavour, and aroma, is the best example (Protecting Darjeeling Tea: The Tea from the Queen of the Hills, 2021).

13.7 CONCLUSION

The term GI denotes a guarantee of quality and uniqueness based on the location of origin. GIs is a major emanate intellectual property field.

Furthermore, in today's globalized world, a GI protects local culture and tradition from being exploited by outsiders who use the GI to deceive buyers into believing that the things they are purchasing are from a specific location and hence have a distinct quality. Intellectual property rights have never been more significant or contentious in terms of economics and politics than they are now. Patents, copyrights, trademarks, industrial designs, and GIs are frequently mentioned in discussions and debates about, among other things, public health, food security, education, trade, industrial policy, traditional knowledge, biodiversity, biotechnology, the Internet, and the entertainment and media industries. In a knowledge-based economy, a grasp of intellectual property rights is unquestionably necessary for informed policymaking in all aspects of human progress.

A GI tag is a source of pride for both the maker and the consumer as a symbol of excellence and a sense of guarantee or originality, as well as the protection of rights for all parties involved in the production process. GI has been a blessing to individuals all around the world, especially poor craftsmen who work hard to maintain a level of quality that is recognized and preserved worldwide. A GI tab is a necessary component for preserving the essence and uniqueness of a product with certain qualities and characteristics.

KEYWORDS

- **geographical indication**
- **TRIPS agreement**
- **agricultural products**
- **intellectual property law**
- **legal protection**

REFERENCES

Chaudhary, R.; Kumar, S.; Kumar, S. Geographical Indications in Indian Agriculture on the Anvil. *J. AgriSearch* **2017**, *6*, 790–816.

Commission on Intellectual Property Rights. *Integrating Intellectual Property Rights and Development Policy*; Commission on Intellectual Property Rights: London, 2004.

Das, K. *Socioeconomic Implications of Protecting Geographical Indications in India*; SSRN, 2009. https://ssrn.com/abstract=1587352 or http://dx.doi.org/10.2139/ssrn.1587352

Anil, W. H. *Geographical Indication: Indian Scenario*, 2006; http://eprints.rclis.org/7878/1/EF6C0C53.pdf (accessed on Sep 7, 2021).

GI in the WTO & Doha Negotiations by Miguel Rodriguez Mendoza Presented in Worldwide Symposium on GI, 2007. www.wipo.int/edocs/mdocs/geoind/en/wipo_geo_bei_07/wipo_geo_bei_07_www_81777.doc (accessed on July 31, 2021).

Intellectual Property India, 2021. https://ipindia.gov.in/faq-gi.htm (accessed on Sep 18, 2021).

Jena, P. R.; Grote, U. *Changing Institutions to Protect Regional Heritage: A Case for Geographical Indications in the Indian Agrifood Sector*, 2007. http://www.pegnet.ifw-kiel.de/activities/pradyot.pdf (accessed on Sep 15, 2021).

List of Geographical Indications [GI Tags] in India, 2021. https://byjus.com/free-ias-prep/geographical-indication-tags-in-india/ (accessed on Sep 20, 2021).

Protecting Darjeeling Tea: The Tea from the Queen of the Hills, 2021. https://www.sourcetrace.com/blog/protecting-gi-darjeeling-tea/ (accessed on Sep 16, 2021).

Sharma, S. *Registration of Geographical Indications in India*, 2021 https://www.mondaq.
com/india/trademark/1043076/registration-of-geographical-indications-in-india (accessed
on Sep 19, 2021).

Sahai, S.; Barpujari, I. *Are Geographical Indications Better Suited to Protect Indigenous
Knowledge? A Developing Country Perspective*, 2007. https://genecampaign.org/
wp-content/uploads/2014/07/Are_Geographical_Indications_Better_Suited.pdf (accessed
on Sep 19, 2021).

Tosato, A. The Protection of Traditional Foods in the EU: Traditional Specialities Guaranteed.
European Law J. **2013,** *19* (4), 545–576. DOI: 10.1111/eulj.12040.

The Protection of Geographical Indication in India, 2009. https://www.altacit.com/resources/
gi/the-protection-of-geographical-indication-in-india/#i_5 (accessed on Sep 15, 2021).

The Protection of Geographical Indications in India: Issues and Challenges, 2013. http://yucita.
org/uploads/yayinlar/diger/makale/8-The_Protection_of_Geographical_Indications_india.
pdf (accessed on Sep 12, 2021).

CHAPTER 14

Farmers' Rights: The Indian Scenario

POOJA JHA MAITY and REETU SHARMA

Department of Botany, Ramjas College, University of Delhi, Delhi, India

ABSTRACT

From ancient times farmers have been conserving, maintaining, improving, and making plant genetic resources accessible for agricultural development. Farmers may have a limited scientific understanding of genetic diversity and breeding, yet their cumulative knowledge and effort in saving, selecting, and maintaining seeds have been the building blocks for crop improvement. Farmers' contribution is as important as that of plant breeders in ensuring food security and sustainability. The two international treaties that regulate plant variety protection and plant breeders' rights are UPOV Convention (International Union for the Protection of New Varieties of Plants) and TRIPs Agreement (Trade-Related Aspects of Intellectual Property Rights). The Government of India enacted the Protection of Plant Varieties and Farmers' Rights (PPV and FR) Act in 2001. This was done to protect the rights of farmers and breeders as intellectual property rights and to encourage development of new plant varieties. This Act is an efficacious *sui generis* system developed by integrating the rights of farmers, breeders, researchers, and local communities. It also provides protection to new plant varieties, extant varieties, and essentially derived varieties. This Act grants the following nine rights to farmers—access to seed; benefit sharing; compensation in case of nonperformance; recognition and reward for contributing to conservation;

Crop Sustainability and Intellectual Property Rights. Soumya Mukherjee, Piyali Mukherjee & Tariq Aftab (Eds)

protection against an infringement action. Proper implementation of PPV and FR Act will not only provide food and nutritional security but also uplift and empower the farmers of the nation.

14.1   INTRODUCTION

Agriculture plays an important role in India's economy and generates a lot of employment. Almost 70% of the population is engaged in agriculture and is a major contributor to the country's economic output (around 25% of GDP). The role of farmers in the conservation of agrobiodiversity and innovation in farming practices is being increasingly recognized both by the Governments and the society (https://www.fni.no/getfile.php/131801-1469869136/Filer/Publikasjoner/FNI-R0606.pdf). Plant Genetic Resources are the basic blocks for crop improvement. They are the "heritage of mankind" that helps in the development of a nutritionally secure society (Brahmi et al., 2004). From ancient times farmers have been conserving, maintaining, improving, and making plant genetic resources accessible for agricultural development. Although most farmers do not have a formal training or knowledge of genetic diversity and breeding, yet their cumulative knowledge and effort in saving, selecting, and maintaining seeds have been the building blocks for crop improvement. Farmers' contribution is as important as that of plant breeders in ensuring food security and sustainability. Most modern crop varieties incorporate genetic content derived either from farmers' traditional varieties or their wild relatives. However, in several instances, small farmers' communities have been deprived of the benefits of cultivating improved varieties. Modern plant breeding is also based on the germplasm provided by the farmers and small farming communities. The contribution of farmers toward crop improvement has been recognized by several countries by granting rights that ensure maximum benefit and support (FAO.org).

14.2   GLOBAL SCENARIO

Agriculture was considered as an area of industrial enterprise for which Property Rights could be secured for the first time at the Paris Convention, 1883. United States of America was probably the first nation to bring out a suitable legislation by enacting The Plant Patent Act in 1930. This Act recognized and protected the rights of developers of new plant varieties (excluding sexually and tuber propagated plants) through patents.

Presently, the two major international systems that regulate the agreement to protect plant varieties and plant breeders' rights under the umbrella of Intellectual Property Rights (IPR) are the UPOV Convention (International Union for the Protection of New Varieties of Plants) and TRIPs Agreement (Trade-Related Aspects of Intellectual Property Rights). UPOV is an independent international intergovernmental organization with headquarters at Geneva (Switzerland) (UPOV.int). UPOV was founded by the International Convention for the Protection of New Varieties of Plants in 1961 at Paris and came into force in 1968. Subsequently, it was revised thrice (1972, 1978, and 1991) to account for scientific advancement in the area of plant breeding (Maurya, 2016). The protection provided by the UPOV is also subject to the United Nations Convention on Biological Diversity. The objectives of UPOV are to provide a legal basis for the protection of new plant varieties as well as to encourage their development for the welfare of society. The UPOV convention gives rights only to breeders of the member countries. Farmers, agriculturists, cultivators, and conservers of biodiversity are given no protection under this convention (Jain, 2015). UPOV provides protection of new plant varieties (commercialized for up to 1 year only), which are Distinct, Uniform, and Stable (DUS) (https://www.upov.int/overview/en/conditions.html). DUS is the set of criteria for which new varieties are tested. A variety is considered "Distinct" if it distinguishable from all other known varieties by one or more significant characteristics. It is "Uniform" if the plant attributes remain consistent amongst several individuals within the variety. It is "Stable" if the plant attributes are genetically fixed and thus persist unchanged over the subsequent generations, or after one reproduction cycle in the case of hybrid varieties (Blakeney, 2012).

As per the UPOV convention, all signatory nations must enact legislation to protect breeders' rights and the new plant varieties. In compliance with these obligations, several nations have brought in such laws, notable being the Plant Variety and Seeds Act, 1964 of the United Kingdom and the Plant Variety Protection Act, 1970 of the United States of America. These were intended to encourage the development of new, sexually reproduced plant varieties and hybrids (Jain, 2015).

The TRIPs Agreement of World Trade Organization specifies minimum standards for protection of various forms of intellectual property. According to the TRIPs provision outlined in Article 27.3 (b), all signatories are required to provide protection for plant varieties through patents or a *sui generis* system or any combination thereof. TRIPs Agreement was signed by India in 1995, opting for a *sui generis* regime over the patent mode. The Protection of Plant Varieties and Farmers' Rights (PPV and FR) Act was enacted in 2001. The reason to opt for this regime was that The Patents Act, 1970 of India excludes agricultural and horticultural methods from the scope of patentability. Moreover, a *sui generis* regime gives greater flexibility to frame legislations suited to the country's system (Brahmi et al., 2004).

Most of the international treaties like UPOV and TRIPs for establishing IPRs in agriculture focus only on breeders' rights. The idea of farmers' rights was first conceptualized in international negotiations of the Food and Agriculture Organization in 1986 and developed further through the International Treaty on Plant Genetic Resources for Food and Agriculture in 2001 (http://www.fao.org/plant-treaty/en/). International Treaty on Plant Genetic Resources for Food and Agriculture acknowledges farmers' rights and binds the member countries to promote and protect these rights. As per this treaty farmers' rights are basically about enabling farmers to continue their work as stewards and innovators of agricultural biodiversity, and about recognizing and rewarding them for their contribution to the global pool of genetic resources" (https://www.farmersrights.org/index.html). It is widely accepted that farmers' rights go a long way toward conservation of genetic resources and ensure sustainable development. These also ensure food security at the global level and are pivotal in the effort to eliminate rural poverty in developing nations.

14.3 INDIAN SCENARIO

14.3.1 THE SEEDS ACT, 1966

The Seeds Act was implemented by the Government of India in 1966 to give a legal framework for seed certification and to ensure the availability of good quality seeds to the farmers. The rules were notified in 1968 and systematic seed certification was initiated in 1969 in the country. This Act also provided for the establishment of a Central Seed Committee. This Committee advises the states in seed-related matters. The Act has provision for the establishment of Seed Certification Boards, Seed Certification Agencies, and Seed Testing Laboratories by the states as well (https://seednet.gov.in/PDFFILES/Seed_Act_1966.pdf).

14.3.2 THE PROTECTION OF PLANT VARIETIES AND FARMERS' RIGHTS (PPV AND FR) ACT, 2001

In order to meet the obligation as a signatory to the TRIPs Agreement, the PPV and FR Act was passed by the Government of India in 2001 (https://www.plantauthority.gov.in). The Rules were framed in 2003. India is one of the pioneer nations in the world to enact such a legislation. This Act provides protection to the rights of farmers and breeders in the broad realm of intellectual property rights and promotes the development of new plant varieties. India's PPV and FR Act, 2001 is special and unique as it aims to protect the rights of farmers, breeders, researchers, and small indigenous communities as well.

The PPV and FR Act promotes agricultural development, protects plant varieties, and encourages investment in the development of new plant varieties. This in turn facilitates the advancement of the seed industry and guarantees the availability of high-quality seed and plant material to the farmers (http://www.plantauthority.gov.in/about-authority.htm). This Act has 11 chapters and 97 sections. The first chapter outlines 27 important definitions used in the Act. The next nine chapters are about the PPV and FR Authority, plant variety registration, benefit sharing, farmers' rights, compulsory license, appellate tribunal, financial matters, infringement, surrender, and revocation of certificate. The final chapter deals with miscellaneous matters (Brahmi et al. 2004).

14.3.2.1 OBJECTIVES OF PPV AND FR ACT, 2001

a. To establish an effective system for the protection of plant varieties, the rights of farmers and plant breeders, and to encourage the development of new varieties of plants.

b. To recognize and protect the rights of farmers in respect of their contributions made at any time in conserving, improving, and making available plant genetic resources for the development of new plant varieties.

c. To accelerate agricultural development in the country, protect plant breeders' rights; stimulate investment for research and development both in public and private sector for the development of new plant varieties.

To facilitate the growth of seed industry in the country which will ensure the availability of high-quality seeds and planting material to the farmers (http://www.plantauthority.gov.in/about-authority.htm).

14.3.2.2   REGISTRABLE PLANT VARIETIES IN INDIA

The following three types of plant varieties can be registered under the Act.

a. **New Variety**—A variety developed by any agency or individual that has not been in trade or in use as a breeding line for more than 1 year if it is from India and for more than 4/6 years (in case of trees/vines only) if it is from any other country, from the date of submission of application.

b. **Extant Variety**—A plant variety already in existence but not for more than 15 years in India at the time of its notification. The extant variety category facilitates those varieties to be put under protection that are not novel, and are represented by any one of the following:

   ➢ **Variety of Common Knowledge**: An existing variety known to be in trade for 1–15 years in India;
   ➢ **Variety notified under The Seeds Act, 1966**: Registration valid from the date of notification to a total of 15 years;
   ➢ **Farmers' Variety**: A variety that has been traditionally culti-vated and evolved by the farmers in their fields; or is a wild relative or landrace of a variety about which the farmers possess the common knowledge;
   ➢ **Any other variety**: A variety that is in public domain.

c. **Essentially Derived Variety (EDV)**—A variety is considered to be essentially derived from such initial variety when

   ➢ it is predominantly derived from such initial variety while retaining the expression of the essential characteristics that are due to the genotype of such initial variety;
   ➢ it is clearly distinguishable from such initial variety;
   ➢ it conforms to such initial variety that results from the genotype of such initial variety.

(Ahuja 2013 and http://plantauthority.gov.in/pdf/FinalNewFAQ23.02.2021.pdf)

14.3.3 MAIN PROVISIONS OF THE PPV AND FR ACT 2001

14.3.3.1 AUTHORITY

Section 3 of the Act provides for the establishment of PPV and FR Authority to implement the provisions of the Act. This Authority was established in 2005 and comprises of a chairperson and 15 members from concerned ministries, departments, industries, and organizations).

General Functions of the Authority (Section 8)

1. Registration of new plant varieties, EDV, extant varieties;
2. Developing DUS test guidelines for new plant varieties;
3. Developing characterization and documentation of registered varieties;
4. Compulsory cataloging facilities for all varieties of plants;
5. Documentation, indexing, and cataloging of farmers' varieties;
6. Recognizing and rewarding farmers, communities of farmers, particularly tribal and rural communities engaged in conservation and improvement;
7. Preservation of plant genetic resources of economically important plants and their wild relatives;
8. Maintenance of the National Register of Plant Varieties;
9. Maintenance of National Gene Fund.

Ensuring that seeds of the registered variety are available to the farmers. (http://www.plantauthority.gov.in/pdf/G_Brochure_New_English.pdf)

14.3.3.2 REGISTRATION OF VARIETIES

A Plant Variety Registry has been established under Section 12 and a National Register of Plant Variety is maintained at the Registry. The names of all the registered varieties and other details of the breeder are entered in the Register. The criteria for registration (Section 15) of a new variety are Novelty (not required for an extant variety), Distinctiveness, Uniformity, and Stability. So far, 157 crop species have been notified for the purpose of registration. The process of registration of a new variety involves application, DUS testing, acceptance, advertisement, call for opposition, registration upon fulfillment of required formalities, and Issuance of a certificate of registration (Section 18) (http://www.plantauthority.gov.in/pdf/G_Brochure_New_English.pdf).

14.3.3.3   TERM OF PROTECTION

The registration certificate of the new varieties (Section 24) is initially valid for 9 years in the case of trees and vines and 6 years in the case of other crops. It can be renewed for the total period of validity which is 18 years in case of trees and vines and 15 years in case of other crops, from the date of registration. For new hybrid varieties, the period of protection is 15 years for annuals and 18 years for perennials if none of the parents involved is previously registered. If one or more of the parents is previously registered, the validity period of the hybrid is that of the period of protection of the earliest registered parent (http://plantauthority.gov.in/pdf/FinalNewFAQ23.02.2021. pdf).

14.3.3.4   RIGHTS GRANTED BY THE ACT

The PPV and FR Act recognizes the contribution towards crop improvement not only of plant breeders but also of researchers, communities, and farmers and gives them the following rights (Brahmi et al. 2004 and http://www. plantauthority.gov.in/pdf/G_Brochure_New_English.pdf).

14.3.3.4.1   Breeders' Rights

Under Section 28 breeders or their successors have exclusive rights to:

   a.   produce, sell, market, distribute, import or export the protected and registered varieties;
   b.   appoint agents/licensees;

seek remedies in case of infringement. (http://www.plantauthority.gov.in/ pdf/G_Brochure_New_English.pdf)

14.3.3.4.2   Researchers' Rights

Under Section 30 researchers have the following rights:
   a.   access to protected varieties for bonafide research;
   b.   use of any of the registered varieties for conducting experimental or research work; use of a registered variety as an initial source for

the purpose of developing another variety (repeat use needs prior permission of the registered breeder). (http://www.plantauthority.gov.in/pdf/G_Brochure_New_English.pdf)

14.3.3.4.3   Communities' Rights

Under Section 41 local communities are entitled to compensation for their contribution in the evolution of new varieties registered under the Act. Claims for such a contribution can be filed at any notified center in the country by any person/group of persons/governmental or nongovernmental organizations on behalf of any local community. (http://www.plantauthority.gov.in/pdf/G_Brochure_New_English.pdf)

14.3.3.4.4   Farmers' Rights

Under Section 39 farmers have the following rights:

a. entitlement to registration and other protection for farmers involved in breeding or development of new varieties, similar to the rights of breeders;
b. entitlement to registration of farmers' varieties;
c. entitlement to recognition and reward from the National Gene Fund for farmers involved in the conservation of plant genetic resources; entitlement to save, use, sow, resow, exchange, share and sell farm produce of a protected variety, except sale under a commercial marketing arrangement (branded seeds). (http://www.plantauthority.gov.in/pdf/G_Brochure_New_English.pdf)

Farmers' rights are discussed in detail in Section 3.4.

14.3.3.5   COMPULSORY LICENSE

Section 47 of the PPV and FR Act provides for the grant of compulsory license under certain circumstances. Authority can grant compulsory license to any person interested in the production, distribution, and sale of the seeds or other propagating material of the plant variety. However, this can be done only after the completion of 3 years of registration. The grant of compulsory license is subject to the following conditions:

a. if the requirement for seeds/propagating material of the variety is not being met;
b. if it is being marketed at an unreasonably high price.

Compulsory license is granted only after hearing the registered breeder and granting extension of time for him to produce adequate quantities of registered varieties to meet the demands of public (Brahmi et al. 2004 and http://plantauthority.gov.in/pdf/FinalNewFAQ23.02.2021.pdf).

14.3.3.6   BENEFIT SHARING

PPV and FR Act 2001 (Section 26) provides benefits to the farmers/individuals/organizations who make available the genetic material used in the development of new varieties. The PPV and FR Authority publishes the certificates of registration in the *Plant Varieties Journal of India* (PVJI) for inviting claims for benefit sharing. These may be submitted by any Indian citizen/firms/nongovernmental organizations formed in India. The claims have to be filed within six months from the date of advertisement in the PVJI. The breeder has to deposit an amount in the National Gene Fund proportionate to the extent and nature of the use of genetic material of the claimant in the development of the new variety, its commercial utility and demand in the market. The amount deposited by the breeder is transferred to the claimant from National Gene Fund (http://www.plantauthority.gov.in/pdf/G_Brochure_New_English.pdf).

14.3.3.7   NATIONAL GENE FUND

Under Section 45 of the Act the National Gene Fund has been constituted by the Government of India. It is an aggregation of funds from the following sources:

a. the benefit sharing amount received from the breeder;
b. the royalty component of the Annual Fee paid by the registered breeder/his agent;
c. the compensation money paid by the defaulter registered breeder/agent/licensee that is granted to the farmer/community by the Authority;

d. contribution made by any national or international organization and other sources.

The Gene Fund is to be utilized for the following:

a. disbursing shares to benefit claimants (individuals/organizations) and for compensation to village communities;
b. for the support of conservation and sustainable use of genetic resources;
c. enhancing the capabilities of the panchayat for promoting conservation measures (Brahmi et al. 2004).

14.3.3.8 *PLANT VARIETY PROTECTION APPELLATE TRIBUNAL (PVPAT)*

Section 54 of the Act provides for the establishment of the PVPAT. The Tribunal consists of a chairman, judicial members, and technical members. The Tribunal addresses appeals relating to orders or decisions of the Authority regarding:

a. registration of a variety;
b. registration as agent/licensee,
c. benefit sharing, revocation of compulsory license, and payment of compensation.

The PVPAT is required to settle the appeal in 1 year's time. The ruling of the PVPAT may be challenged in the High Court. (https://www.plantauthority.gov.in/).

14.3.4 *FARMERS' RIGHTS IN THE INDIAN PPV AND FR ACT, 2001*

The PPV and FR Act seeks to bring the rights of farmers at par with those of the breeders. The Act takes cognizance of the various roles played by farmers in conserving, cultivating, selecting, and making Plant Genetic Resources available for creating new plant varieties. The Act also acknowledges the value added by farmers through the identification and selection of useful attributes of traditional varieties and wild species. Farmers are granted the following specific rights described below (Bala Ravi, 2013).

RIGHT 1: ACCESS TO SEED (SECTION 39)

Farmers are entitled to save, use, sow, re-sow, exchange, share, or sell their farm produce, including seed of protected varieties except branded seeds. The quantity of seed of a protected variety that can be saved by a farmer from his own cultivation is not specified in the Act.

RIGHT 2: BENEFIT-SHARING (SECTION 26)

Entitlement of all Indian farmers/individuals/organizations providing Plant Genetic Resources to breeders for developing new varieties to receive a reasonable share of the commercial gains of the registered varieties. Thus, PPV and FR Act is the first Act in India that integrates a provision for rights to access and benefit-sharing for the farmers along with the rights of the plant breeders. Legal acquisition of the plant genetic resource to be used in breeding is not covered in the PPV and FR Act but falls under the provisions of The Biological Diversity Act, 2002.

RIGHT 3: COMPENSATION (SECTION 39)

Farmers are entitled to claim compensation through the PPV and FR Authority if a variety or the seed/propagating material does not meet the expected performance, as had been claimed by the breeder. The expected performance of a variety is to be revealed by the breeder to the farmers at the time of sale of seed/propagating material.

RIGHT 4: REASONABLE SEED PRICE (SECTION 47)

Farmers are entitled to access the seed of registered varieties at a reasonable price under the PPV and FR Act. The breeder's right over the variety may be suspended under the provision of Compulsory Licensing if the condition of a reasonable price is not met. The breeder is then bound to license the seed production, distribution, and sale of the variety to an authorized legal entity. These provisions assure the farmers of an adequate seed supply at a reasonable price.

RIGHT 5: FARMERS' RECOGNITION AND REWARD FOR CONTRIBUTING TO CONSERVATION (SECTION 39 AND 45)

Farmers are entitled to recognition and reward from the National Gene Fund for contributing toward conservation of genetic resources in the form of landraces and wild relatives of economically important plants and their improvement, if their material is used as donor of genes.

RIGHT 6: REGISTRATION OF FARMERS' VARIETIES (SECTION 39)

Farmers are entitled to the registration of Farmers' varieties that satisfy the requirements of distinctness, uniformity, and stability. However, the Act does not include the requirement of novelty for the registration of Farmers' variety. Once their varieties are registered, farmers have the same rights as those of the plant breeders.

RIGHT 7: PRIOR AUTHORIZATION FOR THE COMMERCIALIZATION OF ESSENTIALLY DERIVED VARIETIES (SECTION 28)

Farmers are entitled to the negotiation of terms of authorization with any third party (regarding royalties, one-off payments, benefit-sharing, etc.) that wishes to use a Farmers' variety as source material for the development of an EDV. Also, prior authorization from the farmer is required for its commercialization.

RIGHT 8: EXEMPTION FROM REGISTRATION FEES FOR FARMERS (SECTION 44)

To ease the burden of expenditure the Act provides for the exemption to the farmers from

- a. payment of any fees/payment payable for variety registration;
- b. DUS testing;
- c. other services rendered by the Authority;
- d. legal expenses with regard to infringement or appeals etc.

RIGHT 9: PROTECTION FROM INNOCENT INFRINGEMENT (SECTION 42)

In view of the traditional rights enjoyed by the farmers over the cultivation of all varieties and inadequate legal awareness, the Act provides that infringement by a farmer will be deemed as innocent infringement if the farmer can prove that he/she was unaware of the existence of these rights. Under this provision the farmer will not be charged for infringement. (http://www.plantauthority.gov.in/pdf/F_Brochure_English.pdf, Bala Ravi, 2013).

14.3.5 HOW INDIAN LEGISLATION IS EMPOWERING THE FARMERS

The PPV and FR Act, 2001 is an effective *sui generis* system, which provides protection to new varieties, extant varieties, and essentially derived varieties of plants. By not opting for a patent-based legislation for plant variety protection India has been able to incorporate provisions of benefit sharing, compensation, rewards, and recognition for the farmers and local communities. Considering the high percentage of the population engaged in farming in our country, PPV and FR Act ensures benefits to a large number of people. This Act aims at balancing the rights and interests of farmers, breeders, researchers, and local communities. The Act also ensures the availability of good quality seeds of registered varieties at reasonable prices to farmers. It also aids the cause of conservation and sustainable use of landraces and traditional varieties by the local communities. To ensure that benefits of such a unique legislation flow down to each and every farmer and farming community of the country, the government has linked institutions like Krishi Vigyan Kendras, Agricultural Universities, Technology Development and Management Centers and Research Institutions with the PPV and FR Authority (http://www.nbpgr.ernet.in). This effort not only goes a long way in making the farmers aware of their rights over plant varieties but also in their commercialization.

KEYWORDS

- **farmers' rights**
- **plant genetic resources**
- ***sui generis***
- **plant varieties**
- **PPV and FR Act**

REFERENCES

Ahuja, V. K. *Law Relating to Intellectual Property Rights*; 2nd ed.; LexisNexis: Gurgaon, 2013.

Bala Ravi, S. P. Farmers' Rights, Their Scope and Legal Protection in India. In *Community Biodiversity Management: Promoting Resilience and the Conservation of Plant Genetic Resources*; de Boef, W. S., Subedi, A., Peroni, N., Thijssen, M., 'Keeffe, E.; Routledge: London, 2013.

Blakeney, M. Patenting of Plant Varieties and Plant Breeding Methods. *J. Exp. Bot.* **2012,** *63* (3), 1069–1074.

Brahmi, P.; Saxena, S.; Dhillon, B. S. The Protection of Plant Varieties and Farmers' Rights Act of India. *Curr. Sci.* **2004,** *86* (3), 392–298.

http://plantauthority.gov.in/pdf/FinalNewFAQ23.02.2021.pdf

http://www.fao.org/3/X0255E/x0255e03.htm#:~:text=Their%20purpose%20is%20to%20 encourage,landraces)%20or%20wild%20crop%20relatives

http://www.nbpgr.ernet.in/Training_Management_PGR/Compendium/06_Protection_of_ Farmers_and_Community_Rights_RC_Agrawal.pdf

http://www.plantauthority.gov.in/pdf/F_Brochure_English.pdf

http://www.plantauthority.gov.in/pdf/G_Brochure_New_English.pdf

https://www.farmersrights.org/index.html

https://www.fni.no/getfile.php/131801-1469869136/Filer/Publikasjoner/FNI-R0606.pdf

https://www.upov.int/overview/en/conditions.html

https://seednet.gov.in/PDFFILES/Seed_Act_1966.pdf

Jain, A. K. *Intellectual Property Laws-II,* 3rd ed.; Ascent Publications: Delhi, 2015.

Maurya, A. K. *Basic Intellectual Property Rights Law*; Book Age Publications: Delhi, 2016.

CHAPTER 15

From Green Revolution to Green Innovation: How IP and Trademarks Catalyze Commercialization of Agriculture and Plant Products?

BIBIN GEORGE VARUGHESE

Subject Matter Expert—Trademarks and Copyrights; Product Marketing/ Brand Consultant for Consumer Brands; Faculty, Digital Marketing—T.A. Pai Management Institute, Manipal, Karnataka, India

ABSTRACT

This chapter primarily explores various aspects of achieving sustainable agriculture, that is, the gamut of factors that influence commercialization or creating scalable economic value for plant products. It further discusses an important catalyst that can power such innovation—Intellectual property Rights (IPR) and trademarks. Plant products have long been underrated for multiple reasons, thus having realized only a fraction of its potential value. In the first part, we try to understand why this value has been realized only for a select variety of plants and products derived from them. Drawing references from data and precedents, we build a strong case for the existence of a potential green mine; one that can be commercially exploited to greatly improve the economy and therefore the per capita income of people in those countries. And such economic growth can in turn ensure the sustenance of agriculture itself. The next section of this chapter seeks to educate the readers the two concerns of scaling up—Firstly, the question of incentive for

Crop Sustainability and Intellectual Property Rights. Soumya Mukherjee, Piyali Mukherjee & Tariq Aftab (Eds)

© 2023 Apple Academic Press, Inc. Co-published with CRC Press (Taylor & Francis)

hundreds of hours spent on R&D in terms of academic and monetary credit. Secondly, the imminent skepticism among native people that the fruits of such opportunities, economic and beyond, will move away from its rightful owners to fulfill vested interests of bigwigs and developed countries. The chapter, in its final leg, compiles the existing IPR and different classes of trademarks around the world (among other regulations). In addition, we also discuss their role in giving fillip for the journey from green revolution to green innovation and further catalyzing sustainability.

15.1 INTRODUCTION

Ever wondered why agriculture, being such an important driver of the economy, still fails to scale up? What makes agriculture so lucrative as a contributor to the economy in western countries, whereas the very sector is still unorganized and is yet to realize even half of its potential in a country as diverse in soil and climate such as India? Further, why do we as nations across the globe struggle to get the right form of nutrition from the food we consume?

As an economics enthusiast at heart, I tend to draw connections between any aspect of science or art and its contribution to the economy of a country. So keeping that in mind I would love to shed some light on the market size and contribution of the agricultural sector across various countries and the role of exports and high quality farm produce as a driver for the same.

Further in, we explore the various aspects that contribute to the flourishing of agriculture while delving deep into the most important factor—use of scientific approach toward realizing such superior output and the downsides of various traditional, unorganized and unscientific repetitive practices. We will then explore the various developments and stories in recent times with the opportunities and challenges that reveal themselves at each stage. We will also take the help of various data points from around the world to add a strong bedrock to our contentions.

As we progress through the chapter, we will discuss Intellectual property Rights (IPRs), it's various types and how each law helps in various stages from produce to purchase. We then move on to discuss why patent filing is not very actively pursued and the ones filed don't see acceptance as much as they should.

To top it all, we will discuss an important catalyst that drives the amplification of such scientific practices and how it is inevitable to keep such a catalyst aside.

I should caution you that too many technical details and data or charts have been avoided in this chapter as the primary intent is to cater to all proficiencies of readers.

It is 2021, and the agriculture sector is playing a crucial role in the world GDP. From 2018 to 2020, the agricultural value-added up to $3.4 trillion, an almost 68% hike between 2 years. The sector employed approximately 27% of the world population while adding 4% of the total GDP of the world in 2018.

In the last few years, the radical transformation in the agriculture sector is supported by advanced technology and artificial intelligence. According to research by McKinsey, in the next decade, the agriculture sector will be among the seven crucial sectors contributing $2–$3 trillion in added value to world GDP.

The recent trends in the agriculture sector are evident and powered by data science, genetic engineering, and advanced machines. However, the advancement is not parallel to every part of the world. The scenario is uneven compared with developed countries, developing countries, and underdeveloped countries.

15.2 CONTRIBUTION OF AGRICULTURE SECTORS IN THE WORLD ECONOMY

According to the Statistics Times, China was ranked first followed by India in the GDP contribution in the agriculture sector in the year 2018. The developed countries like America have only 0.9% contribution, but their productivity was much higher than China or India.

However, the top nine countries where agriculture dominates the overall GDP of the countries are Sierra Leone, Gibraltar, Macau, Monaco, and Singapore.

Precisely, the latest trend in technology and modern tools is changing the whole scenario in the next decade. Moreover, in the last 10 years, investment in agriculture technology has witnessed high growth. Almost $6.7 billion investment has been made in the past 5 years, and the number will be going to be whooping high in upcoming years.

But agriculture today is a lot more than just planting and harvesting. Agricultural production as a system has developed over the years. Technological advancements have led to many new methods being tried, tested, approved, and used around the globe to maximize agriculture. Thanks to improved farming practices, crops both food and nonfood, can now be grown in previously unsuitable areas.

And over a period of time there has been a strong shift of trends to keep up with the inconveniences (if I may call it so) caused by some of our earlier reckless practices.

For one, pollution has destroyed the fertility of soils. Also, due to deforestation and industrialization, reduced farming land, and shifting the paradigm of agriculture to the new local-based food system, broadening the horizon of agriculture is essential.

And which obviously calls for groundbreaking techniques that have evolved the way farmers cultivate worldwide. Some instances that we are most familiar with are

(a) Hydroponics: Hydroponics allows farmers to grow the crop without using soil. It can be in water or using an oxygen container, materials such as vermiculture, peat moss, coconut fiber, etc., or using daily light are integral.

(b) Precision Agriculture: Using IT in farming and get the correct information about crops and soil is the main feature of precision agriculture. Also known as satellite agriculture, it assists farmers to do site-specific crop management.

(c) Vertical Urban Farming: Since 55% of the world population are from cities, vertical farming is the futuristic way to produce large quantities of food in door. It saves resources and also ensures year-round production.

Apart from these trends, farmers around the globe are adhering to other practices such as genetic modification, 3D printing, bioplastic, algae feedstock, seawater farming, and desert farming. In addition, any new forms of technology such as drones, data analytics, use of the internet, artificial intelligence, nanotechnology, blockchain, etc. are also playing a vital role in helping farmers secure a higher and good quality yield. For some of our readers, the fact that these innovations are actually put into practice may come as a surprise since most of these are theories that we believe exist all in research papers (Or so I suppose).

15.3 HURDLES OF IT REVOLUTION IN AGRICULTURE

Although the internet and data transfer are more accessible than before, many farmers have still not adopted the digital tool for agriculture. In the USA alone, only one-quarter of farms are using data for deciding the prominent issues. Due to bad networks or the lack of availability of these resources, technology, especially those that use the internet, has less credibility.

Current IoT technologies that support 3G and 4G data are not state-of-the-art nowadays. Among other things, Telco in the USA is dismantling its low-band IoT networks and moving toward 3G and 4G. But these are only helpful in limited cases such as monitoring of crops and livestock. Also, awareness of these technologies remains a challenging task in many countries.

Alone in India, the overall level of mechanization is very low compared with the developed countries. The country is far behind the world average in terms of farm mechanization. Although farmers are using technologies to get assistance in harvesting and threshing, the application of such is almost negligible in seeding and planting. Almost 90% of stakeholders of agriculture use mechanization in developed countries, but the rate is <50% in India. This means that Indian farmers are deprived of the most basic of technology-driven tools and practices that help important steps like irrigation. As these technologies are either expensive or unavailable to the farmers in rural areas. The poor condition of basic infrastructure in most regions of India poses logistics challenges. Also, the farmers in India have an inertia in adapting to the technological changes and rely heavily on existing yet less efficient manual labor to plant and harvest crops. These and many other tariffs/nontariff issues have lessened the contribution of India in agricultural exports when compared with the developed countries.

15.4 TRENDS OF AGRICULTURAL EXPORTS AROUND THE WORLD

Food availability in a country is either by its production or by importing agricultural products, including crops and noncrops from other countries. Over the last decade, agricultural trade has seen a steady rise worldwide due to increased transport facilities, storage, and other technological advancements. The global economy is changing rapidly and agricultural trade influences a significant part of the global GDP.

The last 70 years have witnessed significant growth in agricultural trade. Export trends over the past decade have soared high due to uneven crop growth and the production of agricultural products in different countries. There was a rise seen in such exports during the 1970s, which came to a temporary halt in the 1980s only to kick off again later.

Since early 1970, the export of fishing and forest produce has also grown by about 75% and its value increased from $148 to $580 sometime by 1997.

Low-income or underdeveloped countries depend greatly on agricultural trade and exports from other countries to sustain themselves. In return, they have other resources to trade, which may or may not be agricultural. Their foreign exchange revenue solely depends on that. Today, Latin America and the Caribbean are a major exporter of agricultural commodities among developing countries, while Asia and Africa are net importers. Australia, France, the United States, etc. on the other hand gain majorly out of the export earnings.

Western Europe and its countries conduct one-third of the global agricultural trade just among themselves. North America and Western Europe are leading export countries while Asia is one of the major importers. The Common Agricultural Policy of the EU launched in 1962 proved successful over the next few years leading to a decline in the import market in Western Europe and an increase in the export of commodities such as cereals and sugar.

It is also important to note at this point that there are two significant dimensions of the agricultural trade, which we often see: Horizontal diversification by the product and vertical diversification by the stage of processing of the product. And therefore it is implicit that only a careful coherence between both will help amplify the trade.

Moving to more recent times, the export of various rice varieties, sugar, and other spices increased during the last few years worldwide. Growing agricultural trade affects the overall development of a nation. However, an imbalance of trade is a negative developmental outcome in countries predominantly dependent on export earnings. Which meant that innovation in agriculture and the increased and improved use of technology and scientific methods right from tilling to selling (these produce) was key to have robust crop production, protection from pests utilizing sustainable farming techniques.

So if technology and innovation are such a key factor in how agriculture can scale up whether it is for subsistence or cash crops, the next part is but an indispensable part in helping further the cause.

15.5 INTELLECTUAL PROPERTY (IP): PATENTS AND TRADEMARKS IN THE AGRICULTURAL SECTOR

The biggest of any innovation is the right of ownership and the pleasure of carrying forward that legacy. Well that surely means income as well but I will keep remuneration aside for the time being; a by-product of ownership and legacy. Which means every innovation whether by an individual or by an organization needs to be backed by enough legislative right so there is no unauthorized access/tamper/infringement/adaptations by others in the ecosystem. So how far has this been implemented with regard to agriculture? And what are the boons and banes that such rights have harnessed for agricultural innovation in practice?

IPRs are rights given to people over the creations of their minds. They usually give the creator an exclusive right to use his/her creation for a certain period of time. IPR can be broadly divided into patents, copyrights, trademarks, or even trade secrets. In some countries, geographical indications are also very relevant in being able to market certain indigenous varieties of crops/products.

But we will explore only two classifications of IPR—patents and trademarks as these are most relevant in our context.

As per most patent bodies across the globe, patents for an invention can be defined as the grant of a property right to the inventor." In the US, a patent filed with the government is valid for a period of 20 years from the date of filing the same. But what does it cover and what are its limitations? Broadly, speaking, anyone who "invents or discovers any new and useful process, machine, manufacture, or composition of matter, or any new and useful improvement thereof, may obtain a patent and is covered under the ambit of this law."

Trademarks on the other hand, are "words/phrases/symbols/designs, or even combination of the former; one which distinguishes the source of the goods of one business from those of the others." For instance, your business' name, logo, or your company tagline can be trademarked under the various classes of business you operate in. The trademark also serves as a unique brand identifier for your company/products. What's even better? You can now trademark images, slogans, and colors making it much helpful during the sale of products and specifically with regard to a brand, a logo, or a certain kind of packaging as in the example of Coco-Cola bottles.

IP Protection in agriculture was only limited to the Plant Breeders' Rights. It was far from sufficient even at the time. The way it worked was that the department of agriculture in a country's ministry issued a Plant Breeders' Rights certificate to a seed owner that restricted others from selling seeds of the same variety in that country. However, they could have the same seeds as part of breeding programs to safeguard their unique varieties. The United States of America broadened the horizon of IP wherein regular patents could be filed for plants, seeds and other biological plant components, making the laws more restrictive than Plant Breeders' Rights.

In 1980, the Bayh-Dole Act moved towards stricter laws that would allow universities to obtain patents of their inventions and further monetize them under government grants. Although this proved to be beneficial for quite some time, later there began to occur clashes among universities to use each other's patented inventions for research purposes. As a result, laws changed a little to foster technological advancements in agriculture by incorporating patents from nonprofit institutions.

15.5.1   WHO BENEFITS FROM THE IP IN THE AGRICULTURAL SECTOR?

Agricultural innovations are necessary to sustain the ever-rising global population and to ensure food security worldwide. For the innovations to be successful and safe, patenting is the priority. But who tends to benefit from the patents?

Many stakeholders are involved in agriculture like farmers, their agents, suppliers, food distributors, environmental protection agencies, etc. In the initial stage, due to the patent, the prices of the seeds may encounter a sudden hike, making it difficult for farmers to buy them. So, it might seem like they would be less likely to purchase a particular variety. However, in the long run, the high prices could coax the competitors to develop their seed hybrids and increase the demand in the market.

15.5.2   INDIAN SCENARIO

IP protection act in agriculture, especially biotechnology, has a serious implication on stakeholders and agriculture innovation. The act has helped developed countries having an IP monopoly on the agriculture sector. As it enables giant agro-based companies and undermines plant breeders' rights over yields, it has been criticized by farmers and plant breeders.

In India, there is a dual form of IP protection in agriculture. It includes agriculture biotechnology under protection of plant variety and Farmers Rights Act 2001. Also, the Patents Act 1970 protects the genetic traits of plants. Also, the trademark law in India is applicable within the Indian Territory. It means that any trademark registered in India is effective in India alone and, for its protection outside, it needs to be registered in another country too.

It is also pertinent to understand what are some of the different types of IP rules and statutes that have been in place.

15.5.3 DIFFERENT LAWS AROUND IP AND TRADEMARK

According to the definition of World Intellectual Property Organization (WIPO), a trademark is any sign that distinguishes the products of one organization from its competitors. It can be a word, phrase, or symbol that symbolizes a particular organization. The Olympics has five rings while Sony Corp. Japan owns the trademark "AIWA." Some of the other trademark laws are:

(a) **Trade-Related Intellectual Property (TRIPS) Agreement**
   In 1994, the TRIPS agreement was signed in Uruguay by most nations to safeguard their plant varieties. Thus providing fillip to companies across the world to sell seeds without the risk of being copied.

(b) **UPOV treaties**
   The UPOV treaty is a sui generis system of protection (a unique system, or of its kind), especially tailored to the needs of plant breeders. The USA and Europe together have assorted agricultural patents. Similarly, China has patented its famous hybrid rice.

15.6 HOW DIFFERENT COUNTRIES ARE HANDLING PATENT LAW

According to the WIPO study of ASEAN countries' reaction to patent Law, the Republic of Korea is at par with G5 countries such as France, Germany, Japan, and the UK, followed by Singapore. In 2008, Singapore applied for 250 patents, whereas in India, the resident share of the total patent application was only 20%. However, among ASEAN countries, Malaysia and India are leading the patent applications.

15.6.1 LIMITATION OF INDIAN PATENT LAW

Patent Law many times protects monopoly in the market. To curb this, several countries have adapted several forms of exceptions/limitations against IPRs that ensure sustainable development. India, after signing to

TRIPS also evolved some exceptions and limitations as directed by TRIP Article 30 of TRIPS. The "Exceptions to Rights Conferred" allows member countries the rights to some extent where they can confer the limitations if it is not prejudicing the legitimate interest of any third party.

Section 48 of the Indian Patent Act 1970 provides the rights of the patentee. Section 84, Section 85, and Section 90 of the Indian Patent Act 1970 talk about exceptions.

15.6.2   A GLANCE AT THE EXCEPTIONS AND LIMITATIONS

1. Private and noncommercial use exception prevents monopoly over commercial activity. Through a facility called compulsory license, the government allows the third party to use the patented invention. It is helpful when the patented invention is either not available in India or has an unreasonable price to use.
2. Experimental and scientific use exception lies under section 47 of the act and allows scientists or inventors to use the patented process or products merely of experiments. It allows researchers to do their activities without infringing the patent holder's legal rights.
3. Regulatory or prior use exemption under Section 107A offers a trade-off between innovators and consumers.

15.6.3   PEPSICO VS. GUJARAT PLANT BREEDER

In India, the government discourages the patent of agricultural products to safeguard the rights of plant breeders and farmers. The Plant Variety Protection and Farmers Rights 2001 is the big step to help agriculture stakeholders to work freely without any fear of monopoly in the market. The famous case of PepsiCo vs. Gujarat farmers is a relevant example of this. PepsiCo has filed a suit against nine potato farmers in Gujarat and asked `10.5 million from each farmer stating the violation of infringement on a variety of FC5 potato. Ultimately PepsiCo withdrew the case as India does not support patents in plant variety.

15.6.4   MONSANTO VS. BT COTTON

The second interesting case was Monsanto, the American seed company against Gujarat farmers. Monsanto is a giant agriculture-based company having a monopoly on agriculture globally. In 2008, the Indian government granted patent number 214436 to Monsanto for "Monsanto technology." Later in 2015, many seed companies collaborated with Monsanto through sublicense agreements to produce seeds, using Monsanto technology.

Later, Monsanto canceled the agreement in a year and sued Nuziveedu and other seed companies in India claiming illegal use of its technology to produce Bt cotton seeds. Bt cotton seed is immune to pink bollworm as this genetically modified seed can produce an enzyme to resist the pink bollworm. However, Nuzuveedu and other seed companies protested as the patent was against section 3 (j) of the Indian Patent Act. The section prohibits all plants and animals other than microorganisms, including seeds, varieties, and species, to be patented. The act also restricts all essential biological processes that help production and propagation of animals and plants against the patent.

The courts overruled Monsanto claims stating it cannot assert any patent rights on the seeds that have been into production through an essentially biological process.

15.6.5   *TRENDS IN FILING OF PATENT APPLICATIONS*

For instance, India witnessed a gradual increase in patenting activity between 2005 and 2012. A total of 3718 patents were applied, and 1041 patents were granted, as per the data available on the Indian Patent Office weblink. A significant increase in patenting activity was highlighted during this period in different sectors of agriculture.

On the top of the leaderboard in patent applications, with a contribution of 60%, are Biocides, pest repellants or attractants, and plant growth regulators. This was followed by new plants or processes for obtaining them, at a distant second place (9.35%), animal husbandry, silk rearing or breeding new animal breeds coming third (7.48%) and horticulture, cultivation, forestry (5.91%) completing the lot.

A noticeable amount of diversification was also addressed during 2005–2012, as dairy products and animal husbandry registered a 19.0 and 5.78 times rise in patenting activity, compared with the 1995–2004 period. The development of new plants and processes (10.87 times) and horticulture and cultivation forestry increased (5.87 times) also rose significantly during this period. You may also observe in Table 15.1 the number of patents granted in agriculture as a percentage of the total.

TABLE 15.1 The Number of Patents Granted (in Total) and in Agriculture in the Last Decade.

Year	Patents granted	Agricultural patents
2005–2006	4320	29
2006–2007	7539	33
2007–2008	15,316	244
2008–2009	16,061	343
2009–2010	6168	111
2010–2011	7509	150
2011–2012	4280	74
Total	61,193	984

And such a trend could be due to multiple reasons:

(a) Big corporations and developed countries holding the reins to most of the innovations, which prevents anybody from doing research in a related space.
(b) The laws are too straining and individuals who do not have access to robust legal teams may get stuck, often having applied but giving up during the time of objections and defense.
(c) The sheer number of years it takes to get a patent approved specially in countries like India (now 4–5 years and earlier almost 10 years) is not much of an encouragement either.

15.6.6 PATENT TRENDS FROM THE WORLD

In recent times, the number of patents in various areas of basic agricultural research has grown exponentially. In the USA alone, the number of patents related to rice remained below 100 per year before 1995. The numbers increased dramatically during 1999 and 2000, with more than 600 patents registered annually.

The numbers were even greater for crops like corn, given the greater commercial interest of the crop in the west. And what may come with surprise is that, about three-quarters of plant DNA is held by private firms, as per a survey report published in Nature. And almost half of these patents are with 14 multinational firms.

China, for instance, has seen a significant surge in the number of such applications over the past two decades.

According to a report released by the Chinese Academy of Agricultural Sciences, more than 612,000 agricultural patents were applied by Chinese institutes and enterprises between 2015 and 2019. This was the largest in the world and the number of patent applications grew by 7.6% on an average annually during the 5 years. The research was carried out among 22 countries on agricultural patent competitiveness. It found that China alone accounted for 62% of the total agricultural patent applications of all 22 countries, including the United States, the United Kingdom, Japan, and Canada.

Of all patent applications of China, about 11% became authorized patents. Patents are mostly rejected on the grounds of nonfulfilment of patentability criteria–invention, nonobviousness, and industrial applicability. The other three criteria of patent rejection are morality, public order, or human rights considerations of each country.

15.6.7 ISSUES WITH PATENT APPLICATION: IS IT THAT DIFFICULT TO GET ONE APPROVED?

Well having worked with organizations to help file trademarks and copyrights, I can only imagine the process gets tougher as we cross over to patents. Simply because there is a lot of nitty gritties to be considered and supporting documents to be presented compared with the former.

Inventors usually make three common patent application mistakes, which keeps them waiting. Any of these is sufficient to result in their application rejection in the United States Patent and Trademark Office (USPTO) Application Assistant Unit's initial review. This rules out the possibilities of the patent not being able to reach before the examiner.

The three common mistakes are:

1. Mistaking discoverability of patents: This one mistake sometimes puts researchers in a dark spot because they would not realize that the patent they filed for was not something new that was created but just the latest discovery of something, which already existed
2. Missing some of the requirements that the appropriate Patents office of the country provides for every inventor including drawings related to the invention.

3. A concise description of the drawings that comply with the MPEP prescriptions is not included, which makes the process difficult.
4. The inventor has not communicated the fees they are ready to pay to the USPTO, and/or does not pay the stated fees.
5. Application of patents in a foreign country

Other rare mistakes are:

1. Putting an invention on the wrong table may result in the dearth of making a product permanently unpatentable. This happens because no patent or legal protection is in place and when the inventor discloses an invention publicly, they have 1 year to file a patent application.
2. Failure to include enough details in the patent application is the most onerous mistake. In the case of New Railhead Mfg., L.L.C. vs. Vermeer Mfg. Co., the applicant had failed to explain the particular nature of one of the components in their provisional application. This, along with previous sales that caused the public announcement of the device, rendered the invention unpatentable. It is important that the inventor describes each part of the invention in detail and by accurately using exact wordings.
3. One must consult a patent lawyer before publishing the patents as they might guide through the nitty-gritties of patent filing making the patent filing experience smooth.
4. The inventor must be able to explain their invention to the examiner. Examiners are mostly interested in understanding the structure of the invention.

15.6.8 HOW CAN A LEGAL TEAM HELP IN PATENT APPLICATION?

Having the help of a legally compliant person to prosecute patents is an added advantage. A person having specific technical as well as legal knowledge is capable of drafting patents efficiently. The knowledge of both domains helps in fulfilling the obligations of patent prosecution. In India, a patent can be prosecuted through a registered Indian patent agent.

As per Section 127 of the Indian Patents Act, 1970, a patent agent is qualified for patent prosecution and is entitled to practice before the Controller. This involves preparing the necessary documents, conducting all business transactions and completing all such functions as prescribed in accordance with any proceeding before the controller within the ambit of this Act.

15.6.9 WHAT MAKES PATENTING AN ARDUOUS TASK IN INDIA AND THE WORLD?

India, in 2005, had brought about significant changes in the Patents Act to bring it on par with the World Trade Organization requirements. And at this point the count of pending patent applications stood at around 56,171. What's interesting (or rather concerning) is that the number of such applications by resident Indian researchers has not risen significantly since then. On the other side, India's patent office has a stockpile of pending applications, which have not been processed.

Picture this in a country that is the fifth largest economy in the world with over 1.2 million businesses and the largest exporter of talent in science and technology. The high technology exports were only a meager 0.81% of total exports (as of 2014) and the fact is that India still relies heavily on other countries in terms of agriculture exports.

India does not have robust IP laws and just like most other laws, creating ones that deliver the required result is another complex problem. One of the major challenges in doing so is to strike a balance between the needs of a large population and the rights of patent holders. The situation intensifies as foreign companies file more patent applications in India than the local, Indian-grown companies. In 2018–2019, foreign applicants filed 32,304 patent applications, while Indian residents filed 15,550 applications.

India has been lagging immensely in its efforts to boost R&D and innovation. Patenting activity in the country makes it evident, as India does not figure in the top five patent offices of the world. However, much smaller countries like Japan and South Korea shine in this space.

IP Applications in India (Last 5 years)				
Application	2014–2015	2015–2016	2016–2017	2017–2018	2018–2019
Patent	42,763	46,904	45,444	47,854	50,659
Trade mark	210,501	283,060	278,170	272,974	323,798

In the Global Intellectual Property Index, India ranks 40th out of 53 countries. However, over the past few years, a gradual increase can be seen in India's IP applications. India's overall score increased from 36.04% in 2019 to 38.46% in 2020.

15.7 THE GLOBAL PICTURE

The forum of the world's five largest IP offices is known as IP5. To improve the efficiency of the examination process for patents worldwide the forum paves the way for having a global patent system by sharing the patent data.

The members of IP5 are

1. The European Patent Office.
2. The Japan Patent Office.
3. The Korean Intellectual Property Office.
4. The National Intellectual Property Administration of the People's Republic of China.
5. The USPTO.

Forum working group members are drawn from these IP5 offices. WIPO undertakes quality surveys, handles financial issues, observes key metrics, and coordinates with IP5 offices.

Around 80% of the world's total patent applications come to the IP5 offices alone. A whopping 95% of all work is carried out under Patent Cooperation Treaty.

Almost US $1.7 trillion was spent globally on R&D, an all-time high. Intriguingly, 80% of that spending is accounted for by the top 10 countries. China's R&D expenditures reach $372 billion annually, with more than 77% coming from the private sector, according to the UNESCO Institution for Statistics. During the same period in India, R&D investments totaled $52 billion, with only 37% of that coming from private sources.

The list of top 50 filers globally did not have even a single Indian company, while 13 Chinese companies were featured in the list.

15.8 PATENTS ON PLANTS

Companies have been making continuous property claims on plants and also with full vigor. Take for instance, *the PepsiCo vs. Gujarat Plant Breeder case* has been evident of the dramatic consequences it can have on farmers around the globe.

There's an ongoing global trend of companies making property rights on plants or generic materials of plants. Companies are privatizing the resources that were available in the market to mankind as a community.

In the 1980s, the United States first initiated the idea of patenting living materials, followed by the Western countries. In 1990, the number of plant patents worldwide was a mere 120, which is more than 12,000 today. About 3500 of these patented plants are registered in Europe, as per the data available from a European initiative No-Patents-On-Seeds.

Patenting a plant reserves the exclusive rights of breeding, growing and selling the product. It restricts the rights of a farmer from sowing, planting, harvesting or breeding that variety without permission.

There have been instances when patented pollens were spread by the wind causing an accidental intermixing of genetic material found in different fields. Such an event could call for legal action by the patent holder.

In 2004, Canadian farmer Percy Schmeiser was sued by the multinational company Monsanto as he replanted saved soybeans after one harvest without the company's permission. As per the claims made by the farmer, his field had been contaminated years ago by genetically modified pollen. The court's decision came in favor of Monsanto. However, the court did not ask Schmeiser to pay any damages to the company as the patented genes were found in a small amount in his crops ruling out any possible advantages.

15.9 WHAT IS BIOPIRACY?

The agrochemical company Monsanto argues that having patent laws is necessary as they guarantee finances for new inventions. The development of newer and better technologies will be affected without protecting them.

According to critics, farmers' access to genetic material is blocked by patents, reducing biodiversity, species diversity, and increasing farmers' dependence on seed producers.

The companies involved in bio innovations say that farmers have the liberty to buy from whichever supplier and use the products of a particular supplier only if they see a clear advantage.

Several years ago in Europe, a case became media headlines that involved Monsanto and a certain breed of melon. Monsanto discovered an Indian variety of melon was naturally resistant to a specific virus. It successfully applied for a patent in Europe after breeding into other melons.

By developing this trait, Monsanto became the sole owner of all melon varieties as well as the original Indian melon variety. Organizations having an otherwise view on bio patenting calls this practice biopiracy.

Later on, EU institutions revoked the patent stating that the natural trait cannot be considered an invention.

15.9.1 IS IT A THREAT TO FOOD SECURITY?

Mordor Intelligence, an India-based market research agency, suggests that revenue in the seed sector will grow from $60 billion in 2018 to $90 billion by 2024. Bayer-Monsanto, DuPont, and Syngenta hold over 50% of the worldwide market share.

As outlined by Bram de Jonge from Oxfam Netherlands, the ever-growing number of plant patents and the rapid changes in the seed industry is putting farmers' rights on threat as the UN outlines how to store, use, exchange and sell seeds or crops from their own harvest.

A report published by the UN--The right to food, has raised concerns about food security. It issues a warning stating that it could increase food prices and deprive the poorest of this basic necessity.

An NGO German watch has stated that most of the seed-producing industries are from Global North while 90% of agricultural products and natural materials come from Global South.

With patenting laws more restrictive in the Global South region, big companies tend to find loopholes for economic gains.

15.9.2 THE ROAD FORWARD FOR DEVELOPING COUNTRIES

A developing country's social and economic development depends upon the use of science and technology. The rice–wheat rotation technique in South Asia has made a significant contribution to the improvement of food production.

Attracting financial investments is crucial to developing countries. Paying more attention toward fundamental research brings world recognition and encourages FDI. A country that is capable of making fundamental discoveries is slated to have an exponential manpower and having great manpower is one of the basics of a business's success.

Countries must look for public-private partnerships in harnessing research and development. Taking a cue from the developed countries, developing countries must make policy changes that must be in sync with common and developmental goals.

To create an innovative society, schools of higher quality, research incentives for universities, and research that is relevant to the private sector are central to the supply side of knowledge.

15.9.3   WHY IS INNOVATION IMPORTANT FOR DEVELOPING AND EMERGING COUNTRIES?

Innovation has played a pivotal role in economic development over the past two decades. Successful developing countries have long emphasized the importance of innovation capacities in their growth dynamics. In order to address the local challenges, these countries need to develop innovation capacity at the local level. In order to foster growth, a successful development strategy must build extensive innovation capacity.

15.9.4   ROLE OF GOVERNMENT LAWS, PUBLIC PRIVATE PARTNERSHIPS IN AMPLIFYING GREEN INNOVATION IS IMMENSE

Over the years, we have seen the role of government laws and intervention has helped make or break any scientific progress. And this means right from cropping to commercialization of farm produce, progress is best effected and amplified with government support. Some instances of such support being (1) encouraging laws that regulate the process, product, and the markets, (2) grants for mass adoption of new technology whether it be machinery, or (3) newer varieties of seeds, subsidies on newer varieties of fertilizers, seeds and other resources.

Because besides the monetary support it also gives the stakeholders (farmers, industries, middlemen to name a few) the notion that the government is also poised towards the same goals as they are. And such like every other change in behavior, this too gets picked up. Just like how digital payments have come to the forefront in many countries mostly because people see government acceptance to the same.

So why can not the government do it all and is there a need to involve private players? Well as history has shown us, especially in recent times the best results of infra, innovation, tech show us that the prowess of capitalist mindset (to get an incentive for every goal) combined with the trust that people place on their governments and the laws of the land has resulted in winning partnerships in the last 50 years.

Well that leads us to explore what are some of the areas in agriculture where PPPs can best function.

I have compiled a list of what I think some of the top areas of collaboration (1) Joint R&D initiatives from produce to purchase (with joint IPR); (2) Proliferating of the public research output to private sector for commercialization via a licensing model; (3) Public sector infrastructure for testing of private-sector products for agronomic performance and bio-safety; (4) Committees, chairs and fellowships hosted and sponsored by Private sector in public institutions; (5) Establishment of incubation centers, biotechnology parks before commercialization is highly desirable to promote such partnership; and (6) Incentivizing adoption of newer technologies especially the indigenously created ones into the existing products, businesses to ensure they experimented in practice and not just in labs Having said that, every PPP effort poses inevitable and commonly seen challenges in terms of common visions, operational overheads, change of government, and consequent focus areas to state a few.

The adoption of PPP policies in these areas is subject to high investment, accomplished resources, training, and change in agricultural practice patterns for the poor farmers. Such upgrades become infeasible in terms of the real-life socioeconomic conditions of the farmers.

Consequently, PPP policies are limited to high-value produce only. It creates a significant competitive gap in the agro market and leaves scanty scope to the poor farmers in receiving any benefits from it. For this reason, PPP growth in agro-industries is comparatively slower than in other commercial areas. For example, in India, 0.6% is the current GDP portion from agriculture invested in agricultural R&D and education. This percentage needs to increase to at least 1.0% (as per Planning Commission 2011) to get the desired benefit in the global agro market.

But pinning our hopes on the success of PPPs in Singapore with the completion of the Tuas Desalination Plant in 2005 and its eventual development; Or the Australian PPP to revamp their transport system, especially their railways, which have been world class. Furthering the strategy over to developing nations, the example of the services sector in India receiving the largest FDI inflow of US $87.06 billion from April 2000 to March 2021. The sector has played a crucial role in India's economic growth. In the financial year 2020, the sector contributed a massive 55.39% to the nation's Gross Value Added. According to the Reserve Bank of India, India's service exports in February 2021 stood at US $21.17 billion and imports at US $10.61 billion during the same time.

From a comparison of India's GDP growth to that of the U.S. and China, it becomes clear that the contribution of the Service sector to employment and income distribution is unique. Eventually, the strength of the manufacturing and agricultural sectors in India leads to the growth of the service sector. The Service sector has become extremely important to our economy, both as a growth engine and as a source of job creation.

Therefore, I am bullish about the results that Public-Private Partnership can bring in the field of agricultural innovation making countries around the world advance in scientific innovations, higher quality crops and therefore drive the journey from green revolution to green innovation.

KEYWORDS

- **agricultural trademarks**
- **research**
- **marketing agricultural products**
- **agricultural IPs**
- **state of agricultural produce around the world**
- **benefits of IP in agriculture**

REFERENCES

Agricultural patent analysis during 2005-2012 in India (https://www.researchgate.net/publication/264835795_Agricultural_patent_analysis_during_2005-2012_in_India) accessed on 4.1.2023

Agricultural trade, trade policies and the global food system (http://www.fao.org/3/y4252e/y4252e11.htm) accessed on 4.1.2023

Bayh Dole Act (https://en.wikipedia.org/wiki/Bayh%E2%80%93Dole_Act) accessed on 4.1.2023

Future how technology can yield new growth (https://www.mckinsey.com/industries/agriculture/our-insights/agricultures-connected-futurehow-technology-can-yield-new-growth) accessed on 4.1.2023

https://www.fastscience.tv/insights/india-innovation-patents/) accessed on 4.1.2023

Issues With Patent Applications (https://www.obrienpatents.com/common-patent-mistakes/) accessed on 4.1.2023

List of Countries by GDP Sector Composition The World Fact Book (https://statisticstimes.com/economy/countries-by-gdp-sector-composition.php) accessed on 4.1.2023

Nation now leader in agricultural patent applications (https://www.chinadaily.com.cn/a/202011/21/WS5fb84c57a31024ad0ba95877.html) accessed on 4.1.2023

Patent life cycle (https://www.uspto.gov/patents/basics#heading-2) accessed on 4.1.2023

Patenting Agriculture (https://issues.org/barton/) accessed on 4.1.2023

Patents on Plants Threaten farmers Tim Schauenberg (https://www.dw.com/en/patents-on-plants-is-the-sellout-of-genes-a-threat-to-farmers-andglobal-food-security/a-49906072) accessed on 4.1.2023

Salient trends in world agricultural production, demand, trade and food security (http://www.fao.org/3/x2996e/x2996e.htm#P67_7344) accessed on 4.1.2023

Sustainability and environmental aspects of agriculture (http://www.fao.org/3/cb1329en/online/cb1329en.html#chapter-4) accessed on 4.1.2023

The Strategic Use of Intellectual Property to Enhance Competitiveness in Select Industries in ASEAN (https://www.wipo.int/edocs/pubdocs/en/intproperty/953/wipo_pub_953.pdf) accessed on 4.1.2023

The What and Why of Hydroponic Farming (https://www.verticalroots.com/the-what-and-why-of-hydroponic-farming/) accessed on 4.1.2023

Who Benefits from IP Rights in Agricultural Innovation? Catherine Jewell (https://www.wipo.int/wipo_magazine/en/2015/04/article_0003.html) accessed on 4.1.2023

CHAPTER 16

Delineating the Legal Application and Protection of Plant Patents: A Critical Study

PIYALI MUKHERJEE

Advocate, West Bengal Bar Council, Kolkata, West Bengal, India

ABSTRACT

Patent law is important in intellectual property law, especially when it comes to innovative innovations and distinctive creations. It is a monopoly granted by the government to a patentee in exchange for the benefit that provides to society through the application of his creative faculties. Though the specific origins of patent law are uncertain, the idea that man's intellectual mind may have some type of property stretches back to ancient Greece. This patented procedure can safeguard the owner's legal rights to invention. Patent rights are integrated under private law in many nations, and the patent holder must sue anyone for infringement or enforce their rights. Prior to TRIPS, India's patent policy excluded life and its components. TRIPS' provision that all disciplines of technology be safeguarded resulted in biotechnology being protected in India. This chapter summarizes the grant of plant patent, procedure, application, and legal protection of the patent in agricultural aspects.

Crop Sustainability and Intellectual Property Rights. Soumya Mukherjee, Piyali Mukherjee & Tariq Aftab (Eds)

© 2023 Apple Academic Press, Inc. Co-published with CRC Press (Taylor & Francis)

16.1 INTRODUCTION

Patent law plays a pivotal role in the intellectual property law that includes unique creation and new inventions. It divaricates throughout the world by giving the owner's legal rights for their innovations. It concedes a property right by a supreme authority to an inventor. In exchange for a comprehensive disclosure of the innovation, the inventor receives exclusive rights to the patented process, design, or invention for a specified length of time. They are a type of intangible property right. This patented procedure, on the other hand, can preserve the owner's specific legal rights to creation. Patent rights are integrated under private law in many nations, and the patent holder must sue anyone for infringement or enforce their rights (WIPO Intellectual Property Handbook: Policy, Law and Use, 2013).

The United States government will grant a plant patent to an inventor who discovers a distinct and new type of plant or a plant found in an uncultivated unusual state. A plant patent protects the part features of a novel and unique plant from being copied, marketed, or utilized by others. By restricting competitors from using the plant throughout the patent protection term, a plant patent can help an innovator earn more money (Plant Patent, 2019).

In India, any part of any plants is prohibited under patent law. Any methods through a genetically modified organism (GMOs) are also purely biological. The only noteworthy exception was microorganisms. Therefore, according to Article 27.3 of the TRIPS Agreement allowed governments to exclude biological objects, although microorganisms and nonbiological processes were not exempted from the patents law.

A patent is a monopoly granted by the state to reward the use of the Patentee's creative faculties for the benefit of the society. The capacity to profit from new ideas is at the heart of endogenous growth theories, and patent protection is frequently cited among a set of institutions that are crucial for long-term development (Romer, 1990; Acemoglu and Robinson, 2012).

Although most discoveries have been protected by patents since the founding of the United States, agricultural variants, such as seeds, have been refused because living organisms and genetic material are not considered patentable subject matter. (Kloppenburg, 2005). This chapter looks into the impact, legal application, and protection of plant and patent rights on technology development, productivity, and technique around the world.

16.2 EARLY HISTORICAL BACKGROUND AND PROTECTION OF PATENT

Plant biotechnology was granted patent protection in the United States in 1985, and it had varying effects on crops based on their reproductive architecture. The idea that there might be some type of property in man's intellectual labor extends back to the ancient Greek period, although the specific origin of patent law is uncertain (Fisher, 2007). The first formal system for granting patents for "new and inventive things" was created in 1474 by a Venetian legislation (Ashok, 2015).

The Venetian Statute of 1474 is frequently regarded as the commencement of patent history and patent rights (Schippel, 2001). Patents were frequently granted in Venice in the year 1450. They did, however, establish a legislation requiring the Republic to be notified of new and inventive equipment to receive legal protection against infringement of rights. The shield was in place for 10 years (Frumkin, 1945). The English patent system evolved from its early gothic origins to become the first modern patent system to recognize restorative inventions as intellectual property rights. The legal implementation of these rights could usher in the Industrial Revolution.

Patents for inventions were awarded and documented in the States General of the United Provinces of the Netherlands' deed books beginning in 1589. There were also "patents for trademarks and manufacturers' trademarks," such as the right to sell brooches made of a green piece of paper with an angel's image on it (History of the Patent System).

In the Netherlands, the first Patent Act was passed in 1817. Patent rights could be valid for 5, 10, or 15 years under this Act. It was necessary to file detailed descriptions of the innovation. When the applicant arrived to "collect" the patent, that is, when the applicant paid the fee and obtained a certified copy of the patent, these became public. The Act was repealed in 1869, and the Netherlands gained a reputation as a free-spirited nation. Following that, in 1883, over 140 countries convened in Paris to compare national acts covering industrial property (patents, trademarks, and designs) in order to create uniformity. Later, the Berne Convention in 1886 was one of numerous international accords for the protection of intellectual property rights.

16.3 INTERNATIONAL FRAMEWORK ON PLANT PATENT

In 1930, the United States became the first country to give plant patents. In 1931, Henry Bosenberg got the first plant patent for his ever-blooming climbing rose. According to the patent law, the inventor is the person who first appreciates a plant's unique characteristics and reproduces it asexually. To put it another way, a plant can be created through breeding or grafting (Stim, 2021).

A plant identified in a cultivated region that is owned can be copyrighted if it is discovered by anyone in that cultivated area. An inventor or the inventor's heirs who have invented or discovered asexually reproduced a separate and novel variety of plants, other than a tuber propagated plant or an uncultivated state of a plant, are issued a plant patent by the US government. This grant preserves the patent owner's right to prevent anyone from asexually reproducing the plant for 20 years from the date of filing the application. It is from using, offering for sale, or selling the plant thus reproduced, or any other plant parts, within the United States, or from importing the plant so reproduced.

The Plant Patent Act of 1930 is a federal statute in the United States that allows for the patenting of new plant types, excluding sexual and tuber-propagated plants. "The UPOV 1961 treaty and the enactment of the US Plant Variety Protection Act of 1970 occurred during the development of plant breeders' rights, which coincided with controversy over intellectual property and its relationship to human health, food security, and environmental development, the UPOV 1961 treaty and the enactment of the US Plant Variety Protection Act of 1970 (Halbert, 2005)." After April 24, 2019, the US Department of Agriculture will accept applications for plant variety protection for industrial hemp (Cannabis sativa) (Agricultural Marketing Service, 2019). Yet, none has been issued; thus breeders have turned to the Plant Patent Act of 1930 for intellectual property protection, such as PP31918 Cannabis plant entitled "RAINBOW GUMMEEZ" on June 30, 2020 (Bosse, 2020).

16.3.1 PROVISION AND LIMITATION OF PLANT PATENT

Plant patents, which are stable and reproduce asexually, are rare as well as other edible tuber reproduced plants are provided by Title 35 US Code, Section 161, which states: "A patent may be obtained by whoever invents or discovers and asexually reproduces any distinct and new variety of plant, including cultivated sports, mutants, hybrids, and newly discovered seedlings, other than a tuber propagated plant or a plant discovered in an uncultivated state, subject to the conditions and requirements of this title."

"Except as otherwise stated, the provisions of this section related to patents for inventions apply to patents for plants."

The plant patent must also satisfy the broad standards of plant patentability, according to the statute. The application's subject matter would be a plant that the inventor has invented or discovered and has proven to be stable through asexual reproduction as follows:

- The plant was invented or found in a cultivated form and asexually reproduced to be patented.
- The plant is not a statutory exclusion, such as a potato, where the part of the plant employed for asexual reproduction is not a tuber food part.
- The person who actually originated the claimed plant, that is, the person who discovered, developed, recognized, or isolated the plant and asexually reproduced it, must be mentioned as the inventor in a plant patent application.
- With certain exceptions, the plant was not patented, in public use, for sale, or otherwise available to the public prior to the effective filing date of the patent application.
- The plant has not been described in a U.S. patent or published patent application, with a few exceptions.
- At least one distinguishing characteristic of the plant must be shown, which is more than a difference caused by growing conditions, fertility levels, or other factors.
- As of the effective filing date of the claimed plant invention, the invention must not have been obvious to a person of ordinary skill in the art.

"A knowledgeable patent practitioner should be consulted before filing an application for a plant patent if there is any uncertainty about the patentability of a specific plant (USPTO, 2021)."

16.3.2 APPLICATION PROCEDURE AND PATENT GRANT IN UNITED STATES

The systematic criteria for a plant patent application are outlined below. Anyone seeking a patent should, however, check the USPTO website just before filing an application to ensure that no additional requirements have been added and that the fees filed with the application represent the current amount payable, as these are subject to change. The USPTO may reject an incomplete application, resulting in the loss of intellectual property rights. Loss of rights may also occur if applications are not filed in a formal manner. Prospective applicants should be aware that plant patent applications filed under 35 U.S.C. 161 are currently not permitted to be filed electronically.

Applicants for plant patents should be aware that they have the option of hiring a registered patent agent to assist them with their applications. A patent practitioner must be hired if the applicant is the assignee (legal entity). While the USPTO cannot assist with attorney or agent selection, anyone pursuing a plant patent should utilize a USPTO-registered attorney or agent.

In the last three decades, several plant patent applications have been submitted in Unites States Patent & Trademark Office (USPTO) by patent holders as follows (U.S. Patent Statistics Chart, 2021):

TABLE 16.1   Percentage of Plant Patent Application Submitted and Plant Patent Grant.

Year	Plant patent application submitted	Plant patent Grant	Plant patent Grant (%)
1991–2000	575.5	428.8	74.5%
2001–2010	1089.1	986.9	90.6%
2011–2020	1152.7	1110.3	96.3%

In Table 16.1, here is the increasing number of plant patent applications submitted and plant patent application granted over 30 years.

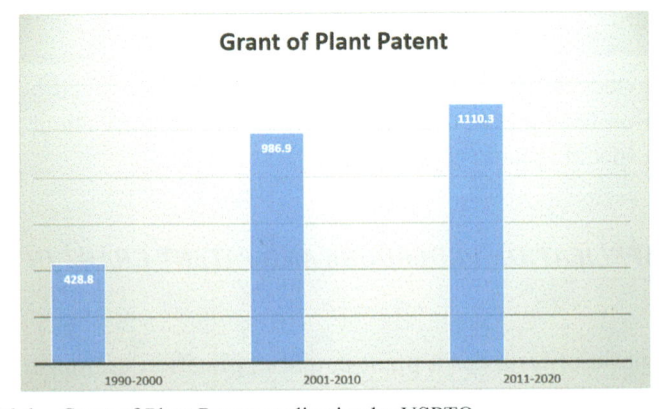

FIGURE 16.1   Grant of Plant Patent application by USPTO.

In the year of 2020–2021, most of the plant patent applications are granted by USPTO compared with other two previous decades and there is also an increase in the number of patent granting. In the current scenario, extensive increase in granting plant patent is seen.

The World Intellectual Property Organization (WIPO) released an annual report titled, "World Intellectual Property Indicators," which contains a wide range of indicators covering all elements of intellectual property. The data for these indicators comes from a variety of national and regional IP offices. Since 2009, the reports have been issued annually until now (World Intellectual Property Indicator, 2018).

16.3.3   PLANT PATENTABILITY AND CLAIMING PROCEDURE

Each country is required to establish such systems under the Agreement on Trade-Related Aspects of Intellectual Property Rights (TRIPS), albeit there are local variances that are explained below:

16.3.3.1   PATENT CLAIMS ON PLANTS

In some jurisdictions (such as the United States, Australia, and Europe), patent claims can cover plants, if the patent applications meet all of the country's patentability standards and restrictions. Under the Trade-Related Aspects of Intellectual Property Agreement (TRIPS), which binds World Trade Organization members, member countries that choose not to provide such mechanisms for plants under their national patent system must provide an alternative way for an entity to claim a legal right to intellectual property in plants and plans, to the partial exclusion of others.

Utility Patents: The terminology "utility patent" is used to differentiate between patents and other types of intellectual property claims that can be found in some jurisdictions.

Plants and inventions directed to plants or plant products (e.g. seed) are not patentable in most countries, but some (such as the United States and Australia) allow an entity to claim a time-limited right to exclude others from using plants and plant products if the legal criteria for patentability are met. While each jurisdiction's patent legislation differs, the conditions and privileges granted to patentees are quite similar (A Green Industry Guide to Plant Patents and Other Intellectual Property Rights, 2020).

Patent grants in practically all European countries can be obtained by filing at the European Patent Office and registering the approved claims in national patent offices, in addition to getting a patent through the patent offices of specific European countries. The European Patent Convention regulates this method of patenting (EPC). As a result, each member country must bring its national legislation into agreement with the Directive, even if some may not be entirely compliant yet.

(i) **Types of claims that are permissible:** Exclusionary rights in various forms of plant can be claimed using utility and standard patents as follows:

- new varieties of plants (U.S. only),
- transgenic plants,
- plant groups,
- individual plants and their descendants,
- particular plant traits,
- plant parts,
- plant components (e.g. specific genes or chromosomes),
- plant products (e.g. fruit, oils, pharmaceuticals),
- plant material used in industrial processes (e.g. cell lines used in cultivation methods),
- reproductive material (e.g. seeds or cuttings),
- plant culture cells,
- plant breeding methodologies, and
- vectors and processes involved in the production of transgenic plants.

(ii) **Inclusions and Exclusions:** Natural source material is not original because it should not be covered by patents as exclusionary rights in any country around the world. Individual plant varieties are patentable in the United States and Australia. Individual plant varieties are not patentable in Europe. A plant that is distinguished by a single gene (rather than its entire genome) does not fall under the concept of a plant variety and is thus patentable.

Transgenic plants are patentable in Europe if they are not limited to a single plant variety but rather represent a broader plant grouping. Plant cells are considered "microbiological products" by the European Directive and therefore patentable (Plant patent, 2013).

16.3.3.2   *INTERNATIONAL ASPECTS FOR PROTECTION OF PLANTS*

Looking at the various national laws, we can observe that the United States has one of the oldest laws preserving plant types. The Plant Patent Act of 1930 allows for the patenting of all novel plant types, with the exception of sexual and tuber-propagated plants. For a period of 20 years from the date of filing for the patent, the Act grants the patentee the right to prohibit others from asexually reproducing the plant, from using, offering for sale, or selling the plant so reproduced in the United States, or from importing it into the United States (United States Patent & Trademark Office, 2021). As a result, plant patents have a fairly similar scope and duration of protection as other patents. In 1970, the Plant Varieties Protection Act was enacted, which supplemented this regulation. This Act gives "breeders of sexually reproduced or tuber propagated plant varieties an exclusive right to exclude others from selling, offering for sale, reproducing, importing, exporting, or using it in producing a hybrid or different variety over his new, distinct, uniform, and stable variety for a period of 25 years for trees and vines and 20 years for other varieties." The Act also includes exemptions such as research exemptions, the right to keep seeds, and other provisions that appear to be missing from the 1930 Act. As can be observed, this Act is very similar to the UPOV Convention of 1991. These Acts are reinforced by the Patent Act of 1956's judicial interpretations.

16.3.3.3   *PROTECTION OF PLANT PATENT*

In 1930, the United States Congress passed the Plant Patent Act. It was created largely to help the horticulture sector by fostering plant breeding and increasing genetic diversity among plants.

(i) Limited types of plants are eligible for protection: The Plant Patent Act was amended on October 27, 1998, to extend the exclusive right to plant parts obtained from protected varieties; however, the amendment is not retroactive. All asexually reproduced plants are protected by the Plant Patent Act of 1930 (35 U.S.C. 161). There are also the following exceptions:

(1) tuber-propagated plants, and (2) plants discovered in their natural state.

Plant patents cover newly discovered plant types, cultivated spores, mutants, hybrids, and newly discovered seedlings that reproduce asexually. Any reproductive process that does not include the union of persons or germ cells is referred to as asexual reproduction. It refers to the process of multiplying a plant without the usage of genetic seeds. Grafting, bulbs, apomictic seeds, rhizomes, and tissue culture are all examples of asexual reproduction in plants.

Tuber-propagated plants and plants occurring in an uncultivated state are specifically excluded from protection under the Plant Patent Act.

(ii) Plant patents have less severe patentability requirements: The conditions for acquiring a Plant Patent are identical to the requirements for obtaining a Utility Patent. Plant patents, for example, are not rejected or invalidated if the plant is not completely described, according to the Plant Patent Act. It is also not required that the claimed plant be deposited with an official depositary.

Plant patents are exempt from the usual "non-obviousness" criteria. This is owing to the fact that developing a new plant variety that is stable and can be preserved through asexual reproduction is incredibly difficult.

(iii) Exclusionary right: For a period of 20 years, the proprietor of a plant patent has the right to restrict anyone from asexually reproducing, using, selling, offering for sale, or importing the reproduced plant (or any of its parts) into the United States.

Plant patents, unlike utility patents, only protect a single plant or genome, and the protection they provide is relatively restricted. It does not protect plant traits, mutations, or cultivation technology related to the patented plant. Because plant patents cover the entire plant, each plant patent can only have one claim.

(iv) Dual protection is allowed: To protect the same plant, both a utility patent and a plant patent can be obtained.

In the United States, it is possible to get protection for the same plant that may be patented under both a utility and a plant patent at the same time, as long as both patents are valid.

16.3.3.4 LEGAL IMPLEMENTATION OF PLANT PATENT IN THE INDIAN CONTEXT

The Indian Patent Act of 1856, which was based on the British Patent Law of 1852, was enacted as a result of Colonial control. It can be used to learn about India's patent history. The system evolved into the current Act, which was passed in 1970 and most recently updated in 2012. Despite this long history, one fundamental aspect that has remained constant is that a patent can only exist with the consent of the sovereign concerned. As can be seen, unlike copyright, which is an automatic grant of protection, certain prescribed conditions must be met before a patent may be recognized, and these restrictions vary by nation. There was no international agreement on what these standards should be in the beginning. This is evident from the fact that despite the fact that the Paris Convention of 1883 stipulates that patents must be protected, the question of what constitutes a patentable innovation has been left unaddressed. Because there was no worldwide mandate, countries had a lot of leeway in deciding what kinds of patents they wanted to award and under what terms they wanted to give them. India is a classic example of a country that exclusively recognizes process patents for pharmaceutical items. With the entry into the force of TRIPS, which established binding international criteria for what constitutes a patentable innovation, this spectacle came to an end. TRIPS stipulates a number of requirements that member countries must meet as a minimum. For example, a patent is only available for an item, including product and process, if it is new, entails advancement above existing knowledge that is not evident, and is valuable (Ashok, 2015). On January 1, 1995, India joined the World Trade Organization (Understanding the WTO: The Organization, 2003). As a member, it was required to adhere to the Trade Related Aspects of Intellectual Property Systems (TRIPS) agreement (Barnes, 2003). TRIPS requires member countries to "provide adequate standards and principles concerning the availability, scope, and use of intellectual property rights, as well as effective means for the enforcement of these rights" (Barnes, 2003). (De Gaulle, 2000) The Patents Act of 1970 governed India's patent system prior to TRIPS (Patents Act). The Patents Act was revised in 1999 and 2002 to begin aligning India's legal framework with TRIPS. In addition, by January 1, 2005, India must make other improvements to become fully TRIPS compliant.

Prior to TRIPS, India's patent policy excluded life and its components. TRIPS' provision that all disciplines of technology be safeguarded resulted in biotechnology being protected in India. Biotechnology makes use of biological materials and their integration with technology to improve human existence. Products or processes of gene engineering technologies, methods of producing organisms, methods of isolating microorganisms from culture medium, methods of mutation, cultures, mutants, transformants, plasmids, processes for making monoclonal antibodies, cell lines for making monoclonal antibodies, and other biotechnological innovations are all examples of biotechnological inventions. These technologies have prompted severe questions about notions such as novelty, utility, and obviousness, as well as a variety of moral and ethical difficulties.

India had not issued a single patent relating to (a) natural or artificial living entities, (b) biological materials or other materials with replicating properties, (c) substances derived from such materials, or (d) any procedures for the manufacturing of living substances, including nucleic acids, until 2002.

The scope of biotechnology in relation to plant patents was further broadened by a 1970 amendment to the Patent Act that included biotechnological and microbiological processes for the purposes of patent grant. This was accomplished by specifically exempting certain sorts of inventions from patent protection. According to the section, microorganisms are particularly included in the patentable subject matter, provided that they meet other conditions of patentability such as novelty, inventive step, and utility. The following subject matter is specifically excluded from the patentable subject matter: (a) entire or partial plants, (b) whole or partial animals, (c) seeds, (d) plant and animal kinds and species, and (e) a biological process for producing or propagating plants and animals. Thus, claims relating to important biological processes of growing plants, seed germination, and developmental stages of plants and animals will be opposed under Section 3 (j) of the Act. This has been accomplished by adopting a distinct statutory procedure and allowing appropriate exemptions to vulnerable groups such as farmers who are not knowingly preserving plant protection outside of the patent framework, which shows little consideration for India's farmers, who remain the country's economic backbone.

16.4   CASE STUDIES

There are several case studies in India as follows:

(1) Turmeric Patent

Turmeric is a tropical herb found on the east coast of India. Turmeric powder is used in India as a medicine, a food ingredient, and a dye, to name a few uses. It is used as a blood purifier, a cold treatment, and an antiparasitic for a range of skin issues, among other things. It is also a key ingredient in the production of many Indian dishes. In 1995, the United States granted a patent on turmeric to the University of Mississippi Medical Center for its wound-healing qualities. The ostensible subject was the use of "turmeric powder and its administration" for wound healing, both topically and orally. An exclusive selling and distribution licence has been granted. The Indian Council for Scientific and Industrial Research (CSIR) has filed a patent objection and provided extensive prior art evidence to the United States Patent and Trademark Office (USPTO). Though it was common knowledge that turmeric had been used in Indian households for centuries, it was a tremendous endeavor to uncover written literature on the use of turmeric powder for wound healing via oral and topical routes.

As a result, the USPTO cancelled the application, claiming that the claims in the invention were obvious and anticipated, and admitting that using turmeric to treat wounds was an ancient art. As a result, India's traditional knowledge was safeguarded in the Turmeric case.

(2) Neem Patent

The first patent for Neem was filed with the European Patent Office by W.R. Grace and the United States Department of Agriculture. The patent in question is for a technique of controlling fungi on plants that involves contacting the fungi with a Neem oil formulation. India has launched a lawsuit to prevent the patent from being granted. A legal challenge to this patent was launched by the Research Foundation for Science, Technology, and Ecology (RFSTE) in New Delhi, in partnership with the International Federation of Organic Agriculture Movements (IFOAM) and Magda Aelvoet, a former green MEP. The Neem tree has a variety of medicinal properties from its roots to the aerial parts that contain powerful compounds, most notably azadirachtin, a chemical contained in its seeds. It is utilized as an astringent in a variety of applications. The neem tree's bark, leaves, blossoms, and seeds are used to treat a variety of illnesses, including leprosy, diabetes, skin problems, and ulcers. Since the dawn of time, neem twigs have been employed as antibacterial teeth brushes. The opponents' evidence included old Indian ayurvedic literature that noted that hydrophobic neem seed extracts had been recognized and used in the past in India for ages, both for treating dermatological problems in humans and for preserving agricultural plants. The patent was cancelled by the EPO due to a lack of novelty, inventive step, and perhaps relevant prior art. Aside from that, multiple US patents for Neem-based emulsions and solutions were recently granted (India: Traditional Knowledge and Patent Issues: An Overview Of Turmeric, Basmati, Neem Cases, 2017).

(3) Basmati patent

The US patent office granted RiceTec a patent for a Basmati rice strain, an aromatic rice that has been cultivated in India and Pakistan for decades. Rice is a staple cuisine in most Asian countries, including India and Pakistan. Over a hundred thousand different types of rice have been developed, fostered, and protected by farmers in this region for generations to suit diverse tastes and demands. Ricetec noted this in its patent application in 1997 "Basmati rice of high grade is historically sourced from northern India and Pakistan. In fact, in certain nations, the name "basmati rice" refers specifically to the rice farmed in India and Pakistan." However, the business went on to say that it had developed "new" Basmati lines and grains "that enable the production of high-quality, higher-yielding Basmati rice over the world." The Indian government only disputed three of the 20 claims in RiceTec Inc.'s original patent application. Only certain claims concerning the properties of basmati rice were challenged. It is important to note that the WTO Agreement does not obligate states to use patents to protect plant varieties. All that is required is for countries to enact legislation to protect plant variety in some other way (not necessarily through patents). The patent application was granted, however, because the United States is a strong proponent of patent protection for plant varieties. RiceTec has been given patent protection for its three strains, allowing them to name their rice "Superior Basmati Rice" (India: Traditional Knowledge and Patent Issues: An Overview of Turmeric, Basmati, Neem Cases, 2017).

As a result, in the Basmati case, RiceTec changed the strain by crossing it with a Western grain strain and successfully claiming it as their own, and the case is an example of this TRIPS' challenges with patenting biotechnological methods.

16.5 CONCLUSION

Agriculture marketing today has experienced a significant move away from conventional agricultural processes, relying instead on contemporary technology, the most important of which being genetically modified plants. The Patents Act lays forth the requirements for patents, including the rights and terms of patents, as well as the types of subject matter that cannot be patented. The Act was not always TRIPS compatible, but the World Trade Organization's Dispute Settlement Body found it to be so. The number of patents on plants and plant components has risen dramatically in many regions of the world during the previous few decades. However, much research has concentrated on industrialized countries, specifically the United States and European Union member states while little is known about the extent to which plants are patented in other areas of the world. Plant patentability is prohibited in at least 51 countries, utilizing one of the key flexibilities provided by Article 27.3 of the TRIPS Agreement (b). Plant varieties and primarily biological procedures for obtaining them, rather than plants themselves, have been excluded by the majority of countries, following the European approach. Plants, seeds, plant varieties, and essentially biological procedures for their production are all exempt from patentability in India, according to the country's laws. Although the patent office's rules urge strict application, a number of patents have been uncovered that suggest considerable latitude in how these requirements are handled. Developing countries may choose a patent system that facilitates rather than restricts access to the building blocks of life to feed a growing global population and best respond to the challenges of climate change.

KEYWORDS

- **patent law**
- **intellectual property rights**
- **legal application**
- **plant patent agreements**

REFERENCE

A Green Industry Guide to Plant Patents and Other Intellectual Property Rights, 2020. https://extension.tennessee.edu/publications/Documents/PB1882.pdf (accessed on Sept 3, 2021).

Acemoglu, D.; Robinson, J. A. *Why Nations Fail: The Origins of Power, Prosperity, and Poverty*; Crown Books, 2012.

Agricultural Marketing Service; U.S Department of Agriculture, 2019. https://www.ams.usda.gov/content/usda-now-accepting-applications-seed-propagated-hemp-plant-variety-protection (accessed on Aug 20, 2021).

Ashok, A. *Plant Patent and Indian IP Regime*; SSRN Electronic Journal, 2015. DOI: 10.2139/ssrn.3086994.

Barnes, S. Pharmaceutical Patents and TRIPS: A Comparison of India and South Africa, 91 KY. L.J. 911, 917, 2003.

Bosse, J. Before the High Court: The Legal Systematics of Cannabis. *Griffith Law Review* 2020. DOI: 10.1080/10383441.2020.1804671

De Gaulle, C. Intellectual Property and Patent in India. *News Time*, 2000. http://www.foxmandal.com/publication/oct2000/oct2kca.html (accessed on Sept 3, 2021).

Fisher, M. *Fundamentals of Patent Law: Interpretation and Scope of Protection*; Hart Publishing: North America, 2007; p 27.

Frumkin, M. The Origin of Patents. *J. Patent Office Soc.* March 1945, *XXVII* (3), 143 et Seq.

Graham v. John Deere Co. of Kansas City, 383 U.S. 1 (1966).

Halbert, D. J. *Resisting Intellectual Property*; Routledge, 2005; pp 87–163. ISBN 9780415429641.

History of the Patent System. https://english.rvo.nl/topics/innovation/patents-other-ip-rights-topic/patent-law/history-patent-system (accessed on March 2, 2021).

India: Traditional Knowledge And Patent Issues: An Overview Of Turmeric, Basmati, Neem Cases, 2017. https://www.mondaq.com/india/patent/586384/traditional-knowledge-and-patent-issues-an-overview-of-turmeric-basmati-neem-cases (accessed on Sept 4, 2021).

Kloppenburg, J. R. *First the Seed: The Political Economy of Plant Biotechnology*; University of Wisconsin Press, 2005.

Online at Library of Congress. A Century of Lawmaking for a New Nation: U.S. Congressional Documents and Debates, 1774–1875. First Congress, Session II, Chapter VII, 1790.

Plant Patent. 2013. https://support.lens.org/help-resources/biological/plant-patents/ (accessed on Sept 2, 2021).

Plant Patent. 2019. https://www.investopedia.com/terms/p/plant-patent.asp (accessed on Feb 26, 2021).

Romer, P. M. Endogenous Technological Change. *J. Polit. Econ.* 1990, *98* (5, Part 2), S71–S102.

Schippel, H. Die Anfänge des Erfinderschutzes in Venedig. In Uta Lindgren (Hrsg.): Europäische Technik im Mittelalter. 800 bis 1400. Tradition und Innovation, 4. Aufl., Berlin, 2001; pp S.539–550. ISBN 3-7861-1748-9

Sichelman, T.; O'Connor, S. Patents as Promoters of Competition: The Guild Origins of Patent Law in the Venetian Republic, 49 San Diego L. ReV. 1267 (2012).

Stim, S. Can I Patent a Plant, Fruit, Seed, or Other Growing Thing? https://www.nolo.com/legal-encyclopedia/plant-patents.html (accessed on July 23, 2021).

Understanding the WTO: The Organization. 2003. http://www.wto.org/english/thewto_e/whatis_e/tif_e/org6_e.htm (accessed on June 30, 2021).

United States Patent & Trademark Office (USPTO); General Information About 35 U.S.C. 161 Plant Patents. 2021. https://www.uspto.gov/patents/basics/types-patent-applications/general-information-about-35-usc-161 (accessed on Aug 27, 2021).

WIPO Intellectual Property Handbook: Policy, Law and Use; Chapter 2: Fields of Intellectual Property Protection, 2013.

World Intellectual Property Indicator. 2018. https://www.wipo.int/publications/en/details.jsp?id=4369 (accessed on Sept 2, 2021).

CHAPTER 17

IP Protection of Traditional Knowledge (TK) and Traditional Cultural Expressions (TCE) in Regions of North-Eastern India: An Ecological Context

SUNANDA BHARTI

University of Delhi, Delhi, India

ABSTRACT

Mahatma Gandhi once remarked when some of his followers in South Africa complained to him about African nudity—"when a large society follows a particular custom, it is quite possible that the custom is harmless enough even if it seems highly improper to the members of another society. We would do well to approach the tribal custom with humility and respect and not judge it by our own standards" (Singh, 2021). One needs to investigate if India has been able to achieve the above for the tribes of Arunachal Pradesh.[1]

For the tribes residing in the north-eastern part of India, there have been two main administrative approaches—First has been the policy of "leave them alone," which was practised by the British government during colonial times. They were inclined, overall to leave the tribesmen alone, partly because the task of administration (especially in the wild border areas) was difficult and unrewarding and "partly because a number of officers sincerely held the view that the people were better and happier as they were" (Elwin, 2009).

---

[1] Arunachal Pradesh is the biggest of the seven north-eastern states of India.

Crop Sustainability and Intellectual Property Rights. Soumya Mukherjee, Piyali Mukherjee & Tariq Aftab (Eds)

The second policy is that of "detribalisation" through which the Christian missionaries and the social reformers interacted with the tribals intending to uplift them and bring them out from their drudgery. "Supporters of this policy still take a poor view of tribal life and maintain that their vices and superstitions should go. The third policy lies between these two extremes of *doing too little* and *doing too much*. The present chapter would analyze this in detail.

In the attempt, the author points out that much needs to be achieved in the context of Indian tribal culture and practices of Northeast India. To begin with, the focus needs to shift to documenting traditional practices and rewarding those who are involved in preserving it through the ages. It has been a travesty that most of the focus, presently, goes on bioprospecting and biopiracy. One must not forget that besides the plant resources, the unique ecological practices of the people of that region play an important role in forming what they are. It is an inseparable component of their unique way of life. Hence, it deserves to be protected and promoted. Instead, the authors' submission is that all the governmental efforts till now have been made to change such indigenous ways/practices in the name of introducing "civilization" to that part of the world.

17.1 INTRODUCTION

Mahatma Gandhi once remarked when some of his followers in South Africa complained to him about African nudity—"when a large society follows a particular custom, it is quite possible that the custom is harmless enough even if it seems highly improper to the members of another society. We would do well to approach the tribal custom with humility and respect and not judge it by our own standards" (Singh, 2021). One needs to investigate if India has been able to achieve the above for the tribes of Arunachal Pradesh.[2]

As per Verrier Elwin, who is credited to have undertaken extensive research in the North Eastern part of India, there have been two main approaches to the tribes of India. First is the policy of "leave them alone," which was practised by the British government during colonial times. They were inclined, overall to leave the tribesmen alone, partly because the task of administration (especially in the wild border areas) was difficult and unrewarding and "partly because a number of officers sincerely held the view that the people were better and happier as they were " **(Elwin, 2009).**

---

[2]Arunachal Pradesh is the biggest of the seven north-eastern states of India.

The second policy is that of "Detribalization" and is in sharp contrast to the British policy. It is through this second policy that both the Christian missionaries and the social reformers interacted with the tribals intending to uplift them and bring them out from their drudgery. "Supporters of this policy still take a poor view of tribal life and maintain that their vices and superstitions should go." Tribal dress is considered as a mark of inferiority which mandates that it should be replaced by something "modern" (Elwin, 2009).

The third policy lies between these two extremes of *doing too little* and *doing too much*. It is this third policy with which the name of the first Indian prime minister Pandit Jawaharlal Nehru is associated. This may be summarized as one that approaches tribal life and culture with deserving respect and value, thus eliminating the breeding of any feelings of inferiority among them.

Pandit Jawaharlal Nehru cautioned that civilization could have a disastrous effect on the tribes of India and their indigenous ways of life if it is left unchecked. He feared that the so-called civil society might just uproot them in various ways, make them feel inferior for what they are, and leave them frustrated and unhappy in return. Pandit ji maintained that these tribes are innocent and too fragile to bear cultural shocks and onslaughts and might just succumb to it all. Hence, our efforts at modernizing them should be well thought out and minimalistic (Elwin 2009).

From the perspective of preservation of traditional knowledge (TK) and traditional cultural expressions (TCE) of the State of Arunachal Pradesh (known as NEFA-the North East Frontier Agency till 1972), India is on the brink of letting the above fears materialize.

The present chapter is an attempt to bring forth the urgent need to legally recognize (by creating an authentic inventory) the TK and TCE as intellectual property of the tribes of Arunachal Pradesh.

As this chapter was in the process of being finalized, a ray of hope appeared in the form of inauguration of the State's first formal indigenous language and knowledge system school in East Kameng district. Though the profile of the tasks to be accomplished by the school is yet to come to the fore, one is optimistic that this "first-of-its-kind school" that goes under the name of "*Nyubu Nyvgam Yerko*" will help in promoting and preserving indigenous traditions, culture, and language (PTI, 2021). Literally, it translates into *Nyubu* (priest), *Nyvgam* (person having wisdom and knowledge), and *Yerko* (learning institution).

Overall, however, the author submits that insufficient attention and inadequate measures have been taken by the Central and State Governments till now toward securing its TK/TCE. It would not be an exaggeration to maintain that particularly governmental administrative efforts have more been toward changing the indigenous ways/practices of the north east hill tribes in the name of introducing "civilization" to that part of the world (Elwin, 2009).

This chapter brings to light the ad-hoc and inconsistent fashion in which TK and TCEs have been dealt with by India. Some of them have been accommodated somehow under the Geographical Indications of Goods Act, 1999 (GIGA, 1999) or Patents Act, 1970—in case novelty is met, or the Plant Varieties Act. 2001. IP rights have not adequately been extended to the holders of TK, through a separate law. Consequently, TK/TCEs have either been neglected or "considered a part of the public domain with no protections or benefits for the knowledge holders or expropriation for the financial gains of others" (Hansen and Van Fleet, 2007).

Also, within the realm of TK that has been so absorbed, attention has been paid only to those practices that generate revenue. IPRs are not only about economics. Had it been the case, we would not have had the concept of moral rights within the IP fold at all. The basic idea behind the moral rights of integrity and paternity, under Copyright law, is preservation.[3] TK/TCEs deserve protection under the same spirit.

Traditional Knowledge (TK) is that organic element of knowledge that is passed on from generation to generation within a community. In this context, the skills and practices related to agriculture that are developed, sustained, and passed generationally within a community, often forming part of its cultural or spiritual identity, need to be investigated.

This paper would attempt to trace some traditional practices of a few main tribes of Arunachal Pradesh in relation to agriculture, fisheries, water conservation, and related aspects. It would also briefly discuss why protection to such practices is required under the IP umbrella. Also, throughout this paper, the ethos that is underscored is that the mentioned traditional practices are *ecological.* They are human actions that create and maintain such harmonious and sustainable conditions as are necessary for survival, development, and prosperity of their habitat; and for this alone their protection is critical.

The author has had the good fortune to briefly look at the life and culture of a few tribes of the State by spending a few years in the region, intermittently

---

[3]*See*, Copyright Act, 1957, s 57.

between 1979 and 1994.[4] Many of the remarks in the chapter come from personal observation and experience.

The tribes considered for the paper are *Adi*, *Apa Tani*, *Monpa*, and *Idu Mishmi*. The broad parts of the Chapter are as follows:

Part I: Concept of TK and TCE

Part II: Traditional Knowledge embedded in Eco-Agricultural Practices of wet-rice cultivation, bamboo grove and forest management, traditional irrigation, etc.

Part III: Traditional Cultural Expressions embedded in the old craft of basketry, cane and bamboo works, wood carving, dye making, loom weaving, etc.

Part IV: What has been achieved and what remains in the context of TK/TCE.

17.2 WHAT ARE TRADITIONAL KNOWLEDGE AND TRADITIONAL CULTURAL EXPRESSIONS?

The United Nations (UN) documents club a range of the diverse and vague descriptions under the category of TK. For instance, TK has been characterized as "complex bodies and systems of knowledge...[and] practices"(SUNPFII, 2006). This is a vast term that may include "content or substance of knowledge held by traditional communities (SWIPO, 2005)," which form a part of an "intangible cultural heritage" of indigenous people, and a part of "the heritage of humanity" (UNDG, 2008).

As per the World Intellectual Property Organisation (WIPO), TK is that organic element of knowledge that is passed on from generation to genera- tion within a community, not necessarily through any institutionalised or formal means of teaching.

[4]The author's father, an IAS officer from the AGMUT Cadre was posted in various parts of Arunachal Pradesh. From 31 July 1979 to May 1980, he was posted in Changlang, an independent subdivision of Tirap district; from June 1980 to September 1982—East Kameng District, with headquarters at Seppa, locally called as "Seppla". West Kameng is primarily a Buddhist region with a few major tribes being *Monpa* and *Sherdukpen*. East kameng area was primarily dominated by *Bangni* (now *Dafla* or *Nyishi* tribes). The area, in common parlance, was referred to as the *Dafla* region. He was then sent on central deputation by MHA from 1st May 1987 to 31st Aug 1992. & then from October 1992 to September 1995 as Divisional Commissioner of East Division of Arunachal Pradesh covering 5 districts of Tirap including Changlang, Lohit, Anini including Roing sub division like Changlang, East Siang & West Siang.

WIPO's program on TK also addresses Traditional Cultural Expressions (TCEs). TCEs are considered synonymous with folklore. Section 2 of the WIPO-UNESCO Model Provisions for National Laws on the protection of Expressions of Folklore against Illicit Exploitation and other Prejudicial Actions, 1985, defines Folklore.

Broadly speaking, it maintains that TCE can be taken as that specific subset of TK which deals with traditional artistic expressions.

17.2.1 TRADITIONAL KNOWLEDGE VS. "INDIGENOUS" KNOWLEDGE

While writing this chapter, the author was mindful of the overlap of a few terms/concepts with the discipline of Anthropology. For instance, the expression "indigenous knowledge" finds its place in Anthropological studies and literature, whereas intellectual property law has grown more comfortably with the word "Traditional Knowledge."

Stephen B. Brush, while quoting Arun Agarwal, notes that, "while 'traditional knowledge' and indigenous knowledge are not synonymous, they share many attributes, such as being unwritten, customary, pragmatic, experiential, and holistic. The terms are frequently used in the same context to distinguish the knowledge of traditional and indigenous communities from other types of knowledge, such as the knowledge of scientific and industrial communities"(Agrawal, 1995).

"Indeed, the primary distinction between traditional and indigenous knowledge pertains to the holders rather than the knowledge per se. Traditional knowledge is a broader category that includes indigenous knowledge as a type of traditional knowledge held by indigenous communities" (Mugabe, 1999).

17.2.2 PROBLEM WITH POPULAR ATTITUDE TOWARD TK

TK, as given by WTO and the UN, appears to be a category that is wide open for some incognito and un-investigated forms of knowledge. It is notable that various traditional eco-agricultural practices of the tribes of Arunachal Pradesh may fit into that definition. As for TCE, the myriad and mesmerising crafts of the region would easily fit the description.

Despite the wide definitions attempted for TK, as noted above, the concept of traditional knowledge has been used in the IP context mostly to understand "technical" know-how and knowledge related to or associated with biodiversity conservation, medicine, genetic resources, and other similar areas. In the context of Indian tribal culture and practices of North East India, it is a travesty that most of the focus goes on bioprospecting and biopiracy. Even the Traditional Knowledge Digital Library (TKDL) centralizes its attention only on "protection of Indian traditional *medicinal* knowledge and prevention of its misappropriation at International Patent Offices." It is as if all misappropriation is possible only in respect of potentially patentable medicinal traditional knowledge (TK).

The author asserts that the research and hence discourse, over the decades, has been lop sided and partial leading to a substantial neglect of other TK and TCEs that might not have much to do with laws of patents, trade secrets, and biodiversity. The traditional eco-agricultural practices (wet-rice cultivation, irrigation techniques, forest management etc, elaborated later in the chapter[5]) are the examples of this neglect.

17.2.3 AGRICULTURAL PRACTICES AND TWO ASPECTS OF TK

Traditional knowledge, in relation to agricultural practices, is not just confined to plant varieties, breeding, seed maintenance, and herbal/medicinal cures. It has two aspects—"Biological" and "Perceptible."

Crop germplasm through close observations of how plants species behave, their election, exchange, and maintenance constitute the first aspect, while the second is reflected in the traditional cultural practices of farmers. The boundaries of both forms are not clearly defined. Both aspects of traditional knowledge have fuzzy boundaries because of their versatile and adaptable nature (Brush, 2005).

Traditional knowledge has been described to include numerous farming systems (Bray, 1986), and its value is evident in such specific activities as designing and managing irrigation (Bray, 1986), coping with marginal farming environments (Sanchez, 1976), enhancing production with local inputs, and developing crop diversity (Brush, 2005).

The author maintains that for recognition as IP, while the latter two have been concentrated upon, focus on the first few has been largely neglected. One must not forget that besides the plant resources, the unique ecological practices of the people of any region play an important role in forming what they are and in shaping the region to successfully battle the adversities thrust by climate change.6 Such practices are an inseparable component of their unique way of life. Hence, they merit unequivocal legal protection and promotion.

---

[5]*See*, Part-II of the Chapter.
[6]*See*, para 3.0.

17.2.4 THE WRONG PROJECTION

What is most burlesque is that in most of the available literature that the author read on the topic, so much space was devoted to writing about the ill-effects of *Jhum* or shifting cultivation,[7] hunting practices etc of the tribes belonging to the State of Arunachal Pradesh (Madegowda, 2009).[8]

This overstretched flogging of the region in the name of *Jhum* and hunting game should stop. Conceding that perhaps some of the ecological practices of the tribes (that were relevant in the earlier times) *have* turned exploitative in the current context, but there is no reason why the administration should dedicate their energies only into shunning those and not documenting, promoting, and protecting the ones that have stood the test of times and have been determined as relevant by experts.

There is a dire need that such traditional ecological practices and knowledge of resources management and conservation are not tinkered with in the name of progress—any such measure would not only be counterproductive, in the sense of creating a disturbance in the social as well as ecological setting of the area, but also lead to the destruction of valuable IP (TK and TCE).

It is maintained by environmentalists that modernization and cultural infestation can lead to loss of environmentally sustainable traditional ecological knowledge forever. Traditional practices not only promote the integration and maintenance of diverse land use types, but also ensure the continuity of diverse species and varieties within each component... (Dollo et al., 2009)'

It cannot hence be denied that these traditional practices are valuable. In fact, one of the unique aspects of such practices is embedded in the fact that they are a result of centuries of informal experimentations with local environment and persistent human adaptations to maintain a symbiotically sustainable lifestyle. It is but natural that they work the best for the area by effectively tapping the local ecosystem.

In Part II, some light would be thrown at the types of such practices that are prevalent in Arunachal Pradesh.

---

[7]Shifting cultivation is a traditional agricultural practice—common in the regions of north-eastern India. Land is cleared of all vegetation, except tall trees, through burning and sowing is then done on it. After a few years of usage and harvest, the land is kept fallow to allow the soil to replenish the nutrients. A typical fallow period lasts about 10 years, sometimes less.

[8]"[While] some ecologists have suggested that *Jhum* may increase biodiversity because it creates new habitats, others see it as a system that threatens the ecology."

Part II

17.3   THIS PART IS CONCERNED WITH TK IN AGRICULTURAL PRACTICES/METHODS IN THE CONTEXT OF ARUNACHAL PRADESH

Arunachal Pradesh is an Indian state in the Eastern part of the Himalayas with a rich ethnic and cultural diversity. The State contains around 26 major tribes and 100 subtribal groups distributed throughout the area who speak around 50 main dialects.

17.3.1   APATANIS—*THE MASTER AGRICULTURISTS AND RESOURCE MANAGERS*

"*Apatani*" is a tribal group that occupies the Ziro valley. The community takes the credit of possessing valuable traditional ecological knowledge of natural resources handling and tending, acquired over the centuries through informal enterprise. Agricultural practices include *jhum* or shifting cultivation, fallow systems, sedentary systems such as rice cultivation etc and the use is sustainable—benefitting both man and nature. Indigenous integration of pisciculture in valley rice cultivation is a distinct characteristic of *Apatani* agro-ecosystem, which has borne rich results for the local economy.

17.3.1.1   WET-RICE CULTIVATION

The *Apatani* farmers routinely indulge in rice-cum-fish cultivation with finger millet on the bund (risers).[9]

"Experts have highlighted that the *Apatani* paddy-cum-fish culture system is one of the most well organised and systematic crop production systems and have further encouraged the *Apatani* farmers to continue their long-established practices" (Dollo et al., 2009). The practice is not only beneficial for sustaining environment and human life but is conducive for the economy; as integrating fish with rice cultivation assures higher economic productivity and year round employment opportunities for farmers (Dollo et al., 2009).

---

[9]The exact date and time of initiation of wet-rice cultivation is still untraceable but through mythology, it is believed that it might have been started centuries ago.

17.3.1.2   WET-RICE FIELD IRRIGATION

As against the popular conception that farming in the so-called remote and difficult areas of India is unorganized and ad hoc, the *Apatani* tribe displays phenomenal synchronization and cooperation. The farmers are divided into informal groups, with each group being responsible for a defined task. Among these, the *"Bogo"* (Dollo, 2009)[10] is most important in matters of irrigating the rice fields.

The Bogo are tasked with construction and maintenance of water supply system. Water being scarce in the area needs to be efficiently regulated. This entire paraphernalia surrounding wet-rice fields irrigation is a masterpiece of water management. "Numerous streams that originate in the wet forests are first diverted into a single canal. From there, each agriculture field is supplied water through bamboo or pinewood pipe" (Dollo, et al., 2009). It is worth noting that the use of plastic and other nondegradable material is absent.

17.3.1.3   MAINTENANCE OF BAMBOO PLANTATION

In the year 2000, the G.B. Pant Institute of Himalayan Environment and Development studied the *Apatani* community of Arunachal Pradesh. The still relevant study report specifically noted the unparalleled contribution of the *Apatanis* in managing the bamboo-plantations. Right from the plantation of stems to timely weeding, young shoots pruning, and finally sensible harvesting, the tribe has been maintaining this tradition devotedly since ages.

The Report notes that, "traditional construction techniques that use bamboo in flooring, roofing, as post and beam, and in the false ceiling and fencing remain largely undocumented" (Sundriyal, et al., 2002).

17.3.1.4   DEVELOPMENT POTENTIAL OF APATANI KNOWLEDGE

From the IP perspective, it is well accepted that in the development of any new technology, most of the money goes into research and development. In such a scenario, if time tested, completely ecological and sustainable solutions are being provided by the traditional eco-agricultural practices of

---

[10]A farmer group sharing the common water sources. The group manager leads all the activities. Posts can be held for 1–3 year(s) and are selected/elected from within the group. Group size is between 3 and 600 households depending on village size.

Apatanis, one can only imagine the economic and developmental potential that such knowledge holds for "designing new technologies for sustainable management of valuable natural resources and efficient ways of resource conservation" (Dollo et al., 2009).

It is notable that barring financial support for things such as erosion control, drainage maintenance, and fencing/boundary creation, the farmers do not look for any other technological interventions or assistance from outside agencies (Dollo et al., 2009). Such self-initiated and traditionally maintained bamboo plantation management, involving maintenance of a balance between usage and replenishment, organic nutrient management through zero reliance on chemical fertilizers,[11] and soil and water conservation are some examples of traditional ecological knowledge which supports the cause of its legal protection through the IP regime, as valuable TK.

17.3.1.5   NEED TO DEVELOP INVENTORY OF TK IN REGISTER

The traditional wisdom of the preceding generations should not be doubted, because it has stood the test of times; and neither should it be polluted by needless intervention of modern methods. It merits to be preserved and nurtured with care. This is also one of the reasons why such ecological practices and their conservers must be supported through legal protection mechanisms—in addition to what is provided by environmental laws. In case of *Apatanis*, despite their legendary skills in sustainable living that ensures a win–win for the environment and human life, they find little mention in law books as preservers of this traditional knowledge. Time has come to correct this historical legal/administrative error.

Times have proven that these traditional agro-ecosystems are favorably linked with nature and gel completely with the local environmental conditions and cultural needs. However, when it comes to protection, the question of documentation is presented and that of the identification of beneficiaries.[12] The author submits that their methods and practices should be recognized in ***their*** names (communities) in the proposed Register containing the inventory of TK. TK Register[13] would serve as an official collection of documentation that describe the TK/TCE. It would be a most authentic record that would

---

[11]For example, the plots of land on which rice-cum-fish culture is practised are mainly fed on organic manure with a variety of animal excreta and plants waste such as rice husk, local beer, and compost such as decomposed straws and weeds. This is done to increase the fertility of the soil and for increasing fish food.

[12]*See*, Para 6.0 onward.

[13]*See*, Para 5.0.

act as a defensive disclosure. It is also expected to effectively plug credit-mongering and usurpation of fame by unassociated third parties.

To this end, several detailed suggestions, including the application of "Domain Public Payant" scheme, have been made toward the end.[14]

17.3.2 MONPAS *OF WEST KAMENG*

The "*Monpa*" who reside in the western region of Arunachal Pradesh[15] are among the major ethnic groups of the State. Being one of the coldest regions of the state, and in proximity with Tibet and Bhutan, their lifestyle has generated a different league of TK. Some examples follow:

17.3.2.1 NATURAL DYE MAKING

The *Monpas* paint their religious temples and monasteries with bright colored traditionally made natural dyes. The dyes are made with tedious effort. They have been engaged in extracting, producing, and processing natural colors from tree barks, leaves, fruits, and roots. Some processes are bizarre wherein insects, animal fat, and even animal urine are used to produce fast colors (Mahanta and Tiwari, 2005).

Natural dyes are colorants having several applications in textiles, cosmetics, inks, wall decoration etc. Organic dyeing not only helps preserve the traditional art of weaving and design, but also provides employment. Surely, the vast treasure of indigenous methods developed by this ethnic tribe in the preparation of the eco-friendly dyes calls for proper documentation, and legal protection as TK.

17.3.2.2 ANCIENT TIERED FOREST MANAGEMENT

Another ingenious aspect of *Monpa* living, which reflects on the remarkable interdependence that the tribe has with forests, lies in their masterful technique/s of forest management.

The distribution of forests is dependent on the altitude—and there are three notional rings or circular divisions that dictate human presence and usage of resources within that zone (Mizuno, 2016).

To elaborate the "*Soeba Shing*" is closest to human dwelling. *Soeba Shing* is the local name for forests used for collecting fallen oak leaves. The

---

[14]*See*, Para 12.2.
[15]*See*, map.

oak leaves from an individual's own land are collected by that person and spread over agricultural fields as fertilizer.

Most noteworthy in the management of *Soeba Shing* is that nobody, not even the owner, is allowed to cut down live oak trees. Only dead trees, fallen trees, and branches can be used. In the event of unlawful logging, an offender is reported to the village leader and made to pay the fine, usually of a cow. Most agricultural households own a section of *Soeba Shing*.

After the *Soeba Shing* ring, comes another type of forested area called "*Borong*" (deciduous forest whose timber is used as fuel), which is further encircled by the third type of forest type called "*Mun*" (dense forests for gathering construction timber and for hunting), which are at some distance from settlements.[16]

FIGURE 17.1: *Monpa* Forest Divisions

Within the above divisions, deliberately curated, traditionally perfected, and environmentally sensitive mechanisms operate to enable human access. It means management of forest resources has traditionally been institutionalized by *Monpas*. This is a) to ensure equitable distribution of fuel timber, construction timber, bamboo, oak leaves, and other forest resources, b) to offset the mounting population pressures, and c) to prevent overexploitation of forests. For example, a traditional informal rural social institution called

[16]Kindly see the diagram-Monpa Forest Divisions.

Chhopa, consisting of a dozen male members and headed by the village elderly (*Gaon Bura*), decides how much of forest resources can be used at any given point in time (Singh, 2013).[17]

17.3.2.3 MONPA FISHING

The *Monpa* tactics of fishing are ecological and traditional to the core. Not only is the activity strictly regulated by the *Chhopa*, but it is also expected to be carried out by using bamboo made tools.

Crushed leaves of walnut are placed in the water to stupefy the fish. Use of chemicals like DDT or any unethical means of fishing is penalized by the *Chhopa*.

In a way, the *Monpas* have been efficiently using the present-day game theory philosophy of John Nash, traditionally.[18]

17.3.2.4 MONPA PAPER MAKING

Paper making is a distinctive craft known to the *Monpas*. They obtain the pulp from a local tree known as the *Sukso* or "paper-tree" (Elvin, 1959).

The traditional knowledge involved in paper making is considered to be approximately 1,000 years old. *Monpas* have been using the bark of a local shrub called "*Shugu sheng*" (scientifically Daphne papyracea) for making paper for religious scriptures used in Buddhist monasteries, manuscripts, prayer flags, and even as a part of flag poles and prayer wheels and for making exquisite gift items.

The process of turning the dry bark into paper is lengthy and tedious[19] but is totally ecological because it does not involve cutting down trees, but just using the bark of a live tree/shrub.

While most artisans have abandoned the skill because of it being laborious and uneconomical, the forests surrounding Mukto, a village in Tawang district, grows the *Shugu sheng* shrub, and a few artisan families still practice this craft (Strengthening Bonds, 2021).

---

[17]For example, the *Chhopa* of Dirang circle restricts fish catch to not more than 1–2 kg/family/day. Nearby the Tawang District, village elders have banned access to forest during November to support regeneration of forest species.

[18]Give and Take—theory of John Nash.

[19]The bark of the shrub is beaten into a pulp, then boiled, beaten, dried and the paper is finally cut—all by hand. The entire hand-crafted process results in paper that has a unique texture. The smoothness of the paper surface depends on drying—generally, one side of the paper is like cloth and the other is slightly rough with long fiber-like texture. In a day, around 100 sheets of paper can be prepared and dried if the weather condition is suitable. Mon shugu paper is sold in the local area at INR 15 to 25 per sheet.

As recent as December 2020, the Khadi Village Industries Corporation (KVIC) partnered with a local NGO, Youth Action for Social Welfare, to commercialize paper-making in the Tawang area.[20]

17.4   CLIMATE CHANGE AND RELEVANCE OF TRADITIONAL ECO-AGRICULTURAL PRACTICES

One remarkable aspect of tribal life (of Arunachal Pradesh) is their blending with and understanding of nature. Knowledge worth eons of ages enables them to adapt to the ever-changing natural environment around them. Despite their defencelessness to the negative effects of climate change, they have found the potential to completely adapt to these changes by carrying on with their age-old practices and by relying on traditional knowledge.

While noting the contributions of indigenous communities, Mirian Masaquiza Jerez elaborates their achievements in reprevention of deforestation; mineral, oil, and gas extraction in their ancestral lands; their incessant protests against monocrop plantations, and their promotion of all forms of traditional knowledge (Jerez, 2021).

The above elaboration might well fit the *Apatanis* of Arunachal Pradesh. Given their proven expertise as sustainable (as against exploitative) agriculturists, if they are included in discussions (along with scientists, meteorologists, climate experts, policymakers, administration/civil servants, and other relevant actors) in "sharing of sustainable strategies to overcome risks and strengthen resilience to climate change" (Jerez, 2021), it might bear fruitful results for the entire north-east region of India.

Traditional eco-agricultural modus operandi, such as wet-rice irrigation, division of forests into zones and collective regulation of each zone, extensive propagation, and use of bamboo and cane instead of plastic etc are practices that reduce the per head carbon footprint and increase the gross health of the forest and jungle ecosystems. They are hence critical strategies for diminution of global warming. These strategies, as fine-tuned through constant adaptations over time, are still practiced by the abovementioned tribes of Arunachal Pradesh.

It is clear that these indigenous peoples have masterfully controlled and used the natural resources so that their future stock and conservation is secured. This way, the indigenous people have fruitfully contributed toward mitigation

---

[20]The Tawang-based voluntary organization, Youth Action for Social Welfare, has been working toward preserving this traditional knowledge and craft. A lawyer by profession, Maling Gombu, is chairman of the NGO.

of climate change and espousal of adaptation strategies (Jerez, 2021). Given the above-proven benefits of such TK, it is more important that legal protection be given to the custodians of this knowledge or to entire communities or peoples against its misuse and misappropriation (Corpuz, 2003).

The above sentiment has been echoed by Mirian Masaquiza Jerez in the UN Policy Brief 101, which suggests the dire need to ensure that policy formulation is in sync with the UN Declaration on the Rights of Indigenous Peoples. Only such measures, it is underscored, are likely to protect blatant plundering of indigenous traditional knowledge and encourage the recognition of indigenous peoples as legitimate holders of their knowledge (Jerez, 2021).

17.5   NEED TO ENFOLD TK AS VALUABLE IP

Traditional forms of creativity and innovation can be protected to some degree by the conventional IP system—through GIs,[21] and Plant Variety Act etc. But there is a residuum of intellectual creativity that does not find protection in the current IP regime. What has been discussed above present a classic example of such practices/methods. These activities cannot be separated from the everyday lives of the tribes. It is to address the felt need to incorporate such practices into the fold of IP that there has been an ongoing discussion at WIPO and elsewhere on developing new forms of IP protection for TK (and TCEs, to which Part III is dedicated).[22]

Confining the discourse presently to traditional knowledge, the author submits that within TK, there is a strict need to define the scope of the same under two heads. One would concern with TK that deals with biodiversity, traditional medicine (for which the TKDL was essentially established), trade-secrets and patents; and the second subset that would be reserved for TK imbued in indigenous agricultural practices, traditional water and forest management systems/methods, sustainable fishing practices etc. It is here that defensive protection is being suggested.

17.5.1   DEFENSIVE PROTECTION AND RELATED SUBMISSIONS

All the strategies aimed at ensuring that no third party extracts illegitimate or unfounded rights over any concerned IP can be clubbed as "Defensive

---

[21]*Idu Mishmi* Textiles and Arunachal Orange are the only two registered GIs from the State of Arunachal Pradesh.
[22]*See*, Part III of this Chapter.

protection." Classically, the IP regime's response to protecting TK (and even TCE—to be discussed in Part III) has manifested in discussions around how their intangible intellectual components can be defended against some form of unauthorized use by third parties.23 In addition to this kind of protection, "preservation," is also a function of the IP system, especially in case of TK. The author submits that this is not a radical or "deviating" revelation, as the IP system has been achieving this substantially (in the Indian context) through Geographical indications and Plant varieties—there, the idea is not just to protect the rights but also to ensure that the concerned IP is preserved and survives the passage of time/technological changes. Also, as stated elsewhere, the concept of moral rights that is deeply imbued in the Copyright system finds its very rationale in the fact of preservation of creative expression. Finally, if business methods can merit patent protection, the traditional methods mentioned above also should. Both are creations of the mind having immense utility. Them being in the public domain should not defeat their claim, as it does not in case of GIs.

17.6 TK CALLS FOR TITLE-BASED REGISTRATION PROTECTION

It is desirable that in the forms of TK explained in Part II—particularly those that are clearly attributable to a certain community receive defensive IP protection. It is a knowledge that should be safeguarded against being poached by others in their name. To this effect the author suggests that title based registration protection should be put in place, where in the Register of TK should legally identify a certain practice/methodology by the name of the community, like "*Apatani* wet rice cultivation," *Monpa* Natural Dyes etc. The operation of such a scheme (it is proposed) should be a blend of "moral rights" concerning attribution and integrity and GI (in being community specific).

17.6.1 TK CENSUS SHOULD BE CONDUCTED

A policy challenge that this approach poses is that any such recognition would entail an indefinite term of protection. Another would be the question as to who would "own" and/or manage the rights that flow from such defensive IP protection of the concerned TK? This task, it is proposed, should be

---

23In fact, the TKDL (Traditional Knowledge Digital Library) in India is playing an important role in protecting the misappropriation of medicinal TK from exploitation (through bio-piracy) and unethical patents by employing defensive protection.

undertaken by the State of Arunachal Pradesh (perhaps through a specially appointed authority), as an essential part of preserving the common heritage of the area.

A TK Census is proposed to be conducted across Arunachal Pradesh with an aim of identifying them and the community (common) beneficiaries.

The general objective should be to ensure that 1) the TK is recorded and hence process of preservation initiated, 2) any usage or improvisation in the concerned practice is duly acknowledged, and 3) the benefits, if any, in the form of royalties etc should flow to the appropriate tribal communities. Existing or new collective management organizations already tasked with looking after the welfare of such communities could play a role in managing the rights for the direct benefit of the relevant communities.

17.7   CAN BENEFIT SHARING WORK IN CASE OF TK/TCES

The traditional technical know-how that might be in the public domain but is attributable to a certain tribal community, or traditional ecological practices, and various aspects of everyday lifestyle may all merit protection against misuse and credit mongering by unrelated third parties.

The role that the IP regime is expected to perform for TK/TCE should largely be that of a sentinel –meant to safeguard the concerned knowledge against third-party exploitation through all possible means.

Conventionally, the Access and Benefit Sharing (ABS) mechanism has been one of the many ways through which genetic resources have been approached and utilized for the common good. ABS refers to the mechanism of permitting access and use of genetic resources to give benefits to the users and providers. Some level of control is always exercised by the right owner here.

The author submits that there is no reason why this mechanism cannot be applied to TK, albeit in a suitably modified form. In case of TK of indigenous people of Arunachal Pradesh, as mentioned above, the agricultural/ fishing/forest management practices are not a secret and can be largely said to be in the public domain (barring in-depth experiential aspects that might not be widely known by the outside world). So, access to the same is available. But that should not deprive the concerned tribes from being accredited and compensated if a third party[24] wants to make use of the process/practice or improvise the same. The author maintains that, this also deserves to be compensated for, if any real/tangible protection is meant to be given to such

---

[24] "Third party" here would mean any person outside the community (if TK is vested in the community).

TK, thus meaning a "BS" or Benefit Sharing Mechanism should be instituted. (Since A=Access of such TK is not a problem-it is in the public domain) This can give such tribes the right to control the use of TK.

17.7.1 BENEFIT SHARING TO CATER TO COMMUNITIES, NOT INDIVIDUALS

It is often asserted that since TK is held collectively, it would be arduous to identify exclusive proprietors; and lack of documentation makes this task even more difficult. The author concedes to this.

But this difficulty can be surpassed once we stop thinking of Benefit Sharing only as a personal-rewarding system. When negotiating on benefit-sharing in regard TK, since there is no single proprietor, the needs of the entire community should/must be taken into consideration. This can be done by:

1. Creating a detailed database of communities and register their TC/ TCEs in the TK Register, as suggested in the preceding paragraphs. Help may be taken from the Census data of 2011.

2. Allowing the concerned community to operate through a representative who could be village/community elders[25]—who are, mostly, already an important part of the social structure of every tribal community in Arunachal Pradesh.

3. Once the terms of BS have been agreed upon, the concerned benefits/ money in the form of royalties can go to a Community Development Fund, and later be used for the benefit of the community by looking at common interests/needs.

4. A few representatives of the civil administration can also be a part of this negotiation exercise. This is purely to ensure that the negotiation process and the agreement between the parties is not ambiguous or exploitative etc. The process should be transparent and understandable to enable representatives of indigenous tribes to make an informed decision.

5. Any such arrangement should clearly specify the rights and duties of each party.[26]

---

[25]Gaon Bura is a village elder who is identified by his distinctive red clothes. All important decisions of a tribe are taken in consultation with the Gaon Bura.
[26]See Para 12.0 of the Chapter.

Part III

17.8 TRADITIONAL CULTURAL EXPRESSIONS AND INTELLECTUAL PROPERTY

For TCEs, the coverage in case of Arunachal Pradesh may focus on myriad tribe specific handicrafts like basketry, hats and jewellery making, loin loom woven apparel and carpet weaving, designs and traditional motifs, pottery, wood carving, painting or any other aspect of everyday tribal life that has technical as well as aesthetic qualities.[27] It is worth mentioning that many of these may be eligible for protection as GIs, provided there is state support or some entity motivates/interests itself in managing the rights of the concerned producers of a certain craft/TCE by bringing them together. Until that happens, it is unwise to leave them unprotected.

A few examples of TCE of (mentioned) Arunachal tribes are being detailed below (Fig. 17.2).

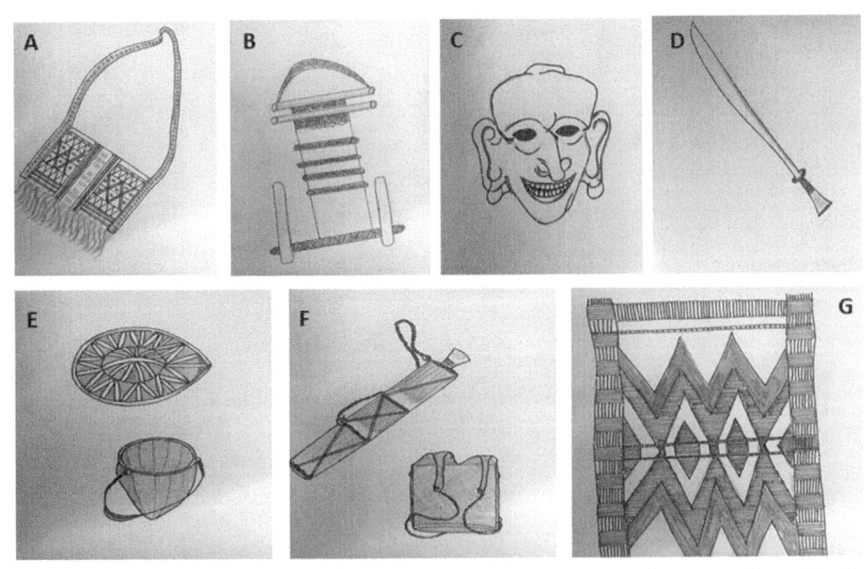

FIGURE 17.2   A few examples of TCE of Arunachal tribes: (A) The Monpa Bag; (B) The Loin Loom; (C) Monpa Wooden Mask; (D) Dao; (E) Apatani hat of cane and bamboo and Beautiful sturdy basket of cane and bamboo; (F) Dao in sheath and Haversack made of cane and bamboo strips; (G) Adi "Galle" or Women's skirt strip, vertically woven.

---

[27]An inventory of the same should be prepared by the government of Arunachal Pradesh.

17.8.1   TCE IN BAMBOO, CANE, AND WOOD-WORK OF TRIBES

17.8.1.1   APATANIS

Using bamboo for various activities is never seen as a craft by the tribes of north east India. It is an extension of their way of life, as almost all products of daily use are made at home.[28] Traditionally, the older male members of the household made bamboo and cane articles for home use.[29] Traditional *Apatani* bamboo and cane work is characterized by unique shapes and intricately woven patterns in some extremely attractive designs (Sundriyal et al., 2002).

17.8.1.2   ADIS

The author has had the opportunity to look at the lives of a subgroup within the main *Adi* tribe. They are known as *Pae-Libos*. Hence, mention has been made of their TK and TCE in the Chapter.

Pae-Libo or Libo is the name of a small tribe of the "Adi" group of Siang district. The *Pae-Libos* have a long history and tradition of using ecological and sustainable methods of living their lives like their ancestors.

A *Pae-Libo* house does not require any furnishings. Their needs are less and whatever furniture is there, is made of bamboo, cane, or wood. All the utensils are made of bamboo and are exceptionally long lasting (Kumar, 1979). This eliminates the use of plastic altogether. For example, *"Kubu"* is a small cylindrical bowl of bamboo used for curry. It is made from a piece of bamboo of appropriate diameter. A piece of suitable diameter with two joints at each end is taken and held longitudinally. A fine strip is slashed with the *Daou*, thus making the mouth of the bowl.

Similarly, *"Lobum-Duki"* is a cylindrical bamboo salt container. *"Isi-Udu"* is a water container of various sizes. Some utensils are made from wood like *"Ekku"* it is a rectangular plate with a handle use for serving or even eating rice. It is hewn from a suitable piece of wood.

Apart from utensils, there are baskets of all suitable shapes and sizes (Kumar, 1979).30 Every article made from bamboo and cane is simple in

---

[28]Though now, some influence of urban city life is visible reflecting in the use of plastic wares.

[29]*See,* image 5: Classical haversack made of cane and bamboo strips; image 6: *Adi* hat of cane and bamboo; image 7: beautiful sturdy basket of cane and bamboo.

[30]*"Ekken"* is a very big basket meant for use by men only because it is too huge and very heavy when loaded. It is used for carrying loads of maize and finger millet from the fields to the granary. It is closely knit to avoid leakage of grains of finger millet. It is fitted with a strong chain strap with which it is lifted and carried. *"Ebar,"* a huge basket conical in form, *"Igin"* made of cane for lighter loads, *"Tirdu"* a big basket for storing beer ferment cakes.'

design but remarkably fine in finish. The area is abound in bamboo and cane and the *Pae-Libos* have diligently used these materials.

A possession worth mentioning is the *Oyok-Dao*.[31] Almost all tribes in the State possess a version of the *Dao*, which is a hybrid between a long knife and a sword/machete. "It is a multipurpose object of maximum utility—a weapon in war and an implement in peace" (Kumar, 1979). The creativity displayed in creating the sheath or "*Chobuk*" of the *Dao* is remarkable. The sheath is made from cane and is bound by thin bamboo strips. The carving on hilt is also fascinating.[32]

17.8.1.3   MONPAS

The *Monpa* community of Tawang and West Kameng districts builds bamboo suspension bridges of twisted split bamboo cables and woven walkway (Sundriyal, 2002). Apart from big construction works, the *Monpas* are adept at working with materials like cane and bamboo, making baskets and accessories etc. Masterfully designed bottles of cane are crafted with dexterity and made sustainable through rubber linings to hold the liquid for decent duration.

They are good carpenters and produce several wall hangings, wooden masks, bowls, cups, plates, and saucers, which are often painted with a number of indigenous, often religious designs.[33]

17.8.2   TCE IN WEAVING

An amazing contraption is the "loin loom"[34] which is entirely made of wood and bamboos. In many parts of Arunachal, which are designated as remote areas garments are still commonly woven on this loin-loom. Some commonly woven garments are loincloth skirts, headwear for women, blouses, and belts etc. Both yarn and wool or animal hair, and sometimes even human hair are used.

Adis and *Apatanis* both have their own unique design repository, and a keen eye can distinguish between them just by a look.[35] *Monpas* weaving patterns are bright and colorful, which they create through their *Monpa* loom. This tribe historically has had relations with Tibet and Bhutan and the influence is visible in their creations—making them visibly different.

---

[31]*See*, image 4, *Dao* or *Daou* in sheath.
[32]*See*, image 4, Sheath of *Dao* or *Daou*.
[33]*See*, image 3, *Monpa* wooden mask.
[34]*See*, image 2, Loin Loom made of bamboo and cane.
[35]*See*, image 9, Vertical design woven on loin loon for *Adi* "*Galle*" or women's skirt.

Cotton, wool, and bark fiber is used for weaving on looms. They mainly produce shawls, sashes, an array of bags, and coats. They are also good carpet weavers.[36]

Idu Mishmi weaving, and design patterns are much more elaborate and decorative. No wonder the textile boasts of being one of the two registered GIs of the state. Their black "war-coat" ornamented with a white pattern is a remarkable piece of work woven with a mixture of cotton, bark-fibre, and human hair. It is said to work as an armour and is projected to be strong enough to turn aside a hostile arrow (Elvin, 1959).

17.9 VERRIER ELWIN WRITINGS VINDICATE CLAIM FOR PROTECTION OF TCES

What is listed above in the context of artwork can be taken as substantially true for almost all tribes of Arunachal Pradesh, with small to sometimes significant deviations in the craft involved. The point being that entire State of Arunachal Pradesh is a hub of sustainable creativity, which merits recognition and respect under the IP laws.

Verrier, as early as 1950s, noted that the Hill people have an impeccable taste in color and in mixing-matching colors. Some of them display an extraordinary talent in creating patterns for weaving. In fact, he records that old documentations and evidence go to prove how many of the designs of present times had already been developed by the tribes 50 years ago. They continue to create newer patterns by getting influenced by contemporary art or their own natural zest for creation (Elvin, 1959).

Regarding cane/bamboo and wood carving expertise, Verrier notes that "the wood-carving [of hill people of NEFA] often reveals vitality and strength; the cane-work reaches a high standard of technical perfection" (Elvin, 1959).

For jewellery, Verrier records that even the most remotely situated tribes of NEFA display a high degree of taste, innovativeness, and proficiency when it comes to the art of personal adornment. Their instinct for ornamentation is unmatched whether it is regarding armlets made of ivory/ polished wood, cane leggings, conch shells and bead strings and a variety of ear-ornaments—some even made of seeds! The collection of hats, baskets, monkeys' skulls, and other distinctive and splendidly made artifacts –all speak volumes of their sense of grace. All this, he maintains, should not be allowed to die out (Elvin, 1959).

---

[36]*See*, image 1, *Monpa* bag woven on loin loom.

Long ago, Verrier had cautioned that tribal art might decline if over-exposed to external influence or reduced to crudity through improper transfer of knowledge. For the latter, he records, "training [in Cottage Industries] in different crafts is being given without proper regard to the art and culture of the tribes. The tribesmen have a great tradition of artistic and cultural self-reliance; they have impeccable taste in colour form and design. It is essential therefore that articles which are tribal in their background should be produced by the trainees and the aim should be to preserve and develop the present art and culture of the hill people" (Elwin, 1967).

The abovementioned dangers in regard TCEs exist even in the present times and have only multiplied in intensity. A huge responsibility lies on the state government to ensure that the creators of these traditional art forms continue to take pride in their works. An equal responsibility lies on the purchasers; "it is essential that they should refuse to buy inferior articles and those which show a falling-off from the high standards of tradition" (Elvin, 1959) .

PART IV

17.10   STATE RESPONSIBILITY TO GET TCE PROTECTED AS IP

A concrete step in the direction of legally protecting TCEs, through the IP system, would be to prepare a detailed inventory of TCE and recognize the crafts of different indigenous Arunachal tribes, in one comprehensive Register of TK/TCE. Simultaneously, aggressive efforts must be made by the state government to get a GI tag for them.

A suitable signage or logo/symbol akin to that of "*Atulya Bharat ki Amulya Nidhi*" used for Indian origin GI goods should be developed for such documented TK/TCE. This legal recognition coupled with the symbol would act as an essential information and educate the casual buyer about the authenticity of the product and the associated do's and dont's, if any. At the same time, it would give the indigenous people a reason to be proud of their heritage.

Since most of the TCEs, as mentioned above, have immense commercial potential, they are especially vulnerable and susceptible to unauthorized commercialization. This makes the need to document them and also legally protect them, *imminent*. Within the limited scope of the present chapter, the author suggests that the policy makers look at providing "positive protection" (which entails active exercise of IP rights in the concerned IP) for TCEs such as basketry, carpet weaving, and tribal designs.

A few examples from foreign, culturally rich countries which explain the stated urgency to protect TCEs are 1) some traditional musical instruments have been remodeled into modern ones, baptized anew and commercialized. They have also been used by nontraditional persons who operate in the music community across the globe. Some are exploited to promote for tourism (such as the steel pan of the Caribbean region (Sandler, 2001). 2) A few unique musical instruments, such as drums and the singularly unique "didgeridoo"—which is a wind instrument, have also been unauthorizedly and blatantly subjected to mass-production as mementos and souvenir items. In short, objects that are clearly made outside Australia are imported into the country by unscrupulous persons and are shamelessly passed off as locally produced, genuine items (Janke, 2003).

17.11 HAS LAW DONE ANYTHING TOWARD THE NEEDS AND EXPECTATIONS OF TCE CUSTODIANS?

The complex nature of TCEs mandates the use of more than one IP strategy to effectively protect the concerned knowledge. Toward this end, several fact-finding missions and consultations have been conducted by WIPO since 1998. Primarily, the three approaches that have been encountered are as follows (WIPO, 2003):

(a) *IP protection to buttress development of economy,* wherein communities that desire to exercise their IP in their tradition-based creations and innovations are so enabled. In the Indian context the protection accorded to some forms of traditional cultural expressions (that exists in the form of artistic creations, handicrafts, and foodstuffs etc) has been given through the Geographical Indications of Goods Act, 1999. The ones that manage to fulfil the requirement of novelty, inventive step etc may be protected by the Patents Act, 1970.

(b) *IP protection to stave-off unwanted use of protectable knowledge by others*, wherein communities are equipped to prevent the use and commercialization of their cultural heritage by others. "Uses which may wish to be prevented could include, for example: (i) uses that falsely evince an association with a community; (ii) derogatory, libellous, defamatory or fallacious uses; and (iii) uses of TCEs that are considered sacred and/or secret by communities (WIPO, 2003).'

For instance, a tribal community in Arunachal Pradesh would want to protect their unique tribal apparel motifs and designs. They may use IP protection to stop the use of such design by an outside commercial manufacturer, but at the same time, the community can also use the same protection as the basis of their own commercial enterprise. In such a scenario, these indigenous-tribal-traditional-communities should have legal means under IP laws to be able to protect their hand-woven or hand-made designs, textiles, weavings, and garments from being copied and commercialized by nonindigenous persons.

Perhaps the protection offered by the PPVFRA, 2001 would fit in this category. Because aspects like breeding/generational preservation and experimentation with plant varieties etc find protection under the Protection to Plant Varieties and Farmers Rights Act, 2001.

In both above approaches, owners, and custodians of TCEs wish to protect their TCEs by actively asserting IP rights. This could be termed "positive protection."

(c) **Defensive strategies to protect** TCEs: A third approach is to put defensive protection strategies into action. These are targeted to prevent others from getting or maintaining IP over adaptations and derivations of TCEs. Those adopting this approach usually are not looking for monetary gains through their IP, but are more interested in safeguarding their knowledge, heritage, and cultural expressions from unauthorized use. To that end, they believe that no such rights should be procured by anyone over them. For example, throughout the North Eastern states (Arunachal Pradesh being no exception) the beak and feathers of the Great Indian Hornbill (*Dichoceros Bicornis*) are given utmost regard for its (believed) magical powers. Additionally, it is also a representation of courage and magnificence. In many tribes, wearing of a hornbill feather in uniquely crafted hats of cane is considered the prerogative of a man; it is strictly taboo for anyone else to wear it. It is natural for the tribes NOT to want such deeply cultural artifacts being commercialized or otherwise slighted.

17.12   WHETHER ATTENTION RECEIVED BY INDIAN TK/TCE AT NATIONAL AND INTERNATIONAL FORUMS

The WIPO Intergovernmental Committee on Intellectual Property and Genetic Resources, Traditional Knowledge and Folklore (IGC) is an international forum mandated by WIPO's Member States to comprehensively discuss and address policy development in relation to IP, including TK and

TCEs. WIPO's member states and several accredited observers participate in the process, including many indigenous organizations.

The IGC has held 40 sessions so far. A range of policy, legal, and practical questions along with intense discussions have been witnessed. The axis of all that has ensued is to deliberate on the expanse to which IP is relevant to meeting the needs of indigenous and local communities. A related aspect of usefulness of the present IP system for protecting TK and TCEs has also been considered at length. Many views and questions have been put forth, but consensus has been elusive.

It is also a matter of concern that the presentations, that representatives of the member states make at the IGC sessions, India does not bother to mark its attendance in most of the sessions. This attitude is reflective of the extent to which the country takes its TK and TCE seriously. In fact, as per available resources, it is first in March 2011 that the Government of India marked any tangible presence on the world forum, in relation to TK. That was when the government and the WIPO Secretariat coorganized an International Conference on Utilisation of the Traditional Knowledge Digital Library (TKDL) as a prototype for protection of TK. Experts from the Patent Offices of several countries such as the European Patent Office (EPO), United States Patent and Trademark Office (USPTO), and the Indian Patent Office (IPO) exchanged notes on their experiences on databases.

The Indian presentations were focused only on TKDL, which, as already mentioned in the text above, has a lop-sided objective, as it focuses only on traditional medicines—that is, diverse health practices that help maintain wellbeing (AYUSH). There is no mention of other forms of TK.

Further, in 2015, 2016, and 2017, the WIPO Secretariat organized several seminars on Intellectual Property and GIs/TK. The Indian speakers yet again shared their experiences only on medicine-based TK and the limited scope within which the TKDL works currently.

At the national level, an ambitious suggestion regarding TK and TCEs was made in the National IPR Policy-May 2016—that "the ambit of Traditional Knowledge Digital Library (TKDL) should be expanded to include other fields besides Ayurveda, Yoga, Unani and Siddha." Also, it was decided that all oral TK should be documented, as it was declared to be necessary to safeguard the latter's integrity. All this was given out as inextricably linked and imperative to preserve the traditional ways of life of communities.

A promise was made to the traditional knowledge holders that all the necessary support and incentive would be provided to them so that their age-old knowledge system can be used for better future. However, nothing tangible has been done in regard the above in the past 5 years since the announcement of the Policy.

Coming back to the international stage, only time will tell on how WIPO's Member States will eventually address the protection of TK and TCEs. Presently, the IGC is examining, among other documents, draft provisions for a sui-generis protection of TK and TCEs.

17.13   RIGHTS THAT ANY MODEL IP LAW ON TK/TCE SHOULD RECOGNIZE

Presently, the mentioned TK and TCEs of indigenous people are being used/or are susceptible and vulnerable to be exploited by others without a compensation mechanism/without acknowledging their contribution. A *sui generis* (IP) protection will offer support. The proposed system, however, must guarantee certain rights to the holders and preservers of TK/TCE.

Terri Janke, who has worked extensively on Aboriginal rights, in "Our Culture: Our Future—Report on Australian Indigenous Cultural and Intellectual Property Rights (Janke, 1998)" has listed the rights that IP law should recognize in favor of indigenous people. They are (Deer, 2013):

1.   Right to own and control indigenous cultural and intellectual property.
2.   Right to define what constitutes indigenous cultural and intellectual property.
3.   Right to control the commercial use and to benefit commercially.[37]
4.   Right to full and proper attribution.[38]
5.   Right to be recognized as the primary guardians and interpreters of their cultures.
6.   Right to protect sacred and significant sites/symbols/objects.
7.   Right to prevent derogatory, offensive, and fallacious use.
8.   Right to maintain secrecy.
9.   Right to have a say in preservation and care.[39]
10.   Right to control use of traditional knowledge.[40]

These rights are important because they would help the indigenous communities to stake a claim on what is rightfully theirs, instead of being

---

[37]*See*, para 10.0.
[38]*See*, para 5.0.
[39]*See*, para 3.0.
[40]*See*, para 6.0.

at the mercy of the system. It would also assist the governments to regulate and monitor the use of the concerned TK/TCE against all unauthorized use.

17.13.1 DOMAINE PUBLIC PAYANT

Domaine Public Payant (Paying for Public Domain) for TK/TCE in the public domain—royalty that is payable by public for commercial or noncommercial use of works in the public domain, whether it is improvisation of the forest management systems, natural dye making processes, basket weaving, loom textile designs etc.

Royalty can be collected by a collecting society created for the purpose. The proceeds then can go to the government coffers and later, invested to support and encourage the concerned TK/TCE in different ways. The idea is to create a system very similar to that of a compulsory licence regime—where the use is dependent upon whether the prescribed fee has been paid to the collecting society or not.

This system has been adopted by many countries for copyright works that have fallen in the public domain. "According to the system, a work that has fallen into the public domain may be used without restriction, subject however to the payment of a fee calculated as a percentage of the receipts produced using the work or its adaptations."

The author proposes a model provision along similar lines for TK/TCE which may read as follows:

Domaine Public Payant: (1) The user shall pay to the competent authority percent (a quantum to be decided and fixed) of the receipts produced using TK/TCE-based practices and in the public domain, or their adaptation

(2) The sums collected shall be used by a competent authority for the following purposes:

- ➢ (i) to promote further research and development in the concerned TK/TCE field, including extensive documentation of the same.
- ➢ (ii) to protect the TK/TCE by educating the public, and disseminating information, particularly about its genesis, paternity, and integrity through all means of publicity at national and international levels.
- ➢ (iii) to vigilantly act as a global monitor in regard the TK/TCE to prevent and plug its misuse/abuse.

17.13.2   TEN TANGENTS THAT PROPOSED LAW SHOULD COVER

Whether it is a *sui generis* legislation or an administrative ruling, any system proposing to protect TK/TCE must ALSO look at the following tangents, for it to be effective:

(i)   whether the policy objective of the protection has been identified—does it propose to adopt the positive protection strategy or the concerned TK/TCE merits only defensive protection.

(ii)   what is the subject matter of protection—whether it is traditional agricultural and related practices or cultural expressions like basketry, jewellery making, loin loom weaving etc or both.

(iii)   what are the criteria that the concerned subject matter **must** meet as a condition for its protection—this might involve an authentic survey or inquiry into how old the traditional practice or cultural expression has been. Wherever, new methods have been adopted to supplement and improvise on the traditional techniques, the same should be noted, recorded, and made known as such. This would bring in the necessary clarity about the traditional practices and help trace their evolution over the period.

(iv)   whether owners of the rights have been identified—in case of Arunachal, most of the practices and expressions are community/tribe specific; so the concerned tribe to which the practice can be attributed needs to be ascertained. Efforts should be made to document these through administrative surveys and census activity. Help of NGOs, already working in the area, may be taken.

(v)   whether the scope of those rights has been defined, including exceptions—workshops and outreach programmes should be held with the help of grassroot NGO workers and original knowledge holders or village/community elders to acquaint the locals about the value of what they have conserved over the ages and why there is a need to protect it from over-exposure through a right-based system.

(vi)   whether there are any procedures and formalities that need to be completed by the proposed right holders for the acquisition and maintenance of the rights conferred—like getting yourself registered as a member of a certain tribe etc.

(vii)   what are the consequences against infringement of the concerned TK/TCE—the extent of penalties that would be imposed for unauthorized use, malafides etc.

(viii) what is the duration of protection—protection to such TK and TCEs that are recognized to vest in indigenous tribal communities of Arunachal must vest in them till perpetuity, as the nature of this IP demands so.

(ix) what is the level of interaction or overlap, if any, of the concerned legislation with the existing branches of intellectual property—for example, what would happen is a certain TCE (say, natural dyes and its making) is recognized and registered as a GI.

(x) whether due process has been followed in governmental/nongovernmental projects involving TK and TCE—the author submits that it should be made mandatory for the government not to undertake projects that are likely to interfere/tinker with these ecological traditional practices in the area, unless a go-ahead has been given by independently conducted research, involving private and public agencies and neutral (preferably foreign-based) scientists.

17.14 CONCLUSION

One thing is clear—the free or easy/unprotested availability of TK/TCE cannot be understood as a natural permission for its unconditional commercial use and exploitation. Also, that any effective management of TK/TCEs must include an *exclusive and ethically appropriate policy of use/exploitation/art of sharing.*

The TK and TCEs of the kind mentioned in the chapter have allurement quotient to them. The outside world is attracted to them, which is welcome. Many of the eco-agricultural practices mentioned have been there, shared willingly by the originator and preserver community as "common heritage of mankind." But that does not translate into any tangible benefit for the community. The *"Vasudhaiv Kutumbkam"* philosophy has clearly not worked, as knowledge is being usurped by unscrupulous fame peddlers. Moreover, the indigenous tribal communities now live on the margins of civilized society –the balance of the world order is skewed against them. It is hence critical that their TK in eco-agricultural practices be preserved and protected through IP rights.

Same is the case with TCEs mentioned in the Chapter. They define an entire culture and are an extension of the tribal lifestyle. But the moment they are unauthorizedly used or commercially exploited by the outer world, they no longer retain that aesthetic quotient, and everything degenerates into a money-making venture of unworthy and substandard imitations.

KEYWORDS

- **traditional cultural expressions**
- **traditional knowledge**
- **IPR**
- **intellectual property rights**
- **Arunachal Pradesh Tribes**
- **arts and living of indigenous people**
- **TK**
- **TCE**
- **benefit sharing**

REFERENCES

Arun, A. Dismantling the Divide Between Indigenous and Scientific Knowledge. *Dev. & Change* **1995**, *26*, 413 and 422.

Assemblies of the Member States of WIPO, Fifty-Ninth Series of Meetings, Report on the Intergovernmental Committee on Intellectual Property and Genetic Resources, Traditional Knowledge and Folklore (IGC); 30 September to 9 October 2019, Agenda Item 20. https://www.wipo.int/export/sites/www/tk/en/igc/pdf/igc_mandate_2020-2021.pdf   (accessed June 20, 2021).

Birender Pal, S. *Gandhian Perspective on Tribal Resources and the Modern State*, https://www.mkgandhi.org/articles/tribalresources_modernstate.html (accessed June 20, 2021).

Bray, F. *The Rice Economies*; University of California Press, 1986; p 68.

Brush, S. B. Protecting Traditional Agricultural Knowledge. *Wash. U. J. L. & Policy* **2005**, *17*, 59; https://openscholarship.wustl.edu/law_journal_law_policy/vol17/iss1/5 (accessed June 20, 2021).

Corpuz, V. T. *Biodiversity, Traditional Knowledge and Rights of Indigenous Peoples*; Third World Network, 2003.

Deer, K. Indigenous ICT Taskforce: Managing Traditional Knowledge in the Information Society – From Indigenous Customary Law to Global Internet Governance. In *Traditional Knowledge & Indigenous Peoples*; Gosart, Ulia Popova. Ed., 2013.

Dollo, M.; Samal, P. K.; Sundriyal, R. C.; Kumar, K. Environmentally Sustainable Traditional Natural Resource Management and Conservation in Ziro Valley, Arunachal Himalaya, India. *J. Am. Sci.* **2009**, *5*(5) 41, 47-para 3.4, 51-para 4.

Dollo, M.; Samal, P. K; Sundriyal, R. C; Kumar, K. Environmentally Sustainable Traditional Natural Resource Management and Conservation in Ziro Valley, Arunachal Himalaya, India. *J. Am. Sci.* **2009**, *5*(5), 42, para 1.

Dollo, M.; Samal, P. K; Sundriyal, R. C; Kumar, K. Environmentally Sustainable Traditional Natural Resource Management and Conservation in Ziro Valley, Arunachal Himalaya, India. *J. Am. Sci.* **2009**, *5*(5), 47, Table.

Dollo, M.; Samal, P. K; Sundriyal, R. C; Kumar, K. Environmentally Sustainable Traditional Natural Resource Management and Conservation in Ziro Valley, Arunachal Himalaya, India. *J. Am. Sci.* **2009,** *5*(5), 44, para 3.1, Table

Dollo, M.; Samal, P. K.; Sundriyal, R. C; Kumar, K. Environmentally Sustainable Traditional Natural Resource Management and Conservation in Ziro Valley, Arunachal Himalaya, India. *J. Am. Sci.* **2009,** *5*(5), 48, para 3.4.

Dollo, M.; Samal, P. K.; Sundriyal, R. C; Kumar, K. Environmentally Sustainable Traditional Natural Resource Management and Conservation in Ziro Valley, Arunachal Himalaya, India. *J. Am. Sci.* **2009,** *5*(5), 51, para 1.

Domaine Public Payant https://sckool.org/the-domaine-public-payant.html (accessed June 20, 2021).

Elvin, V. *India's North East Frontier in the 19th Century*; OUP 1959; p 33.

Elvin, V. *The Art of North East Frontier of India*; Shillong, 1959; p 14.

Elvin, V. *The Art of North East Frontier of India*; Shillong, 1959; pp 13–14.

Elvin, V. *The Art of North East Frontier of India*; Shillong, 1959; p 70.

Elvin, V. *The Art of North East Frontier of India*; Shillong, 1959; p 124.

Elvin, V. *The Art of North East Frontier of India*; Shillong, 1959; p 187.

Elwin, V. *A Philosophy for NEFA*, New Edition (first published 1957); Isha Books, 2009; p 41.

Elwin, V.; Shastri, B.; Simon, I. Eds. *Important Directives on Administration of NEFA*; NEFA Administration-Shillong 1967; Ch 6, p 77.

Give and Take—theory of John Nash.

González, R. J. *Zapotec Science: Farming and Food in the Northern Sierra of Oaxaca*; University of Texas Press: Austin, 2001; p 130.

Hansen, S. A.; Van Fleet, J. W. Issues and Options for Traditional Knowledge Holders in Protecting Their Intellectual Property. In *Intellectual Property Management in Health and Agricultural Innovation: A Handbook of Best Practices*; Krattiger, A., Mahoney, R. T., Nelsen, L. et al., Eds.; MIHR: Oxford, and PIPRA: Davis; 2007; p 1523; www.ipHandbook. org.

Home Page < About TKDL> (accessed June 20, 2021).

IPR_2019.pdf (arunachalipr.gov.in) (accessed June 20, 2021).

Idu Mishmi Textiles and Arunachal Orange are the only two registered GIs from the State of Arunachal Pradesh.

In 1976 WIPO published the Tunis Model Law on Copyright for Developing Countries, whose Section 17 included a *Domaine Public Payant clause*: (WIPO 1976) www.wipo.int/edocs/pubdocs/en/copyright/120/wipo_pub_120_1976_07-08.pdf (accessed June 20, 2021).

Janke, T. Case Studies on Intellectual Property and Traditional Cultural Expressions. WIPO, *Minding Culture*: Study No. 1, 2003, pp 37–40; http://www.wipo.int/globalissues/studies/cultural/minding-culture/index.html (accessed June 20, 2021).

Janke, T. *Our Culture, Our Future* 1998; https://www.terrijanke.com.au/our-culture-our-future (accessed June 20, 2021).

Jerez, M. M. Challenges and Opportunities for Indigenous Peoples' Sustainability. UN. Department of Economic and Social Affairs, Division for Inclusive Social Development: UN-DESA Policy Brief No 101, 2021; www.un.org/development/desa/publications/ (accessed June 20, 2021).

John, M. Intellectual Property Protection and Traditional Knowledge. In *Intellectual Property and Human Rights*; WIPO, 1999; p 97.

Kumar, K. District Research Officer, Kameng, *The Pailibos*; Research Department, Govt. of Arunachal Pradesh, Shillong Press, 1979, 31.

Kumar, K. District Research Officer, Kameng. *The Pailibos*; Research Department, Govt. of Arunachal Pradesh, Shillong Press, 1979; Ch 2, pp 17–55.

Kumar, K. District Research Officer, Kameng. *The Pailibos*; Research Department, Govt. of Arunachal Pradesh, Shillong Press, 1979; Ch 2, p 45.

Madegowda, C. *Traditional Knowledge and Conservation*. Economic and Political Weekly, 23–29 May 2009, 44(21), pp 65–69. https://www.jstor.org/stable/40279037 (accessed June 20, 2021).

Mahanta, D.; Tiwari, S. C. Natural Dye-Yielding Plants and Indigenous Knowledge on Dye Preparation in Arunachal Pradesh, Northeast India. *Curr. Sci.* **2005,** *88*(9), 1474–1480; https://www.jstor.org/stable/24110717 (accessed June 20, 2021).

Microsoft Word - Arunachal_HLST_RGI.doc (censusindia.gov.in), (accessed June 20, 2021).

Mihin, D.; Samal, P. K.; Sundriyal, R. C; Kumar, K. Environmentally Sustainable Traditional Natural Resource Management and Conservation in Ziro Valley, Arunachal Himalaya, India. *J. Am. Sci.* **2009,** *5*(5), 47, para 3.4.

Mizuno, K. The Distribution and Management of Forests in Arunachal Pradesh, India. In *Environmental Geography of South Asia*; Singh, R. B., Prokop, P., Eds.; Springer: Japan, 2016.

PTI. Arunachal gets First-Formal-Indigenous-Knowledge-System. *The Hindu*, 21 March 2021; https://www.thehindu.com/news/national/arunachal-gets-first-formal-indigenous-knowledge-system-school/article34123214.ece (accessed June 20, 2021).

Pal Singh, B. *Gandhian Perspective on Tribal Resources and the Modern State*, https://www.mkgandhi.org/articles/tribalresources_modernstate.html (accessed June 20, 2021).

Sanchez, P. A. *Properties and Management of Soils in the Tropics*; Wiley Inter-science Publication, orig. University of Michigan, 1976; digitised 2010.

Sandler, F. Music of the Village in The Global Marketplace – Self-Expression, Inspiration, Appropriation, or Exploitation? Ph.D. Dissertation, University of Michigan, 2001, pp 35–38.

Secretariat of the United Nations Permanent Forum on Indigenous Issues. Report of the International Technical Workshop on Indigenous Traditional Knowledge; UN Document E./C.19/2006/2, New York: United Nations Permanent Forum on Indigenous Issues, 2006; https://digitallibrary.un.org/record/565533?ln=en (accessed June 20, 2021).

Secretariat of the World Intellectual Property Organisation. Information Note. In *Proceedings of the Workshop on Traditional Indigenous Knowledge*; Panama: United Nations Permanent Forum on Indigenous Issues, 2005; http://www.un.org/esa/socdev/unpfii/documents/workshop_TK_WIPO.pdf (accessed June 20, 2021).

WIPO. Traditional Cultural Expressions; https://www.wipo.int/tk/en/folklore/ (accessed June 20, 2021).

WIPO. Traditional Cultural Expressions; https://www.wipo.int/tk/en/folklore/ (accessed June 20, 2021).

The Biological Diversity Act 2002, s 21.

See, map: https://www.researchgate.net/figure/Tribal-communities-of-Arunachal-Pradesh_fig1_237475024.

Singh, R. K. Ecoculture and Subsistence Living of Monpa Community in the Eastern Himalayas: An Ethnoecological Study in Arunachal Pradesh. *Indian J. Tradit. Knowl.* **2013,** *12*(3), 441–453.

Strengthening Bonds, Indian PM Visit to Bangladesh. *India Perspectives* **2021,** *35*(2), 56; https://www.indembastana.gov.in/docs/India%20Perspective.pdf (accessed June 20, 2021).

Sundriyal, R. C.; Upreti, T. C.; Varuni, R. Bamboo and Cane Resource Utilisation and Conservation in the *Apatani* plateau, Arunachal Pradesh, India: Implications for Management. *J. Bamboo Ratt.* **2002,** *1*(3), 205.

Sundriyal, R. C.; Upreti, T. C.; Varuni, R. Bamboo and Cane Resource Utilisation and Conservation in the *Apatani* plateau, Arunachal Pradesh, India: Implications for Management. *J. Bamboo Ratt.* **2002,** *1*(3), 226.

Sundriyal, R. C.; Upreti, T. C.; Varuni, R. Bamboo and Cane Resource Utilisation and Conservation inthe *Apatani* plateau, Arunachal Pradesh, India: Implications for Management. *J. Bamboo Ratt.* **2002,** *1*(3), 205–246.

The National IP Policy; https://dipp.gov.in/sites/default/files/national-IPR-Policy2016-14October2020.pdf (accessed June 20, 2021).

The presentations can be accessed from the following links: https://www.wipo.int/edocs/mdocs/tk/en/wipo_iptk_ge_2_15/wipo_iptk_ge_2_15_presentation_usha_rao.pdf; https://www.wipo.int/edocs/mdocs/tk/en/wipo_iptk_ge_2_16/wipo_iptk_ge_2_16_presentation_12javed.pdf; https://www.wipo.int/edocs/mdocs/tk/en/wipo_iptk_ge_16/wipo_iptk_ge_16_presentation_13dhar.pdf (accessed June 20,2021).

Their presentations are available at https://www.wipo.int/meetings/en/details.jsp?meeting_id=22423 (accessed June 20, 2021).

Under the Biodiversity Act, 2001 and the PPVFRA, 2002.

United Nations Development Group. UNDG Guidelines on Indigenous Peoples' Issues. Office of the United Nations High Commissioner for Human Rights, 2008; http://www2.ohchr.org/english/issues/indigenous/docs/guidelines.pdf (accessed June 20, 2021).

Verrier, E. *A Philosophy for NEFA*, New Edition; Isha Books, 2009, p 41.

Verrier, E. *A Philosophy for NEFA*, New Edition; Isha Books, 2009, pp 1–13.

Verrier, E. *A Philosophy for NEFA*, New Edition; Isha Books, 2009, pp 9–10.

WIPO. Intellectual Property and Traditional Cultural Expressions/Folklore; Booklet no. 1; WIPO publication No. 913(E). https://www.wipo.int/edocs/pubdocs/en/tk/913/wipo_pub_913.pdf (accessed June 20, 2021).

About Us: Traditional Knowledge Digital Library. http://www.tkdl.res.in/tkdl/langdefault/common/Abouttkdl.asp?GL=Eng (accessed June 20, 2021).

WIPO; Consolidated Analysis of the Legal Protection of Traditional Cultural Expressions/Expressions of Folklore; WIPO Publication 785, 2003. https://www.wipo.int/edocs/pubdocs/en/tk/785/wipo_pub_785.pdf (accessed June 20, 2021).

WIPO; Intergovernmental Committee on Intellectual Property and Genetic Resources, Traditional Knowledge and Folklore (IGC) https://www.wipo.int/meetings/en/topic.jsp?group_id=110 (accessed June 20, 2021).

CHAPTER 18

International Legal Issues and Plant Variety Protection Rights in Agriculture

RASHMITA DASGUPTA[1] and PIYALI MUKHERJEE[2]

[1]*Calcutta High Court, Kiran Sankar Roy Road, Kolkata, West Bengal, India*

[2]*Department of Law, Brainware University, Kolkata, West Bengal, India*

ABSTRACT

Plant varieties are live self-reproducible material and it is critical to preserve the breeder's intellectual property. Plant innovation intellectual property protection differs from country to country. Plant variants are not patented in Europe. Plant Variety Protection grants the breeder complete control over the variety, as well as rights to use it, and fosters the production of new plant varieties. Plant Breeders' Rights were recognized on an international level after the formation of the International Union for the Protection of New Varieties of Plants (UPOV). If a new plant variety is distinct (D) from any other existing variety, uniform (U), and stable, it can be protected (S). Individual species-specific DUS examination guidelines are developed by UPOV. This chapter includes legal issues, application, plant variety protection rights, and related intellectual property rights in agricultural aspects.

18.1   INTRODUCTION

The United Nations Organization formed the World Intellectual Property Organization in 1967 to encourage the development of intellectual property protection around the world through cooperation among countries and other international organizations, as well as to harmonize national IP legislation. WIPO works with governments, nongovernmental organizations, and individuals to manage intellectual property (IP) for social and economical improvement. WIPO consists of 26 international treaties concerning a wide variety of IP issues; protection of plant variety is one of them.

Intellectual is a person with a sufficient ability to understand or solve the problems and is capable of thinking, resolving, and making solution of any matter through his knowledge, experience, experiment, and intellect. Intellectual property is one of the kinds of property that includes creativity by the action of human mind. It is actually an intangible invention of human intellect, but it does not have any proper definition of IP even in the established convention of the WIPO.

- Intellectual property rights (IPRs) are legal rights and these give the creators or inventors or breeders an exclusive right to protect his or her creation for a certain limited period of time, if anyone wants to get benefited, he should buy it or grants license from the creator, if he does not do the same then it will be treated as infringement of IPR of the creator or inventor or the breeder.
- We get various products from different kinds of plants or trees. These help us in many ways,, for example, we get different types of fruits, flowers, leaves, spices, medicines, edible vegetables, crops, woods, extracts, etc., from different types of plants. These also encourage to protect the soil during a flood, give shelters to birds and other animals, and also try to maintain the balance of the carbon dioxide gas in the air by inheriting it and producing oxygen that helps our Earth to be inhabitable for the human being and the all other living things.
- The plant breeders are trying to create a new species of the plant after numerous research and experiment for the benefit of the people, that is why they need to protect their creativity, their newly created, extraordinary plant, and if not anybody with malafide intention can take undue advantage or can take fame of the creativity which has done by the said breeder.

The plant varieties protection is required for giving protection of the rights of the breeder of the newly created plant, also it gives protection to the newly, unique, distinct, stable plants. Many pirates dare to commit black-marketing of the rare plants that affect the plant breeder's right, so that the protection of plant variety is necessary.

- Plant variety protection provides a way to protect the IPRs of the breeders for their creativity of a new variety of plants, and also it protects the variety of plants in the worldwide which are endangered.
- The organization namely UPOV is on the power to look after the matter related to the protection of a newly creative variety of plants.
- Plant variety depends upon the geographical indication.

18.2 ROLE OF INTERNATIONAL UNION FOR THE PROTECTION OF NEW VARIETIES OF PLANTS

The International Union for the Protection of New Varieties of Plants (UPOV) is an intergovernmental organization that was founded in 1961 in Paris and has been updated thrice since then: in 1972, 1978, and 1991. This organization's headquarters are in Geneva (Switzerland). This Convention had 74 parties or countries as members as of October 2, 2015.

The International Union for the Protection of New Varieties of Plants was duly established by the International Convention for giving protection of all the new plant varieties by an IPRs, developing new distinct, unique varieties of plants and also monopolistic control over the variety by the breeder who creates it on an international basis. It maintains a distinct path to protect newly creative varieties of plants, that is, Sui generics system for the protection of new plant varieties. A new plant variety has to pass the DUS examination; then the variety will be protected under the plant variety protection. The new plant variety must be differentiated (D), uniform (U), and stable (S) in comparison with other plant types.

Requirement of the Plant Variety Protection Under the UPOV Convention

Plant variety protection is one of the forms of intellectual property proposed by the UPOV in 1961 to protect the infringement of the breeders' investment in developing new, distinct plant types with consistent features. It is an exclusive right of a breeder over business purposes and commercialization of the protected plant variety's reproductive or vegetative propagation material, such as seeds.

IPRs also include the protection of new plant varieties by granting breeders of the new plant kinds an exclusive right for a short time. As previously said, new plant varieties must pass the DUS test to obtain protection, so there are some specific criteria that must be met or passed by new plant kinds.

The protection of a new variety of plants is very much important to provide safety or security of the plant breeders, for the sake of development by improvising plant varieties, makes new varieties with different initiative that helps the world to improve in agriculture, medicinal plants, horticulture, forest, etc.

For the purpose of progressing the quality and getting more benefits from the new variety of plants, the breeders have to do a lot of research and experiment, which is very expensive and time-oriented. I must spur production of a new plant variety is the result of whole dedication of the plant breeder's intellect, time, raw materials, money, uniqueness, labor, etc., to make it successful.

IPRs provide security and safety by giving an exclusive right to the plant breeders upon the new variety. In that order breeders consider the reasonable opportunity to recover the monetary investment by enforcing various forms of IPRs such as patent right and protection of new varieties of plants.

Plant breeders' rights help to achieve the goals that they are entitled to get and prevent others from taking advantage by multiplying seeds of breeders or other propagating material and selling the variety without considering breeder's permission as well as without providing reparation, for the commercial business purpose.

Key features of the UPOV convention:

The UPOV's members are bound by the organization's values, which include granting plant breeders' rights to new plant varieties and ensuring international harmony. The basic principles of the UPOV convention were adopted in the 1961 Act and thereafter amended in the 1978 and 1991 Acts; the revision was done with that effect as if everyone can equip effectively in the time of the 21st century. There are some guidelines of the UPOV convention toward the members mentioned below

- give protection of genera and species;
- common agreement on essential ideas, concepts, and actions regarding a plant variety and the breeder;

- eligibility for grant of protection based on novelty, distinctness, uniformity, and stability;
- provide a minimum scope and duration of protection;
- describe limits of action to cancel the breeder's rights.

As per Article 1(iv) of the UPOV Convention breeder means "the person who bred, or discovered and developed, a variety, the person who is the employer of the aforementioned person or who has commissioned the latter's work, where the laws of the relevant contracting party so provide, or the successor in title of the first or second aforementioned person, as the case may be."

As per Article 1(vi) of the UPOV Convention variety means "a plant grouping within a single botanical taxon of the lowest known rank, which grouping, irrespective of whether the conditions for the grant of a breeders' right are fully met, can be

- defined by the expression of the characteristics resulting from a given genotype or combination of genotypes,
- distinguished from any other plant grouping by the expression of at least one of the said characteristics, and
- considered as a unit with regard to its suitability for being propagated unchanged."

Denomination of variety:

According to Article 20 of the UPOV Convention, it is chosen by the breeder but it must satisfy the criteria as mentioned below

- it must encourage the variety to be recognized;
- it must be different and unique from all other Union member's denomination for the same or close enough species;
- it must not be responsible for any misleading or creating confusion regarding the characteristic, value, or identity of the variety or identity of the breeder;
- prior right of the third party must not be affected even if required change the variety denomination;
- no rights in the designation registered as the denomination of the variety, even after expiration of the breeder's right, shall hamper the free use of the variety denomination;
- it may not consist solely of figures except this is an established practice."

Exceptions to the breeder right:

The following acts do not consider an infringement of the breeder's right:

- if the act is done for private usage and it is not used for commercial purposes;
- if the act is done for research and experimental purposes;
- if the act is done for breeding other varieties.

18.3   ROLE OF PATENT RIGHTS IN THE FIELD OF PLANT VARIETY PROTECTION

Inventions that are new, original, and helpful are protected by patents. Although some kinds of inventions are eligible to be protected by patent and some of the other common exceptions, to be patentable, innovations must meet specific criteria in terms of novelty and other characteristics. There are three substantive criteria that would need to be met for an invention for patentability provided by The Agreement on Trade Related Aspect of Intellectual Property Rights (TRIPS Agreement).

First, it must be new or novel, meaning that nobody has made it before, carried it out before, or used it before.

An inventive step must be presented there so as to convince its worthy features being patentable.

The third one is it must be capable of industrial application.

Patent is an agreement between public and the patent holder. International treaty provides the time period of patent, that is, at least 20 years from the filling date of the patent application.

We can recognize a plant patent through a patent number on the tag, if it is applied for then it has been shown as PPAF (plant patent applied for), and if it is still in processing then it has been indicated as patent pending.

A plant patent is one kind of patents so it is also an IPR, which protects the stable characteristics of a new, unique plant from being replicated, and concerning about exploitation of the protected plant from being misused or taken advantages by others. A patent protects the invention from everyone except when the patent-holder licenses others to propagate and sell the plant on behalf of him. In fact, a plant patent can assist the patent holder earn more financial profits during the time period of the patent right by preventing others from using illegally or without taking permission the plant and the propagating material including seeds.

United States Government grants a patent for a plant which the inventor has invented or discovered and asexually reproduced the variety, not by tuber-propagated new plants after the parent plants died off.

The TRIPS Agreements states the WTO Members must protect new plant varieties either by a suing generis system or by patents or by some combination thereof. Patent gives protection to all the inventions.

Trade related aspects of IPRs agreement is one of the intellectual property treaties which provides IPR protection for plant varieties. It was implemented in 1994 by the World Trade Organization (WTO).

TRIPS is the first and sole IPR treaty that procures to set up worldwide, minimum engagements of protection all over the major parts of intellectual property, including trademarks, copyrights, industrial designs, patents, trade secret. It dedicates least attention in the matter of plant variety protection or in the plant breeders' right, but it helps to develop the legal protection of the plant varieties international levels.

TRIPS Agreement encourages protection of plant varieties through the connection of various sources. A link to additional international trade agreements is to spread over a vast region in both the developed and developing worlds, unique enactment, review, and resolution of disputes, necessity, and review under Article 27.3(b) of TRIPS.

WTO members are offered TRIPS to provide and improve national laws related to plant variety protection in their states.

First, the Council received responses from 17 states and the European Communities and its member states. Later it got more response from another six states regarding providing national laws that help to furnish the protection of plant varieties in the member states.

Connection with other International Treaty Agreements:

We can say that the TRIPS agreement is connected with a huge group of trade-related agreements related to matters such as trade-in goods and services, textiles, agriculture, health-related restrictions on import. During the Uruguay Round of trade negotiations, which took place between 1988 and 1994, the World Trade Organisation accepted all of these agreements.

Review and Remedy:

Meanwhile, the influence of the TRIPS agreement spread out all over the world about providing effective protection in the field of IPRs. The unique provisions of the TRIPS agreement furnish a connection with the enforcement of IPRs amalgamate with national laws, as well as the review of the said national laws generated by the TRIPS council and also the remedies related to the settlement of disputes between two or more states, who has become a threat of trade sanctions.

TRIPS agreement requires brace of additional effective provision of national statutes by the WTO members, by the application of which the inventors or owners of the intellectual property products can authorize to utilize their rights against those who infringe them by committing illegal acts.

According to the TRIPS agreement provision of remedies has been explained in the following articles:

"Article 41 contains general obligations, Articles 42 to 49 deal with civil and administrative procedures and remedies, Article 50 explains provisional measures, Articles 51 to 60 deal with special requirements related to border measures, and Article 61 contains criminal procedures."

Any of the remedies can be applied to the plant breeder of a new plant variety, the new plant variety of whose illegally sold in a commercial market with the interest of getting profits without that breeder's permission must be able to institute a civil judicial action by seeking an injunction to stop the conduct of the unauthorized sale as well as the breeder is entitled to get compensation from the wrong-doer.

"According to Article 27.3(b) member may also exclude from patent-ability: plants and animals other than micro-organisms, and essentially biological processes for the production of plants or animals other than non-biological and microbiological processes. However, Member shall provide for the protection of plant varieties either by patents or by an effective sui generis system or by any combination thereof."

A debate, regarding this Article, has been arisen among WTO Members and non-governmental organizations with distinct opinion related to IPR protection for plant varieties.

The summary of the debate among both WTO Members and NGOs after analyzing Article 27.3(b) is mentioned later

- There has been no reference of any pre-existing intellectual property agreement, like the 1978 and 1991 UPOV Acts, provided by TRIPS related to plant varieties.

To manage these unfurnished situations that affect also the rest of other intellectual properties such as trademarks, patents copyrights, industrial design, TRIPS suggests WTO Members to follow the provisions of protection in accordance with the pre-existing IPR agreements, such as the Paris

Convention that protects industrial property, Barne Convention that protects literary and artistic works. To manage this fault that influences the rest of the intellectual properties such as trademarks, patents copyrights, industrial designs, TRIPS suggests WTO members to follow the provisions of protection by the pre-existing IPR agreements, such as the Paris Convention that protects industrial property, Barne Convention that protects literary and artistic works. Due to this fault, being a member there have been no such obligations under TRIPS to become a member of UPOV or to enact national laws provided by UPOV. TRIPS wants to serve a standard of protection related to the plant varieties.

- "According to Article 27.3(b) WTO members can protect the plant varieties either by patents or by an effective sui generis system or by a combination of aspects from both, so that, it allows members discretion to choose one of these three approaches by which they can protect plant varieties in their states. It is so natural that different state has a different approach of using this Article in the matter related to plant variety protection."

Some states, including the United States, Australia, the United Kingdom, Japan, etc., have taken advantage of this opportunity in the matter of protection of plant varieties by permitting plant breeders to implicate patent rights in new varieties provided that the criteria for the patent have been met.

"According to Article 33 of the TRIPS agreement, patent is available for the terms of 20 years from the date of filing application and fees are also required to maintain the applicability of patent."

18.4   PLANT VARIETY PROTECTION ACT

The process of enacting national laws related to the Plant Variety Protection Act differs from country to country. Different countries have different plant variety protection rules and regulations based on their circumstances, and after studying the field of plant variety protection, governments have a better understanding of how to implement what kind of rules and regulations to protect new plants under their national laws. The plant variety protection systems of the United States and Australia will be discussed.

18.4.1   PLANT VARIETY PROTECTION IN THE UNITED STATES

In the United States, the Plant Variety Protection Act of 1970 governs intellectual property rights. PVPA grants plant breeders exclusive rights to innovative, distinct, original, and stable sexually reproduced or tuber-propagated plant varieties for up to 25 years. By using patents, the PVPA protects plant breeders' rights and new plant varieties, although patents do not cover tuber-propagated plants.

"Plant Variety Protection Act 1970 confers the breeder who has reproduced that any sexually reproduced or tuber propagated plant variety (other than fungi or bacteria) or the successor in interest of the breeder, shall be entitled to plant variety protection until the requirements for getting protection have been met."

After satisfying the four requirements, the plant variety is eligible for getting a certificate under the PVPA.

"According to Section 42 of the Plant Variety Protection Act 1970 the plant must be new it means that propagating or harvested material of the variety has not been sold or arranged from other sources for the purposes of exploitation for more than one year in the United States prior to the date of filing, or more than four years in outside of the United States, or six years in the case of a tree or vine prior to date of filing."

- The plant variety must be distinct; it implies that the variety is undoubtedly distinguishable from any other publicly launched variety.
- The plant variety must be uniform that signifies that any variations are describable, predictable, and commercially acceptable.
- The plant variety must be stable; it denotes that the variety, when reproduced, will remain unchanged considering its essential and distinctive characteristics within a reasonable degree of commercial reliability.

"According to Section 83 of the said Act it gives exclusive rights to the breeder for selling it or proposing it for sale, or reproducing it or importing/ exporting it, or using it in producing a hybrid or distinct variety."

The Plant Variety Protection Act confers three exemptions:

First, the "Section 44 of the PVPA securing 'public interest in wide usage' which allows the Secretary to declare an otherwise protected variety open to utilize on a basis of equitable remuneration to the owner, on condition that not more than two years authorising compulsory license is necessary relating to protected variety in order to insure an adequate supply of fibre, food or feed, and that the owner is unwilling or unable to supply public needs at the rate of reasonable as well as fair price."

Second, "Section 113 contains crop exemption, which permits a farmer to save seeds from protected plant varieties and to use the saved seeds in the reproduction of the crop without infringement."

Third, "Section 114 of the said Act allows research exemption which means the use and reproduction of a protected variety for plant breeding or other bona side research shall not constitute an infringement."

Infringement of law:

If a person is doing an act without being entitled to do that, like stocking the plant variety or selling the plant variety without holding any authorized license by the owner or instigating any activities that are not legally done, then it is considered an infringement of law.

Remedy:

A plant-breeder or an owner of the plant variety shall entitle to get civil remedy for the infringement of plant variety protection under Section 111.

Available remedies are:

Injunction;

Damages;

Attorney fees;

Otherwise, a person who intends to deceive the public by false marking, he or she shall be liable for getting punishment under the Criminal statute of the United States.

Applications of Intellectual property rights in plant variety protection:

Patent rights, trademark rights, and plant breeder's rights are the three categories of intellectual property rights that provide a breeder with exclusive rights related to the protection of new sorts of plant varieties. Breeders with a registered PBR have exclusive commercial rights to a registered plant variety, and they have the legal authority to prohibit others from commercial enjoyment of the variety for a short time.

18.5   PLANT BREEDER'S RIGHTS ACT (PLANT VARIETY PROTECTION IN AUSTRALIA)

The Plant Breeder's Rights Act protects new plant varieties including propagating material (seed, cuttings, divisions, tissue culture) and harvested material (cut flowers, fruit, foliage), along with varieties of trees, shrubs, vines, and vegetables. Moreover, 100 new breeders seek plant protection every year. First, the right was introduced in Australia as the Plant Variety Rights Act in 1987, but later in the year 1994 this Act was repealed and replaced by the Plant Breeder's Rights to collaborate Australia's international obligations including the updated version of the International Convention for the Protection of New Varieties of Plants (UPOV) 1991 relating to intellectual property rights.

PBRs have excluded others by granting the rights to the plant breeder from:

- producing or reproducing the propagating and harvested material
- proposing the material for sale
- selling the propagating or harvesting material
- importing and exporting the material
- stocking the material with the intention of selling, exporting, etc.
- referring to any other plant of the same plant class using the registered name of the plant variety, or a registered synonym of the true name of the variety.

Procedure of establishing PBRs for a plant variety:

There are three requirements to establish the plant breeder's rights for the protection of plant variety.

1. PBRs must be registered with the Intellectual Property Office of Australia. PBRs are not always applicable to a new innovation or concept.
2. Eligibility of getting registration, the variety must be new or recently exploited. The plant variety's breeder must be cautious about how to use the new variety and when it requires protection.

   A plant variety is termed new if it has never been sold before with the breeder's permission. When a variety has been sold for up to 12 months with the breeder's approval, it is considered "recently exploited" in Australia, and this restriction is extended to 4 years outside of Australia, with the exception of trees and vines, where the maximum is extended to 6 years.
3. The new plant variety must have a breeder and the plant variety must pass the DUS test:
   - If a plant variety has not been commercialized for more than 1 year, it is termed new.

- It is considered distinct if it differs in botanical characteristic such as height, color, from other existing varieties.
- If the attribute of the plant variety remains similar from plant to plant within the variety, the plant is uniform.
- If the plant's features are genetically fixed and do not change even when hybrid varieties are reproduced, the variety is stable.

The Plant Breeder's Right Office (the PBRO) has the authority to administer the PBR Act and it is a division of Intellectual Property in Australia. It maintains the PBR register and is liable for substantive examination of PBR applications.

The National office grants plant variety rights to the new plant variety as well as the plant breeder after qualifying the examination. The PBRO gathers seeds as a sample and grows them for one or more seasons to assess the variety's distinctness, uniformity, and stability. After passing this test the plant breeder gets exclusive rights over the plants for a period of 20 years and 25 years for trees and vines. PBRs are only valid for a certain amount of time just like most IPR and after the expiry of the period the rights are enjoyed by the public.

The plant breeder has to pay the maintenance fees to enjoy the rights once the new variety has been registered or certified. If the renewal payment has not been paid within the required period, the breeder may lose all the existing rights to the plant variety.

Exceptions to the rights:

If the protected variety is employed for study or continued breeding, the PBRs Act allows for exceptions to the law. The following actions do not constitute an infringement of the protected variety or the rights of the plant breeder:

- The use of the variety for private and noncommercial purposes, experimental or research purpose, and breeding other varieties.
- A protected variety's propagative substance is used or marketed as food, an ingredient, or a fuel.
- Conditioning and use of farm-saved seed by the farmer.

Remedy:

The Civil Federal Court imposes a fine of not more than $50,000 on individuals and $250,000 on businesses for international and reckless infringement of a breeder's rights. Injunction, damages, or an account of profits, at the plaintiff's option, and further damages are the other civil remedies available.

18.6 CONCLUSION

The use of the Plant Variety Protection System, which is governed by UPOV and TRIPs, has played a significant role in stimulating the development of new plant varieties and their introduction into the agricultural and horticultural fields for the benefit of the society. In any country, the moto of the PVP is to protect the plant variety, protect the farmer's or breeder's right, and provide opportunities to create another developed unique variety through the experiment upon the variety without malafide intention and otherwise all WTO Members want to maintain harmony in the international market among each other.

KEYWORDS

- **plant variety protection act**
- **plant patent**
- **UPOV**
- **intellectual property rights**
- **trips agreements**

REFERENCES

http://www.fao.org/3/y5714e/y5714e03.htm (accessed on Apr 20, 2021).

https://en.m.wikipedia.org/wiki/Plant_Variety_Protection_Act_of_1970 (accessed on Apr 20, 2021).

https://www.upov.int/portal/index.html.en (accessed on Apr 15, 2021).

https://en.m.wikipedia.org/wiki/International_Union_for_the_Protection_of_New_Varieties_of_Plants (accessed on Apr 18, 2021).

Saurabh Bhatia, Randhir Dahiya, in Modern Applications of Plant Biotechnology in Pharmaceutical Sciences, 2015; https://www.sciencedirect.com/topics/agricultural-and-biological-sciences/plant-variety-protection (accessed on Apr 22, 2021).

https://en.m.wikipedia.org/wiki/Plant_Variety_Protection_Act_of_1970#:~:text=The%20Plant%20Variety%20Protection%20Act,or%20tuber%20propagated%20plant%20varieties (accessed on Apr 18, 2021).

https://www.seednet.com.au/plant-breeders-rights (accessed on Apr 23, 2021).

https://en.m.wikipedia.org/wiki/Plant_breeders%27_rights (accessed on Apr 19, 2021).

http://www.fao.org/3/y5714e/y5714e03.htm#:~:text=In%20short%2C%20the%20 UPOV%20treaties,system%20or%20some%20combination%20thereof (accessed on Apr 27, 2021).

https://www.business.qld.gov.au/running-business/ip/types/plant-breeders-rights;   (accessed on Apr 27, 2021).

https://www.seednet.com.au/plant-breeders-rights#:~:text=Exceptions%20to%20the%20 Rights&text=The%20use%20of%20the%20variety,existence%20of%20Plant%20 Breeder's%20Rights (accessed on Apr 28, 2021).

Tripsagreement.pdf; -www.worldtradelaw.ne (accessed on Apr 28, 2021).

Plant Variety Protection Act.pdf- www.ams.usda.gov (accessed on Apr 19, 2021).

Index

U